CW01464312

MACHINE VISION

This book is an accessible and comprehensive introduction to machine vision. It provides all the necessary theoretical tools and shows how they are applied in actual image processing and machine vision systems. A key feature is the inclusion of many programming exercises that give insights into the development of practical image processing algorithms.

The authors begin with a review of mathematical principles and go on to discuss key issues in image processing such as the description and characterization of images, edge detection, feature extraction, segmentation, texture, and shape. They also discuss image matching, statistical pattern recognition, syntactic pattern recognition, clustering, diffusion, adaptive contours, parametric transforms, and consistent labeling. Important applications are described, including automatic target recognition. Two recurrent themes in the book are consistency (a principal philosophical construct for solving machine vision problems) and optimization (the mathematical tool used to implement those methods).

A CDROM containing software and data used in the book is included. The book is aimed at graduate students in electrical engineering, computer science, and mathematics. It will also be a useful reference for practitioners.

WESLEY E. SNYDER received his Ph.D. from the University of Illinois, and is currently Professor of Electrical and Computer Engineering at North Carolina State University. He has written over 100 scientific papers and is the author of the book *Industrial Robots*. He was a founder of both the IEEE Robotics and Automation Society and the IEEE Neural Networks Council. He has served as an advisor to the National Science Foundation, NASA, Sandia Laboratories, and the US Army Research Office.

HAIRONG QI received her Ph.D. from North Carolina State University and is currently an Assistant Professor of Electrical and Computer Engineering at the University of Tennessee, Knoxville.

MACHINE VISION

Wesley E. Snyder
North Carolina State University, Raleigh

Hairong Qi
University of Tennessee, Knoxville

PUBLISHED BY THE PRESS SYNDICATE OF THE UNIVERSITY OF CAMBRIDGE
The Pitt Building, Trumpington Street, Cambridge, United Kingdom

CAMBRIDGE UNIVERSITY PRESS
The Edinburgh Building, Cambridge, CB2 2RU, UK
40 West 20th Street, New York, NY 10011–4211, USA
477 Williamstown Road, Port Melbourne, VIC 3207, Australia
Ruiz de Alarcón 13, 28014 Madrid, Spain
Dock House, The Waterfront, Cape Town 8001, South Africa

http://www.cambridge.org

First published 2004

Printed in the United Kingdom at the University Press, Cambridge

Typefaces Times 10/12 pt. and Akzidenz Grotesk *System* LaTeX 2_ε [TB]

A catalog record for this book is available from the British Library

Library of Congress Cataloging in Publication data
Snyder, Wesley E.
Machine vision / Wesley E. Snyder and Hairong Qi.
p. cm.
ISBN 0 521 83046 X
1. Computer vision. I. Qi, Hairong, 1970. II. Title.
TA1634.S69 2004
006.3′7 – dc21 2003053083

ISBN 0 521 83046 X hardback

To Graham and Robert
WES

To my parents and Feiyi
HQ

Contents

To the instructor

This textbook covers both fundamentals and advanced topics in computer-based recognition of objects in scenes. It is intended to be both a text and a reference. Almost every chapter has a "Fundamentals" section which is pedagogically structured as a textbook, and a "Topics" section which includes extensive references to the current literature and can be used as a reference. The text is directed toward graduate students and advanced undergraduates in electrical and computer engineering, computer science, or mathematics.

Chapters 4 through 17 cover topics including edge detection, shape characterization, diffusion, adaptive contours, parametric transforms, matching, and consistent labeling. Syntactic and statistical pattern recognition and clustering are introduced. Two recurrent themes are used throughout these chapters: Consistency (a principal philosophical construct for solving machine vision problems) and optimization (the mathematical tool used to implement those methods). These two topics are so pervasive that we conclude each chapter by discussing how they have been reflected in the text. Chapter 18 uses one application area, automatic target recognition, to show how all the topics presented in the previous chapters can be integrated to solve real-world problems.

This text assumes a solid graduate or advanced-undergraduate background including linear algebra and advanced calculus. The student who successfully completes this course can design a wide variety of industrial, medical, and military machine vision systems. A CDROM is included with software tools developed by the authors and images to support the homework assignments and projects. The software will run on PCs running Windows or Linux, Macintosh computers running OS-X, and SUN computers running SOLARIS. Software includes ability to process images whose pixels are of any data type on any computer and to convert to and from "standard" image formats such as JPEG.

Although it can be used in a variety of ways, we designed the book primarily as a graduate textbook in machine vision, and as a reference in machine vision. If used as a text, the students would be expected to read the basic topics section of each chapter used in the course (there is more material in this book than can be covered in a single semester). For use in a first course at the graduate level, we present a sample syllabus in the following table.

Sample syllabus.

Lecture	Topics	Assignment (weeks)	Reading assignment
1	Introduction, terminology, operations on images, pattern classification and computer vision, image formation, resolution, dynamic range, pixels	2.2–2.5 and 2.9 (1)	Read Chapter 2. Convince yourself that you have the background for this course
2	The image as a function. Image degradation. Point spread function. Restoration	3.1 (1)	Chapters 1 and 3
3	Properties of an image, isophotes, ridges, connectivity	3.2, 4.1 (2)	Sections 4.1–4.5
4	Kernel operators: Application of kernels to estimate edge locations	4.A1, 4.A2 (1)	Sections 5.1 and 5.2
5	Fitting a function (a biquadratic) to an image. Taking derivatives of vectors to minimize a function	5.1, 5.2 (1)	Sections 5.3–5.4 (skip hexagonal pixels)
6	Vector representations of images, image basis functions. Edge detection, Gaussian blur, second and higher derivatives	5.4, 5.5 (2) and 5.7, 5.8, 5.9 (1)	Sections 5.5 and 5.6 (skip section 5.7)
7	Introduction to scale space. Discussion of homeworks	5.10, 5.11 (1)	Section 5.8 (skip section 5.9)
8	Relaxation and annealing	6.1, 6.3 (1)	Sections 6.1–6.3
9	Diffusion	6.2 (2)	Sections 6A.2
10	Equivalence of MFA and diffusion	6.7 and 6.8 (1)	Section 6A.4
11	Image morphology	7.5–7.7 (1)	Section 7.1
12	Morphology, continued. Gray-scale morphology. Distance transform	7.10 (2)	Sections 7.2, 7.3
13	Closing gaps in edges, connectivity	7.4 (1)	Section 7A.4
14	Segmentation by optimal thresholding		Sections 8.1, 8.2
15	Connected component labeling	8.2 (1)	Section 8.3
16	2D geometry, transformations	9.3 (1)	Sections 9.1, 9.2
17	2D shape features, invariant moments, Fourier descriptors, medial axis	9.2, 9.4, 9.10 (1)	Sections 9.3–9.7
18	Segmentation using snakes and balloons		Sections 8.5, 8.5.1
19	PDE representations and level sets		Section 8.5.2
20	Shape-from-X and structured illumination	9.10 (1)	Sections 9A.2.2, 9A.2.3
21	Graph-theoretic image representations: Graphs, region adjacency graphs. Subgraph isomorphism		Chapter 12
22	Consistent and relaxation labeling	10.1 (1)	Chapter 10
23	Hough transform, parametric transforms	11.1 (2)	Sections 11.1, 11.2, 11.3.3
24	Generalized Hough transform, Gauss map, application to finding holes in circuit boards		Section 11A.3
25	Iconic matching, springs and templates, association graphs	13.2 and 13.3 (1)	Sections 13.1–13.3
26	The role of statistical pattern recognition		

The assignments are projects which must include a formal report. Since there is usually programming involved, we allow more time to accomplish these assignments – suggested times are in parentheses in column 3. It is also possible, by careful selection of the students and the topics, to use this book in an advanced undergraduate course.

For advanced students, the "Topics" sections of this book should serve as a collection of pointers to the literature. Be sure to emphasize to your students (as we do in the text) that no textbook can provide the details available in the literature, and any "real" (that is, for a paying customer) machine vision project will require the development engineer to go to the published journal and conference literature. As stated above, the two recurrent themes throughout this book are consistency and optimization. The concept of consistency occurs throughout the discipline as a principal philosophical construct for solving machine vision problems. When confronted with a machine vision application, the engineer should seek to find ways to determine sources of information which are consistent. Optimization is the principal mathematical tool for solving machine vision problems, including determining consistency. At the end of each chapter which introduces techniques, we remind the student where consistency fits into the problems of that chapter, as well as where and which optimization methods are used.

Acknowledgements

My graduate students at North Carolina State University, especially Rajeev Ramanath, deserve a lot of credit for helping us make this happen. Bilgé Karacali also helped quite a bit with his proofreading, and contributed significantly to the section on support vector machines.

Of course, none of this would have mattered if it were not for my wife, Rosalyn, who provided the encouragement necessary to make it happen. She also edited the entire book (more than once), and converted it from Engineerish to English.

WES

I'd like to express my sincere thanks to Dr. Wesley Snyder for inviting me to coauthor this book. I have greatly enjoyed this collaboration and have gained valuable experience.

The final delivery of the book was scheduled around Christmas when my parents were visiting me from China. Instead of touring around the city and enjoying the holidays, they simply stayed with me and supported me through the final submission of the book. I owe my deepest gratitude to them. And to Feiyi, my forever technical support and emergency reliever.

HQ

1 Introduction

The proof is straightforward, and thus omitted

Ja-Chen Lin and Wen-Hsiang Tsai[1]

1.1 Concerning this book

This is an important observation: This book does NOT have enough information to tell you how to implement significant large systems. It teaches general principles. You MUST make use of the literature when you get down to the gnitty gritty.

We have written this book at two levels, the principal level being introductory. "Introductory" does not mean "easy" or "simple" or "doesn't require math." Rather, the introductory topics are those which need to be mastered before the advanced topics can be understood.

In addition, the book is intended to be useful as a reference. When you have to study a topic in more detail than is covered here, in order, for example, to implement a practical system, we have tried to provide adequate citations to the relevant literature to get you off to a good start.

We have tried to write in a style aimed directly toward the student and in a conversational tone.

We have also tried to make the text readable and entertaining. Words which are deluberately missppelled for humorous affects should be ubvious. Some of the humor runs to exaggeration and to puns; we hope you forgive us.

We did not attempt to cover every topic in the machine vision area. In particular, nearly all papers in the general areas of optical character recognition and face recognition have been omitted; not to slight these very important and very successful application areas, but rather because the papers tend to be rather specialized; in addition, we simply cannot cover everything.

There are two themes which run through this book: ***consistency*** and ***optimization***. Consistency is a conceptual tool, implemented as a variety of algorithms, which helps machines to recognize images – they fuse information from local measurements to make global conclusions about the image. Optimization is the mathematical mechanism used in virtually every chapter to accomplish the objectives of that chapter, be they pattern classification or image matching.

1 Ja-Chen Lin and Wen-Hsiang Tsai, "Feature-preserving Clustering of 2-D Data for Two-class Problems Using Analytical Formulas: An Automatic and Fast Approach," *IEEE Transactions on Pattern Analysis and Machine Intelligence*, **16**(5), 1994.

These two topics, consistency and optimization, are so important and so pervasive, that we point out to the student, in the conclusion of each chapter, exactly where those concepts turned up in that chapter. So read the chapter conclusions. Who knows, it might be on a test.

1.2 Concerning prerequisites

The target audience for this book is graduate students or advanced undergraduates in electrical engineering, computer engineering, computer science, math, statistics, or physics. To do the work in this book, you must have had a graduate-level course in advanced calculus, and in statistics and/or probability. You need either a formal course or experience in linear algebra.

To find out if you meet this criterion, answer the following question: What do the following words mean? "transpose," "inverse," "determinant," "eigenvalue." If you do not have any idea, do not take this course!

Many of the homeworks will be projects of sorts, and will be computer-based. To complete these assignments, you will need a hardware and software environment capable of

You will have to write programs in C (yes, C or C++, not Matlab) to complete this course.

(1) declaring large arrays (256×256) in C
(2) displaying an image
(3) printing an image.

The CDROM which comes with this book contains all the software you need, except for a compiler, editor, etc.

We are going to insist that you write programs, and that you write them at a relatively low level. Some of the functionality that you will be coding is available in software packages like Matlab. However, while you learn something by simply calling a function, you learn more by writing and debugging the code yourself. Exceptions to this occur, of course, when the coding is so extensive that the programming gets in the way of the image analysis. For that reason, we provide the student with a library of subroutines which allow the student to ignore details like data type, byteswapping, file access, and platform dependencies, and instead focus on the logic of making image analysis algorithms work.

You should have an instructor, and if you do, we strongly recommend that you GO to class, even though all the information you really need is in this book. Read the assigned material in the text, then go to class, then read the text material again. Remember:

A hacker hermit named Dave
Tapped in to this course in his cave.
He had to admit
He learned not a bit.
But look at the money he saved.

And now, on to the technical stuff.

1.3 Some terminology

Students usually confuse machine vision with image processing. In this section, we define some terminology that will clarify the differences between the contents and objectives of these two topics.

1.3.1 Image processing

Many people consider the content of this course as part of the discipline of image processing. However, a better use of the term is to distinguish between image processing and machine vision by the intent. "Image processing" strives to make images look better, and the output of an image processing system is an image. The output of a "machine vision" system is information about the content of the image. The functions of an image processing system may include enhancement, coding, compression, restoration, and reconstruction.

Enhancement

Enhancement systems perform operations which make the image look better, as perceived by a human observer. Typical operations include contrast stretching (including functions like histogram equalization), brightness scaling, edge sharpening, etc.

Coding

Coding is the process of finding efficient and effective ways to represent the information in an image. These include quantization methods and redundancy removal. Coding may also include methods for making the representation robust to bit-errors which occur when the image is transmitted or stored.

Compression

Compression includes many of the same techniques as coding, but with the specific objective of reducing the number of bits required to store and/or transmit the image.

Restoration

Restoration concerns itself with fixing what is wrong with the image. It is unlike enhancement, which is just concerned with making images look better. In order to "correct" an image, there must be some model of the image degradation. It is common in restoration applications to assume a deterministic blur operator, followed by additive random noise.

Reconstruction

Reconstruction usually refers to the process of constructing an image from several partial images. For example, in computed tomography (CT),[2] we make a large number, say 360, of x-ray projections through the subject. From this set of one-dimensional signals, we can compute the actual x-ray absorption at each point in the two-dimensional image. Similar methods are used in positron emission tomography (PET), magnetic resonance imagery (MRI), and in several shape-from-X algorithms which we will discuss later in this course.

1.3.2 Machine vision

Machine vision is the process whereby a machine, usually a digital computer, automatically processes an image and reports "what is in the image." That is, it recognizes the content of the image. Often the content may be a machined part, and the objective is not only to locate the part, but to inspect it as well. We will in this book discuss several applications of machine vision in detail, such as automatic target recognition (ATR), and industrial inspection. There are a wide variety of other applications, such as determining the flow equations from observations of fluid flow [1.1], which time and space do not allow us to cover.

The terms "computer vision" and "image understanding" are often also used to denote machine vision.

Machine vision includes two components – measurement of features and pattern classification based on those features.

Measurement of features

The measurement of features is the principal focus of this book. Except for Chapters 14 and 15, in this book, we focus on processing the elements of images (pixels) and from those pixels and collections of pixels, extract sets of measurements which characterize either the entire image or some component thereof.

Pattern classification

Pattern classification may be defined as the process of making a decision about a measurement. That is, we are **given** a measurement or set of measurements made on an unknown object. From that set of measurements with knowledge about the possible *classes* to which that unknown might belong, we make a decision. For

2 Sometimes, CT is referred to as "CAT scanning." In that case, CAT stands for "computed axial tomography." There are other types of tomography as well.

example, the set of possible classes might be men and women and one measurement which we could make to distinguish men from women would be height (clearly, height is not a very good measurement to use to distinguish men from women, for if our decision is that anyone over five foot six is male we will surely be wrong in many instances).

Pattern recognition

Pattern recognition may be defined as the process of assigning unknowns to classes just as in the definition of pattern classification. However, the definition is extended to include the process of making the measurements.

1.4 Organization of a machine vision system

Fig. 1.1 shows schematically, at the most basic level, the organization of a machine vision system. The unknown is first measured and the values of a number of *features* are determined. In an industrial application, such features might include the length, width, and area of the image of the part being measured. Once the features are measured, their numerical values are passed to a process which implements a *decision rule*. This decision rule is typically implemented by a subroutine which performs calculations to determine to which class the unknown is most likely to belong based on the measurements made.

As Fig. 1.1 illustrates, a machine vision system is really a fairly simple architectural structure. The details of each module may be quite complex, however, and many different options exist for designing the classifier and the feature measuring system. In this book, we mention the process of classifier design. However, the process of determining and measuring features is the principal topic of this book.

The "feature measurement" box can be further broken down into more detailed operations as illustrated in Fig. 1.2. At that level, the organization chart becomes more complex because the specific operations to be performed vary with the type of image and the objective of the tasks. Not every operation is performed in every application.

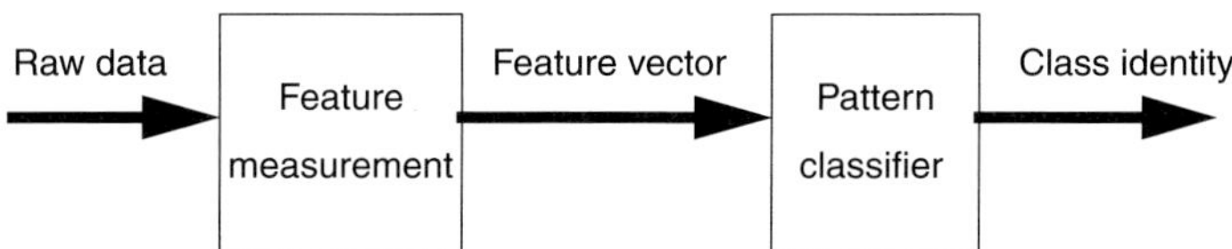

Fig. 1.1. Organization of a machine vision system.

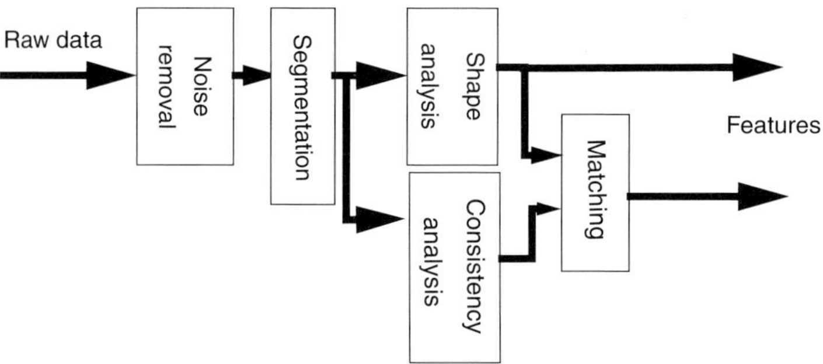

Fig. 1.2. Some components of a feature characterization system. Many machine vision applications do not use every block, and information often flows in other ways. For example, it is possible to perform matching directly on the image data.

1.5 The nature of images

We will pay much more attention to the nature of images in Chapter 4. We will observe that there are several different types of images as well as several different ways to represent images. The types of images include what we call "pictures," that is, two-dimensional images. In addition, however, we will discuss three-dimensional images and range images. We will also consider different representations for images, including iconic, functional, linear, and relational representations.

1.6 Images: Operations and analysis

Some equivalent words.

We will learn many different operations to perform on images. The emphasis in this course is "image analysis," or "computer vision," or "machine vision," or "image understanding." All these phrases mean the same thing. We are interested in making measurements on images with the objective of providing our machine (usually, but not always, a computer) with the ability to recognize what is in the image. This process includes several steps:

- *denoising* – all images are noisy, most are blurred, many have other distortions as well. These distortions need to be removed or reduced before any further operations can be carried out. We discuss two general approaches for denoising in Chapters 6 and 7.
- *segmentation* – we must segment the image into meaningful regions. Segmentation is covered in Chapter 8.
- *feature extraction* – making measurements, geometric or otherwise, on those regions is discussed in Chapter 9.

- *consistency* – interpreting the entire image from local measurements is covered in Chapters 10 and 11.
- *classification and matching* – recognizing the object is covered in Chapter 12 through Chapter 16.

So turn to the next chapter. (Did you notice? No homework assignments in this chapter? Don't worry. We'll fix that in future chapters.)

Reference

[1.1] C. Shu and R. Jain, "Vector Field Analysis for Oriented Patterns," *IEEE Transactions on Pattern Analysis and Machine Intelligence*, **16**(9), 1994.

2 Review of mathematical principles

Everything, once understood, is trivial

W. Snyder

2.1 A brief review of probability

Let us imagine a statistical experiment: rolling two dice. It is possible to roll any number between two and twelve (inclusive), but as we know, some numbers are more likely than others. To see this, consider the possible ways to roll a five.

We see from Fig. 2.1 that there are four possible ways to roll a five with two dice. Each event is **independent**. That is, the chance of rolling a two with the second die (1 in 6) does not depend at all on what is rolled with die number 1.

Independence of events has an important implication. It means that the *joint probability* of the two events is equal to the product of their individual probabilities and the conditional probabilities:

$$Pr(a|b)P(b) = Pr(a)Pr(b) = Pr(b|a)Pr(a) = Pr(a, b). \tag{2.1}$$

In Eq. (2.1), the symbols a and b represent **events**, e.g., the rolling of a six. $Pr(b)$ is the probability of such an event occurring, and $Pr(a \mid b)$ is the **conditional probability** of event a occurring, given that event b has occurred.

In Fig. 2.1, we tabulate all the possible ways of rolling two dice, and show the resulting number of different ways that the numbers from 2 to 12 can occur. We note that 6 different events can lead to a 7 being rolled. Since each of these events is equally probable (1 in 36), then a 7 is the most likely roll of two dice. In Fig. 2.2 the information from Fig. 2.1 is presented in graphical form.

In pattern classification, we are most often interested in the probability of a particular measurement occurring. We have a problem, however, when we try to plot a graph such as Fig. 2.2 for a continuously-valued function. For example, how do we ask the question: "What is the probability that a man is six feet tall?" Clearly, the answer is zero, for an infinite number of possibilities could occur (we might equally well ask, "What is the probability that a man is (exactly) 6.314 159 267 feet tall?"). Still, we know intuitively that the likelihood of a man being six feet tall is higher than the likelihood of his being ten feet tall. We need some way of quantifying this intuitive notion of likelihood.

Sum		Number of ways
0		0
1		0
2	1–1	1
3	2–1, 1–2	2
4	1–3, 3–1, 2–2	3
5	2–3, 3–2, 4–1, 1–4	4
6	1–5, 5–1, 2–4, 4–2, 3–3	5
7	3–4, 4–3, 2–5, 5–2, 1–6, 6–1	6
8	2–6, 6–2, 3–5, 5–3, 4–4	5
9	3–6, 6–3, 4–5, 5–4	4
10	4–6, 6–4, 5–5	3
11	6–5, 5–6	2
12	6–6	1

Fig. 2.1. The possible ways to roll two dice.

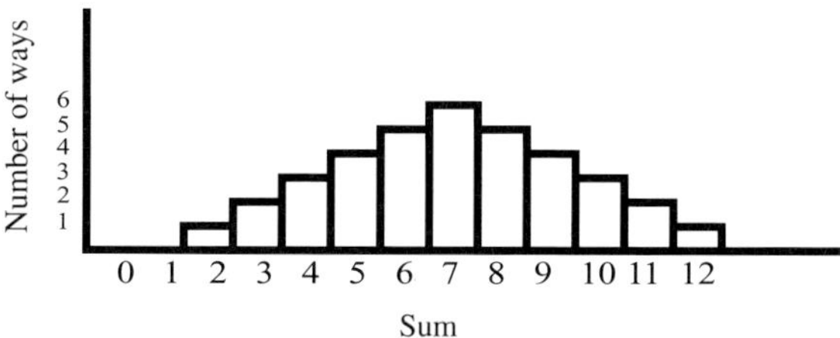

Fig. 2.2. The information of Fig. 2.1, in graphical form.

One question that does make sense is, "What is the probability that a man is **less than** six feet tall?" Such a function is referred to as a **probability distribution** function

$$P(x) = Pr(z < x) \tag{2.2}$$

for some measurement, z.

Fig. 2.3 illustrates the probability distribution function for the result of rolling two dice.

When we asked "what is the probability that a man is less than x feet tall?" we obtained the probability distribution function. Another well-formed question would be "what is the probability that a man's height is between x and $x + \Delta x$?" Such a question is easily answered in terms of the density function:

$$Pr(x \leq h < x + \Delta x) = Pr(h < x + \Delta x) - Pr(h < x) = P(x + \Delta x) - P(x)$$

Dividing by Δx and taking the limit as $\Delta x \to 0$, we see that we may define the **probability density function** as the derivative of the distribution function:

$$p(x) = \frac{d}{dx} P(x). \tag{2.3}$$

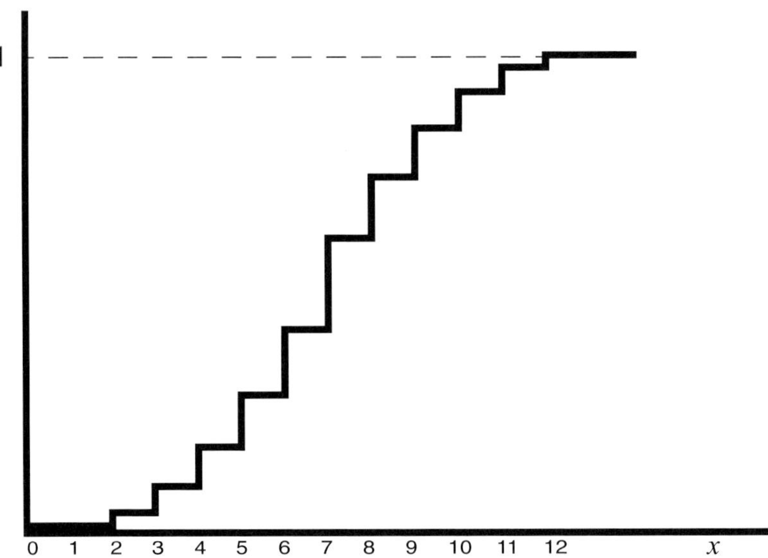

Fig. 2.3. The probability distribution of Fig. 2.2, showing the probability of rolling two dice to get a number LESS than x. Note that the curve is steeper at the more likely numbers.

$p(x)$ has all the properties that we desire. It is well defined for continuously-valued measurements and it has a maximum value for those values of the measurement which are intuitively most likely.

Furthermore:

$$\int_{-\infty}^{\infty} p(x)\, dx = 1, \tag{2.4}$$

which we must require, since **some** value will certainly occur.

2.2 A review of linear algebra

This section will serve more as a reference than a teaching aid, since you should know this material already.

In this section, we very briefly review vector and matrix operations. Generally, we denote vectors in boldface, scalars in lowercase Roman, and matrices in uppercase Roman.

Vectors are always considered to be column vectors. If we need to write one horizontally for the purpose of saving space in a document, we use transpose notation. For example, we denote a vector which consists of three scalar elements as:

$$\boldsymbol{v} = [x_1 \quad x_2 \quad x_3]^{\mathrm{T}}.$$

The inner product of two vectors is a scalar, $\boldsymbol{v} = \boldsymbol{a}^{\mathrm{T}}\boldsymbol{b}$. Its value is the sum of products of the corresponding elements of the two vectors:

$$\boldsymbol{a}^{\mathrm{T}}\boldsymbol{b} = \sum_i a_i b_i.$$

You will also sometimes see the notation $\langle x, y\rangle$ used for inner product. We do not like this because it looks like an expected value of a random variable. One sometimes also sees the "dot product" notation $\boldsymbol{x} \cdot \boldsymbol{y}$ for inner product.

The magnitude of a vector is $|\boldsymbol{x}| = \sqrt{\boldsymbol{x}^{\mathrm{T}}\boldsymbol{x}}$. If $|\boldsymbol{x}| = 1$, $\boldsymbol{x}$ is said to be a "unit vector." If $\boldsymbol{x}^{\mathrm{T}}\boldsymbol{y} = 0$, then $\boldsymbol{x}$ and $\boldsymbol{y}$ are "orthogonal." If $\boldsymbol{x}$ and $\boldsymbol{y}$ are orthogonal unit vectors, they are "orthonormal."

The concept of orthogonality can easily be extended to continuous functions by simply thinking of a function as an infinite-dimensional vector. Just list all the values of $f(x)$ as x varies between, say, a and b. If x is continuous, then there are an infinite number of possible values of x between a and b. But that should not stop us – we cannot enumerate them, but we can still think of a vector containing all the values of $f(x)$. Now, the concept of summation which we defined for finite-dimensional vectors turns into integration, and an inner product may be written

$$\langle f(x), g(x)\rangle = \int_a^b f(x)g(x)\,dx. \tag{2.5}$$

The concepts of orthogonality and orthonormality hold for this definition of the inner product as well. If the integral is equal to zero, we say the two functions are orthogonal. So the transition from orthogonal vectors to orthogonal functions is not that difficult. With an infinite number of dimensions, it is impossible to visualize orthogonal as "perpendicular," of course, so you need to give up on thinking about things being perpendicular. Just recall the definition and use it.

Suppose we have n vectors $\boldsymbol{x}_1, \boldsymbol{x}_2, \ldots \boldsymbol{x}_n$; if we can write $\boldsymbol{v} = a_1\boldsymbol{x}_1 + a_2\boldsymbol{x}_2 + \cdots a_n\boldsymbol{x}_n$, then $\boldsymbol{v}$ is said to be a "linear combination" of $\boldsymbol{x}_1, \boldsymbol{x}_2, \ldots \boldsymbol{x}_n$.

A set of vectors $\boldsymbol{x}_1, \boldsymbol{x}_2, \ldots \boldsymbol{x}_n$ is said to be "linearly independent" if it is impossible to write any of the vectors as a linear combination of the others.

Given d linearly independent vectors, of d dimensions, $\boldsymbol{x}_1, \boldsymbol{x}_2, \ldots \boldsymbol{x}_d$ defined on $\Re^d$, then any vector $\boldsymbol{y}$ in the space may be written $\boldsymbol{y} = a_1\boldsymbol{x}_1 + a_2\boldsymbol{x}_2 + \cdots a_d\boldsymbol{x}_d$.

Since any d-dimensional real-valued vector $\boldsymbol{y}$ may be written as a linear combination of $\boldsymbol{x}_1, \ldots \boldsymbol{x}_d$, then the set $\{\boldsymbol{x}_i\}$ is called a "basis" set and the vectors are said to "span the space" $\Re^d$. Any linearly independent set of vectors can be used as a basis (necessary and sufficient). It is often particularly convenient to choose basis sets which are orthonormal.

For example, the following two vectors form a basis for $\Re^2$

$$\boldsymbol{x}_1 = [0 \quad 1]^{\mathrm{T}} \quad \text{and} \quad \boldsymbol{x}_2 = [1 \quad 0]^{\mathrm{T}}.$$

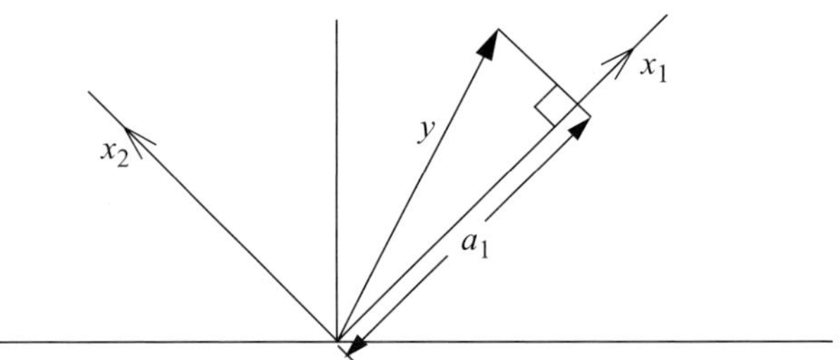

Fig. 2.4. x_1 and x_2 are orthonormal bases. The projection of y onto x_1 has length a_1.

This is the familiar Cartesian coordinate system. Here's another basis set for $\Re^2$

Is this set orthonormal?

$$\boldsymbol{x}_1 = [1 \quad 1]^{\mathrm{T}} \; \boldsymbol{x}_2 = [-1 \quad 1]^{\mathrm{T}}.$$

If $\boldsymbol{x}_1, \boldsymbol{x}_2, \ldots \boldsymbol{x}_d$ span $\Re^d$, and $\boldsymbol{y} = a_1\boldsymbol{x}_1 + a_2\boldsymbol{x}_2 + \cdots a_d\boldsymbol{x}_d$, then the "components" of $\boldsymbol{y}$ may be found by

$$a_i = \boldsymbol{y}^{\mathrm{T}}\boldsymbol{x}_i \tag{2.6}$$

and a_i is said to be the "projection" of $\boldsymbol{y}$ onto $\boldsymbol{x}_i$. In a simple Cartesian geometric interpretation, the inner product of Eq. (2.6) is literally a projection as illustrated in Fig. 2.4. However, whenever Eq. (2.6) is used, the term "projection" may be used as well, even in a more general sense (e.g. the coefficients of a Fourier series).

The only vector spaces which concern us here are those in which the vectors are real-valued.

2.2.1 Linear transformations

What does this say about m and d?

A "linear transformation," A, is simply a matrix. Suppose A is $m \times d$. If applied to a vector $\boldsymbol{x} \in \Re^d$, $\boldsymbol{y} = A\boldsymbol{x}$, then $\boldsymbol{y} \in \Re^m$. So A took a vector from one vector space, $\Re^d$, and produced a vector in $\Re^m$. If that vector $\boldsymbol{y}$ could have been produced by applying A to one and only one vector in $\Re^d$, then A is said to be "one-to-one." Now suppose that there are no vectors in $\Re^m$ that can not be produced by applying A to some vector in $\Re^d$. In that case, A is said to be "onto." If A is one-to-one and onto, then A^{-1} exists. Two matrices A and B are "conformable" if the matrix multiplication $C = AB$ makes sense.

We assume you know the meanings of transpose, inverse, determinant, and trace. If you do not, look them up.

Some important (and often forgotten) properties: If A and B are conformable, then

$$(AB)^{\mathrm{T}} = B^{\mathrm{T}}A^{\mathrm{T}} \tag{2.7}$$

and

$$(AB)^{-1} = B^{-1}A^{-1} \tag{2.8}$$

if A and B are invertible at all.

A couple of other useful properties are

$$\det(AB) = \det(BA) \quad \text{and} \quad \mathrm{tr}(AB) = \mathrm{tr}(BA)$$

which only is true, of course, if A and B are square. If a matrix A satisfies

$$AA^{\mathrm{T}} = A^{\mathrm{T}}A = I \tag{2.9}$$

then obviously, the transpose of the matrix is the inverse as well, and A is said to be an "orthonormal transformation" (OT), which will correspond geometrically to a rotation. If A is a $d \times d$ orthonormal transformation, then the columns of A are orthonormal, linearly independent, and form a basis spanning the space of $\Re^d$. For $\Re^3$, three convenient OTs are the rotations about the Cartesian axes:

Some example orthonormal transformations.

$$R_x = \begin{bmatrix} 1 & 0 & 0 \\ 0 & \cos\theta & -\sin\theta \\ 0 & \sin\theta & \cos\theta \end{bmatrix} R_y = \begin{bmatrix} \cos\theta & 0 & -\sin\theta \\ 0 & 1 & 0 \\ \sin\theta & 0 & \cos\theta \end{bmatrix} R_z = \begin{bmatrix} \cos\theta & -\sin\theta & 0 \\ \sin\theta & \cos\theta & 0 \\ 0 & 0 & 1 \end{bmatrix}$$

Suppose R is an OT, and $\boldsymbol{y} = R\boldsymbol{x}$, then

$$|\boldsymbol{y}| = |\boldsymbol{x}|. \tag{2.10}$$

A matrix A is "positive definite" if

$$y = \boldsymbol{x}^{\mathrm{T}}A\boldsymbol{x} > 0 \; \forall \boldsymbol{x} \in \Re^d, \boldsymbol{x} \neq 0$$

$\boldsymbol{x}^{\mathrm{T}}A\boldsymbol{x}$ is called a *quadratic form.*

The derivative of a quadratic form is particularly useful:

What happens here if A is symmetric?

$$\frac{d}{dx}(\boldsymbol{x}^{\mathrm{T}}A\boldsymbol{x}) = (A + A^{\mathrm{T}})\boldsymbol{x}.$$

Since we mentioned derivatives, we might as well mention a couple of other vector calculus things:

Suppose f is a scalar function of $\boldsymbol{x}$, $\boldsymbol{x} \in \Re^d$, then

$$\frac{df}{d\boldsymbol{x}} = \begin{bmatrix} \frac{\partial f}{\partial x_1} & \frac{\partial f}{\partial x_2} & \cdots & \frac{\partial f}{\partial x_d} \end{bmatrix}^{\mathrm{T}}, \tag{2.11}$$

and is called the "gradient." This will be often used when we talk about edges in images, and $f(\boldsymbol{x})$ will be the brightness as a function of the two spatial directions.

If f is vector-valued, then the derivative is a matrix

$$\frac{d\boldsymbol{f}^{\mathrm{T}}}{d\boldsymbol{x}} = \begin{bmatrix} \frac{\partial f_1}{\partial x_1} & \frac{\partial f_2}{\partial x_1} & \cdots & \frac{\partial f_m}{\partial x_1} \\ \cdots & \cdot & \cdots & \cdots \\ \cdots & \cdot & \cdots & \cdots \\ \cdots & \cdot & \cdots & \cdots \\ \frac{\partial f_1}{\partial x_d} & \frac{\partial f_2}{\partial x_d} & \cdots & \frac{\partial f_m}{\partial x_d} \end{bmatrix}, \tag{2.12}$$

and is called the "Jacobian."

One more: If f is scalar-valued, the matrix of second derivatives

$$\begin{bmatrix} \frac{\partial^2 f}{\partial x_1^2} & \frac{\partial^2 f}{\partial x_1 \partial x_2} & \cdots & \frac{\partial^2 f}{\partial x_1 \partial x_d} \\ \cdots & \cdots & \cdots & \cdots \\ \cdots & \cdots & \cdots & \cdots \\ \cdots & \cdots & \cdots & \cdots \\ \frac{\partial^2 f}{\partial x_d \partial x_1} & \frac{\partial^2 f_2}{\partial x_d \partial x_2} & \cdots & \frac{\partial^2 f}{\partial x_d^2} \end{bmatrix} \tag{2.13}$$

is called the "Hessian."

2.2.2 Derivative operators

Here, we introduce a new notation, a vector containing only derivative operators,

$$\nabla = \begin{bmatrix} \frac{\partial}{\partial x_1} & \frac{\partial}{\partial x_2} & \cdots & \frac{\partial}{\partial x_d} \end{bmatrix}^{\mathrm{T}}. \tag{2.14}$$

It is important to note that this is an OPERATOR, not a vector. We will do linear algebra sorts of things with it, but by itself, it has no value, not even really any meaning – it must be applied to something to have any meaning. For most of this book, we will deal with two-dimensional images, and with the two-dimensional form of this operator,

$$\nabla = \begin{bmatrix} \frac{\partial}{\partial x} & \frac{\partial}{\partial y} \end{bmatrix}^{\mathrm{T}}. \tag{2.15}$$

Apply this operator to a scalar, f, and we get a vector which does have meaning, the gradient of f:

$$\nabla f = \begin{bmatrix} \frac{\partial f}{\partial x} & \frac{\partial f}{\partial y} \end{bmatrix}^{\mathrm{T}}. \tag{2.16}$$

Similarly, we may define the divergence using the inner (dot) product (in all the following definitions, only the two-dimensional form of the del operator defined in

Eq. (2.16) is used. However, remember that the same concepts apply to operators of arbitrary dimension):

$$\operatorname{div} \boldsymbol{f} = \nabla \boldsymbol{f} = \begin{bmatrix} \frac{\partial}{\partial x} & \frac{\partial}{\partial y} \end{bmatrix} \begin{bmatrix} f_1 \\ f_2 \end{bmatrix} = \frac{\partial f_1}{\partial x} + \frac{\partial f_2}{\partial y}. \tag{2.17}$$

We will also have opportunity to use the outer product of the del operator with a matrix:

$$\nabla \times \boldsymbol{f} = \begin{bmatrix} \frac{\partial}{\partial x} \\ \frac{\partial}{\partial y} \end{bmatrix} [f_1 \quad f_2] = \begin{bmatrix} \frac{\partial f_1}{\partial x} & \frac{\partial f_2}{\partial x} \\ \frac{\partial f_1}{\partial y} & \frac{\partial f_2}{\partial y} \end{bmatrix}. \tag{2.18}$$

2.2.3 Eigenvalues and eigenvectors

If matrix A and vector $\boldsymbol{x}$ are conformable, then one may write the "characteristic equation"

$$A\boldsymbol{x} = \lambda \boldsymbol{x}, \lambda \in \Re. \tag{2.19}$$

Since $A\boldsymbol{x}$ is a linear operation, A may be considered as mapping $\boldsymbol{x}$ onto itself with only a change in length. There may be more than one "eigenvalue[1]" λ, which satisfies Eq. (2.19). For $\boldsymbol{x} \in \Re^d$, A will have exactly d eigenvalues (which are not, however, necessarily distinct). These may be found by solving $\det(A - \lambda I) = 0$. (But for $d > 2$, we do not recommend this method. Use a numerical package instead.)

For any given matrix, there are only a few eigenvalue/eigenvector pairs.

Given some eigenvalue λ, which satisfies Eq. (2.19), the corresponding $\boldsymbol{x}$ is called the corresponding "eigenvector."

2.3 Introduction to function minimization

In this book, essentially EVERY machine vision topic will be discussed in terms of some sort of minimization, so get used to it!

Minimization of functions is a pervasive element of engineering: One is always trying to find the set of parameters which minimizes some function of those parameters. Notationally, we state the problem as: Find the vector $\mathbf{x}$ which produces a minimum of some function $H(\mathbf{x})$:

$$\widehat{H} = \min_{\mathbf{x}} H(\mathbf{x}) \tag{2.20}$$

where $\mathbf{x}$ is some d-dimensional parameter vector, and H is a scalar function of $\mathbf{x}$, often referred to as an "objective function." We denote the $\mathbf{x}$ which results in the

1 "Eigen-" is the German prefix meaning "principal" or "most important." These are NOT named for Mr Eigen.

minimal H as $\mathbf{x}$

$$\mathbf{x} = \arg\min_{\mathbf{x}} H(\mathbf{x}). \tag{2.21}$$

The authors get VERY annoyed at improper use of the word "optimal." If you didn't solve a formal optimization problem to get your result, you didn't come up with the "optimal" anything.

The most straightforward way to minimize a function is to set its derivative to zero:

$$\nabla H(\mathbf{x}) = 0, \tag{2.22}$$

where ∇ is the gradient operator – the set of partial derivatives. Eq. (2.22) results in a set of equations, one for each element of $\mathbf{x}$, which must be solved simultaneously:

$$\begin{aligned}
\frac{\partial}{\partial x_1} H(\mathbf{x}) &= 0 \\
\frac{\partial}{\partial x_2} H(\mathbf{x}) &= 0 \\
&\cdots \\
\frac{\partial}{\partial x_d} H(\mathbf{x}) &= 0.
\end{aligned} \tag{2.23}$$

Such an approach is practical only if the system of Eq. (2.23) is solvable. This may be true if $d = 1$, or if H is at most quadratic in $\mathbf{x}$.

EXERCISE

Find the vector $\boldsymbol{x} = [x_1, x_2, x_3]^{\mathrm{T}}$ which minimizes

$$H = ax_1^2 + bx_1 + cx_2^2 + dx_3^2$$

where a, b, c, and d are known constants.

Solution

$$\begin{aligned}
\frac{\partial H}{\partial x_1} &= 2ax_1 + b \\
\frac{\partial H}{\partial x_2} &= 2cx_2 \\
\frac{\partial H}{\partial x_3} &= 2dx_3
\end{aligned}$$

minimized by

$$x_3 = x_2 = 0, x_1 = \frac{-b}{2a}.$$

If H is some function of order higher than two, or is transcendental, the technique of setting the derivative equal to zero will not work (at least, not in general) and we must resort to numerical techniques. The first of these is gradient descent.

In one dimension, the utility of the gradient is easy to see. At a point $x^{(k)}$ (Fig. 2.5), the derivative points **AWAY FROM** the minimum. That is, in one dimension, its sign will be positive on an "uphill" slope.

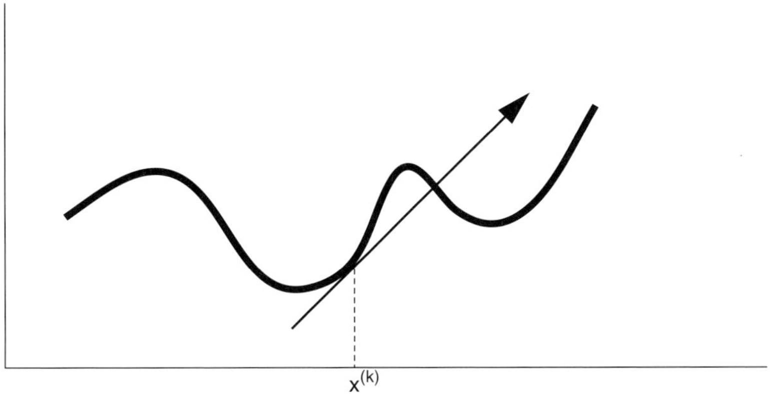

Fig. 2.5. The sign of the derivative is always away from the minimum.

Thus, to find a new point, x^{k+1}, we let

$$x^{(k+1)} = x^{(k)} - \alpha \left. \frac{\partial H}{\partial x} \right|_{x^{(k)}} \tag{2.24}$$

where α is some "small" constant.

In a problem with d variables, we write

$$\mathbf{x}^{(k+1)} = \mathbf{x}^{(k)} - \alpha \nabla H(\mathbf{x})|_{\mathbf{x}^{(k)}}. \tag{2.25}$$

2.3.1 Newton–Raphson

It is not immediately obvious in Eq. (2.25) how to choose the variable α. If α is too small, the iteration of Eq. (2.25) will take too long to converge. If α is too large, the algorithm may become unstable and never find the minimum.

We can find an estimate for α by considering the well-known Newton–Raphson method for finding roots: (In one dimension), we expand the function $H(x)$ in a Taylor series about the point $x^{(k)}$ and truncate, assuming all higher order terms are zero,

$$H\left(x^{(k+1)}\right) = H\left(x^{(k)}\right) + \left(x^{(k+1)} - x^{(k)}\right) H'\left(x^{(k)}\right).$$

Since we want $x^{(k+1)}$ to be a zero of H, we set

$$H\left(x^{(k)}\right) + \left(x^{(k+1)} - x^{(k)}\right) H'\left(x^{(k)}\right) = 0, \tag{2.26}$$

and find that to estimate a root, we should use

$$x^{(k+1)} = x^{(k)} - \frac{H\left(x^{(k)}\right)}{H'\left(x^{(k)}\right)}. \tag{2.27}$$

In optimization however, we are not finding roots, but rather, we are minimizing a function, so how does knowing how to find roots help us? The minima of the function are the roots of its derivative, and our algorithm becomes

Algorithm: Gradient descent

$$x^{(k+1)} = x^{(k)} - \frac{H'\left(x^{(k)}\right)}{H''\left(x^{(k)}\right)}. \tag{2.28}$$

In higher dimensions, Equation (2.28) becomes

$$\mathbf{x}^{(k+1)} = \mathbf{x}^{(k)} - \mathbf{H}^{-1}\nabla H, \tag{2.29}$$

where $\mathbf{H}$ is the Hessian matrix of second derivatives, which we mentioned earlier in this chapter:

$$\mathbf{H} = \left[\frac{\partial^2}{\partial x_i \partial x_j} H(\mathbf{x})\right]. \tag{2.30}$$

EXAMPLE

Given a set of x, y data pairs $\{(x_i, y_i)\}$ and a function of the form

$$y = ae^{bx}, \tag{2.31}$$

find the parameters a and b which minimize

$$H(a, b) = \sum_i (y_i - ae^{bx_i})^2. \tag{2.32}$$

Solution

We can solve this problem with the linear approach by observing that $\ln y = \ln a + bx$ and re-defining variables $g = \ln y$ and $r = \ln a$. With these substitutions, Eq. (2.32) becomes

$$H(r, b) = \sum_i (g_i - r - bx_i)^2 \tag{2.33}$$

$$\frac{\partial H}{\partial b} = 2\sum_i (g_i - r - bx_i)(-x_i) \tag{2.34}$$

$$\frac{\partial H}{\partial r} = 2\sum_i (g_i - r - bx_i)(-1). \tag{2.35}$$

Setting Eq. (2.34) to zero, we have

$$\sum_i g_i x_i - \sum_i r x_i - \sum_i b x_i^2 = 0 \tag{2.36}$$

or

$$r\sum_i x_i + b\sum_i x_i^2 = \sum_i g_i x_i \tag{2.37}$$

and from Eq. (2.35)

$$\sum_i g_i - r\sum_i 1 - b\sum_i x_i = 0 \tag{2.38}$$

or

$$Nr + b\sum_i x_i = \sum_i g_i \tag{2.39}$$

where N is the number of data points. Eqs. (2.37) and (2.39) are two simultaneous linear equations in two unknowns which are readily solved. (See [2.2, 2.3, 2.4] for more sophisticated descent techniques such as the conjugate gradient method.)

2.3.2 Local vs. global minima

Gradient descent suffers from a serious problem: Its solution is strongly dependent on the starting point. If started in a "valley," it will find the bottom of *that* valley. We have no assurance that this particular minimum is the lowest, or "global," minimum.

Before continuing, we will find it useful to distinguish two kinds of nonlinear optimization problems.

- **Combinatorial optimization**. In this case, the variables have discrete values, typically 0 and 1. With $\mathbf{x}$ consisting of d binary-valued variables, 2^d possible values exist for $\mathbf{x}$. Minimization of $H(\mathbf{x})$ then (in principle) consists of simply generating each possible value for $\mathbf{x}$ and consequently of $H(\mathbf{x})$, and choosing the minimum. Such "exhaustive search" is in general not practical due to the exponential explosion of possible values. We will find that simulated annealing provides an excellent approach to solving combinatorial optimization problems.
- **Image optimization**. Images have a particular property: Each pixel is influenced only by its neighborhood (this will be explained in more detail later), however, the pixel values are continuously-valued, and there are typically many thousand such variables. We will find that mean field annealing is most appropriate for the solution of these problems.

2.3.3 Simulated annealing

We will base much of the following discussion of minimization techniques on an algorithm known as "simulated annealing" (SA) which proceeds as follows. (See the book by Aarts and Van Laarhoven for more detail [2.1].)

Algorithm: Simulated annealing

Choose (at random) an initial value of $\mathbf{x}$, and an initial value of $T > 0$.

While $T > T_{\min}$, do

(1) Generate a point $\mathbf{y}$ which is a neighbor of $\mathbf{x}$. (The exact definition of neighbor will be discussed soon.)

(2) If $H(y) < H(x)$ then replace $\mathbf{x}$ with y.

(3) Else compute $P_y = \exp(-\frac{(H(\mathbf{y})-H(\mathbf{x}))}{T})$. If $P_y \geq R$ then replace $\mathbf{x}$ with $\mathbf{y}$, where R is a random number uniformly distributed between 0 and 1.
(4) Decrease T slightly and go to step 1.

How simulated annealing works

Simulated annealing is most easily understood in the context of combinatorial optimization. In this case, the "neighbor" of a vector $\mathbf{x}$ is another vector $\mathbf{x}_2$, such that only one of the elements of $\mathbf{x}$ is changed (discretely) to create $\mathbf{x}_2$.[2] Thus, if $\mathbf{x}$ is binary and of dimension d, one may choose a neighboring $\mathbf{y} = \mathbf{x} \oplus \mathbf{z}$, where $\mathbf{z}$ is a binary vector in which exactly one element is nonzero, and that element is chosen at random, and $\oplus$ represents exclusive OR.

In step 2 of the algorithm, we perform a descent. Thus we "always fall down hill."

In step 3, we provide a mechanism for sometimes making uphill moves. Initially, we ignore the parameter T and note that if $\mathbf{y}$ represents an uphill move, the probability of accepting $\mathbf{y}$ is proportional to $e^{-(H(\mathbf{y})-H(\mathbf{x}))}$. Thus, uphill moves can occur, but are exponentially less likely to occur as the size of the uphill move becomes larger. The likelihood of an uphill move is, however, strongly influenced by T. Consider the case that T is very large. Then $\frac{H(\mathbf{y})-H(\mathbf{x})}{T} \ll 1$ and $P_y \approx 1$. Thus, all moves will be accepted. As T is gradually reduced, uphill moves become gradually less likely until for low values of $T (T \ll (H(\mathbf{y}) - H(\mathbf{x})))$, such moves are essentially impossible.

One may consider an analogy to physical processes in which the state of each variable (one or zero) is analogous to the spin of a particle (up or down). At high temperatures, particles randomly change state, and if temperature is gradually reduced, minimum energy states are achieved. The parameter T in step 4 is thus analogous to (and often referred to as) temperature, and this minimization technique is therefore called "simulated annealing."

2.4 Markov models

A Markov process is most easily described in terms of time, although in machine vision we are primarily interested in interactions over spatial distances. The concept is that the probability of something happening is dependent on a thing that *just recently* happened. We will use Markov processes primarily in the noise removal and segmentation issues in Chapter 6, however, they find applications in a wide variety of problems, including character recognition [16.1].

We start by introducing the simplest kind of Markov model, the Markov chain. This type of model is appropriate whenever a sequence of things can be identified,

[2] Thus the set of neighbors of $\mathbf{x}$ consists of all $\mathbf{x}$'s of Hamming distance = 1.

for example, when a string of symbols is received over a computer network, or a sequence of words is input to a natural language processor. Let the symbols which are received be denoted by $y(t)$, where the argument t denotes the (discrete) time instant at which that symbol is received. Thus, $y(1)$ is received before $y(2)$. Let w denote the class to which a symbol belongs, $w \in \{w_1, w_2, \ldots w_c\}$, if there are c possible classes. As an example the signals could represent 0 or 1, as in a communications system.

We are interested in the probability that $y(t)$ belongs to some particular class. For example, what is the probability that the kth symbol will be a one? We are interested in that probability as a function of history. For example, let w_1 represent the class 1. We formalize our previous discussion by asking what is the probability that $y(t)$ is a 1, given the last N symbols received:

DONT PANIC! It's just notation, and it is explained, right after the equation.

$$P(y(t) \in w_1 | y(t-1) \in w_{t-1}, y(t-2) \in w_{t-2}, \ldots, y(t-N) \in w_{t-N}). \quad (2.40)$$

This is an awkward bit of notation, but here is what it means. When you see the term "$y(t)$" think "the symbol received at time t." When you see the "$\in$," think "is." When you see "w_k" think "1" or think "0" or think "whatever might be received at time k." For example, we might ask "what is the probability that you receive at time k the symbol 1, when the last four symbols received previously were 0110?" Which, in our notation is to ask what is

$$P(y(k) \in w_1 | (y(k-1) \in w_0, y(k-2) \in w_1, y(k-3) \in w_1, y(k-4) \in w_0)).$$

It is possible that in order to compute this probability, we must know all of the history, or it is possible that we need only know the class of the last few symbols. One particularly interesting case is when we need only know the class of the last symbol received. In that case, we could say that the probability of class assignments for symbol $y(t)$, given all of the history, is precisely the same as the probability knowing only the last symbol:

$$P(y(k)|y(k-1)) = P(y(k)|(y(k-1), y(k-2), \ldots)) \quad (2.41)$$

where we have simplified the notation slightly by omitting the set element symbols. That is, $y(k)$ does not denote the fact that the kth symbol was received, but rather that the kth symbol belongs to some particular class. If this is the case – that the probability conditioned on all of history is identical to the probability conditioned on the last symbol received – we refer to this as a *Markov process*.[3]

This relationship implies that

$$P(y(N) \in w_N, \ldots y(1) \in w_1) = \left\{ \prod_{t=2}^{N} P(y(t) \in w_t | (y(t-1) \in w_{t-1})) \right\} P(y(1) \in w_1).$$

[3] To be perfectly correct, this is a *first-order* Markov process, but we will not be dealing with any other types in this chapter.

Suppose there are only two classes possible, say 0 and 1. Then we need to know only four possible "transition probabilities," which we define using subscripts as follows:

$$P(y(t) = 0|y(t-1) = 0) \equiv P_{00}$$
$$P(y(t) = 0|y(t-1) = 1) \equiv P_{01}$$
$$P(y(t) = 1|y(t-1) = 0) \equiv P_{10}$$
$$P(y(t) = 1|y(t-1) = 1) \equiv P_{11}.$$

In general, there could be more than two classes, so we denote the transition probabilities by P_{ij}, and can therefore describe a Markov chain by a $c \times c$ matrix **P** whose elements are P_{ij}.

We will take another look at Markov processes when we think about Markov random fields in Chapter 6.

Assignment 2.1

Is the matrix **P** symmetric? Why or why not? Does **P** have any interesting properties? Do its rows (or columns) add up to anything interesting?

2.4.1 Hidden Markov models

Hidden Markov models (HMMs) occur in many applications, including in particular recent research in speech recognition. In a hidden Markov model, we assume that there may be more than one transition matrix, and that there is an unmeasurable (hidden) process which switches between transition matrices. Furthermore, that switching process itself may be statistical in nature, and we normally assume it is a Markov process. This is illustrated in Fig. 2.6, where the position of the switch determines whether the output $y(t)$ is connected to the output of Markov Process 1 or Markov Process 2. The switch may be thought of as being controlled by a finite

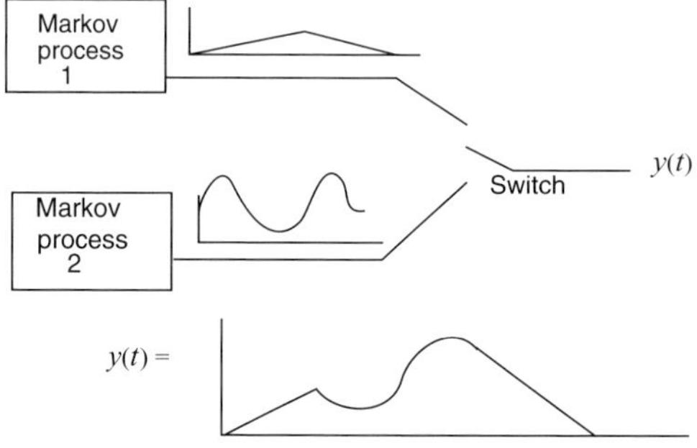

Fig. 2.6. A hidden Markov model may be viewed as a process which switches randomly between two signals.

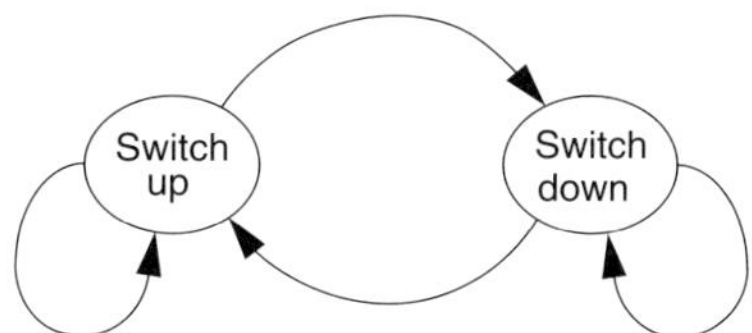

Fig. 2.7. The finite state machine of switches.

state machine (FSM), which at each time instant may stay in the same state or may switch, as shown in Fig. 2.7.

Here is our problem: We observe a sequence of symbols

$$Y = [y(t = 1), y(t = 2), \ldots] = [y(1), y(2), \ldots].$$

What can we infer? The transition probabilities? The state sequence? The structure of the FSM? The rules governing the FSM? Let's begin by estimating the state sequence.

Estimating the state sequence

Let $s(t)\, t = 1, \ldots, N$ denote the state associated with measurement $y(t)$, and denote the sequence of states $S = [s(1), s(2), \ldots s(N)]$, where each $s(t) \in \{s_1, s_2, \ldots s_m\}$. We seek a sequence of states, S, which maximizes the conditional probability that the sequence is correct, given the measurements; $P(S|Y)$.

Using Bayes' rule

$$P(S|Y) = \frac{p(Y|S)P(S)}{p(Y)}. \tag{2.42}$$

We assume the states form a Markov chain, so

$$P(S) = \left[\prod_{t=2}^{N} P_{s(t),s(t-1)}\right] P_{s(0)}. \tag{2.43}$$

Now, let's make a temporarily unbelievable assumption, that the probability density of the output depends only on the state. Denote that relationship by $p(y(t)|s(t))$. Then the posterior conditional probability of the sequence can be written:

$$p(Y|S)P(S) = \left[\prod_{t=1}^{N} p(y(t)|s(t))\right]\left[\prod_{t=2}^{N} P_{s(t),s(t-1)}\right] P_{s(0)}. \tag{2.44}$$

Define $P_{s(1),s(0)} \equiv P_{s(0)}$, and Eq. (2.44) simplifies to

$$p(Y|S)P(S) = \prod_{t=1}^{N} p(y(t)|s(t))P_{s(t),s(t-1)}. \tag{2.45}$$

Now look back at Eq. (2.42). The choice of S does not affect the denominator, so all we need to do is find the sequence S which maximizes

$$E = \prod_{t=1}^{N} p(y(t)|s(t)) P_{s(t),s(t-1)}. \tag{2.46}$$

Surprisingly, there is an algorithm which will solve this maximization problem. It is called the Viterbi algorithm. It has many, many applications. We explain it in the next section, since this seems like a good place to motivate it.

2.4.2 The Viterbi algorithm

This is used for a particular kind of optimization problem, one where each state $s(t)$ has only two neighboring states, $s(t+1)$ and $s(t-1)$. Imposing this "neighboring state" requirement allows us to use an efficient algorithm.

First, it is (almost) always easier to work with a sum than with a product, so let's define a new objective function by taking logs.

$$L \equiv \ln E \equiv \sum_{t=1}^{N} (\Psi(t) + \Theta_{i,j}) \tag{2.47}$$

where $\Psi(t) = \ln p(y(t)|s(t))$ and $\Theta_{i,j} = \ln P_{i,j}$.

Pictorially, we illustrate the set of all possible sequences as a graph as illustrated in Fig. 2.8.

A particular sequence of states describes a *path* through this graph. For example, for a graph like this with $N = 4$, $m = 3$, the path $[s_1, s_2, s_3, s_1]$ is illustrated in Fig. 2.9.

A path like this implies a set of values for the functions. For each node in the graph, we associate a value of Ψ. Suppose we measured $y(1) = 2$, $y(2) = 1$, $y(3) = 2.2$,

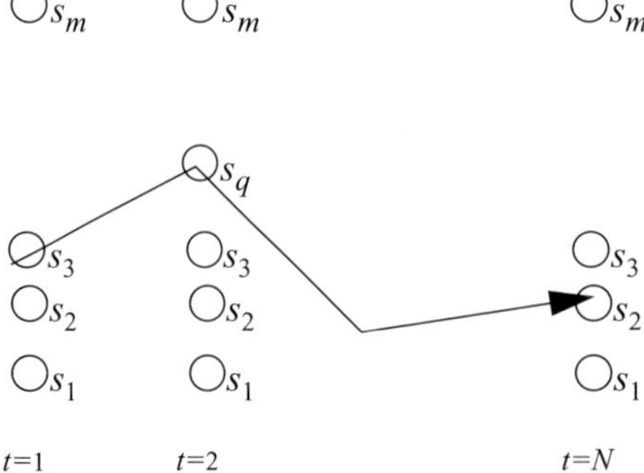

Fig. 2.8. Every possible sequence of states can be thought of as a path through such a graph.

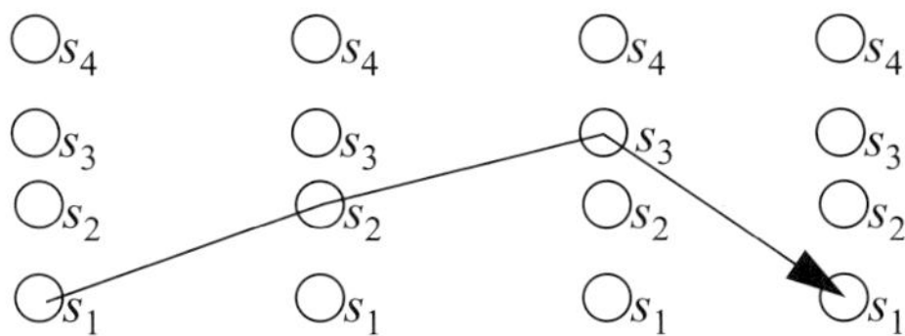

Fig. 2.9. A path through a problem with four states and four time values.

$y(4) = 1$. The function Ψ has value

$$\Psi = \ln p(y(1) = 2|s(1) = s_1) + \ln p(y(2) = 1|s(2) = s_2) \\ + \ln p(y(3) = 2.2|s(3) = s_3) + \ln p(y(4) = 1|s(4) = s_1).$$

There is a value associated with each edge in the graph as well, the function Θ determined by the associated transition probability. So every possible path through the graph has a corresponding value of the objective function L. We describe the algorithm to find the best path as an induction: Suppose at time t we have already found the best path to each node, with cost denoted $\mathrm{LB}_i(t)$, $i = 1, \ldots, m$. Then we can compute the cost of going from each node at time t to each node at time $t + 1$ (m^2 calculations) by

$$L'_{ij}(t + 1) = LB_i(t) + \Psi(y(t + 1)|s_j(t + 1)) + \Theta_{i,j}. \tag{2.48}$$

The best path to node j at time $t + 1$ is the maximum of these. When we finally reach time step N, the node which terminates the best path is the final node.

The computational complexity of this algorithm is thus Nm^2, which is a **lot** less than m^N, the complexity of a simple exhaustive search of all possible paths.

2.4.3 Markov outputs

In the description above, we assumed that the probability of a particular output depended only on the state. We do not have to be that restrictive; we could allow the outputs themselves to be the Markov processes.

Assume if the state changes, the first output depends only on the state, as before. But afterward, if the state remains the same, the outputs obey a Markov chain. We can formulate this problem in the same way, and solve it with the Viterbi algorithm.

2.4.4 Estimating model parameters

One final detail still eludes us: Given an observation Y, how do we estimate the conditional output probabilities $p(y(k)|s(k))$, and those transition probabilities for the states, $P_{i,j}$?

To do this, first, we need to go from a continuous to a discrete representation for the values of $y(t)$, in order to be able to use probabilities rather than densities. Assume $y(t) \in \{y_1, y_2, \ldots y_r\}$. Then, we may define an output probability matrix $\Pi = [\pi_{k,l}], k = 1, \ldots m; l = 1, \ldots r$, which represents the probability of observing output y_l if in state s_k.

Define

$$P_{i,j|Y}(t) = Pr((s(t-1) = i, s(t) = j)|Y). \tag{2.49}$$

That is, given the observation sequence, what is the probability that we went from state i to state j at time t? We can compute that quantity using the methods of section 2.4.2 if we know the transition probabilities $P_{i,j}$ and the output probabilities $\pi_{k,l}$. Suppose we do know those. Then, we estimate the transition probability by averaging the probabilities over all the inputs.

$$P_{i,j} = \frac{\sum_{t=2}^{N} P_{i,j|Y}(t)}{\sum_{t=2}^{N} P_{j|Y}(t)} \tag{2.50}$$

where, since in order to go into state j, the system had to go there from somewhere,

$$P_{j|Y}(t) = \sum_{i=1}^{N} P_{ij|Y}(t). \tag{2.51}$$

Then we estimate the probability of the observation by again averaging all the observations.

$$\pi_{k,l} = \frac{\sum_{t=1, y(t)=j}^{N} P_{i|Y}(t)}{\sum_{t=1}^{N} P_{i|Y}(t)}. \tag{2.52}$$

At each iteration, we use Eqs. (2.50) and (2.52) to update the parameters. We then use Eqs. (2.49) and (2.51) to update the conditional probabilities. The process then repeats until it converges.

2.4.5 Applications of HMMs

Hidden Markov models have found many applications in speech recognition and document content recognition [17.29].

Assignment 2.2

(Trivia question) In what novel did a character named Markov Chaney occur?

Assignment 2.3

Find the OT corresponding to a rotation of 30° about the z axis. Prove the columns of the resulting matrix are a basis for $\Re^3$.

Assignment 2.4

Prove Eq. (2.10). Hint: Use Eq. (2.7); Eq. (2.9) might be useful as well.

Assignment 2.5

A positive definite matrix has positive eigenvalues. Prove this. (For that matter, is it even true?)

Assignment 2.6

Does the function $y = xe^{-x}$ have a unique value x which minimizes y? If so, can you find it by taking a derivative and setting it equal to zero? Suppose this problem requires gradient descent to solve. Write the algorithm you would use to find the x which minimizes y.

Assignment 2.7

We need to solve a minimization problem using gradient descent. The function we are minimizing is $\sin x + \ln y$. Which of the following is the expression for the gradient which you need in order to do gradient descent?

(a) $\cos x + \frac{1}{y}$ (b) $y = -\frac{1}{\cos x}$ (c) $-\infty$

(d) $\begin{bmatrix} \cos x \\ 1/y \end{bmatrix}$ (e) $\frac{\partial}{\partial y}\sin x + \frac{\partial}{\partial x}\ln y$

Assignment 2.8

(a) Write the algorithm which uses gradient descent to find the vector $[x, y]^T$ which minimizes the function

$z = x\exp(-(x^2+y))$. (b) Write a computer program which finds the minimizing x, y pair.

Assignment 2.9

Determine whether the functions $\sin x$ and $\sin 2x$ might be orthonormal or orthogonal functions.

References

[2.1] E.H.L. Aarts and P.J.M. van Laarhoven. *Simulated Annealing: Theory and Applications*, Dordrecht, Holland, Reidel, 1987.

[2.2] R.L. Burden, J.D. Faires, and A.C. Reynolds. *Numerical Analysis*, Boston, MA, Prindle, Weber and Schmidt, 1981.

[2.3] G. Dahlquist and A. Bjorck. *Numerical Methods*, Englewood Cliffs, NJ, Prentice-Hall, 1974.

[2.4] B. Gottfried and J. Weisman. *Introduction to Optimization Theory*, Englewood Cliffs, NJ, Prentice-Hall, 1973.

3 Writing programs to process images

Computer Science is not about computers any more than astronomy is about telescopes

E. W. Dijkstra

One may take two approaches to writing software for image analysis, depending on what one is required to optimize. One may write in a style which optimizes/minimizes programmer time, or one may write to minimize computer time. In this course, computer time will not be a concern (at least not usually), but *your* time will be far more valuable. For that reason, we want to follow a programming philosophy which produces correct, operational code in a minimal amount of programmer time.

The programming assignments in this book are specified to be written in C or C++, rather than in MATLAB or JAVA. This is a conscious and deliberate decision. MATLAB in particular hides many of the details of data structures and data manipulation from the user. In the course of teaching variations of this course for many years, the authors have found that many of those details are precisely the details that students need to grasp in order to effectively understand what image processing (particularly at the pixel level) is all about.

3.1 Image File System (IFS) software

The objective of quickly writing good software is accomplished by using the image access subroutines in IFS. IFS is a collection of subroutines and applications based on those subroutines which support the development of image processing software. Advantages of IFS include the following.

- IFS supports any data type including char, unsigned char, short, unsigned short, int, unsigned int, float, double, complex float, complex double, complex short, and structure.
- IFS supports any image size, and any number of dimensions. One may do signal processing by simply considering a signal as a one-dimensional image.
- IFS is available on most current computer systems, including Windows on the PC, Linux on the PC, Unix on the SUN, and OS-X on the Macintosh.[1] Files written on

[1] Regrettably, IFS does not support Macintosh operating systems prior to OS-X.

one platform may be read on any of the other platforms. Conversion to the format native to the platform is done by the read routine, without user intervention.

- A large collection of functions are available, including two-dimensional Fourier transforms, filters, segmenters, etc.

3.1.1 The IFS header structure

All IFS images include a header which contains various items of information about the image, such as the number of points in the image, the number of dimensions for the image, the data format, the units and scan direction of each dimension, and so on. Also associated with the image is the actual data for the image. The image header includes a pointer to the image data.

The user manipulates an image by calling some function in the IFS library; one of the arguments to the function will be the address of the header. From the information in the header, the IFS library functions automatically determine where the data is and how to access it. In addition to accessing data in images, the IFS routines automatically take care of allocating space in memory to store data and headers. Everything is totally dynamic in operation; there are no fixed-dimension arrays. This relieves the user of the difficulties involved with accessing data in arrays, when the arrays are not of some fixed size.

The header structure for an image is defined in the file ⟨ifs.h⟩, and is known by the name IFSHDR. To manipulate an image, the user merely needs to declare a pointer to an image header structure (as IFSHDR *your_image; or IFSIMG your_image;). Then, the user simply calls some IFS function to create a new image, and sets the pointer to the value returned from that function.

3.1.2 Some useful IFS functions

Now, you should look at the IFS manual. Read especially carefully through the first two example programs in the front of the manual.

You may read images off disk using *ifspin*. This subroutine will calculate the size of the image, how much memory is required, and determine what sort of computer wrote the image. It will do all necessary data conversions (byte swapping, floating point format conversions, etc.) and read it into your computer in the format native to the machine you are using. You do not need to know how to do any of those data conversion operations. Similarly, you can write images to disk using *ifspot*. This subroutine will write both the IFS header and the associated actual image data from memory to the disk.

You may access the image using *ifsigp* or *ifsfgp* and *ifsipp* or *ifsfpp*. These subroutine names stand for IFS Integer Get Pixel, IFS Floating Get Pixel, IFS Integer Put Pixel, and IFS Floating Put Pixel. The words Integer and Floating refer to the data type returned or being written. If you say, for example,

v = ifsfgp(img,x,y)

then a floating point number will be returned *independent of the data type of the image*. That is, the subroutine will do data conversions for you. Similarly, *ifsigp* will return an integer, no matter what the internal data type is. This can, of course, get you in trouble. Suppose the internal data type is float, and you have an image consisting of numbers less than one. Then, the process of the conversion from float to int will truncate all your values to zero.

For some projects you will have three-dimensional data. That means you must access the images using a set of different subroutines, *ifsigp3d*, *ifsfgp3d*, *ifsipp3d*, and *ifsfpp3d*. For example,

y = ifsigp3d(img,frame,row,col)

3.1.3 Common problems

Two common problems usually occur when students first use IFS software.

(1) **ifsipp(img,x,y,exp(-t*t))** will give you trouble because *ifsipp* expects a fourth argument which is an integer, and exp will return a double. You should use *ifsfpp*.

(2) **ifsigp(img,x,y,z)** is improperly formed. *ifsigp* expects three arguments, and does not check the dimension of the input image to determine number of arguments (it could, however; sounds like a good student project...). To access a three-dimensional image, either use pointers, or use *ifsigp3d(img,x,y,z)*, where the second argument is the frame number.

3.2 Basic programming structure for image processing

Images may be thought of as two-dimensional arrays. They are usually processed pixel-by-pixel in a raster scan. In order to manipulate an image, two-nested for-loops is the most commonly used programming structure, as shown in Fig. 3.1.

```
......
int row, col;
......
for ( row = 0; row < 128; row++)
{
    for( col = 0; col < 128; col++)
    {
        /* pixel processing */
        ......
    }
    ......
}
```

Fig. 3.1. Basic programming structure: two-nested for-loops.

```
......
int row, col, frame;
......
for (frame = 0; frame < 224; frame++)
{
    for ( row = 0; row < 128; row++)
    {
        for( col = 0; col < 128; col++)
        {
            /* pixel processing */
            ......
        }
        ......
    }
}
```

Fig. 3.2. Basic programming structure: three-nested for-loops.

In this example, we use two integers (*row* and *col*) as the indices to the row and column of the image. By increasing *row* and *col* with a step one, we are actually scanning the image pixel-wise from left to right, top to bottom.

If the image has more than two dimensions, (e.g. hyperspectral images) a third integer is then used as the index to the dimensions, and correspondingly, a three-nested for-loops is needed, as shown in Fig. 3.2.

3.3 Good programming styles

You should follow one important programming construct: **all programs should be written so that they will work for any size image**. You are not required to write your programs so that they will work for any number of dimensions (although that is possible too), or any data type; only size. One implication of this requirement is that you cannot declare a static array, and copy all your data into that array (which novice programmers like to do). Instead, you must use the image access subroutines.

Another important programming guideline is that except in rare instances, **global variables are forbidden**. A global variable is a variable which is declared outside of a subroutine. Use of such globals is poor programming practice, and causes more bugs than anything else in all of programming. Good structured programming practice requires that everything a subroutine needs to know is included in its argument list.

Following these simple programming guidelines will allow you to write general-purpose code easily and efficiently, and with few bugs. As you become more skilled you can take advantage of the pointer manipulation capabilities to increase the run speed of your programs, if this is needed later.

```
for (<cond>) {        for (<cond>)          for (<cond>)          for (<cond>)
    <body>            {                         {                     {
}                         <body>                <body>                    <body>
                      }                         }                     }
```

(a) the K&R style (b) the Allman style (c) the Whitesmiths style (d) the GNU style

Fig. 3.3. The four commonly used indenting styles.

You will lose major points if your code does not follow these guidelines.

Besides the aforementioned programming guidelines, it is very important for students to also follow the *indenting*, *spacing*, and *commenting* rules in order to make your code "readable."

There are four commonly used indenting styles, the K&R style (or kernel style), Allman style (or BSD style), Whitesmiths style, and the GNU style, as shown in Fig. 3.3. In this book, we use the Allman indenting style. Adding some space (e.g. a blank line) between different segments of your code also improves readability. We emphasize the importance of the comments. However, do not add too many comments since that will break the flow of your code. In general, you should add a block comment at the top of each function implementation, including descriptions of what the function does, who wrote this function, how to call this function, and what the function returns. You should also add a description for each variable declaration.

3.4 Example programs

Take another look at the IFS manual. It will help you to follow these example programs. Also pay attention to the programming style we use. The comments might be too detailed, but they are included for pedagogical purpose.

A typical program is given in Fig. 3.4, which is probably as simple an example program as one could write. Fig. 3.5 lists another example which implements the same function as Fig. 3.4 does but is written in a more flexible way such that it can handle images of different size.

Both these examples make use of the subroutine calls *ifsigp*, *ifsipp*, *ifsfgp*, and *ifsfpp* to access the images utilizing either integer-valued or floating point data. The advantage of these subroutines is convenience: No matter what data type the image is stored in, *ifsigp* will return an integer, and *ifsfgp* will return a float. Internally, these subroutines determine precisely where the data is stored, access the data, and convert it. All these operations, however, take computer time. For class projects, the authors strongly recommend using these subroutines. However, for production operations, IFS supports methods to access data directly using pointers, trading additional programmer time for shorter run times.

```
/* Example1.c
     This program thresholds an image. It uses a fixed image size.
     Written by Harry Putter, October, 2006
*/
#include <stdio.h>
#include <ifs.h>
main( )
{
     IFSIMG img1, img2;          /* Declare pointers to headers */
     int len[3];                 /* len is an array of dimensions, used by ifscreate */
     int threshold;              /* threshold is an int here */
     int row,col;                /* counters */
     int v;

     /* read in image */
     img1 = ifspin("infile.ifs");    /* read in file by this name */

     /* create a new image to save the result */
     len[0] = 2;                 /* image to be created is two dimensional */
     len[1] = 128;               /* image has 128 columns */
     len[2] = 128;               /* image has 128 rows */
     img2 = ifscreate("u8bit",len,IFS_CR_ALL,0);          /* image is unsigned 8 bit */
     threshold = 55;             /* set some value to threshold */

     /* image processing part - thresholding */
     for (row = 0; row < 128; row++)
          for (col = 0; col < 128; col++)
          {
               v = ifsigp(img1,row,col);                /* read a pixel as an int */
               if (v > threshold)
                    ifsipp(img2,row,col,255);
               else
                    ifsipp(img2,row,col,0);
          }

     /* write the processed image to a file */
     ifspot(img2, "img2.ifs");       /* write image 2 to disk */
}
```

Fig. 3.4. Example IFS program to threshold an image using specified values of dimensions and predetermined data type.

3.5 Makefiles

You really should use makefiles. They are far superior to just typing commands. If you are doing your software development using Microsoft C++, Lcc, or some other compiler, then the makefiles are sort of hidden from you, but it is helpful to know how they operate. Basically, a makefile specifies how to build your project, as illustrated by the example makefile in Fig. 3.6.

The example in Fig. 3.6 is just about as simple a makefile as one can write. It states that the executable named *myprogram* depends on only one thing, the object module myprogram.o. It then shows how to make myprogram from myprogram.o and the IFS library.

Similarly, *myprogram.o* is made by compiling (but not linking) the source file, myprogram.c, utilizing header files found in an “include” directory on the CDROM, named hdr. Note: To specify a library, as in the link step, one must specify the library

```
/* Example2.c
    Thresholds an image using information about its data type and the dimensionality.
    Written by Sherlock Holmes, May 16, 1885
*/
#include <stdio.h>
#include <ifs.h>
main( )
{
    IFSIMG img1, img2;           /* Declare pointers to headers */
    int *len;                    /* len is an array of dimensions, used by ifscreate */
    int frame, row, col;         /* counters */
    float threshold, v;          /* threshold is a float here */

    img1 = ifspin("infile.ifs");  /*read in file by this name*/
    len = ifssiz(img1);           /* ifssiz returns a pointer to an array of image dimensions*/
    img2 = ifscreate(img1->ifsdt,len,IFS_CR_ALL,0);
                                  /* output image is to be the same type as the input */
    threshold = 55;               /* set some value to threshold */

    /* check for one, two or three dimensions */
    switch (len[0]) {
    case 1:                                     /* 1d signal */
        for (col = 0; col < len[1]; col++)
        {
            v = ifsfgp(img1,0,col);             /* read a pixel as a float */
            if (v > threshold)
                ifsfpp(img2,0,col,255.0);       /* write a float */
            else                                /* if img2 not float, will be converted*/
                ifsfpp(img2,0,col,0.0);
        }
        break;
    case 2:                                     /* 2d picture */
        for (row = 0; row < len[2]; row++)
            for (col = 0; col < len[1]; col++)
            {
                v = ifsfgp(img1,row,col);       /* read a pixel as a float */
                if (v > threshold)
                    ifsfpp(img2,row,col,255.0); /* store a float */
                else
                    ifsfpp(img2,row,col,0.0);
            }
        break;
    case 3:                                     /* 3d volume */
        for (frame = 0; frame < len[3];frame++)
            for (row = 0; row < len[2]; row++)
                for (col = 0; col < len[1]; col++)
                {
                    v = ifsfgp3d(img1,frame,row,col);        /* read a pixel as a float */
                    if (v > threshold)
                        ifsfpp3d(img2,frame,row,col,255.0);
                    else
                        ifsfpp3d(img2,frame,row,col,0.0);
                }
        break;
    default:
        printf("Sorry I cannot do 4 or more dimensions\n");
    } /* end of switch */

    ifspot(img2, "img2.ifs"); /* write image 2 to disk */
}
```

Fig. 3.5. **An example IFS program to threshold an image using number of dimensions, size of dimensions, and data type determined by the input image.**

```
myprogram: myprogram.o
    cc -o myprogram myprogram.o /CDROM/Solaris/ifslib/libifs.a

myprogram.o: myprogram.c
    cc -c myprogram.c -I/CDROM/Solaris/hdr
```

Fig. 3.6. An example makefile which compiles a program and links it with the IFS library.

```
CFLAGS=-Ic:\lcc\include -g2 -ansic
CC=c:\lcc\bin\lcc.exe
LINKER=c:\lcc\bin\lcclnk.exe
DIST=c:\ece763\myprog\lcc\
OBJS=c:\ece763\myprog\objs\
LIBS= ifs.lib -lm
# Build myprog.c
myprog:
    $(CC) -c $(CFLAGS) c:\ece763\myprog\mysubroutine1.c
    $(CC) -c $(CFLAGS) c:\ece763\myprog\mysubroutine2.c
    $(CC) -c $(CFLAGS) c:\ece763\myprog\myprog.c
    $(LINKER) -subsystem console -o myprog.exe myprog.obj mysubroutine1.obj
mysubroutine2.obj $(LIBS)
```

Fig. 3.7. An example WIN32 makefile.

name (e.g. libifs.a), but to specify an include file (e.g. ifs.h), one specifies only the directory in which that file is located, since the file name was given in the #include preprocessor directive.

In WIN32 the makefiles look like the example shown in Fig. 3.7. Here, many of the symbolic definition capabilities of the make program are demonstrated, and the location of the compiler is specified explicitly.

The programs generated by IFS are (with the exception of ifsview) console-based. That is, you need to run them inside an MSDOS window on the PC, inside a terminal window under Linux, Solaris, or on the Mac, using OS-X.

Assignment 3.1 Learning how to use the tools

Windows users:

If you are using a WIN32 computer, open an MSDOS session and cd to the directory where you installed all the IFS files. What program would you use to add Gaussian random noise to an image? (Hint: Look in INDEX). Apply that program to the image named angio128.ifs. Display the result using ifsview. Figure out how to print the result. Hand in the answers to the questions and the image.

Assignment

3.2 Learn how to use IFS

The intent of this assignment is to get you to use the computer and to begin to write programs.

Use ifsview to view the following images:

images/echo1
images/echo2
images/echo3

Convert each of these to TIFF, using any2any. (For usage, type "any2any -h"). For example, you might use

```
any2any echo1.ifs echo1.tif tiff
```

Important observation: echo1 is three-dimensional. TIFF is a two-dimensional image format. When you use any2any as above, you only get the zeroth frame!

You will see that the three echo images are similar. Written assignment, to be handed in: Describe in one paragraph what the images look like. We recognize that you do not know the anatomy. Just describe what you see; include an image (figure out how to print a file – you know this already, we hope.) If it happens that you do know the anatomy, your instructor will be impressed.

4 Images: Formation and representation

Computers are useless. They can only give us answers

Pablo Picasso

In this chapter, we describe how images are formed and how they are represented. *Representations* include both mathematical representations for the information contained in an image and for the ways in which images are stored and manipulated in a digital machine. In this chapter, we also introduce a way of thinking about images – as surfaces with varying height – which we will find to be a powerful way to describe both the properties of images as well as operations on those images.

4.1 Image representations

In this section, we discuss several ways to represent the information in an image. These representations include: iconic, functional, linear, probabilistic, spatial frequency, and relational representations.

4.1.1 Iconic representations (an image)

An iconic representation of the information in an image is an image. "Yeah, right; and a rose is a rose is a rose." When you see what we mean by functional, linear, and relational representations, you will realize we need a word[1] for a representation which is itself a picture. Some examples of iconic representations include the following.

y

$f(x, y)$

x

- 2D brightness images, also called luminance images. The things you are used to calling "images." These might be color or gray-scale. (Be careful with the words "black and white," as that might be interpreted as "binary"). We usually denote the brightness at a point $\langle x, y\rangle$ as $f(x, y)$. Note: x and y could be integers (in this case, we are referring to discrete points in a sampled image; these points are called "pixels," short for "picture elements"), or real numbers (in this case, we are thinking of the image as a function).

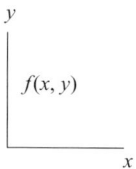

[1] "Iconic" comes from the Greek word meaning picture.

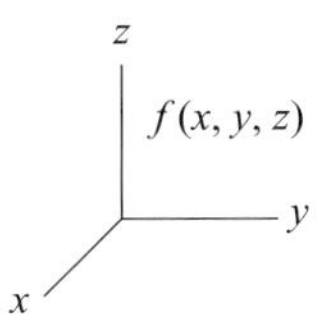

- 3D (density) images. These sorts of data structures usually occur in medical images, like CT (computed tomography), MRI (magnetic resonance imaging), ultrasound, etc. In the typical 3D image, each pixel represents the density at a point, (technically, we cannot measure the density AT a point, we can only measure the density in the *vicinity* of a point – but the meaning should be clear). We usually denote the density at a point $\langle x, y, z \rangle$ as $f(x, y, z)$.

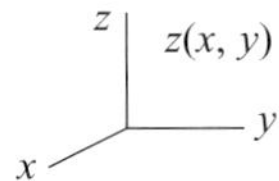

- $2\frac{1}{2}$D (range) images. In a range image, the value at each point on a surface represents a distance, usually the distance from the camera or the distance along a normal to a plane containing the camera. Thus, in a range image, we denote the position of the surface point as $z(x, y)$. Suppose, for example, that the sensor is a laser, measuring time-of-flight. Then the "brightness" is actually proportional to the time required for a pulse of light to travel from the laser to a surface, bounce off, and return to a detector located at the source. In truth, such a sensor is measuring range as a function of two deflection angles, $r(\theta, \varphi)$, and this produces a *range image*. However to transform from these coordinates to $z(x, y)$ is straightforward, and normally, any data you get (in this course anyway) will already have this correction performed to produce an *altitude image*.

You have a homework involving files named (e.g.) site1.c.ifs. The c indicates that correction has been done.

4.1.2 Functional representations (an equation)

We can fit a function to any set of data points. When we measure an image, it always consists of a discrete and finite set of measurements, $f(i, j)$, where i and j are the discrete pixel coordinates (integers) in the x and y directions. By performing some operation such as minimum-least-squares, we can find a continuous function which best fits this set of data. We could thus represent an image, at least over a small area, by an equation such as a biquadratic:

$$z = ax^2 + by^2 + cxy + dx + ey + f \tag{4.1}$$

or a quardic:

$$ax^2 + by^2 + cz^2 + dxy + exz + fyz + gx + hy + iz + j = 0. \tag{4.2}$$

Implicit and explicit representations

The form given in Eq. (4.1), in which one variable is defined in terms of the others, is often referred to as an *explicit* representation, whereas the form of Eq. (4.2) is an *implicit* representation [4.23], which may be equivalently represented in terms of the *zero set*, $\{(x, y, z): f(x, y, z) = 0\}$. Implicit polynomials have some convenient properties. For example consider a point (x_0, y_0) which is not in the zero set of $f(x, y)$, that is, the set of points x, y which satisfy

$$f(x, y) \equiv x^2 + y^2 - R^2 = 0. \tag{4.3}$$

If we substitute x_0 and y_0 into the equation for $f(x, y)$, we know we get a nonzero result (since we said this point is not in the zero set); if that value is negative, we know that (x_0, y_0) is inside the curve, otherwise, outside [4.3]. This inside/outside property holds for all closed curves (and surfaces) representable by polynomials.

4.1.3 Linear representations (a vector)

Throughout this book, we use the transpose whenever we write a vector on a row, because we think of vectors as column vectors.

5	10
6	4

We will discuss this in more detail later, but the basic idea is to unwind the image into a vector, and then talk about doing vector–matrix operations on such a representation. For example, the 2×2 image with brightness values as shown could be written as a vector $f = [5\ 10\ 6\ 4]^{\mathrm{T}}$.

4.1.4 Probabilistic representations

In Chapter 6 we will represent an image as the output of a random process which generates images. We can in that way make use of a powerful set of mathematical tools for estimating the best version of a particular image, given a measurement of a corrupted, noisy image.

4.1.5 The spatial frequency representation

Let your eye scan across the two images in Fig. 4.1, say from left to right, and decide in which of them the brightness changes more rapidly. You define what "more rapidly" means. We think you will agree that the right-hand image has more rapid changes in brightness. We can quantify that concept of brightness change by characterizing it in cycles per inch of paper, where one cycle is a transition from maximum brightness to minimum and back. Of course, if you scan the image at a fixed rate, the brightness variation seen by your eye, measured in cycles per second (hertz) is proportional to the spatial frequency content of the image you are scanning.

(a)

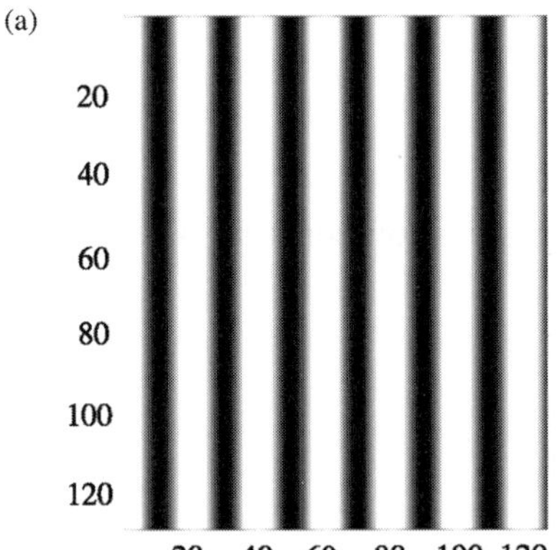

(b)

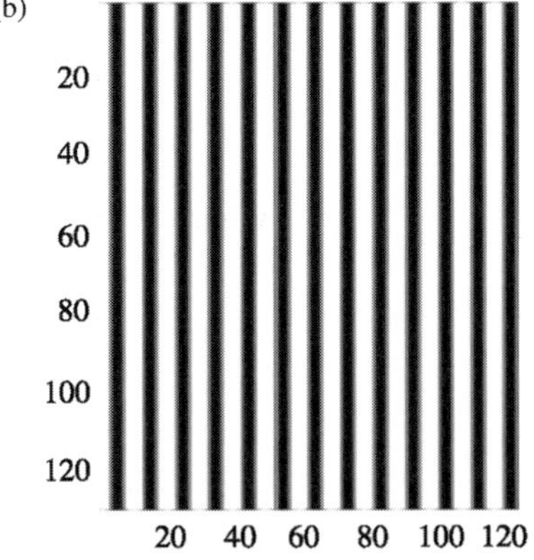

Fig. 4.1. (a) An image with lower horizontal frequency content. (b) An image with higher horizontal frequency content.

Fig. 4.2. (L) An image. (R) A low-frequency iconic representation of that image. The right-hand image is blurred in the horizontal direction only, a blur which results when the camera is panned while taking the picture. Notice that horizontal edges are sharp.

The spatial frequency content of an image can be modified by filters which block specific frequency ranges. For example, Fig. 4.2 illustrates an original image and an iconic representation of that image which has been passed through a *low-pass* filter. That is, a filter which permits low frequencies to pass from input to output, but which blocks higher frequencies. As you can see, the frequency response is one way of characterizing *sharpness*. Images with lots of high-frequency content are perceived as sharp.

Although we will make little use of frequency domain representations for images in this course, you should be aware of a few aspects of frequency domain representations.

First, as you should have already observed, spatial frequencies differ with direction. Fig. 4.1 illustrates much more rapid variation, higher spatial frequencies in the vertical direction than in the horizontal. Furthermore, in general, an image contains many spatial frequencies. We can extract the spatial frequency content of an image using the two-dimensional Fourier transform, given by

$$F(u, v) = \frac{1}{K} \sum_x \sum_y f(x, y) \exp(-i2\pi(ux + vy)) \tag{4.4}$$

where K is a suitable normalization constant. The Fourier transform is properly defined using integrals, and the form in Eq. (4.4) is defined only for discrete two-dimensional grids, such as a two-dimensional sampled image. It is important to note that the Fourier transform converts a function of x and y into another function, whose arguments are u and v. We think of u and v as the spatial frequency content in the x and y directions respectively. If $F(u_0, v_0)$ is a large number, for some particular values u and v, then there is a lot of energy at that particular spatial frequency. You may interpret the term *energy* as "lots of pixels," or "very large variations at that frequency."

The second observation made here is that spatial frequencies vary over an image. That is, if one were to take subimages of an image, one would find significant variation in the Fourier transforms of those subimages.

Third, take a look at the computational complexity implied in Eq. (4.4). The Fourier transform of an image is a function of spatial frequencies, u and v, and may thus be considered as an image (its values are complex, but that should not worry you). If our image is $N \times N$, we must sum over x and y to get a SINGLE u, v value – a complexity of N^2. If the frequency domain space is also sampled at $N \times N$, we have a total complexity of N^4 to compute the Fourier transform. BUT, there exists an algorithm called the fast Fourier transform which very cleverly computes a single u, v value in $N \log_2 N$ rather than N^2, resulting in a significant saving. Thus it is sometimes faster to compute things in the frequency domain.

Finally, there is an equivalence between convolution, which we will discuss in the next chapter and multiplication in the frequency domain. More on that in section 5.8.

4.1.6 Relational representations (a graph)

Finally, graphs are a very general way to think about the information in an image. We will talk a lot more about graphs in Chapter 12.

4.2 The digital image

Now suppose the image is sampled; that is, x and y take on only discrete, integer values, and also suppose f is quantized (f takes on only a set of integer values). Such an image could be stored in a computer memory and called a "digital image."

4.2.1 The formation of a digital image

The imaging literature is filled with a variety of imaging devices, including dissectors, flying spot scanners, vidicons, orthicons, plumbicons, CCDs (charge-coupled devices), and others [4.6, 4.8]. Currently, CCDs dominate the market. Imaging devices differ both in the ways in which they form images and in the properties of the images so formed. However, all these devices convert light energy to voltage in similar ways. Since our intent in this chapter is to introduce the fundamental concepts of image analysis, we will choose one device, the CCD, and discuss the way in which it acquires a digital image. More precise details are available in the literature, for example, for CCDs [4.19].

Image formation with a silicon device

A lens is used to form an image on the surface of the CCD. When a photon of the appropriate wavelength strikes the special material of the device, a quantum of charge is created (an electron–hole pair). Since the conductivity of the material is quite low, these charges tend to remain in the same general area where they were created. Thus, to a good approximation, the charge, q, in a local area of the CCD follows

$$q = \int_{0}^{t_f} i \, dt$$

where i is the incident light intensity, measured in photons per second. If the incident light is a constant over the integration time, then $q = i t_f$, where t_f is called the *frame time*.

In vidicon-like devices, the accumulated (positive) charge is cancelled by a scanning electron beam. The cancellation process produces a current which is amplified and becomes the video signal. In a CCD, charge is shifted from one cell to the next synchronously with a digital clock. The mechanism for reading out the charge, be it electron beam, or charge coupling, is always designed so that as much of the charge is set to zero as possible. We start the integration process with zero accumulated charge, build up the charge at a rate proportional to local light intensity, and then read it out. Thus, the signal measured at a point will be proportional to both the light intensity at that point and to the amount of time between read operations.

Since we are interested only in the intensities and not in the integration time, we remove the effect of integration time by making it the same everywhere in the picture. This process, called *scanning*, requires that each point on the device be interrogated and its charge accumulation zeroed, repetitively and cyclically. Probably the most straightforward, and certainly the most common way in which to accomplish this is in a top-to-bottom, left-to-right scanning process called *raster scanning* (Fig. 4.3).

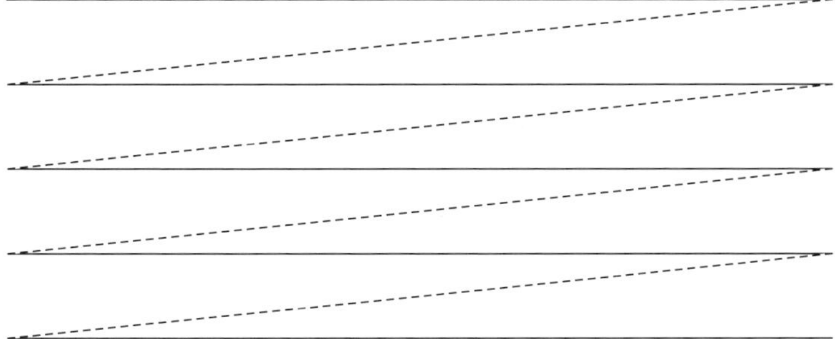

Fig. 4.3. Raster scanning: Active video is indicated by a solid line, blanking (retrace) by a dashed line. In an electron beam device, the beam is turned off as it is repositioned. Blanking has no physical meaning in a CCD, but is imposed for compatibility. This simplified figure represents noninterlaced scanning.

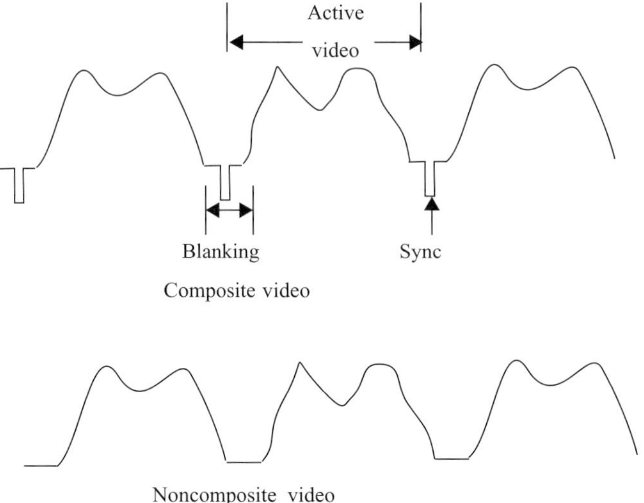

Fig. 4.4. Composite and noncomposite outputs of a television camera, voltage as a function of time.

To be consistent with standards put in place when scanning was done by electron beam devices (and there needed to be a time when the beam was shut off), the television signal has a pause at the end of each scan line called *blanking*. While charge is being shifted out from the bottom of the detector, charge is once again built up at the top. Since charge continues to accumulate over the entire surface of the detector at all times, it is necessary for the read/shift process to return immediately to the top of the detector and begin shifting again. This scanning process is repeated many times each second. In American television, the entire faceplate is scanned once every 33.33 ms (in Europe, the frame time is 40 ms).

To compute exactly how fast the electron beam is moving, we compute

$$\frac{1\text{s}}{30\,\text{frame}} \div \frac{525\,\text{lines}}{\text{frame}} = 63.5\ \mu\text{s/line}. \tag{4.5}$$

Using the European standard of 625 lines and 25 frames per second, we arrive at almost exactly the same answer, 64 μs per line. This 63.5 μs includes not only the active video signal but also the blanking period, approximately 18 percent of the line time. Subtracting this *dead time*, we arrive at the active video time, 52 μs per line.

Fig. 4.4 shows the output of a television camera as it scans three successive lines. One immediately observes that the raster scanning process effectively converts a picture from a two-dimensional signal to a one-dimensional signal, where voltage is a function of time. Fig. 4.4 shows both *composite* and *noncomposite* video signals,

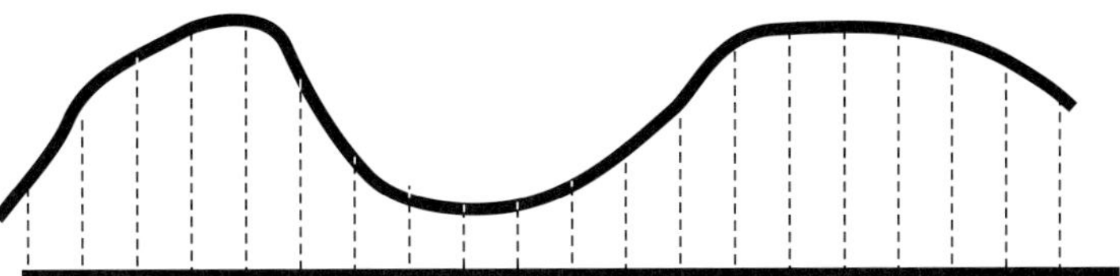

Fig. 4.5. Video signal, showing sampling times.

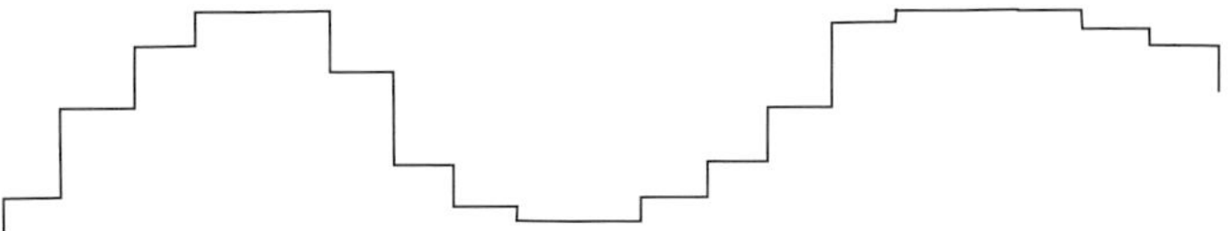

Fig. 4.6. Sampled video signal.

that is, whether the signal does or does not include the sync and blanking timing pulses.

The sync signal, while critical to operation of conventional television, is not particularly relevant to our understanding of digital image processing at this time. The blanking signal, however, is the single most important timing signal in a raster scan system. *Blanking* refers to the time that there is no video. There are two distinct blanking events: *horizontal blanking*, which occurs at the end of each line, and *vertical blanking*, which occurs at the bottom of the picture. In a digital system, both blanking events may be represented by pulses on separate digital wires. Composite video is constructed by shifting these special timing pulses negative and adding them to the video signal.

Now that we recognize that horizontal blanking signifies the beginning of a new line of video data, we can concentrate on that line and learn how a computer might acquire the brightness information encoded in that voltage.

The sampling process

The charge on the detector is converted to a voltage by passing through a resistor and then amplified. That signal is converted to a digital representation using an analog-to-digital converter. An analog-to-digital converter performs two functions simultaneously – sampling and quantization.

The sampling process approximates the video signal by measuring it at specific times. At each discrete time, the video signal is measured and that value is remembered until the next sampling time. Fig. 4.5 shows an analog voltage represented as a function of time, and Fig. 4.6 shows the same voltage after sampling.

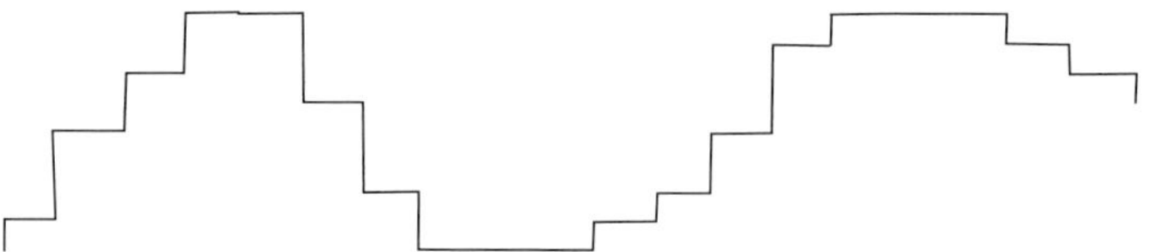

Fig. 4.7. Sampled quantized video signal.

Resolution

The number of samples on a single line defines the *horizontal resolution* of a video system. Similarly, the number of lines in a single image defines the *vertical resolution*. It is interesting to note that European television, with 625 lines per picture, has a greater vertical resolution than American. This is why viewers observe that European TV has a "better picture" than American.

The term *resolution* may also refer to the physical size of the smallest thing the imaging system can clearly image. For example, the resolution of mammographic x-ray film is around 50 microns; meaning that a dot on the film as small as that may be discovered.

For computer monitors, there are many resolution standards, and we will not list them all here. However the approach to calculating clock rates is the same.

Dynamic range

The sampled analog signal is converted to digital form by the *quantization* process, as shown in Fig. 4.7. The digital representation of any signal can have only a finite number of possible values, which are defined by the number of bits in the output word. Video signals are often quantized to 8 bits of accuracy, thus allowing a signal to be represented as one of 256 possible values.

One definition of the *dynamic range* of the imaging system is the number of bits of the digital representation. An alternative definition specifies the dynamic range as the range of input signal over which a camera successfully operates. Both meanings are accepted and are in common use, but they differ according to the context.

Since a digital image is raster scanned and sampled, there is a one-to-one relationship between time and space. That is, if we refer to the *sampling time*, we must speak of it relative to the top-of-picture signal (vertical blanking). That timing relationship identifies a unique position on the screen.

The sampling theorem

An interesting question arises if we wish to sample and store an analog signal and we wish to reconstruct that signal exactly from the sampled version. It can be shown

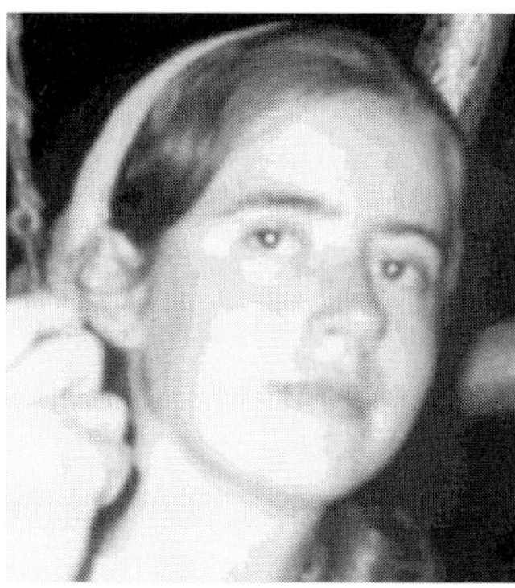

Fig. 4.8. Image of a face represented using 16 shades of gray (4 bits) on left, and with eight shades (3 bits) on right.

that exact reconstruction requires a sampling rate of at least twice the highest frequency in the signal.

In machine vision, we are usually not concerned with exact reconstruction of the most subtle details of the image, but wish to extract just the information we need to accomplish the task at hand.

Quantization error is the term used to refer to the fact that information is lost whenever the continuously valued analog signal is partitioned into discrete ranges. Quantization error is often observed as *contouring*, as illustrated in Fig. 4.8.

4.2.2 Formation of range images

There are two dominant ways to acquire a range image – using stereopsis and structured lighting.

Stereopsis

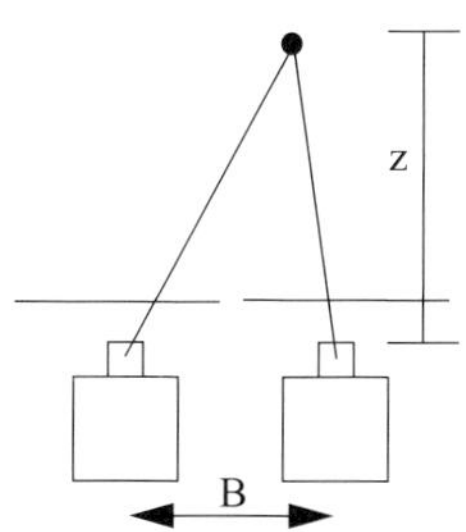

Fig. 4.9. From the images extracted from two cameras, it is possible to compute the location in 3-space of any point, provided we can solve the correspondence problem.

Most animals have two eyes, and we know from experience that with two views, we can extract three-dimensional information. It is not hard to work that out geometrically (Fig. 4.9).

If we know the distance between the cameras, the angle of observation of each camera (in most stereopsis systems, the cameras are set up so their midlines are parallel), and we can measure where particular points in the scene appear in both images, then we can calculate distance to the object, which will be referred to as "range." If we can do this in the general case, that is, if we can always identify which point in the left image corresponds to which point in the right image, we have solved the *correspondence problem*.

The correspondence problem is one of the fundamental problems of machine vision! Some people would say it is THE problem in machine vision.

An often-used simplifying assumption is that the two cameras are set up so that they are exactly parallel, and if a point occurs on a particular line in the left image then it will appear on the same "epipolar line" in the right image. In other words, the epipolar line connects a point and its correspondence. This assumption makes it possible to reduce the complexity of the correspondence problem dramatically.

The literature is filled with papers on approaches to the correspondence problem. Most of them focus on point matching. That is, find a point on the epipolar line in the second image which in some way resembles a point in the first image. For example, Bokil and Khotanzad [4.5] extended the work of Marr and Poggio [4.27], which makes use of epipolar assumption. They accomplish point matching by establishing a gray level compatibility matrix (GLCM). The pixel values of the left and right images are labels at the bottom and left of the matrix. The i, j element of the matrix is determined by computing the absolute value of the difference between brightness values in the ith row of the left-hand image and the jth column in the right-hand image. The GLCM values are then normalized. Row-to-row correlations are then established and a best match is selected.

The correspondence problem may be made easier by hierarchical matching [4.26] (two low-level features like epipolar edges correspond only if they belong to regions which correspond).

There are methods for finding curves in 3-space which do not explicitly require a solution to the correspondence problem. For example, Cohen and Wang [4.10, 4.11] solve for the best matching curve rather than for the individual points.

Camera calibration [4.28, 4.37] is important for stereo [4.31], and at the same time, stereopsis can be used for camera calibration, since it establishes a relationship between the two-dimensional image and the three-dimensional world. A great deal of effort has been expended to determine the minimum set of correspondences [4.1, 4.32] or other relationships [4.33, 4.34] required to calibrate the cameras.

A set of correspondences implies a transformation determining the pose of an object (the position and orientation of the object in 3-space is known as the "pose"). In a given scene, there may be multiple sets of correspondences [4.20, 4.38].

There are many special case applications [4.13], including how to obtain stereo information from panoramic cameras [4.30].

We will revisit stereopsis in Chapter 11 after we have learned how the concepts of parametric transformations will help in the solution of the correspondence problem.

Structured illumination

One variation eliminates the correspondence problem – replace one camera with a light source (e.g. a laser beam passing through a cylindrical mirror). However, this is really no longer stereopsis. Instead, it is a method referred to as *structured illumination*. To see how this works, look back at Fig. 4.9, and think of one of

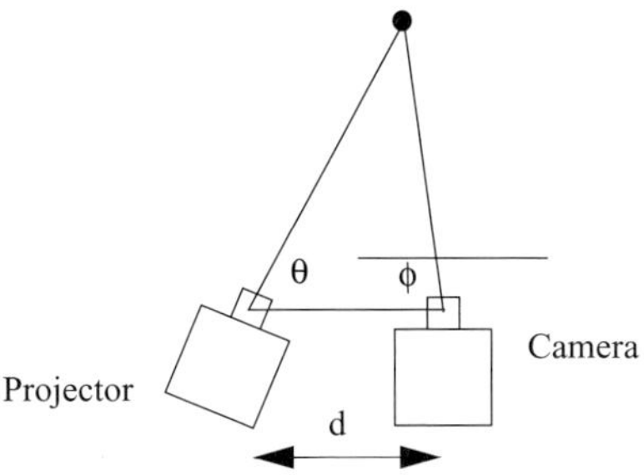

Fig. 4.10. Structured illumination.

those cameras as being replaced by a projector which shines a very narrow, very bright slit on the scene, as illustrated in Fig. 4.10. Now, one angle, θ, is known from the projector; the other angle, ϕ, is measured by finding the bright spot in the camera image, counting over pixels, and knowing the relationship between pixels and angle. Finally, knowledge of the distance between cameras, d, makes the triangle solvable.

One observation seems relevant to this point in describing images. An interesting problem occurs when one uses structured illumination to look at specular reflectors such as metal surfaces. With specular reflectors, either not enough or too much light may be reflected (polarization filters help [4.29]).

We will see more about using structured illumination when we get to the "shape-from-X" sections of this book.

4.3 Describing image formation

Define brightness as a function of two spatial variables. $f(x, y)$ is the brightness at point $x, y; x, y \in \Re$, and f is assumed real-valued $f \in \Re$.

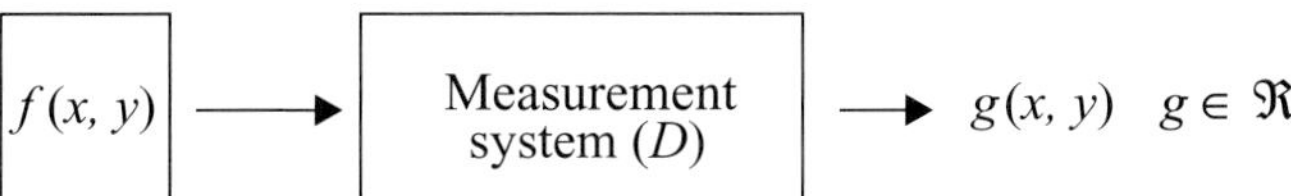

A measurement system corrupts the input image to produce the measured image:

$$g(x, y) = D(f(x, y)), \tag{4.6}$$

where D is some distortion function which typically includes some random noise process.

A distortion of particular interest to us is one where the noise is additive and independent of the signal, and g can be written as

$$g(x, y) = \iint_{\infty}^{\infty} f(\alpha, \beta)h(x - \alpha, y - \beta)\, d\alpha\, d\beta + n(x, y). \tag{4.7}$$

Suppose f is a delta function, then g will be h. That is, the image of a point is h: The "point spread function" (PSF).

This is the convolution integral. We can show that for any distortion operator D, if D is linear and is the same wherever it is applied (space-invariant), then there is a convolution in the form of Eq. (4.7) which produces distortion identical to D. Here is how it happens: The following derivation of the convolution integral is done in one dimension, only for convenience. The extension to two dimensions is trivial. First, observe that any function f, evaluated at a point x, can be written as

$$f(x) = \int_{-\infty}^{\infty} f(x')\delta(x - x')\, dx' \tag{4.8}$$

where $\delta(r)$ represents the delta function which is equal to zero when its argument is nonzero, equal to infinity when its argument is zero, and has an integral of one. Equation (4.8) is not profound – it simply defines the way the delta samples a function. However, now let us suppose our function f is corrupted by some operator which changes f at every point x. Then,

First assumption: d is linear.

$$D(f(x)) = D\left(\int_{-\infty}^{\infty} f(x')\delta(x - x')\, dx'\right). \tag{4.9}$$

Now *if D is a linear operator*, then we can interchange the operator D and the integral to obtain

$$D(f(x)) = \int_{-\infty}^{\infty} D(f(x')\delta(x - x'))\, dx'. \tag{4.10}$$

D is an operator which operates on things that are functions of x, not x'. As far as D is concerned, anything which just depends on x' is a constant. So we can factor the first term outside of the operator to obtain

$$D(f(x)) = \int_{-\infty}^{\infty} f(x')D(\delta(x - x'))\, dx'. \tag{4.11}$$

Now observe that D may depend on x, or it may depend on the difference between x and x', but in any case, it is the distortion operator applied to just the delta function. So any LINEAR distortion of f can be written as an integral of the product of f with a function which is the distortion applied to the delta. Since the delta function is really just a very bright spot, with infinite height and zero width, in one dimension, we call it an impulse, and call $D(\delta(x - x'))$ the *impulse response*. The two-dimensional delta function is a point of light, so we call the result of applying the distortion to it the *point spread function.* The impulse response and the point spread function are precisely the same thing, the only difference is in usage.

Since the impulse response might depend on both x and x', we introduce a new notation which we call h:

Look out! another assumption: Space-invariant.

$$h(x, x') = D(\delta(x - x')). \tag{4.12}$$

If we make another assumption, we can get a simpler expression: Let's assume that D depends not on x, but only on the difference between x and x'. In that case, we can write $h(x, x') = h(x - x')$, and Eq. (4.11) simplifies to

$$g(x) = D(f(x)) = \int_{-\infty}^{\infty} f(x')h(x - x')\,dx' \tag{4.13}$$

where we have introduced g, the output of the system. This, you will come to recognize as the *convolution integral.* This integral is very important for a variety of reasons, including the fact that it can be computed rapidly using the fast Fourier transform (FFT). Even more significant is the observation that ANY distortion of an image (as long as it is linear and space-invariant) can be computed by an integral like this.

4.4 The image as a surface

In this section, we consider the problem of interpreting the image as a surface in a 3-space. Thinking of an image in this way will allow us to conceive of image properties as heights.

4.4.1 Isophotes

Consider the value of $f(x, y)$ as a surface in space, described by $z = f(x, y)$. Then the ordered triple $[x, y, f(x, y)]^{\mathrm{T}}$ describes this surface. For every point, (x, y), there is a corresponding value in the third dimension. It is important to observe that there is just ONE such z value for any x, y pair ($f(x, y)$ is a function). Therefore, z is a surface.

Consider the set of all points satisfying $f(x, y) = C$ for some constant C. If f represents brightness, then this set of points is a set of points, all of which have the same brightness. We therefore refer to this set as an "isophote."

Theorem

Proof of this theorem will be a test question.

At any image point, (x, y), the isophote passing through that point is perpendicular to the gradient.[2]

2 The gradient vector is defined in Eq. 2.11 and elaborated upon in Eq. 5.22.

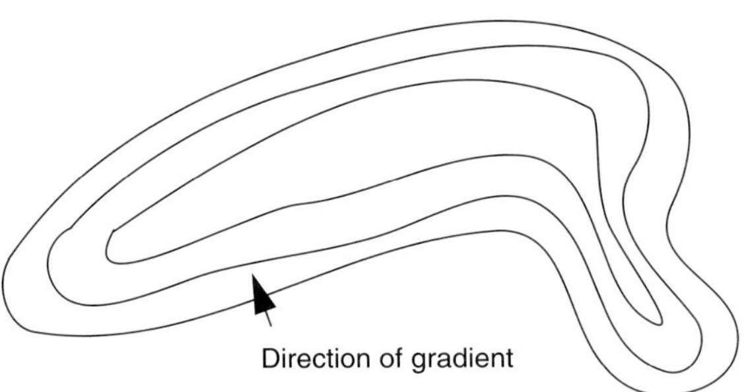

Fig. 4.11. Contour lines on an elevation map are equivalent to isophotes. The gradient vector at a point is perpendicular to the isophote at that point.

4.4.2 Ridges

Now let's think about $z(x, y)$, a surface in space, as a mountain (see Fig. 4.11). If we draw a geological contour map of this mountain, the lines on the map are lines of equal elevation. However, if we think of "elevation" denoting brightness, then the contour lines are isophotes. Stand at a point on this "mountain," and look in the direction of the gradient. The direction you are looking is the way you would go to undertake the steepest ascent.

Look to your right or left and you are looking along the isophote. Note that the direction of the gradient is the steepest direction *at that particular point*. It does not necessarily point at the peak.

Let's climb this mountain by taking small steps in the direction of the local gradient. What happens at the ridge line? How would you know you were on a ridge? How can you describe this process mathematically?

Think about taking steps in the direction of the gradient. Your steps are generally in the same direction, until you reach the ridge, then, the direction radically shifts. So, one useful definition of a ridge is the locus of points which are **local maxima of the rate of change of gradient direction.** That is, we need to find the points where $\partial\theta/\partial v$ is maximized. Here, v represents a derivative taken in the direction of the gradient. In Cartesian coordinates,

A local maximum is any point which does not have a larger neighbor.

$$\frac{\partial\theta}{\partial v} = \frac{2f_x f_y f_{xy} - f_y^2 f_{xx} - f_x^2 f_{yy}}{\left(f_x^2 + f_y^2\right)^{3/2}}. \tag{4.14}$$

Maintz *et al.* [4.24] point out that it is essentially equivalent to a slightly simpler formulation based on simply the second derivative of brightness in the v direction, which leads to maximizing

$$\frac{f_y^2 f_{xx} - 2f_x f_y f_{xy} + f_x^2 f_{yy}}{\left(f_x^2 + f_y^2\right)}, \tag{4.15}$$

where the subscript denotes "partial derivative with respect to." In three-dimensional data, the concepts of ridges are the same, just harder to visualize. In that case, the gradient is a 3-vector, pointing in the direction of increasing density. Isophotes are surfaces instead of curves. In that same paper [4.24], Maintz *et al.* also consider the concept of ridges in three-dimensional data; check it out if you have to implement such things.

4.4.3 Binary images and the medial axis

Compared to the number of pixels in the image, the number where the gradient is strong, the boundaries of regions, is a very small percentage. In fact, as the resolution goes up, the percentage gets smaller and smaller. This is called a "set of measure zero."

The medial axis is used to characterize a region inside an image. It is especially effective when dealing with a binary image where the gradient vector is not useful, since it is zero almost everywhere. Inside the region, find a place where you can draw a circle that touches the boundary at exactly two points. The set of centers of all such circles is the *medial axis.*

The medial axis is easy to define, but tough to compute. We will deal with that in Chapter 9 after morphology and distance transforms have been introduced.

It is possible to relate ridges and the medial axis by using a scale-space representation of the image. This will be covered in section 9.7.1.

4.5 Neighborhood relations

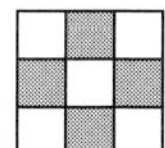

Fig. 4.12. The 4-neighbors of the center pixel are shaded.

We may define neighborhoods in a variety of ways, but the most common and most intuitive way is to say that two pixels are neighbors if they share a side (4-connected) or they share either a side or a vertex (8-connected). The neighborhood of a pixel is the set of pixels which are neighbors (surprise!). The 4-neighbors of the center point are illustrated in Fig. 4.12. Denote the neighborhood of a point s by $\aleph_s$. Later we will discuss operations on sets of points, neighborhoods of points, and on sets of neighborhoods. For example, let A and B be sets of points in the image, and let s be a point in the set A. We may define the aura [4.15] of a set A with respect to a set B, for a neighborhood structure $\aleph_s$ by

$$O_B(A) = \bigcup_{s \in A} (\aleph_s \cap B).$$

The neighbors of a pixel are usually adjacent to that pixel, but there is no fundamental requirement that they be. We will see this again when we talk about "cliques."

That is, the aura of a set of pixels A relative to a set B is the collection of all points in B that are neighbors of pixels in A, where the concept of "neighbor" is given by a problem-specific definition. Fig. 4.13 illustrates an image containing (a) a set A (defined to be the set of shaded pixels), a set B (defined to be the set of blank pixels), a neighborhood relation given by (b), and the aura of set B with respect to set A in (c) [4.15]. We will see more about relationships like this when we discuss morphology.

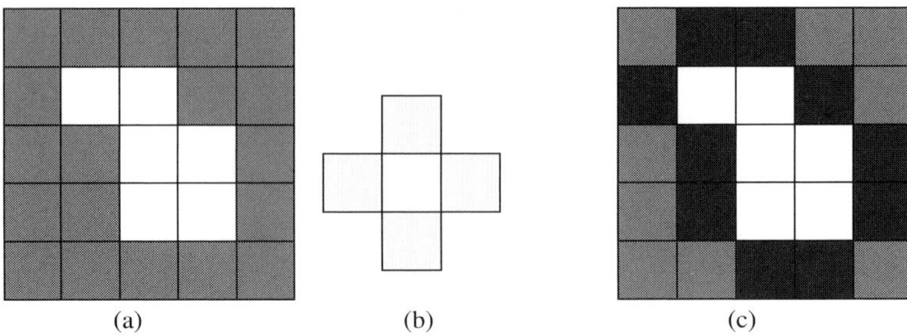

Fig. 4.13. (a) Set A (shaded pixels) and set B (blank pixels); (b) a neighborhood relation. The shaded pixels are neighbors by definition of the center pixel. (c) The aura of the set of white pixels in (a) relative to the set of shaded pixels is given by the dark pixels. It is important to observe that this example uses the standard 4-connected definition of neighbor, there is no requirement that neighbors even be spatially adjacent.

Digression: The connectivity paradox

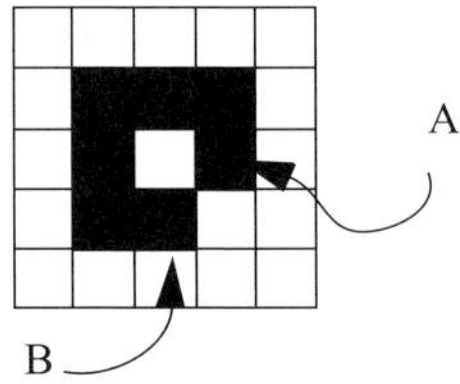

Here is an image. Foreground pixels in black, background in white. We have learned about 4-connectedness and 8-connectedness. Let's apply what we have learned. The foreground in this image is a ring, perhaps a very low-resolution image of a washer. Is this ring closed? That is, can we start at one point, walk around the ring, without retracing our steps, and passing only from one pixel to a connected neighbor? (Incidentally, we just defined a property of connectivity called a *path*.) If we can walk all the way around this region like that, we say the region is closed. Use 4-connectedness. Is it closed? If you said no, good! In a 4-connected system, we cannot get from the pixel A to pixel B. If it is not closed, it must be open. Right? If it is open, the inside and outside must be connected, right? But, using a 4-connected definition, the inside and outside are separate regions.

Well, guess we have to give up on 4-connectedness, since it leads to a paradox. Let's go to 8-connected. Now, the foreground is closed, right? (say right). But now! The inside and outside are also connected! How can that be? If the region is closed, the inside and outside must (logically) be separate. So 8-connected does not work either.

For this particular paradox, there is a fix, at least sort of a fix, which is to use one definition (8- or 4-connected) for the foreground and the other for the background. This fix works for binary images, but when we get to images with more than two levels of brightness, these discretization problems occur again. We present this example to illustrate that in digital images, intuition is not always right. There are lots of other examples of similar problems involving, for example, the measure of the perimeter of regions. Just be aware that weird things happen and intuition does not always hold.

The hexagonal representation discussed in section 4A.1 solves the connectivity paradox. It may solve other problems as well.

Assignment **4.1 Processing and viewing range images.**
The purpose of this exercise is to get you thoroughly used to using the computer and the software packages we will need for this course.

(1) Look on your CDROM in the "images" directory (Unix users, look on your system disk in the appropriate place) and in the "leadhole" subdirectory, you will notice a number of images with the name "site . . . ." View these with whatever tool you wish and copy your favorite to your directory. These are tiny images, and will not take up much space.

(2) Run **ifs viewpoint** on each of the images you just copied (to get the arguments, type "ifs viewpoint -h"). This program accepts range images as input, and allows you to view those images from a viewpoint other than straight up and down. When viewed from another angle, you can see the three-dimensionality of the range image. To select your movement, respond to the "enter transform" with a command, something like

Having trouble with this? Does it seem to hang? It is probably waiting for you to type "end".

movex − 10 movey 10 roll 30 pitch 0 yaw 5 movex 3 movez 10 end

(3) You can enter these on successive lines if you wish. It continues to read motion command until it gets the "end". Roll, pitch, and yaw are rotations, and their arguments are in degrees. Play with this until you can answer the following questions.
   (a) Does movex move the viewpoint or the image, and in which direction (rows or columns)?
   (b) Roll rotates about the *z* axis. About what axis does pitch rotate?

When using ifs2avi, you can choose the name of the output file. To make your PC work correctly, name the file with a .avi extension.

(4) Now, take this image and create six new images, each resulting from a different roll (0, 30, 60, 90, 120, and 150 degrees). Did the roll move any of them out of the image? Come up with a transform that fixes this problem (Hint: Consider moving the

origin to the center of the image using a move command before you do the roll). Use the program **ifs stack** to convert these six two-dimensional images into a single three-dimensional image. If you are using Unix, view that image using imp, and demonstrate how to use imp to display the rotation as a "movie" (Hint: Use the "volume" button). If you are using a PC, you may convert the three-dimensional image produced by stack into an AVI image. For this, use **ifs2avi**. The AVI image can be viewed by any of a large collection of PC programs. A double click on the icon of the .avi image should do the job.

(5) Now, learn how[3] to use the program *ifs spin*. Demonstrate that you can use it to generate a sophisticated movie. NOTE: *ifs spin* actually runs *viewpoint*. The Unix version will generate quite a large set of temporary files, which it deletes when done. Be aware of a need for temporary disk space.

Write up your results, and show your instructor a demo.

Hint: On your CDROM, in the "leadhole" directory, you will find an image name spinout.avi. The image you produced should vaguely resemble the output.

4.6 Conclusion

In this chapter, you have been introduced to a variety of ways to represent images, and the information in images. In subsequent chapters, we will build on these representations, developing algorithms which extract and categorize that information.

4.7 Vocabulary

You should know the meanings of the following terms.

Correspondence problem
Curvature

[3] To learn how to use an IFS program, either type program_name -h or look it up in the manual.

Dynamic range
Functional representation
Graph
Iconic representation
Isophote
Linear system
Medial axis
Probabilistic representation
Quantization
Range image
Raster scan
Ridge
Resolution
Sampling
Spatial frequency
Stereo
Structured illumination

Topic 4A Image representations

4A.1 A variation on sampling: Hexagonal pixels

In a number of papers [4.36], imaging sensors have been described which use hexagonally-organized pixel arrays. Hexagons are the minimum-energy solution when a collection of tangent circles with flexible boundaries is subjected to pressure. The beehive is the best known such naturally occurring organization, but many others occur too, including the organization of cones in the human retina.

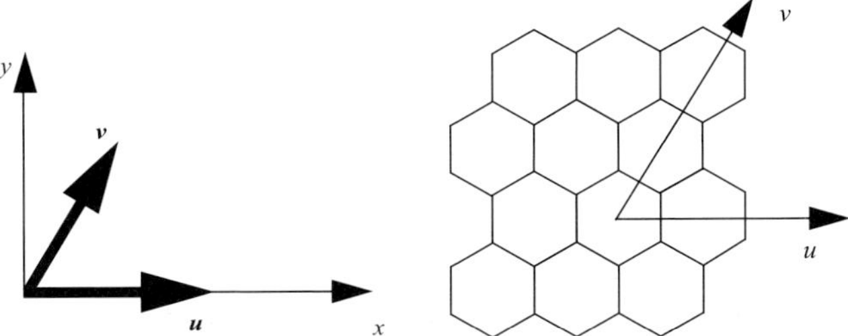

Fig. 4.14. A coordinate system which is natural to hexagonal tessellation of the plane. The *u* and *v* directions are not orthogonal. Unit vectors *u* and *v* describe this coordinate system.

Traditionally, electronic imaging sensors have been arranged in rectangular arrays mainly because an electron beam needed to be swept in a raster-scan way, and more recently because it is slightly more convenient to arrange charge-coupled devices in a rectangular organization. Rectangular arrays, however, introduce an ambiguity in attempts to define neighborhoods. On the other hand, we see no connectivity paradoxes in hexagonal connectivity analysis: Every pixel has exactly six neighbors, foreground, background, or other colors.

Notation

We denote a point in R^2 by $\boldsymbol{p} = u\boldsymbol{u} + v\boldsymbol{v}$, where the unbolded character denotes the magnitude in the direction of the unit vector denoted by the bold character. In the case that we discuss two or more points, we will denote different vectors by using subscripts, with the same subscripts on the components, e.g., $\boldsymbol{p}_i = u_i\boldsymbol{u} + v_i\boldsymbol{v}$.

We will also use column vector notation for such points:

$$P_i = [u_i, v_i]^{\mathrm{T}}. \tag{4.16}$$

In some cases, we will be interested in the location of points in the familiar Cartesian representation, $[x, y]^{\mathrm{T}}$. In this case, we will denote points by subscripts as well, e.g. $\boldsymbol{P}_i = [u_i, v_i]^{\mathrm{T}} = [x_i, y_i]^{\mathrm{T}}$ with corresponding values for u, v, x, and y.

Lemma 1

Any ordered pair $[u, v]$. corresponds to exactly one pair $[x, y]$.

Proof

Using simple trigonometry, and noting that the cosine of 60 degrees is $1/2$, it is straightforward to derive that

$$x = u + \frac{v}{2} \quad \text{and} \quad y = \frac{\sqrt{3}v}{2}. \tag{4.17}$$

Lemma 2

Any ordered pair of Cartesian coordinates $[x, y]$ corresponds to exactly one pair $[u, v]$.

Proof

By solving Eq. (4.17) for u and v, we find

$$u = x - \frac{y}{\sqrt{3}} \quad \text{and} \quad v = 2\frac{y}{\sqrt{3}}. \tag{4.18}$$

A set of vectors $\boldsymbol{b}_1, \boldsymbol{b}_2, \ldots \boldsymbol{b}_d$, is said to be a *basis* for the vector space $\Re^d$ if any vector in $\Re^d$ can be written as a linear combination of $\boldsymbol{b}_1, \boldsymbol{b}_2, \ldots \boldsymbol{b}_d$. If the $\boldsymbol{b}_i$s are orthonormal ($\boldsymbol{b}_i^{\mathrm{T}}\boldsymbol{b}_j = 0$ if $i \neq j$ and $\boldsymbol{b}_i^{\mathrm{T}}\boldsymbol{b}_i = 1$), this is sufficient to claim that these vectors constitute a basis. However, being orthonormal is not a necessary condition. In the case of $\boldsymbol{u}$ and $\boldsymbol{v}$ which are normalized but not orthogonal, they still form a basis. The nonorthogonality is shown in Eq. (4.19), where the inner product of $\boldsymbol{u}$ and $\boldsymbol{v}$ in Cartesian coordinates does not equal zero.

$$\boldsymbol{u} = \boldsymbol{x}, \boldsymbol{v} = \frac{1}{2}\boldsymbol{x} + \frac{\sqrt{3}}{2}\boldsymbol{y}, \quad \text{so} \quad \boldsymbol{u}^{\mathrm{T}}\boldsymbol{v} = [1, 0]\begin{bmatrix} \frac{1}{2} \\ \frac{\sqrt{3}}{2} \end{bmatrix} = \frac{1}{2}. \tag{4.19}$$

Theorem

The vectors $\boldsymbol{u}$ and $\boldsymbol{v}$ form a basis for $\Re^d$.

Proof

Since $\boldsymbol{x}$ and $\boldsymbol{y}$ obviously are a basis for $\Re^d$, we can write any point $\boldsymbol{p}$ in $\Re^d$ as an ordered pair $p = [x, y]^{\mathrm{T}} = x\boldsymbol{x} + y\boldsymbol{y}$. But from Eq. (4.19), we have

$$\boldsymbol{p} = x\boldsymbol{u} + y\frac{(2\boldsymbol{v} - \boldsymbol{u})}{\sqrt{3}} = \left(x - \frac{y}{\sqrt{3}}\right)\boldsymbol{u} + \frac{2y}{\sqrt{3}}\boldsymbol{v}.$$

Thus, any point in R^2 may be written as a weighted sum of $\boldsymbol{u}$ and $\boldsymbol{v}$. QED.

4A.1.1 Identifying the neighbors of a pixel

Given a pixel with coordinates u, v (assumed integer), the coordinates of the neighbors are illustrated in Fig. 4.15.

We can use the following loop to efficiently access all six neighbors of pixel u, v. We observe that no "if" statements are required to determine if the center pixel is on an even- or odd-numbered row.

```
ou = {−1, −1, 0, 1, 1, 0}
ov = {0, 1, 1, 0, −1, −1}
for i = 1 to 6
   nu = u + ou[i]
   nv = v + ov[i]
   value = image[nu][nv]
```

We note that this method of accessing the neighbors of a pixel is also useful in rectangular grids for 8-neighbors and is more efficient than the doubly indexed loop

```
   for i = −1 to 1
   for j = −1 to 1
   if((i! = 0) or (j! = 0))
      value = image[u + i][v + j]
```

It is also interesting to note that ou and ov are circulations of each other.

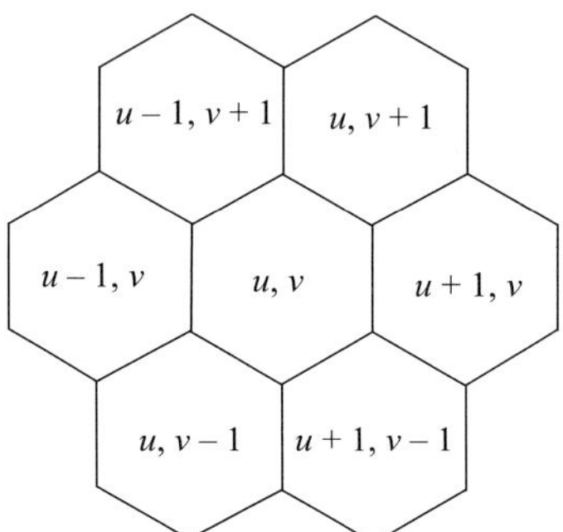

Fig. 4.15. The neighborhood of a pixel is absolutely symmetric, with no ambiguity of 4- or 8-neighborhoods.

4A.2 Other types of iconic representations

So far, we have only spoken of images as if they were always brightness. Well actually, we did mention $2\frac{1}{2}$ D images, which range as a function of x and y. However, there are other things which we could compute to represent image properties. Curvature is one.

4A.2.1 Curvature

The computation of local curvature could be performed at every point in an image. For $2\frac{1}{2}$ D images (surfaces), the curvature cannot be described adequately by a single scalar, but rather takes the form of a matrix. (See doCarmo's book [4.12] or other texts on differential geometry for details.)

$$K = \begin{bmatrix} E & F \\ F & G \end{bmatrix}^{-1} \begin{bmatrix} e & f \\ f & g \end{bmatrix} \tag{4.20}$$

where

$$E = 1 + \left(\frac{\partial z}{\partial x}\right)^2, F = \frac{\partial z \partial z}{\partial x \partial y}, G = 1 + \left(\frac{\partial z}{\partial y}\right)^2,$$
$$e = \left(\frac{\partial^2 z}{\partial x^2}\right)^2 \Big/ H, f = \left(\frac{\partial^2 z}{\partial x \partial y}\right)^2 \Big/ H, g = \left(\frac{\partial^2 z}{\partial y^2}\right)^2 \Big/ H$$

and finally,

$$H = \sqrt{\left(\frac{\partial z}{\partial x}\right)^2 + \left(\frac{\partial z}{\partial y}\right)^2 + 1}.$$

The principal curvatures K_1 and K_2 are defined as the two eigenvalues of the matrix K, and the corresponding eigenvectors determine the directions of the curvature.

For many of our purposes, we will need scalar measurements of curvature which are invariant to viewpoint. Two such scalars are easily defined, the mean curvature

$$K_m = \frac{1}{2}(K_1 + K_2) = \frac{1}{2}\,\mathrm{Tr}\,(K) \tag{4.21}$$

and the Gauss curvature

$$K_G = K_1 K_2 = \det\,(K). \tag{4.22}$$

Since it is a product, the Gauss curvature is zero whenever either of the two principal curvatures is zero, a condition which routinely occurs with industrial parts. For this reason, we seldom use the Gauss curvature.

4A.2.2 Texture

Texture is one of those words that everybody seems to know, but knows without a definition. There are at least two different definitions of texture – "natural" textures, which are best characterized by random process descriptions, and "regular" textures, which are best characterized by frequency–domain representations.

Haralick and Shapiro [4.18] describe textures as "having one or more of the properties of fineness, coarseness, smoothness, granulation, randomness, lineation, or as being mottled, irregular, or hummocky." In detecting that clusters of pixels are different, many features may be used including moments of the power spectrum [4.4, 4.14], fractal dimension [8.12], and the cepstrum [4.35]. Texture segmentation involves representing [4.17, 4.21] an image in a way which incorporates both spatial and spatial–frequency information, and then using that information to identify regions with similar characteristics [4.7, 4.14, 4.16].

The fact that textures can be effectively represented by self-similar (fractal) processes is addressed in a number of papers, the first of which was presented in the classic work by Mandelbrot and Van Ness [4.25]. Kaplan and Kuo [4.22] point out that true textures do not necessarily keep the same exact textures over scale, and the concept of self-similarity should be modified.

Assignment 4.A1

Suppose the image $f(x, y)$ is describable by $f(x, y) = x^4/4 - x^3 + y^2$. At the point $x = 1$, $y = 2$, which of the following is a unit vector which points along the isophote passing through

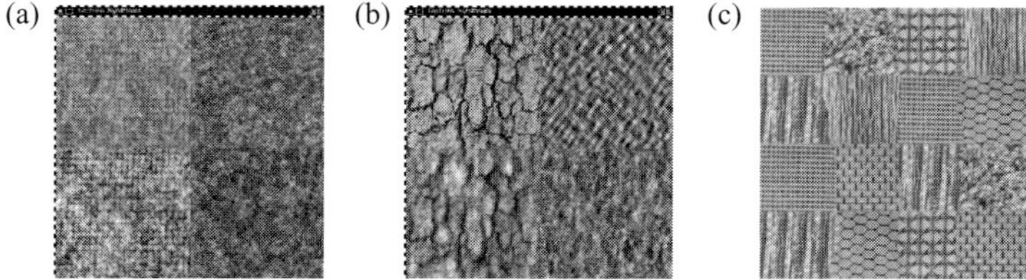

Fig. 4.16. (a) Examples of wool textures [4.2]. (b) Examples of tree bark textures [4.2]. (c) Examples comparing natural and regular textures [4.4]. Used with permission.

that point?

(a) $\left[\frac{2}{(\sqrt{5})} \quad \frac{1}{\sqrt{5}}\right]^T$ (c) $\left[\frac{-1}{(\sqrt{5})} \quad \frac{2}{\sqrt{5}}\right]^T$ (e) $[2 \quad 1]^T$

(b) $\left[\frac{1}{(\sqrt{5})} \quad \frac{2}{\sqrt{5}}\right]^T$ (d) $[-2 \quad 4]^T$ (f) $\left[-\frac{2}{\sqrt{5}} \quad \frac{1}{\sqrt{5}}\right]^T$

Assignment 4.A2

Imagine you are standing on a surface. You cannot see the entire surface, but you can see a fairly large portion. If you measure the curvature at all the points you can see, you find that one of the two principal curvatures is zero. The other principal curvature varies monotonically in one direction. You cannot measure it precisely, but you suspect that variation of curvature is linear in that one direction. On what type of surface are you standing?

References

[4.1] T. Alter, "3-D Pose from 3 Points Using Weak-perspective," *IEEE Transactions on Pattern Analysis and Machine Intelligence*, **16**(8), 1994.

[4.2] D. Badler, J. JáJá, and R. Chellappa, "Scalable Data Parallel Algorithms for Texture Synthesis and Compression using Gibbs Random Fields," *IEEE Transactions on Image Processing*, **4**(10), 1995.

[4.3] R. Bajcsy and F. Solina, "Three Dimensional Object Representation Revisited," *International Conference on Computer Vision*, London, May, 1987.

[4.4] J. Bigün and J. du Buf, "N-folded Symmetries by Complex Moments in Gabor Space and Their Application to Unsupervised Texture Segmentation," *IEEE Transactions on Pattern Analysis and Machine Intelligence*, **16**(1), 1994.

[4.5] A. Bokil and A. Khotanzad, "A Constraint Learning Feedback Dynamic Model for Stereopsis," *IEEE Transactions on Pattern Analysis and Machine Intelligence*, **17**(11), 1995.

[4.6] K. Castleman, *Digital Image Processing*, Englewood Cliffs, NJ, Prentice-Hall, 1996.

[4.7] J. Chen and A. Kundu, "Rotation and Gray Scale Transformation Invariant Texture Identification using Wavelet Decomposition and Hidden Markov Models," *IEEE Transactions on Pattern Analysis and Machine Intelligence*, **16**(2), 1994.

[4.8] R. Chien and W. Snyder, "Hardware for Visual Image Processing," *IEEE Transactions on Circuits and Systems*, **22**(6), 1975.

[4.9] D. Clausi, "Texture Segmentation Example," Web publication, http://www.eng.uwaterloo.ca/~dclausi/texture.html, Spring 2001.

[4.10] F. Cohen and J. Wang, "Part I: Modeling Image Curves Using Invariant 3-D Object Curve Models – A Path to 3-D Reconstruction and Shape Estimation from Image

Contours Using B-Splines, Shape Invariant Matching and Neural Network," *IEEE Transactions on Pattern Analysis and Machine Intelligence*, **16**(1), 1994.

[4.11] F. Cohen and J. Wang, "Part II: 3-D Object Recognition and Shape Estimation from Image Contours," *IEEE Transactions on Pattern Analysis and Machine Intelligence*, **16**(1), 1994.

[4.12] M. doCarmo, *Differential Geometry of Curves and Surfaces*, Englewood Cliffs, NJ, Prentice-Hall, 1976.

[4.13] U. Dhond, and J. Aggarwal, "Stereo Matching in the Presence of Narrow Occluding Objects using Dynamic Disparity Search," *IEEE Transactions on Pattern Analysis and Machine Intelligence*, **17**(7), 1995.

[4.14] D. Dunn, W. Higgins, and J. Wakeley, "Texture Segmentation using 2-D Gabor Elementary Functions," *IEEE Transactions on Pattern Analysis and Machine Intelligence*, **16**(2), 1994.

[4.15] I. Elfadel and R. Picard, "Gibbs Random Fields, Co-occurrences, and Texture Modeling," *IEEE Transactions on Pattern Analysis and Machine Intelligence*, **16**(1), 1994.

[4.16] H. Greenspan, R. Goodman, R. Chellappa, and C. Anderson, "Learning Texture Discrimination Rules in a Multiresolution System," *IEEE Transactions on Pattern Analysis and Machine Intelligence*, **16**(9), 1994.

[4.17] M. Gürelli and L. Onural, "On a Parameter Estimation Method for Gibbs–Markov Random Fields," *IEEE Transactions on Pattern Analysis and Machine Intelligence*, **16**(4), 1994.

[4.18] R. Haralick and L. Shapiro, *Computer and Robot Vision*, Volume I, Reading, MA, Addison-Wesley, 1992.

[4.19] G. Healey and R. Kondepudy, "Radiometric CCD Camera Calibration and Noise Estimation," *IEEE Transactions on Pattern Analysis and Machine Intelligence*, **16**(3), 1994.

[4.20] Y. Hel-Or and M. Werman, "Pose Estimation by Fusing Noisy Data of Different Dimensions," *IEEE Transactions on Pattern Analysis and Machine Intelligence*, **17**(2), 1995.

[4.21] A. Jain and K. Karu, "Learning Texture Discrimination Masks," *IEEE Transactions on Pattern Analysis and Machine Intelligence*, **18**(2), 1996.

[4.22] L. Kaplan and C. Kuo, "Texture Roughness Analysis and Synthesis via Extended Self-similar (ESS) Model," *IEEE Transactions on Pattern Analysis and Machine Intelligence*, **17**(11), 1995.

[4.23] D. Keren, D. Cooper, and J. Subrahmonia, "Describing Complicated Objects by Implicit Polynomials," *IEEE Transactions on Pattern Analysis and Machine Intelligence*, **16**(1), 1994.

[4.24] J. Maintz, P. van den Elsen, and M. Viergever, "Evaluation of Ridge Seeking Operations for Multimodality Medical Image Matching," *IEEE Transactions on Pattern Analysis and Machine Intelligence*, **18**(4), 1996.

[4.25] B. Mandelbrot and J. Van Ness, "Fractional Brownian Motions, Fractional Noises, and Applications," *SIAM Review*, **10**, October, 1968.

[4.26] S. Marapan and M. Trivedi, "Multi-primitive Hierarchical (MPH) Stereo Analysis," *IEEE Transactions on Pattern Analysis and Machine Intelligence*, **16**(3), 1994.

[4.27] D. Marr and T. Poggio, "Cooperative Computation of Stereo Disparity," *Science*, **194**, pp. 283–287, October, 1976.

[4.28] P. McLauchlan and D. Murray, "Active Camera Calibration for a Head-eye Platform Using Variable State-dimension Filter," *IEEE Transactions on Pattern Analysis and Machine Intelligence*, **18**(1), 1996.

[4.29] N. Page, W. Snyder, and S. Rajala, "Turbine Blade Image Processing System," In *Advanced Software for Robotics*, ed. A Danthine, Amsterdam, North-Holland, 1984.

[4.30] S. Peleg, M. Ben-Ezra, and Y. Pritch, "Omnistereo: Panoramic Stereo Imaging," *IEEE Transactions on Pattern Analysis and Machine Intelligence*, **23**(3), 2001.

[4.31] L. Quan, "Invariants of Six Points and Projective Reconstruction from Three Uncalibrated Images," *IEEE Transactions on Pattern Analysis and Machine Intelligence*, **17**(1), 1995

[4.32] A. Shashua, "Projective Structure from Uncalibrated Images: Structure from Motion and Recognition," *IEEE Transactions on Pattern Analysis and Machine Intelligence*, **16**(8), 1994.

[4.33] A. Shashua, "Algebraic Functions for Recognition," *IEEE Transactions on Pattern Analysis and Machine Intelligence*, **17**(8), 1995.

[4.34] A. Shashua and N. Navab, "Relative Affine Structure: Canonical Model for 3D From 2D Geometry and Applications," *IEEE Transactions on Pattern Analysis and Machine Intelligence*, **18**(9), 1996.

[4.35] P. Smith and N. Nandhakumar, "An Improved Power Cepstrum Based Stereo Correspondence Method for Textured Scenes," *IEEE Transactions on Pattern Analysis and Machine Intelligence*, **18**(3), 1996.

[4.36] W. Snyder, H. Qi, and W. Sander, "A Hexagonal Coordinate System," *SPIE Medical Imaging: Image Processing*, Pt. 1–2, pp. 716–727, February, 1999.

[4.37] G. Wei and S. Ma, "Implicit and Explicit Camera Calibration: Theory and Experiments," *IEEE Transactions on Pattern Analysis and Machine Intelligence*, **16**(5), 1994.

[4.38] X. Zhuang and Y. Huang, "Robust 3-D – 3-D Pose Estimation," *IEEE Transactions on Pattern Analysis and Machine Intelligence*, **16**(8), 1994.

5 Linear operators and kernels

Now I see through a glass darkly, but then, face to face

Paul of Tarsus

In this chapter,[1] we investigate linear operations on images. We first consider the derivative, probably the most common linear operator. That discussion is extended into edge detection, and we consider a variety of methods for accomplishing this objective.

5.1 What is a linear operator?

Suppose D is an operator which takes an image f and produces an image g. If D satisfies

$$D(\alpha f_1 + \beta f_2) = \alpha D(f_1) + \beta D(f_2), \tag{5.1}$$

where f_1 and f_2 are images, α and β are scalar multipliers, then we say that D is a "linear operator."

A gedankenexperiment

Consider the image operator D

$$g = D(f) = af + b \qquad a, b \in \Re$$

Is D a linear operator?

We suggest you work this out for yourself before reading the solution. It certainly LOOKS linear. Multiplication by a constant followed by addition of a constant. If f were a scalar variable, then D describes the equation of a line, which SURELY is linear (isn't it?)! OK. Let's prove it. Using Eq. (5.1), we evaluate

$$D(\alpha f_1 + \beta f_2) = a(\alpha f_1 + \beta f_2) + b = a\alpha f_1 + a\beta f_2 + b$$

1 The authors are grateful to Bilge Karacali, Rajeev Ramanath, and Lena Soderberg for their assistance in producing the images used in this chapter.

and check to see if this is the same as

$$\begin{aligned}\alpha D(f_1) + \beta D(f_2) &= \alpha(af_1 + b) + \beta(af_2 + b)\\ &= a\alpha f_1 + \alpha b + a\beta f_2 + \beta b\\ &= a\alpha f_1 + a\beta f_2 + b(\alpha + \beta)\end{aligned}$$

so unless $\alpha + \beta = 1$, D is NOT a linear operator! This seems counter-intuitive, doesn't it? We'll have another look at this later and see if we can figure out why. In the remainder of this chapter, we will look at image operators which are linear.

5.2 Application of kernel operators in digital images

Since f is now digital, many authors choose to write f as a matrix, f_{ij}, instead of the functional notation $f(x, y)$. However, we prefer the x, y notation, for reasons which shall become apparent later. We will find it more convenient to use a single subscript f_i, at several points later, but for now, let's stick with $f(x, y)$ and remember that x and y take on only a small range of integer values, e.g. $0 < x < 511$.

Think about a one-dimensional image named f with five pixels and another one-dimensional image, which we will call a *kernel* named h with three pixels, as illustrated in Fig. 5.1.

Place the kernel down so its center, pixel h_0, is over some pixel of f, say f_2; we get $g_2 = f_1 h_{-1} + f_2 h_0 + f_3 h_1$, which is a sum of products of elements of the kernel and elements of the image. With that understanding, let us consider the most common example of Eq. (5.2), the case we will come to call *application of a* 3×3 *kernel*:

$$g(x, y) = \sum_{\alpha} \sum_{\beta} f(x + \alpha, y + \beta) h(\alpha, \beta). \tag{5.2}$$

This case occurs when both α and β take on only the values -1, 0, and 1. In this case, Eq. (5.2) expands to

Remember, $y = 0$ is at the TOP of the image and y goes up as you go down in the image.

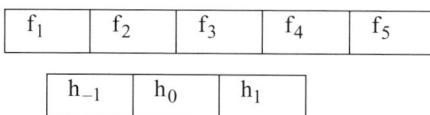

$$\begin{aligned}g(x, y) = {} & f(x-1, y-1)h(-1,-1) + f(x, y-1)h(0,-1)\\ & + f(x+1, y-1)h(1,-1) + f(x-1, y)h(-1, 0)\\ & + f(x, y)h(0, 0) + f(x+1, y)h(1, 0) + f(x-1, y+1)h(-1, 1)\\ & + f(x, y+1)h(0, 1) + f(x+1, y+1)h(1, 1).\end{aligned} \tag{5.3}$$

f_1	f_2	f_3	f_4	f_5

h_{-1}	h_0	h_1

Fig. 5.1. A one-dimensional image with five pixels and a one-dimensional kernel with three pixels. The subscript is the x-coordinate of the pixel.

Note the order of the arguments here, x (column) and y (row). Sometimes the reverse convention is followed.

Table 5.1. *Values of the elements of a kernel.*

$h(-1,-1)$	$h(0,-1)$	$h(1,-1)$
$h(-1,0)$	$h(0,0)$	$h(1,0)$
$h(-1,1)$	$h(0,1)$	$h(1,1)$

To better capture the essence of Eq. (5.3), let us write h as a 3×3 grid of numbers (yes, we used the word "grid" rather than "array" intentionally), as in Table 5.1.

Now we imagine that we place this grid down on top of the image so that the center of the grid is directly over pixel $f(x, y)$; then each h value in the grid is multiplied by the corresponding point in the image. We will refer to the grid of h values henceforth as a "kernel."

5.2.1 On the direction of the arguments: Convolution and correlation

Let's restate two important equations. First the equation for a kernel operator, recopied from Eq. (5.2), and then the equation for two-dimensional, discrete convolution.

$$g(x, y) = \sum_{\alpha} \sum_{\beta} f(x + \alpha, y + \beta) h(\alpha, \beta) \tag{5.4}$$

$$g(x, y) = \sum_{\alpha} \sum_{\beta} f(x - \alpha, y - \beta) h(\alpha, \beta). \tag{5.5}$$

Mathematically, convolution and correlation differ in the left–right order of coordinates.

The observant student will have noticed a discrepancy in order between Eqs. (5.4) and (5.5). In formal convolution, as given by Eq. (5.5), the arguments reverse: the right-most pixel of the kernel (h_1) is multiplied by the left-most pixel in the corresponding region of the image (f_2). However, in Eq. (5.4), we think of "placing" the kernel down over the image and multiplying corresponding pixels. If we multiply corresponding pixels, left–left and right–right, we have correlation. There is, unfortunately, a misnomer in much of the literature – both may be called "convolution." We advise the student to watch for this. In many publications, the authors use the term "convolution" when they really mean "sum of products." In order to avoid confusion, in this book, we will avoid the use of the word "convolve" unless we really do mean the application of Eq. (5.5), and instead use the term "kernel operator," when we mean Eq. (5.4).

5.2.2 Using kernels to estimate derivatives

Let us examine this concept via an example – approximating the spatial derivatives of the image, $\partial f / \partial x$ and $\partial f / \partial y$.

We recall from some dimly remembered calculus class,

$$\frac{\partial f}{\partial x} = \lim_{\Delta x \to 0} \frac{f(x + \Delta x) - f(x)}{\Delta x}$$

which would suggest the following kernel could be used (for $\Delta x = 1$):

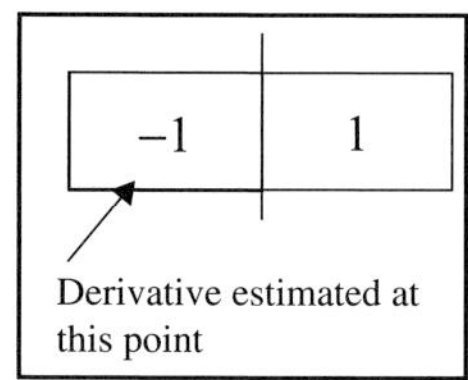

But this kernel is aesthetically unpleasing – the estimate at x depends on the value at x and at $x + 1$, but not at $x - 1$; why? We actually like a symmetric definition better, such as

$$\left.\frac{\partial f}{\partial x}\right|_{x_a} = \lim_{\Delta x \to 0} \frac{f(x_0 + \Delta x) - f(x_0 - \Delta x)}{2\Delta x}.$$

We cannot get Δx smaller than 1, and we end up with this kernel

−1/2	0	1/2

which, for notational simplicity, we write as

$$1/2 \begin{array}{|c|c|c|} \hline -1 & 0 & 1 \\ \hline \end{array}.$$

A major problem with derivatives is noise sensitivity. We compensate for this by taking the difference horizontally, and then averaging vertically – which produces the following kernel:

$$\left.\frac{\partial f}{\partial x}\right|_{x_0} = \frac{1}{6}\left(\begin{bmatrix} -1 & 0 & 1 \\ -1 & 0 & 1 \\ -1 & 0 & 1 \end{bmatrix} \otimes f\right) \tag{5.6}$$

where we have introduced a new symbol, $\otimes$ which will denote the sum-of-products implementation described above. The literature abounds with kernels like this one. All of them combine the concept of estimating the derivative by differences, and then averaging the result in some way to compensate for noise. Probably the best known of these ad hoc kernels is the Sobel:

$$\left.\frac{\partial f}{\partial x}\right|_{x_0} = \frac{1}{8}\left(\begin{bmatrix} -1 & 0 & 1 \\ -2 & 0 & 2 \\ -1 & 0 & 1 \end{bmatrix} \otimes f\right). \tag{5.7}$$

The Sobel operator has the benefit of being center-weighted.

5.3 Derivative estimation by function fitting

This approach presents yet another way to make use of the continuous representation of an image $f(x, y)$. Think of the brightness as a function of the two spatial coordinates, and consider a plane which is tangent to that brightness surface at a point, as illustrated in Fig. 5.2.

In this case, we may write the continuous image representation using the equation of a plane

$$f(x, y) = ax + by + c. \tag{5.8}$$

Then, we may consider the edge strength using the two numbers $\partial f/\partial x = a$, $\partial f/\partial y = b$, and the rate of change of brightness at the point (x, y) is represented by the gradient vector

$$\nabla f = \begin{bmatrix} \frac{\partial f}{\partial x} & \frac{\partial f}{\partial y} \end{bmatrix}^{\mathrm{T}} = [a \quad b]^{\mathrm{T}}. \tag{5.9}$$

The approach followed here is to find a, b, and c given some noisy, blurred measurement of f, and the assumption of Eq. (5.8).

To find those parameters, first observe that Eq. (5.8) may be written as $f(x, y) = A^{\mathrm{T}}X$ where the vectors A and X are $A^{\mathrm{T}} = [a \;\; b \;\; c]$ and $X^{\mathrm{T}} = [x \;\; y \;\; 1]$.

Suppose we have measured brightness values $g(x, y)$ at a collection of points $\aleph \subset Z \times Z$ (Z is the set of integers) in the image. Over that set of points, we wish to find the plane which best fits the data. To accomplish this objective, write the error as a function of the measurement and the (currently unknown) function $f(x, y)$.

A sum-squared error objective function

$$E = \sum_{\aleph}(f(x, y) - g(x, y))^2 = \sum_{\aleph}(A^{\mathrm{T}}X - g(x, y))^2.$$

Expanding the square and eliminating the functional notation for simplicity, we find

$$E = \sum_{\aleph}(A^{\mathrm{T}}X)(A^{\mathrm{T}}X) - 2A^{\mathrm{T}}Xg + g^2.$$

Remembering that for vectors A and X, $A^{\mathrm{T}}X = X^{\mathrm{T}}A$, and taking the summation

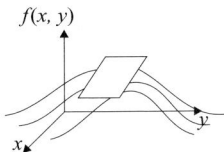

Fig. 5.2. The brightness in an image can be thought of as a surface, a function of two variables. The slopes of the tangent plane are the two spatial partial derivatives.

through, we have

$$E = \sum_{\aleph} A^{\mathrm{T}} X X^{\mathrm{T}} A - 2\sum_{\aleph} A^{\mathrm{T}} X g + \sum_{\aleph} g^2$$
$$= A^{\mathrm{T}} \left(\sum_{\aleph} X X^{\mathrm{T}} \right) A - 2A^{\mathrm{T}} \sum_{\aleph} X g + \sum_{\aleph} g^2.$$

Now we wish to find the A (the parameters of the plane) which minimizes E; so we may take derivatives and set the result to zero.

$$\frac{dE}{dA} = 2\left(\sum_{\aleph} X X^{\mathrm{T}} \right) A - 2\sum_{\aleph} X g = 0. \tag{5.10}$$

Let's call $\sum_{\aleph} X X^{\mathrm{T}} \equiv S$ (it is the "scatter matrix") and see what Eq. (5.10) means: consider a neighborhood $\aleph$ which is symmetric about the origin. In that neighborhood, suppose x and y only take on values of -1, 0, and 1, then

$$S = \sum_{\aleph} X X^{\mathrm{T}} = \sum_{\aleph} \begin{bmatrix} x \\ y \\ 1 \end{bmatrix} [x \quad y \quad 1] = \begin{bmatrix} \sum x^2 & \sum xy & \sum x \\ \sum xy & \sum y^2 & \sum y \\ \sum x & \sum y & \sum 1 \end{bmatrix}$$

which, for the neighborhood described, is

$$\begin{bmatrix} 6 & 0 & 0 \\ 0 & 6 & 0 \\ 0 & 0 & 9 \end{bmatrix}.$$

More detail on how the elements of the scatter matrix are derived. Do not forget the positive direction for y is down.

You do not see where those values came from?

Ok, here's how you get them. Look at the top left point, at coordinates $x = -1$, $y = -1$. At that point, $x^2 = (-1)^2 = 1$. Now look at the top middle point, at coordinates $x = 0$, $y = -1$. At that point, $x^2 = 0$. Do this for all 9 points in the neighborhood, and you obtain $\sum x^2 = 6$. Got it?

Convince yourself that this is true.

Useful observation: If you make the neighborhood symmetric about the origin, all the terms in the scatter matrix which contain x or y to an odd power will be zero.

Be sure you carefully proofread whatever you write!

Also: A common miss steak is to put a 1 in the lower right corner rather than a 9 – be careful!

So now we have the matrix equation

$$2\begin{bmatrix} 6 & 0 & 0 \\ 0 & 6 & 0 \\ 0 & 0 & 9 \end{bmatrix} \begin{bmatrix} a \\ b \\ c \end{bmatrix} = 2\begin{bmatrix} \sum g(x, y)x \\ \sum g(x, y)y \\ \sum g(x, y) \end{bmatrix}.$$

We can easily solve for a:

$$a = \frac{1}{6}\sum g(x, y)x \approx \frac{\partial f}{\partial x}.$$

So, to compute the derivative from a fit to a neighborhood at each of the nine points in the neighborhood, take the measured value at that point, multiply by its x coordinate, and add them up. Let's write down the x coordinates in tabular form:

−1	0	1
−1	0	1
−1	0	1

That is *precisely* the kernel of Eq. (5.6), which we derived intuitively. Now, we have it derived formally. Doesn't it give you a warm fuzzy feeling when theory agrees with intuition?!? (Whoops, we forgot to multiply each term by 1/6, but we can simply factor the 1/6 out, and when we get the answer, we will just divide by 6.)

We accomplished this by using an optimization method, in this case, minimizing the squared error, to find the coefficients of a function $f(x)$ in an equation of the form $y = f(x)$, where f is polynomial. Recall from section 4.1.2 that this form is referred to as an explicit functional representation.

One more terminology issue: In future material, we will use the expression *radius of a kernel*. The radius is the number of pixels from the center to the nearest edge. For example, a 3×3 kernel has a radius of one. A 5×5 kernel has a radius of 2, etc. It is possible to design kernels which are circular, but most of the time, we use squares.

Finding image gradients in hexagonal arrays of pixels

In this section, we find image gradients again, in exactly the same way, but this time with hexagonally arranged pixels. Refer to section 4A.1 for a discussion of the coordinate system. It is a different presentation of the same material, and if you read both presentations carefully, you will understand the concepts more clearly.

To find the gradient of intensity in an image, we will fit a plane to the data in a small neighborhood. This plane will be represented in the form of Eq. (5.8). We then take partial derivatives with respect to u and v to find the gradient of intensity in those corresponding directions. We choose a neighborhood of six points, surrounding a central point, and fit the plane to them. Define the set of data points as z_i, $(i = 1, \ldots, 6)$. Then the following expression represents the error in fitting these six points to a plane parameterized by a, b, and c.

$$E = \sum_{i=1}^{6}(z_i - (au_i + bv_i + c))^2. \tag{5.11}$$

In order to represent E in a form that will be easy to differentiate, we will reformulate the argument of the summation using matrix notation. Define vectors $A = [a \;\; b \;\; c]^{\mathrm{T}}$ and $Z = [u \;\; v \;\; 1]^{\mathrm{T}}$. Then E may be written using

$$E = \sum_{i=1}^{6} (z_i - A^{\mathrm{T}} Z_i)^2 \tag{5.12}$$

$$= \sum_{i=1}^{6} \left(z_i^2 - 2z_i A^{\mathrm{T}} Z_i + A^{\mathrm{T}} Z_i A^{\mathrm{T}} Z_i \right). \tag{5.13}$$

First, we observe that $A^{\mathrm{T}} Z = Z^{\mathrm{T}} A$, and rewrite Eq. (5.13), temporarily dropping the limits on the summation to make the typing easier:

$$= \sum z_i^2 - 2A^{\mathrm{T}} \sum z_i Z_i + A^{\mathrm{T}} \left(\sum Z_i Z_i^{\mathrm{T}} \right) A. \tag{5.14}$$

The term in parentheses in Eq. (5.14) is the scatter matrix, the collection of locations of points at which data exists. We denote this matrix by the symbol S. In order to find the value of vector A which minimizes E^2, we take the partial derivative with respect to A:

$$\frac{\partial E}{\partial A} = -2 \sum z_i Z_i + 2SA. \tag{5.15}$$

Evaluating S, we find

$$S = \begin{bmatrix} \sum u_i^2 & \sum u_i v_i & \sum u_i \\ \sum u_i v_i & \sum v_i^2 & \sum v_i \\ \sum u_i & \sum v_i & \sum 1 \end{bmatrix} = \begin{bmatrix} 4 & -2 & 0 \\ -2 & 4 & 0 \\ 0 & 0 & 6 \end{bmatrix}.$$

(One finds the numerical values by summing the u and v coordinates over each of the pixels in the neighborhood, as illustrated in Fig. 4.15, assuming the center pixel is at location 0, 0.)

Define

$$\sum z_i u_i \equiv \Upsilon_u$$
$$\sum z_i v_i \equiv \Upsilon_v,$$

and set the partial derivative of Eq. (5.15) equal to zero to produce a pair of simultaneous equations,

$$\begin{aligned} 4a - 2b &= \Upsilon_u \\ -4a + 8b &= 2\Upsilon_v \end{aligned} \tag{5.16}$$

with solution

$$b = \frac{1}{6}(2\Upsilon_v + \Upsilon_u). \tag{5.17}$$

Similarly,

$$a = \frac{1}{6}(\Upsilon_v + 2\Upsilon_u). \tag{5.18}$$

Substituting actual values of u and v at each pixel in the six-pixel neighborhood, we determine the gradient vector, since the gradient in the u direction is a and the gradient in the v direction is b.

We rewrite the equation for a, substituting in the definitions of the Υs:

$$a = \frac{1}{6}\left(\sum z_i v_i + 2\sum z_i u_i\right) = \frac{1}{6}\sum z_i(v_i + 2u_i).$$

Now, look at the pixel directly to the right of center, its u, v coordinates are $1, 0$. So, we should multiply the image brightness, Z_i, at that point by $(v_i + 2u_i)$, or $(0 + 2 \times 1)$ which is 2. That is the value that goes into the kernel. We need to remember in using this method, that the answer we get is not quite right . . . It is actually six times the best estimate, and we need to divide by six, if the actual value of the derivative is important, rather than a result proportional to the estimated derivative. Repeating this process at each point, we obtain the kernels shown in Fig. 5.3.

The concepts of fitting described in this section are ubiquitous. They pop up all over the discipline of machine vision. You find a need to fit gray values, surfaces, lines (and we will do that in Chapter 9), curves, etc. There are even techniques for fitting when data has known statistical variations [5.7, 5.40].

5.4 Vector representations of images

Suppose we list every pixel in an image in raster scan order, as one long vector. For example, for the 4×4 image

$f(x, y) =$

1	2	4	1
7	3	2	8
9	2	1	4
4	1	2	3

$$F = [1\ 2\ 4\ 1\ 7\ 3\ 2\ 8\ 9\ 2\ 1\ 4\ 4\ 1\ 2\ 3]^{\mathrm{T}}.$$

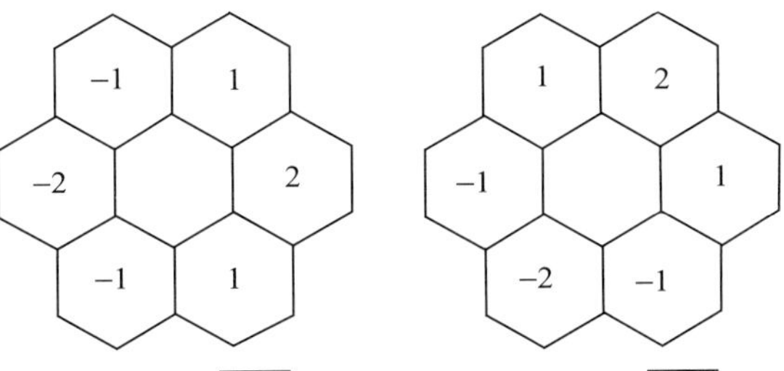

Fig. 5.3. The kernels used to estimate the gradient of brightness in the u direction, and in the v direction.

0, 0 is the UPPER left corner.

This is called the "lexicographic" representation. If we write the image in this way, each pixel may be identified by a single index, e.g., $F_0 = 1$, $F_4 = 7$, $F_{15} = 3$, where the indexing starts with zero.

Now suppose we want to apply the following kernel to this image:

$$h = \begin{bmatrix} -1 & 0 & 2 \\ -2 & 0 & 4 \\ 3 & 9 & 1 \end{bmatrix}$$

at point $x = 1$, $y = 1$, again, starting the indexing at zero.

A dot product computes application of a kernel.

Point (1, 1) corresponds to pixel F_5 in the image. We could accomplish this application of the kernel by taking the dot product of the vector F with the vector

$$H_5 = [-1 \quad 0 \quad 2 \quad 0 \quad -2 \quad 0 \quad 4 \quad 0 \quad 3 \quad 9 \quad 1 \quad 0 \quad 0 \quad 0 \quad 0 \quad 0]^{\mathrm{T}}.$$

Now you try it. Determine what vector to use to apply this kernel at (2, 2). Did you get this?

$$H_{10} = [0 \quad 0 \quad 0 \quad 0 \quad 0 \quad -1 \quad 0 \quad 2 \quad 0 \quad -2 \quad 0 \quad 4 \quad 0 \quad 3 \quad 9 \quad 1]^{\mathrm{T}}.$$

Now try (2, 1) ($x = 2$, $y = 1$):

$$H_6 = [0 \quad -1 \quad 0 \quad 2 \quad 0 \quad -2 \quad 0 \quad 4 \quad 0 \quad 3 \quad 9 \quad 1 \quad 0 \quad 0 \quad 0 \quad 0]^{\mathrm{T}}.$$

Compare H_5 at (1, 1) and H_6 at (2, 1). They are the same except for a rotation. We could convolve the entire image by constructing a matrix in which each column is one such H. Doing so would result in a matrix such as the one illustrated. By producing the product $G = H^{\mathrm{T}}F$, G will be the (vector form of) convolution of image F with kernel H.

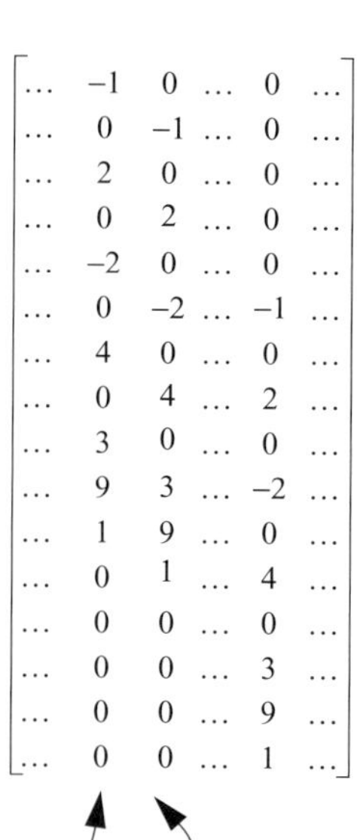

$$\begin{bmatrix} \cdots & -1 & 0 & \cdots & 0 & \cdots \\ \cdots & 0 & -1 & \cdots & 0 & \cdots \\ \cdots & 2 & 0 & \cdots & 0 & \cdots \\ \cdots & 0 & 2 & \cdots & 0 & \cdots \\ \cdots & -2 & 0 & \cdots & 0 & \cdots \\ \cdots & 0 & -2 & \cdots & -1 & \cdots \\ \cdots & 4 & 0 & \cdots & 0 & \cdots \\ \cdots & 0 & 4 & \cdots & 2 & \cdots \\ \cdots & 3 & 0 & \cdots & 0 & \cdots \\ \cdots & 9 & 3 & \cdots & -2 & \cdots \\ \cdots & 1 & 9 & \cdots & 0 & \cdots \\ \cdots & 0 & 1 & \cdots & 4 & \cdots \\ \cdots & 0 & 0 & \cdots & 0 & \cdots \\ \cdots & 0 & 0 & \cdots & 3 & \cdots \\ \cdots & 0 & 0 & \cdots & 9 & \cdots \\ \cdots & 0 & 0 & \cdots & 1 & \cdots \end{bmatrix}$$
H_5
H_6

Some observations about this process:

- The resulting matrix is a "circulant" matrix. Each column is simply a rotation of the adjacent column.
- The fact that we can apply kernels in this way is yet another demonstration of the fact that kernel operators are linear operators.
- This form suggests one approach for dealing with the nasty problem of boundary conditions (you *did* think about that, didn't you?). Specifically, how do you multiply by the data value *above* when you are on the top line? One answer is to rotate around the image and pull from the bottom.
- The matrix H is VERY large. If f is a typical image of 256×256 pixels, then H is $(256 \times 256) \times (256 \times 256)$ which is a large number (although still smaller than the US national debt). Do not worry about the monstrous size of H. Nobody (well, almost nobody) ever computes H and uses it in this way. This form is useful for thinking about images and for proving theorems about image operators – it is a conceptual, not a computational tool.

Finally, multiplication by a circulant matrix can be accomplished considerably faster by using the fast Fourier transform; but more about that later.

5.5 Basis vectors for images

In the previous section, we saw that we could think of an image as a vector. If we can do that for an image, surely we can do the same thing for a small subimage. Consider the nine-pixel neighborhood of a single point. We can easily construct the 9-vector which is the lexicographic representation of that neighborhood.

In Chapter 2, we learned that any vector could be represented as a weighted sum of basis vectors, and we will apply that same concept here. Let's rewrite Eq. (2.6) in the form:

$$V = \sum_{i=1}^{9} a_i \boldsymbol{u}_i$$

where now V is the 9-vector representation of this nine-pixel neighborhood, the a_i are scalar weights, and the $\boldsymbol{u}_i$ are some set of orthonormal basis vectors.

But what basis vectors should we use? More to the point, what basis set *would be useful*? The set we normally use, the Cartesian basis, is

$$\begin{aligned}
\boldsymbol{u}_1 &= [1 \quad 0 \quad 0 \quad 0 \quad 0 \quad 0 \quad 0 \quad 0 \quad 0]^{\mathrm{T}} \\
\boldsymbol{u}_2 &= [0 \quad 1 \quad 0 \quad 0 \quad 0 \quad 0 \quad 0 \quad 0 \quad 0]^{\mathrm{T}} \\
&\vdots \\
\boldsymbol{u}_9 &= [0 \quad 0 \quad 0 \quad 0 \quad 0 \quad 0 \quad 0 \quad 0 \quad 1]^{\mathrm{T}}
\end{aligned}$$

which, while convenient and simple, does not help us at all here. Could another basis be more useful? (answer yes). Before we figure out what, do you remember how many possible basis vectors there are for this real-valued 9-space? The answer is "a zillion".[2] With so many choices, we should be able to pick some good ones. To accomplish that, recall the role of the coefficients a_i. Recall that if some particular a_i is much larger than all the other as, it means that V is "very similar" to $\boldsymbol{u}_i$. Computing the as then allows us a means to find which of a set of prototype neighborhoods a particular image most resembles.

Fig. 5.4 illustrates a set of prototype neighborhoods developed by Frei and Chen [5.12]. Notice that neighborhood ($\boldsymbol{u}_1$) is negative below and positive above the horizontal center line, and therefore is indicative of a horizontal edge, or a point where $\partial f/\partial y$ is large.

Now recall how to compute the projection a_i. The scalar-valued projection of a vector V onto a basis vector $\boldsymbol{u}_i$ is the inner product $a_i = V^{\mathrm{T}}\boldsymbol{u}_i$.

[2] Actually, the correct answer is infinity – a zillion is just an engineering approximation.

1	$\sqrt{2}$	1
0	0	0
−1	$-\sqrt{2}$	−1

u_1

1	0	−1
$\sqrt{2}$	0	$-\sqrt{2}$
1	0	−1

u_2

0	−1	$\sqrt{2}$
1	0	−1
$-\sqrt{2}$	1	0

u_3

$\sqrt{2}$	−1	0
−1	0	1
0	1	$-\sqrt{2}$

u_4

0	1	0
−1	0	−1
0	1	0

u_5

−1	0	1
0	0	0
1	0	−1

u_6

1	−2	1
−2	4	−2
1	−2	1

u_7

−2	1	−2
1	4	1
−2	1	−2

u_8

1	1	1
1	1	1
1	1	1

u_9

Fig. 5.4. Frei-Chen basis vectors.

Do you think you could develop a similar basis set for the seven pixels in a hexagonal neighborhood? Hint: Seven pixels defines a seven-dimensional vector space. You will need to find seven such vectors.

One way to determine how similar a neighborhood about some point is to a vertical edge is to compute the inner product of the neighborhood vector with the vertical edge basis vector. One final question: What is the difference between calculating this projection and convolving the image at that point with a kernel which estimates $\partial f/\partial x$? The answer is left as an exercise to the student. (Don't you wish they were all this easy?)

So now you know all there is to know (almost) about linear operators and kernel operators. Let's move on to an application to which we have already alluded – finding edges.

5.6 Edge detection

Edges are areas in the image where the brightness changes suddenly; where the derivative (or more correctly, *some* derivative) has a large magnitude. We can categorize edges as step, roof, or ramp [5.20], as illustrated in Fig. 5.5.

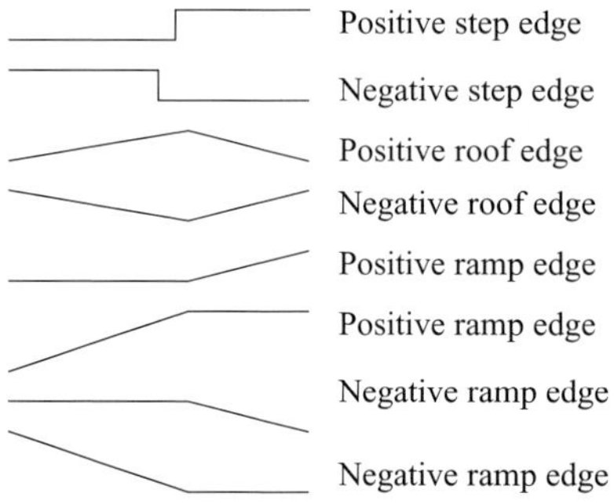

Fig. 5.5. Types of commonly occurring edges. Note that the term positive or negative generally refers to the sign of the first instance of the first derivative.

We have already seen (twice) how application of a kernel such as

$$h_x = \begin{array}{|c|c|c|}\hline -1 & 0 & 1 \\ \hline -1 & 0 & 1 \\ \hline -1 & 0 & 1 \\ \hline \end{array} \tag{5.19}$$

approximates the partial derivative wrt x. Similarly,

Which is correct? It depends on which direction you have chosen as positive y.

$$h_y = \begin{array}{|c|c|c|}\hline -1 & -1 & -1 \\ \hline 0 & 0 & 0 \\ \hline 1 & 1 & 1 \\ \hline \end{array} \quad \text{or} \quad \begin{array}{|c|c|c|}\hline 1 & 1 & 1 \\ \hline 0 & 0 & 0 \\ \hline -1 & -1 & -1 \\ \hline \end{array} \tag{5.20}$$

estimates $\partial f / \partial y$.

Some other forms have appeared in the literature that you should know about for historical purposes.

Remember the Sobel operator?

$$h_y = \begin{array}{|c|c|c|}\hline -1 & -2 & -1 \\ \hline 0 & 0 & 0 \\ \hline 1 & 2 & 1 \\ \hline \end{array} \quad \text{or} \quad \begin{array}{|c|c|c|}\hline 1 & 2 & 1 \\ \hline 0 & 0 & 0 \\ \hline -1 & -2 & -1 \\ \hline \end{array} \tag{5.21}$$

Important

(This will give you trouble for the entire semester, so you may as well start now.) In software implementations, the positive y direction is DOWN! This results from the fact that scanning is top-to-bottom, left-to-right. So pixel (0, 0) is the *upper* left corner of the image. Furthermore, numbering starts at zero, not one. We find the best way to avoid confusion is to never use the words "x" and "y" in writing programs, but instead use "row" and "column" remembering that now 0 is on top.

However, in these notes, we will use conventional Cartesian coordinates in order to get the math right, and to further confuse the student (which is, after all, what Professors are there for. Right?).

Having cleared up that muddle, let us proceed. Given the gradient vector

$$\nabla f = \left[\frac{\partial f}{\partial x} \quad \frac{\partial f}{\partial y} \right]^{\mathrm{T}} \equiv [G_x \quad G_y]^{\mathrm{T}} \tag{5.22}$$

we are interested in its magnitude

$$|\nabla f| = \sqrt{G_x^2 + G_y^2} \tag{5.23}$$

(which we will call the "edge strength") and its direction

$$\angle\nabla f = \text{atan}\left(\frac{G_y}{G_x}\right). \tag{5.24}$$

One way to find an edge in an image is to compute the "gradient magnitude" image, and to threshold that. So go try it: Work homework Assignment 5.5.

While you are at it, you might want to work Assignment 5.6 too.

What did you learn from those experiments? Clearly, several problems arise when we try to find edges by simple kernel operations. During much of the remainder of the course, we will address these issues. First, let us improve our kernel-based edge detection.

5.7 A kernel as a sampled differentiable function

We hope you have realized by now that all the edge detector operators you have used so far are doing two things simultaneously; *smoothing* (read "low-pass filtering," "noise removal," "averaging," or "blurring") and *differentiation* (read "high-pass filtering" or "sharpening"). The kernel of Eq. (5.6) actually takes the vertical average of three derivative estimates. It is actually counter-intuitive, however, since it weights the center pixel the same as the lines above and below.

Consider the result you got on Assignment 5.6. If you did it correctly, the kernel values increase as they are farther from the center. That is even worse, right? Why should data points *farther away* from the point where we are estimating the derivative contribute more heavily to the estimate? Wrong! Wrong! Wrong! It is an artifact of the assumption we made that *all* the pixels fit the same plane. They obviously don't.

So here's a better way – weight the center pixel more heavily. You already saw this – the Sobel operator, Eq. (5.7) does it. But now, let's get a bit more rigorous. Let's blur the image by applying a kernel which is bigger in the middle and then differentiate. We have lots of choices for a kernel like that, e.g., a triangle or a Gaussian, but thorough research [5.28] has shown that a Gaussian works best for this sort of thing. We can write this process as

$$d = \frac{\partial}{\partial x}(g \otimes h)$$

Recall what you learned about linear systems.

where now g is the measured image, h is a Gaussian, and d will be our new derivative estimate image. Now, a crucial point from linear systems theory:

For linear operators D and $\otimes$,

$$D(g \otimes h) = D(h) \otimes g. \tag{5.25}$$

Equation (5.25) means we do not have to do blurring in one step and differentiation in the next; instead, we can pre-compute the derivative of the blur kernel and simply apply the resultant kernel.

Let's see if we can remember how to take a derivative of a 2D Gaussian (did you forget it is a 2D function?).

A d-dimensional multivariate Gaussian has the general form

$$\frac{1}{(2\pi)^{d/2}|K|^{1/2}} \exp\left(-\frac{[\boldsymbol{x}-\boldsymbol{\mu}]^{\mathrm{T}} K^{-1} [\boldsymbol{x}-\boldsymbol{\mu}]}{2}\right) \tag{5.26}$$

where K is the covariance matrix and μ is the mean vector. Since we want a Gaussian centered at the origin (which will be the center pixel) $\boldsymbol{\mu} = 0$, and since we have no reason to prefer one direction over another, we choose K to be diagonal (isotropic)

$$K = \begin{bmatrix} \sigma^2 & 0 \\ 0 & \sigma^2 \end{bmatrix} = \sigma^2 I. \tag{5.27}$$

For two dimensions, Eq. (5.26) simplifies to

Learn this Gaussian blur stuff, and how to use it to compute a first- or second-derivative kernel. It is guaranteed to be on the exam.

$$h(x, y) = \frac{1}{2\pi\sigma^2} \exp\left(\frac{-[x\ y]^{\mathrm{T}}[x\ y]}{2\sigma^2}\right) = \frac{1}{2\pi\sigma^2} \exp\left(-\frac{(x^2+y^2)}{2\sigma^2}\right) \tag{5.28}$$

and

$$\frac{\partial}{\partial x} h(x, y) = \frac{-x}{2\pi\sigma^4} \exp\left(-\frac{(x^2+y^2)}{2\sigma^2}\right). \tag{5.29}$$

If our objective is edge detection, we are done. However, if our objective is precise estimation of derivatives, particularly higher order derivatives, use of a Gaussian kernel, since it blurs the image, clearly introduces errors which can only be partially compensated for [5.39]. Nevertheless, this is one of the most simple ways to develop effective derivative kernels.

For future reference, here are a few of the derivatives of the one-dimensional Gaussian. Even though there is no particular need for the normalizing $\sqrt{2\pi}$ for most of our needs (it just ensures that the Gaussian integrates to one), we have included it. That way these formulae are in agreement with the literature. The subscript notation is used here to denote derivatives. That is,

$$G_{xx}(\sigma, x) = \frac{\partial^2}{\partial x^2} G(\sigma, x),$$

where $G(\sigma, x)$ is a Gaussian function of x with mean of zero and standard deviation σ.

Don't scoff at third derivatives. You never know when you might need one (like finding the maxima of the second derivative).

$$\begin{aligned}
G(\sigma, x) &= \frac{1}{\sqrt{2\pi}\sigma} \exp\left(-\frac{x^2}{2\sigma^2}\right) \\
G_x(\sigma, x) &= \frac{-x}{\sqrt{2\pi}\sigma^3} \exp\left(-\frac{x^2}{2\sigma^2}\right) \\
G_{xx}(\sigma, x) &= \left(\frac{x^2}{\sqrt{2\pi}\sigma^5} - \frac{1}{\sqrt{2\pi}\sigma^3}\right) \exp\left(-\frac{x^2}{2\sigma^2}\right) \\
G_{xxx}(\sigma, x) &= \frac{x}{\sqrt{2\pi}\sigma^5}\left(3 - \frac{x^2}{\sigma^2}\right) \exp\left(-\frac{x^2}{2\sigma^2}\right).
\end{aligned} \tag{5.30}$$

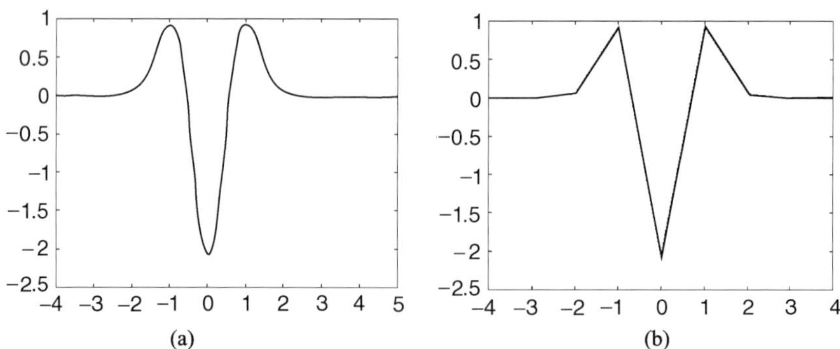

Fig. 5.6. (a) The second derivative of a one-dimensional Gaussian centered at 0. (b) A three-point approximation.

Let's look in a bit more detail about how to make use of these formulae and their two-dimensional equivalents to derive kernels.

The simplest way to get the kernel values for the derivatives of a Gaussian is to simply substitute $x = 0, 1, 2$, etc. along with their negative values, which yields numbers for the kernel. The first problem to arise is "what should σ be?" To address these questions, we will derive the elements of the kernel used for the second derivative of a one-dimensional Gaussian. The other derivatives can be developed using the same philosophy. Take a look at Fig. 5.6(a) and ask, "is there a value of σ such that the maximum of the second derivative occurs at $x = -1$ and $x = 1$?" Clearly there is, and its value is $\sigma = 1/(\sqrt{3})$. Given this value of σ, we can compute the values of the second derivative of a Gaussian at the integer points $x = \{-1, 0, 1\}$. At $x = 0$, we find $G_{xx}(1/\sqrt{3}, 0) = -2.07$, and at $x = 1$, $G_{xx}(1/\sqrt{3}, 1) = 0.9251$. So are we finished? That wasn't so hard, was it? Unfortunately, we are not done. It is very important that the elements of the kernel sum to zero. If they don't, then iterative algorithms like those described in Chapter 6 will not maintain the proper brightness levels over many iterations. The kernel also needs to be symmetric. That essentially defines the second derivative of a Gaussian. The most reasonable set of values close to those given, which satisfy symmetry and summation to zero are $\{1, -2, 1\}$.

Proving this is the correct value of σ is a homework problem.

The elements of any kernel which approximates a derivative must sum to zero.

However, this does not teach us very much. Let's look at a 5×1 kernel and see if we can learn a bit more. We require the following.

- The elements of the kernel should approximate the values of the appropriate derivative of a Gaussian as closely as possible.
- The elements must sum to zero.
- The kernel should be symmetric about its center, unless you want to do special processing.

We can calculate the elements of a five-element one-dimensional Gaussian, and if we do so, assuming $\sigma = 1/\sqrt{3}$, for $x = \{-2, -1, 0, 1, 2\}$, we get $[0.0565, 0.9251, -2.0730, 0.9251, 0.0565]$. Unfortunately, those numbers do not sum to zero. It is very important that the kernel values integrate to zero, not quite so important that the actual values be precise. So what do we do in a case like this? We use constrained optimization. One strategy is to set up a problem to find a second derivative of a Gaussian, which has these values as closely as possible, but which integrates to zero. For more complex problems, the authors use *Interopt* [5.3] to solve numerical optimization problems, but you can solve this problem without using numerical methods. This is accomplished as follows. First, understand the problem (presented for the case of five points given above): We wish to find five numbers as close as possible to $[0.0565, 0.9251, -2.0730, 0.9251, 0.0565]$ which satisfy the constraint that the five sum to zero. By symmetry, we actually only have three numbers, which we will denote $[a, b, c]$. For notational convenience, introduce three constants $\alpha = 0.0565$, $\beta = 0.9251$, $\gamma = -2.073$. Thus, to find a, b, and c which resemble these numbers, we write the mean squared error (MSE) form

$$H_0(a, b, c) = 2(a - \alpha)^2 + 2(b - \beta)^2 + (c - \gamma)^2. \tag{5.31}$$

H is the constrained version of H_0.

Using the concept of Lagrange multipliers, we can find the best choice of a, b, and c by minimizing a different objective function

$$H(a, b, c) = 2(a - \alpha)^2 + 2(b - \beta)^2 + (c - \gamma)^2 + \lambda(2a + 2b + c). \tag{5.32}$$

A few words of explanation are in order for those students who are not familiar with constrained optimization using Lagrange multipliers. The term with the λ in front (λ is the Lagrange multiplier) is the constraint. It is formulated such that it is exactly equal to zero, if we should find the proper a, b, and c. By minimizing H, we will find the parameters which minimize H_0 while simultaneously satisfying the constraint.

To minimize H, take the partials and set them equal to zero:

$$\begin{aligned} \frac{\partial H}{\partial a} &= 4a - 4\alpha + 2\lambda \\ \frac{\partial H}{\partial b} &= 4b - 4\beta + 2\lambda \\ \frac{\partial H}{\partial c} &= 2c - 2\gamma + \lambda. \end{aligned} \tag{5.33}$$

Setting the partial derivatives equal to zero, simplifying, and adding the constraint, we find the following set of linear equations:

$$\begin{aligned} a &= \alpha - \frac{\lambda}{2} \\ b &= \beta - \frac{\lambda}{2} \\ c &= \gamma - \frac{\lambda}{2} \\ 2a + 2b + c &= 0 \end{aligned} \tag{5.34}$$

which we solve to find the sets given in Table 5.2.

In the case of the first derivative, symmetry ensures the values always sum to zero, so no "tweaking" is necessary. Therefore the integer values are just as good as the floating point ones.

Table 5.2. *Derivatives of a one-dimensional Gaussian.*

1st deriv, 3 × 1	[0.2420, 0.0, −0.2420] or [1, 0, −1]
1st deriv, 5 × 1	[0.1080, 0.2420, 0, −0.2420, −0.1080]
2nd deriv, 3 × 1	[1, −2, 1]
2nd deriv, 5 × 1	[0.07846, 0.94706, −2.05104, 0.94706, 0.07846]

0.0261	0	−0.0261
0.1080	0	−0.1080
0.0261	0	−0.0261

Fig. 5.7. The 3×3 first derivative kernel.

We can proceed in the same way to compute the kernels to estimate the partial derivatives using Gaussians in two dimensions.

One implementation of the first derivative with respect to x, assuming an isotropic Gaussian is presented in Fig. 5.7. You will have the opportunity to derive others as homeworks.

In this chapter, we have explored the idea of edge operators based on kernel operators. We discovered that no matter what, noisy images result in edges which are:

- too thick in places
- missing in places
- extraneous in places.

That is just life – we cannot do any better with simple kernels. In Chapter 6, we will explore some approaches to these problems.

As we hope you have guessed, there are other ways of finding edges in images besides simply thresholding a derivative. In later sections, we will mention a few of them.

5.7.1 Higher order derivatives

We have just seen how second or third derivatives may be computed using derivatives of Gaussians. Since the topic has come up, and you will need to know the terminology later, we define here two scalar operators which depend on the second derivatives: the Laplacian and the quadratic variation.

The Laplacian of brightness at a point x, y is

$$\frac{\partial^2 f}{\partial x^2} + \frac{\partial^2 f}{\partial y^2},$$

whereas the quadratic variation of brightness is

$$\left(\frac{\partial^2 f}{\partial x^2}\right)^2 + \left(\frac{\partial^2 f}{\partial y^2}\right)^2 + 2\left(\frac{\partial^2 f}{\partial x \partial y}\right)^2.$$

The Laplacian may be approximated by several kernels, including

−1	2	−1
2	−4	2
−1	2	−1

.

5.8 Computing convolutions

Remembering that the only difference between convolution and a kernel operator is the direction of x and y (section 5.2.1) for any kernel operator, there is an equivalent convolution kernel. Therefore efficient ways to calculate convolution are also efficient ways to apply kernel operators. The convolution operation may be computed directly as discussed above. It is simply a sum of products, calculated in the neighborhood of each pixel. However, it may also be computed by the Fourier transform. The Fourier transform of a convolution is the product of the Fourier transforms of the two arguments. That is (denoting convolution by the operator $\otimes$), we are concerned with computing

$$g(x, y) = f(x, y) \otimes h(x, y).$$

Let the Fourier transforms of the two images and the convolution kernel be defined by

$$\begin{aligned} G(\omega_x, \omega_y) &= F(g(x, y)) \\ F(\omega_x, \omega_y) &= F(f(x, y)) \\ H(\omega_x, \omega_y) &= F(h(x, y)) \end{aligned}$$

where the symbol F denotes the process of taking the Fourier transform. Remember from section 4.1.5, the Fourier transform of an image (a function of two variables) is itself a function of two variables. We refer to those variables, ω_x and ω_y, as the spatial frequencies in the x and y directions, respectively. Then G is the product of F and H.

$$G(\omega_x, \omega_y) = F(\omega_x, \omega_y) \bullet H(\omega_x, \omega_y). \tag{5.35}$$

The "product" of two transforms means, for each spatial frequency value (each combination of ω_x and ω_y), multiply the values of the two functions. (Just in case you do not remember the details, in general, these values are complex numbers.)

The ability to perform point-by-point multiplication is significant because of the computational complexity. Consider the complexity of convolving an $N \times N$ image with an $L \times L$ kernel. Doing it spatially, we have $L \times L$ multiplications for each pixel, totalling $N^2 L^2$. Doing it with the Fourier transform (this assumes you are using a particular algorithm called the fast Fourier transform) works out in the following way. (The details are outside the scope of this section, but we will assume that the Fourier transform of an $N \times N$ image is itself a two-dimensional array of the same size.)

- Transform f: $N^2 \log N$.
- Transform h: $L^2 \log L$.
- Perform appropriate operations, such as padding, to get H and F the same size.
- Multiply H by F: N^2.
- Inverse transform the result: $N^2 \log N$.

If the sum of these four terms is smaller than $N^2 L^2$, it is computationally more effective to go through the (considerable) inconvenience of using the transform domain. But for a particular image size and kernel size, what should we do? Fortunately, the relative efficiency of Fourier and spatial methods for computing convolutions of varying sizes has been analyzed, and the results are illustrated in Fig. 5.8. In that figure, we see that for kernels larger than about 15×15, we should use Fourier methods, for kernels smaller than 7×7, we should use spatial methods. The peculiar variations in the region boundary occur because the FFT requires the image size be a power of two, and images of other sizes introduce additional complications.

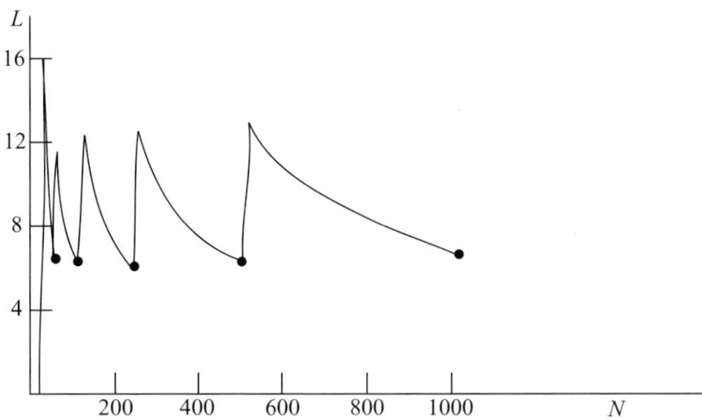

Fig. 5.8. Efficiency of computing a convolution with an $L \times L$ kernel on an $N \times N$ image. Combinations above the curve shown are more efficiently computed using Fourier methods, below the curve, by spatial methods (redrawn from Pratt [5.33]).

5.9 Scale space

"Scale space" is a recent addition to the well-known concept of image pyramids, first used in picture processing by Kelly [5.19] and later extended in a number of ways (see [5.5, 5.8, 5.30, 5.32], and many others). In a pyramid, a series of representations of the same image are generated, each created by a 2 : 1 subsampling (or averaging) of the image at the next higher level (Fig. 5.9).

In Fig. 5.10, a Gaussian pyramid is illustrated. It is generated by blurring each level with a Gaussian prior to 2 :1 subsampling. An interesting question should arise as you look at this figure. Could you, from all the data in this pyramid, reconstruct the original image? The answer is "no, because at each level, you are throwing away high-frequency information."

Usually, when we say "scale-space", we do not mean a pyramid, we mean a varying blur.

Although the Gaussian pyramid alone does not contain sufficient information to reconstruct the original image, we could construct a pyramid that does contain sufficient information. To do that, we use a "Laplacian" pyramid, constructed by computing a similar representation of the image; this preserves the high-frequency information (Fig. 5.11). Combining the two pyramid representations allows reconstruction of the original image.

In a modern scale space representation we preserve the concept that each level is a blurring of the previous level, but do not subsample – each level is the same size as the previous level, but more blurred. Normally, each level is generated by convolving the original image with a Gaussian of variance σ^2, and σ varies from one level to the next. This variance then becomes the "scale parameter." Clearly, at high levels of scale, (σ large), only the largest features are visible. We will see more about scale space later in this chapter, when we talk about wavelets.

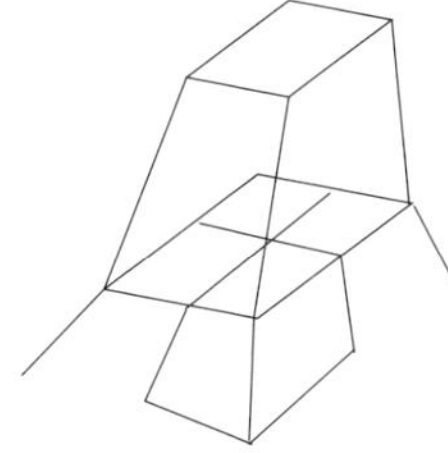

Fig. 5.9. A pyramid is a data structure which is a series of images, in which each pixel is the average of four pixels at the next lower level.

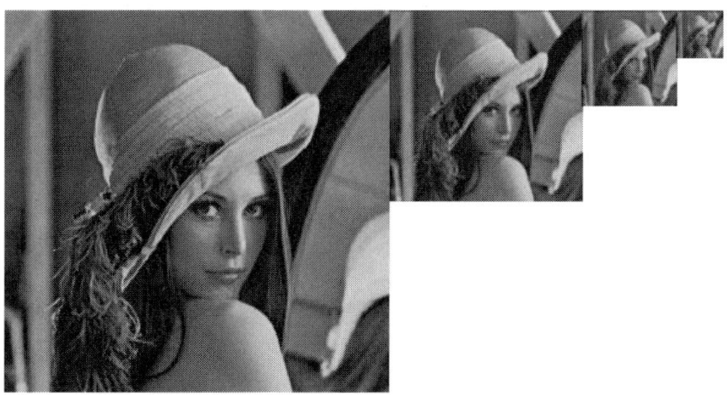

Fig. 5.10. A Gaussian pyramid, constructed by blurring each level with a Gaussian and then 2 : 1 subsampling.

Fig. 5.11. This Laplacian pyramid is actually computed by a difference of Gaussians.

5.9.1 Quad trees

A quad tree [5.21] is a data structure in which images are recursively broken into four blocks, corresponding to nodes in a tree. The four blocks are designated NW (north–west), NE, SW, and SE. The correspondence between the nodes in the tree and the image are best illustrated by an example (see Fig. 5.12).

In encoding binary images, it is straightforward to come up with a scheme for generating the quad tree for an image: If the quadrant is homogeneous (either solid black or solid white), then make it a leaf, otherwise divide it into four quadrants and add another layer to the tree. Repeat recursively until the blocks either reach pixel size or are homogeneous.

It is easy to make a quad tree representation into a pyramid. It is only necessary to keep, at each node, the average of the values of its children. Then, all the information in a pyramid is stored in the quad tree.

If an image has large homogeneous regions, a quad tree would seem to be an efficient way to store and transmit an image. However, experiments with a variety of images, even images which were the difference between two frames in a video

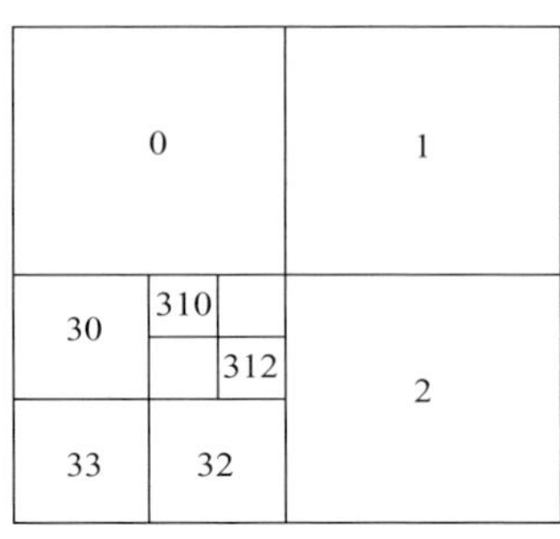

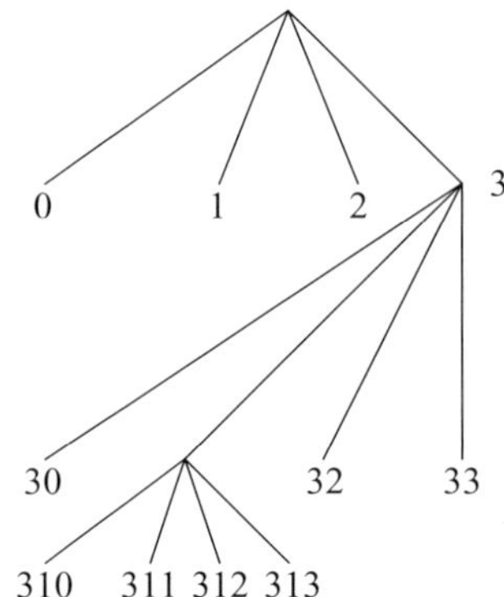

Fig. 5.12. An image is divided into four blocks. Each inhomogeneous block is further divided. This partitioning may be represented by a tree.

sequence, have shown that this is not true. Since the difference image is only nonzero where things are moving, it seems obvious that this, mostly zero, image would be efficiently stored in a quad tree. Not so. Even in that case, the overhead of managing the tree overwhelms the storage gains. So, surprisingly, the quad tree is not an efficient image compression technique. When used as a means for representing a pyramid, it does, however, have advantages as a way of representing scale space.

Another disadvantage of using quad trees is that a slight movement of an object can result in radically different tree representations, that is, the tree representation is not rotation or translation invariant. In fact, it is not even robust. Here, "robust" means a small translation of an object results in a correspondingly small change in the representation. One can get around this problem, to some extent, by not representing the entire image, but instead, representing each object subimage with a quad tree.

The generalization of the quad tree to three dimensions is called an "octree." The same principles apply.

5.9.2 Gaussian scale structures

A good way to remember large and small scale is that at large scale, only large objects may be distinguished.

We know how to blur an image. Here is a *gedankenexperiment* for you: Take an image, blur it with a Gaussian kernel of standard deviation 1. You get a new image. Call that image 1. Now, blur the original image with a Gaussian kernel of standard deviation 2. Call that image 2. Continue until you have a set of images which you can think of as stacked, and the "top" image is almost blurred away. We say the top image is a representation of the image at "large scale." This stack of images is referred to as a "scale space" representation. Clearly, we are not required to use integer values of the standard deviation, so we can create scale space representations with as much resolution in scale as desired. The essential premise of scale space representations is that certain features can be tracked over scale, and how those features vary with scale tells something about the image. A scale space has been formally defined [5.25, 5.26] as having the following properties.

- All signals should be defined on the same domain (no pyramids).
- Increasing values of the scale parameter should produce coarser representations.
- A signal at a coarser level should contain less structure than a signal at a finer level. If one considers the number of local extrema as a measure of smoothness, then the number of extrema should not increase as we go to coarser scale. This property is called "scale space causality."
- All representations should be generated by applications of a convolution kernel to the original image.

The last property is certainly debatable, since convolution formally requires a linear, space-invariant operator. One interesting approach to scale space which violates this requirement is to produce a scale space by using gray scale morphological smoothing (we will discuss this later) with larger and larger structuring elements [5.16].

You could use scale space concepts to represent texture [4.16] or even a probability density function (in which case, your scale space representation becomes a clustering algorithm [5.24]) as well as brightness. We will see applications of scale representations as we proceed through the course.

One of the most interesting aspects of scale space representations is the behavior of our old friend, the Gaussian. The second derivative of the Gaussian (in two dimensions, the Laplacian of Gaussian: LOG) has been shown [5.27] to have some very nice properties when used as a kernel. In particular, the zero crossings of the LOG are good indicators of the location of an edge. One might be inclined to ask, "Is the Gaussian the best smoothing operator to use to develop a kernel like this?" Said another way: We want a kernel whose second derivative never generates a new zero crossing as we move to larger scale. In fact, we could state this desire in the following more general form.

Let our concept of a "feature" be a point where some operator has an extreme, either maximum or minimum. The concept of scale space causality says that as scale increases, as images become more blurred, new features are never created. The Gaussian is the ONLY kernel (linear operator) with this property [5.1, 5.2]. Studies of nonlinear operators have been done to see under what conditions these operators are scale space causal [5.22].

This idea of scale space causality is illustrated in the following example. Fig. 5.13 illustrates the brightness profile along a single line from an image, and the scale space created by blurring that single line with one-dimensional Gaussians of increasing variance. In Fig. 5.14, we see the Laplacian of the Gaussian, and the points where the Laplacian changes sign. The features in this example, the zero crossings (which are good candidates for edges) are indicated in the right image. Observe that as scale increases, feature points (in this case, zero-crossings) are never created as scale increases. As we go from top (low scale) to bottom (high scale), some features disappear, but no new ones are created.

One obvious application of this idea is to identify the important edges in the image first. We can do that by going up in scale, finding those few edges, and then tracking them down to lower scale.

5.10 Quantifying the accuracy of an edge detector

Since there are many options in the design of an edge detection algorithm, we need some objective ways to say that one edge detector works better than another. Pratt [5.33] has suggested a simple formula to address this question. The formula is just

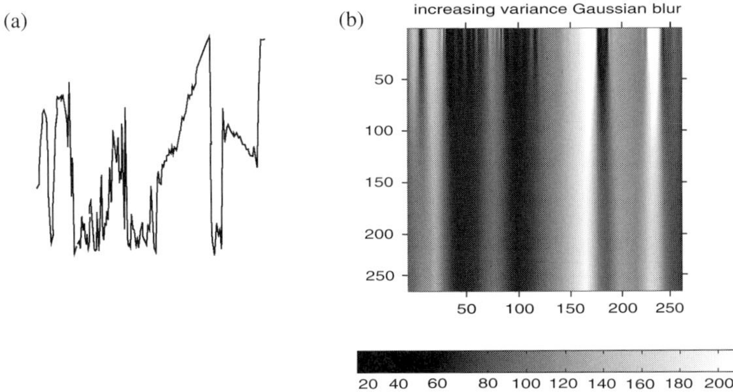

Fig. 5.13. (a) Brightness profile of a scanline through an image. (b) Scale space representation of that scanline. Scale increases toward the bottom, so no new features should be created as one goes from top to bottom.

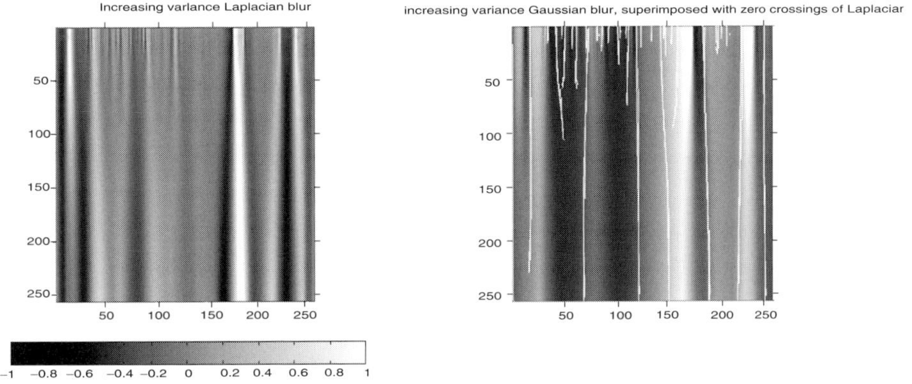

Fig. 5.14. Laplacian of the scale space representation, and the zero crossings of the Laplacian. Since this is one-dimensional data, there is no difference between the Laplacian and the second derivative.

a summation over the edge points

$$R = \frac{1}{I_N} \sum_{i=1}^{I_a} \frac{1}{1 + \alpha d^2} \tag{5.36}$$

where $I_N = \max(I_I, I_A)$, I_I denotes the number of edge points detected and I_A is the number of edge points actually in the image. α is a constant scale factor, and d is a measure of the distance from the edge point detected to the nearest actual edge. This formula requires, of course, some knowledge of where the edge points *should* be. Consequently, it is of primary usefulness when applied to synthetic data, since only in such data can we be absolutely sure of the actual position of edge points (see also [5.4]).

5.11 So how do people do it?

Two neurophysiologists, David Hubel and Thorsten Wiesel [5.13, 5.14] stuck some electrodes in the brains – specifically the visual cortex – first of cats[3] and later of monkeys. While recording the firing of neurons, they provided the animal with visual stimuli of various types. They observed some fascinating results: First, there are cells which fire only when specific types of patterns are observed. For example, a particular cell might only fire if it observed an edge, bright to dark, at a particular angle. There was evidence that each of the cells they measured received input from a neighborhood of cells called a "receptive field." There were a variety of types of receptive fields, possibly all connected to the same light detectors, which were organized in such a way as to accomplish edge detection and other processing. Jones and Palmer [5.17] mapped receptive field functions carefully and confirmed [5.9, 5.10] that the function of receptive fields could be accurately represented by *Gabor functions*, which have the form of Eq. (5.37):

$$G(x, y) = \frac{1}{2\pi\sigma\beta} \exp\left(-\pi\left(\frac{x^2}{\sigma^2} + \frac{y^2}{\beta^2}\right)\right) \exp(i[\xi x + \nu y]). \tag{5.37}$$

The first exponential is a two-dimensional Gaussian whose isophotes form ellipses with major and minor axes aligned with the x and y axes. (If you happen to be dealing with a receptive field which is tilted with respect to x and y, you need to rotate your coordinate system to make this equation still hold.) The second (complex) exponential represents a plane wave. Eq. (5.37) assumes the origin is at the center of the Gaussian. Fig. 5.15 illustrates a Gabor filter.

The following interesting observations have been made [5.23] regarding the values of the parameters in Eq. (5.37), when those parameters are actually measured in living organisms:

- The aspect ratio, β/α, of the ellipse is $2 : 1$.
- The plane wave tends to propagate along the short axis of the ellipse.
- The half-amplitude bandwidth of the frequency response is about 1 to 1.5 octaves along the optimal orientation.

So do we have Gabor filters in our brains? Or Gabor-like wavelet generators? Well, we don't know about YOUR brain, but Young [5.41] has looked at the stimulus–response characteristics of mammalian retina and observed that those same receptive fields that are so nicely modeled with Gabor filters or wavelets may equally well be described in terms of kernels called "difference of offset Gaussians," which is essentially the LOG with an additive Gaussian offset. Fig. 5.16 illustrates a section

[3] We were going to put a dead cat joke here, something like "One of the cats died during the procedure, but its behavior was unchanged," but the publisher told us people would be offended, so we had to remove it.

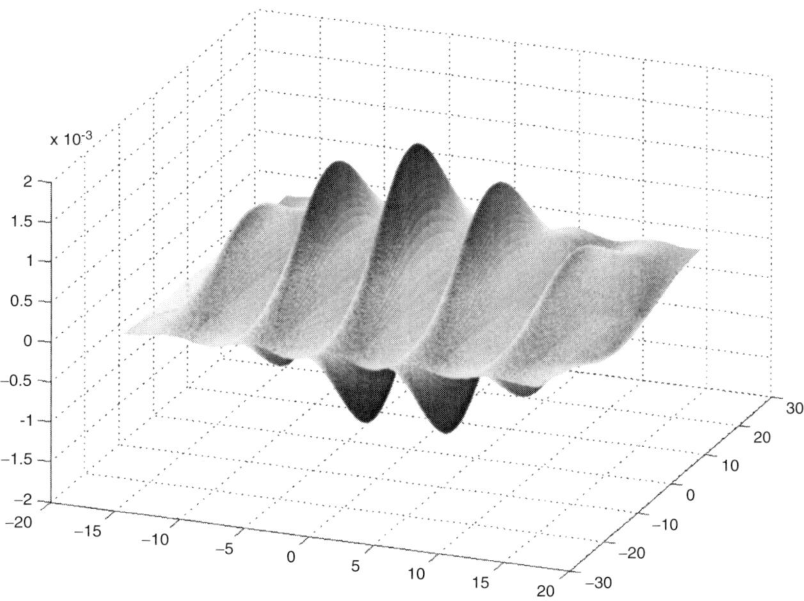

Fig. 5.15. **Gabor filter. Note that the positive/negative response is very similar to those that we derived earlier in this chapter.**

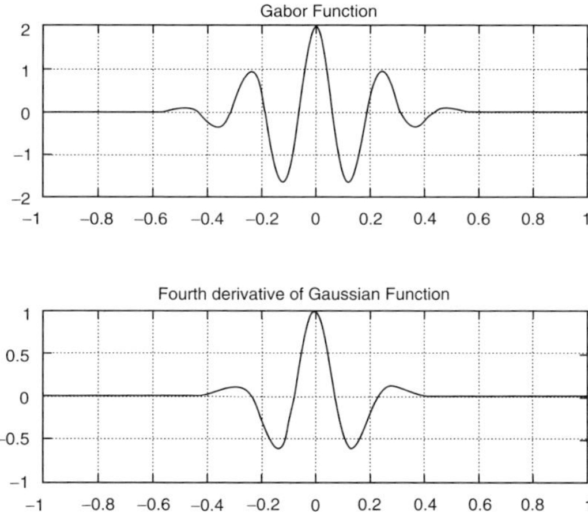

Fig. 5.16. **Comparison of a section through a Gabor filter and a similar section through a fourth derivative of a Gaussian.**

through a Gabor and a fourth derivative of a Gaussian. You can see noticeable differences, the principal one being that the Gabor goes on forever, whereas the fourth derivative has only three extrema. However, to the precision available to neurology experiments, they are the same. The problem is simply that the measurement of the data is not sufficiently accurate, and it is possible to fit a variety of curves to it.

Bottom line: We have barely a clue how the brain works, and we don't really know very much about the retina. There are two or three mathematical models which adequately model the behavior of receptive fields.

5.12 Conclusion

Consistency in edge detection.

Explicit use of consistency has not been made in this chapter. However, in Assignment 10.1, you will see an application of consistency to edge detection. In that problem, you will be asked to develop an algorithm which makes use of the fact that adjacent edge pixels have parallel gradients. That is, if pixel A is a neighbor of pixel B, and the gradient at pixel A is parallel (or nearly parallel) to the gradient at pixel B, this increases the confidence that both pixels are members of the same edge.

In this chapter, we have looked at several ways to derive kernel operators which, when applied to images, result in strong responses for types of edges.

Minimize the sum-squared error.

Constrained optimization and Lagrange multipliers.

- We applied the definition of the derivative.
- We fit an analytic function to a surface, by minimizing the sum squared error.
- We converted subimages into vectors, and projected those vectors onto special basis vectors which described edge-like characteristics.
- We made use of the linearity of kernel operators to interchange the roll of blur and differentiation to construct kernels which are the derivatives of special blurring kernels. We used the constrained optimization and Lagrange multipliers to solve this problem.

5.13 Vocabulary

You should know the meanings of the following terms.

Basis vector
Convolution
Correlation
Gabor filter
Image gradient
Inner product
Kernel operator
Lagrange multiplier
Lexicographic
Linear operator

LOG
Projection
Pyramid
Quad tree
Scale space
Sum-squared error

Assignment **5.1**

The previous section showed how to estimate the first derivative by fitting a plane. Clearly that will not work for the second derivative, since the second derivative of a plane is zero everywhere. Use the same approach, but use a biquadratic

$$f(x, y) = ax^2 + by^2 + cx + dy + e.$$

Then $[a\ b\ c\ d\ e]^T = A$

$$\begin{bmatrix} x^2 & y^2 & x & y & 1 \end{bmatrix}^T = X.$$

Find the 3×3 kernel which estimates $\frac{\partial^2 f}{\partial x^2}$.

Assignment **5.2**

Oh no! Another part of the assignment! (hey, this one is much easier than the one above – a real piece of cake). Using the same methods, find the 5×5 kernel which estimates $\frac{\partial f}{\partial x}$ at the center point, using the equation of a plane.

Assignment **5.3**

Determine whether u_1 and u_2 in Fig. 5.4 are in fact orthonormal. If not, recommend a modification or other approach which will allow all the us to be used as basis functions.

Assignment 5.4

(1) Write a program to generate an image which is 64×64, as illustrated below.

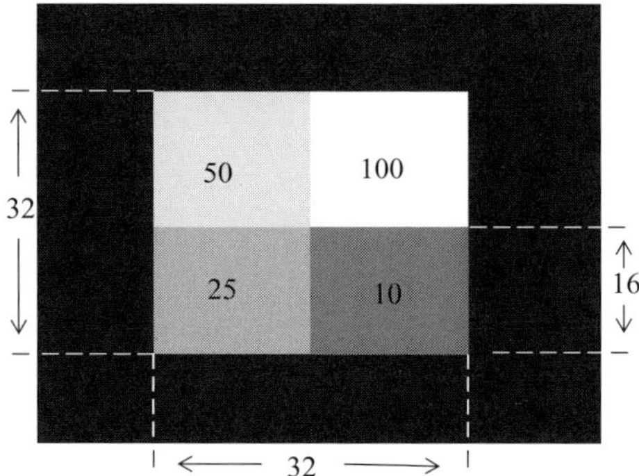

The images should contain areas of uniform brightness, with dimensions and brightness as shown. Save this in a file and call it "SYNTH1."

(2) Write a program which reads in SYNTH1, and applies the blurring kernel

$$\frac{1}{10}\begin{bmatrix} 1 & 1 & 1 \\ 1 & 2 & 1 \\ 1 & 1 & 1 \end{bmatrix}$$

and write the answer to file named "BLUR1."

(3) Add Gaussian random noise of variance $\sigma^2 = 9$ to BLUR1, and write the output to a file "BLUR1.V1."

Assignment 5.5

Write a program to apply the following two kernels (referred to in the literature as the "Sobel operators") to images SYNTH1, BLUR1, and BLUR1.V1.

$$h_x = \begin{bmatrix} -1 & 0 & 1 \\ -2 & 0 & 2 \\ -1 & 0 & 1 \end{bmatrix} \qquad h_y = \begin{bmatrix} -1 & -2 & -1 \\ 0 & 0 & 0 \\ 1 & 2 & 1 \end{bmatrix}$$

To accomplish this, perform the following.

(1) Apply h_x to the input, save the result as a temporary image in memory (remember that the numbers CAN be negative).
(2) Apply h_y to the input, save the result as another array in memory.
(3) Compute a third array in which each point in the array is the sum of the squares of the corresponding points in the two arrays you just saved. Finally, take the square root of each point. Save the result.
(4) Examine the values you get. Presumably, high values are indicative of edges. Choose a threshold value and compute a new image, which is one whenever the edge strength exceeds your threshold and is zero otherwise.
(5) Apply steps (1)–(4) to the blurred and noisy images as well.
(6) Write a report. Include a printout of all three binary output images. Are any edge points lost? Are any points artificially created? Are any edges too thick? Discuss sensitivity of the result to noise, blur, and choice of threshold.

Be thorough; this is a research course which requires creativity and exploring new ideas, as well as correctly doing the minimum required by the assignment.

Assignment 5.6

In Assignment 5.2, you derived a 5×5 kernel. Repeat Assignment 5.5 using that kernel, for $\partial/\partial x$ and the appropriate version for $\partial/\partial y$.

Assignment 5.7

(1) Verify the mathematics we did in Eq. (5.30). Find a 3×3 kernel which implements the derivative-of-Gaussian vertical edge operator of Eq. (5.30). Use $\sigma = 1$ and $\sigma = 2$ and determine two kernels. Repeat for a 5×5 kernel. Discuss the impact of the choice of σ and its relation to kernel size. Assume the kernel may contain real (floating point) numbers.

(2) Suppose the kernel can only contain integers. Develop kernels which produce approximately the same result.

Assignment 5.8

In section 5.7, parameters useful for developing discrete Gaussian kernels were discussed. Prove that the value of σ such that the maximum of the second derivative occurs at $x = -1$ and $x = 1$. Is $\sigma = 1/\sqrt{3}$?

Assignment 5.9

Use the method of fitting a polynomial to estimate $\partial^2 f/\partial y^2$. Which of the following polynomials would be most appropriate to choose?

(a) $f = ax^2 + by + cxy$ (c) $f = ax^3 + by^3 + cxy$

(b) $f = ax^2 + by^2 + cxy + d$ (d) $f = ax + by + c$

Assignment 5.10

Fit the following expression to pixel data in a 3×3 neighborhood: $f(x, y) = ax^2 + bx + cy + d$. From this fit, determine a kernel which will estimate the second derivative with respect to x.

Assignment 5.11

Use the function $f = ax^2 + by^2 + cxy$ to find a 3×3 kernel which estimates $\partial^2 f/\partial y^2$. Which of the following is the kernel which results? (Note: The following answers do not include the scale factor. Thus, the best choice below will be the one proportional to the correct answer.)

(a)

6	4	0
4	6	0
0	0	4

(c)

2	6	2
−4	0	−4
2	6	2

(e)

6	4	0
4	6	0
0	0	1

(b)

1	−2	1
3	0	3
1	−2	1

(d)

1	0	−1
−2	0	2
1	0	−1

(f)

1	2	1
−2	0	−2
−1	−2	−1

Assignment 5.12

Suppose the kernel that estimates $\partial^2 f/\partial x\partial y$ is

2	−1	0
−1	0	1
0	1	−2

(It may not be true that this kernel estimates that derivative, but it does not affect the answer. Assume it is true.) Is it true then that

4	1	0
1	0	1
0	1	4

estimates $(\partial^2 f/\partial x\partial y)^2$? Explain your answer.

Topic 5A Edge detectors

The process of edge detection includes more than simply thresholding the gradient. We want to know the location of the edge more precisely than simple gradient thresholds reveal. Two methods which seem to have acquired a substantial reputation in this area are the so called "Canny edge detector" [5.6] and the "facet model" [4.18]. Here, we only describe the Canny edge detection.

5A.1 The Canny edge detector

The edge detection algorithm begins with finding estimates of the gradient magnitude at each point. Canny uses 2×2 rather than 3×3 kernels as we have, but it does not affect the philosophy of the approach. Once we have estimates of the two partial derivatives, we use Eqs. (5.22) through (5.24) to calculate the magnitude and direction of the gradient, producing two images, $M(x, y)$ and THETA(x, y). We now have a result which easily identifies the pixels where the magnitude of the gradient is large. That is not sufficient, however, as we now need to thin the magnitude array, leaving only points which are maxima, creating a new image $N(x, y)$. This process is called nonmaximum[4] suppression (NMS).

NMS may be accomplished in a number of ways. The essential idea, however, is as follows: First, initialize $N(x, y)$ to $M(x, y)$. Then, at each point, (x, y), look one pixel in the direction

4 This is sometimes written using the plural, nonmaxima suppression. The expression is ambiguous. It could be a compression of suppress every point which is not a maximum, or suppress all points which are not maxima. We choose to use the singular.

of the gradient, and one pixel in the reverse direction. If $M(x, y)$ (the point in question) is not the maximum of these three, set its value in $N(x, y)$ to zero. Otherwise, the value of N is unchanged.

After NMS, we have edges which are properly located and are only one pixel wide. These new edges, however, still suffer from the problems we identified earlier – extra edge points due to noise (false hits) and missing edge points due either to blur or to noise (false misses). Some improvement can be gained by using a dual-threshold approach. Two thresholds are used, τ_1 and τ_2, where τ_2 is significantly larger than τ_1. Application of these two different thresholds to $N(x, y)$ produces two binary edge images, denoted T_1 and T_2 respectively. Since T_1 was created using a lower threshold, it will contain more false hits than T_2. Points in T_2 are therefore considered to be parts of true edges. Connected points in T_2 are copied to the output edge image. When the end of an edge is found, points are sought in T_1 which could be continuations of the edge. The continuation is continued until it connects with another T_2 edge point, or no connected T_1 points are found.

In [5.6] Canny also illustrates some clever approximations which provide significant speedups.

5A.2 Improvements to edge detection

The derivative of $g(x)$ is the second derivative of the image, so look for zero crossings.

Tagare and deFigueiredo [5.34] (see also [5.1]) describe the process of edge detection as follows.

(1) The input is convolved with a filter which smoothly differentiates the input and produces high values at and near the location of the edge. The output $g(x)$ is the sum of the differentiated step edge and filtered noise.
(2) A decision mechanism isolates regions where the output of the filter is significantly higher than that due to noise.
(3) A mechanism identifies the zero crossing in the derivative of $g(x)$ in the isolated region and declares it to be the location of the edge.

A Gaussian low-pass filter followed by finding a zero in the (second) derivative accurately (to subpixel resolution) finds the exact location of the edge, but only if the edge is straight [5.38]. If the edge is curved, errors are introduced. For example, the second derivative in the gradient direction (SDGD) and the Laplacian both make errors in estimating the location of the edge, but interestingly, in opposite directions, which prompts Verbeek and van Vliet [5.38] to suggest an operator which is the sum of the two.

All the methods cited or described in this section perform signal processing in a direction normal to the edge [5.18] in order to better locate the actual edge. Taratorin and Sideman [5.35] present a way to make use of the fact that the images are known to have properties such as positivity and finite support to improve the accuracy with which derivatives may be estimated. Iverson and Zucker [5.15] improve the results of the Canny by adding logical/Boolean reasoning. This improves edge detection over simple thresholding of the derivative, but does not provide results as good as active contours (see Chapter 9) or optimization (see Chapter 6). There are a wide variety of papers on signal processing techniques applied to edge detection [5.36].

Examination of the literature in biological imaging systems, all the way back to the pioneering work of Hubel and Wiesel in the 1960s [5.13] suggests that biological systems analyze images by making local measurements which quantify orientation, scale, and motion. Keeping this in mind, suppose we wish to ask a question like "is there an edge at orientation θ at this point?" How might we construct a kernel that is specifically sensitive to edges at that orientation? A straightforward approach [5.37] is to construct a weighted sum of the two Gaussian first derivative kernels, G_x and G_y, using a weighting something like

$$G_\theta = G_x \cos\theta + G_y \sin\theta. \tag{5.38}$$

Could you calculate the orientation selectivity? What is the smallest angular difference you could detect with a 3×3 kernel determined in this way?

Unfortunately, unless quite large kernels are used, the kernels obtained in this way have rather poor orientation selectivity. In the event that we wish to differentiate across scale, the problem is even worse, since a scale space representation is normally computed rather coarsely, to minimize computation time. Perona [5.31] provides an approach to solving these problems.

5A.3 Inferring line segments from edge points

After we have chosen the very best operators to estimate derivatives, have chosen the best thresholds, and selected the best estimates of edge position, we still have nothing for a set of pixels, some of whom have been marked as probably part of an edge. If those points are adjacent, one could "walk" from one pixel to the next, eventually circumnavigating a region, and there are representations such as the chain code which make this process easy. However, the points are unlikely to be connected the way we would like them to be. Some points may be missing due to blur, noise, or partial occlusion. There are many ways to approach this problem, including relaxation labeling, and parametric transforms, both of which will be discussed in detail later in this book. In addition, there are combination methods, such as the work of Deng and Iyengar [5.11] which combines relaxation and Bayes' methods as well as other methods [5.29] which we do not have space to discuss.

5A.4 Space/frequency representations

Wavelets are very important, but a thorough examination of this area is beyond the scope of this book. Therefore we present only a rather superficial description here, and provide some pointers to literature. For example, Castleman [4.6] has a readable chapter on wavelets.

5A.4.1 Why wavelets?

Consider the image illustrated in Fig. 5.17. Clearly, the spatial frequencies appear to be different, depending on where one looks in the image. The Fourier transform has no mechanism for capturing this intuitive need to represent both frequency and location. The Fourier transform of this image will be a two-dimensional array of numbers, representing the amount of energy *in the entire image* at each spatial frequency. Clearly, since the Fourier transform is invertible, it captures all this spatial and frequency information, but there is no obvious way to answer the question: at each position, what are the local spatial frequencies?

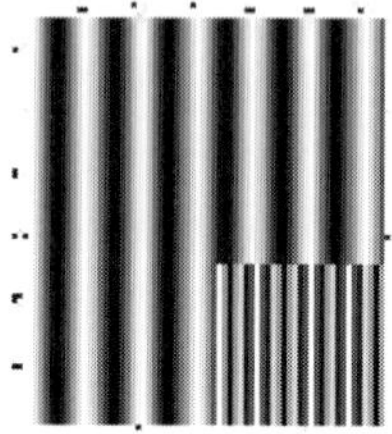

Fig. 5.17. An image in which spatial frequencies vary dramatically.

The wavelet approach is to add a degree of freedom to the representation. Since the Fourier transform is complete and invertible, all we really need to characterize the image is a single two-dimensional array. Instead however, following the space/frequency philosophy as described in section 5.8, we use a three-(or higher) dimensional data structure. In this sense, the space/frequency representation is redundant (or *overcomplete*), and requires significantly more storage than the Fourier transform.

5A.4.2 The basic wavelet and wavelet transform

We define a basic[5] wavelet $\psi(x, y)$ as any function of the two spatial variables x and y, which meets a certain criterion that we need not concern ourselves with here. Basically, we desire a function which is symmetric about the origin and has *almost finite support*. By "almost finite support," we mean that the magnitude of this function drops off to zero rapidly (in a particular way defined by the admissibility criterion) as it goes away from its center. A one-dimensional example basic wavelet is

$$\psi(x) = \frac{2}{\sqrt{3}\sqrt{\pi}}(1 - x^2)\exp\left(-\frac{x^2}{2}\right) \tag{5.39}$$

graphed in Fig. 5.18. From one such wavelet, one may then generate a (potentially infinite) set of similar functions by translation and scaling of the original. That is, we define a translated, scaled version of ψ by (again, in one dimension)

$$\psi_{a,b}(x) = \frac{1}{\sqrt{a}}\psi\left(\frac{x-b}{a}\right). \tag{5.40}$$

The wavelet transform of a function f is computed by the inner products of f with the wavelet at each of the possible values of a and b.

$$W_f(a, b) = \int_{-\infty}^{\infty} f(x)\psi_{a,b}(x)\,dx. \tag{5.41}$$

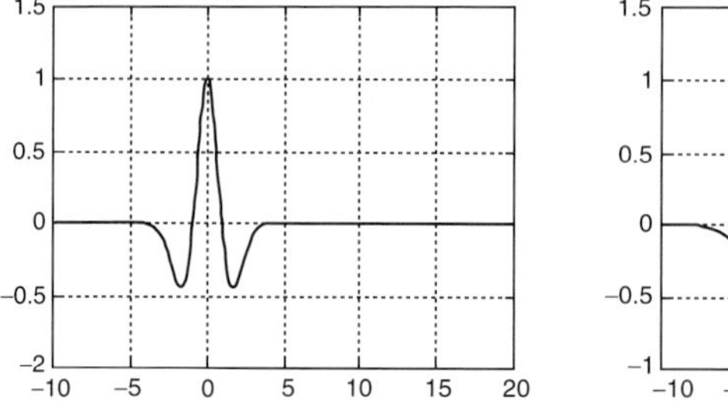

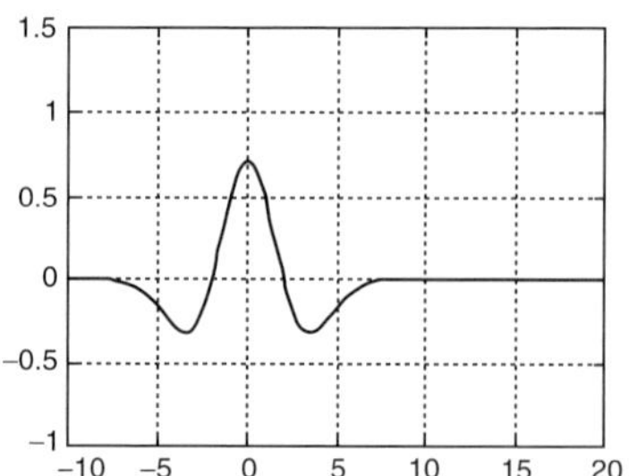

Fig. 5.18. A basic wavelet, and a wavelet generated by scale change.

[5] The basic wavelet is often referred to as the *mother* wavelet.

Fig. 5.19. Original and the result of the inner product of the original and three different two-dimensional wavelets (three slices through the wavelet transform).

Observe that the transform is a function of the scale and the shift. The same idea holds in two dimensions, where the equation for the inner product becomes

$$W_f(a, b_x, b_y) = \int_{-\infty}^{\infty} f(x, y)\psi_{a,b_x,b_y}(x, y)\, dx\, dy. \tag{5.42}$$

Fig. 5.19 illustrates a cross section of W for different values of a. Clearly, this process produces a scale space representation.

Lee [5.23] takes the neurophysiological evidence and derives a mother wavelet of the following form.

$$\psi(x, y) = \frac{1}{\sqrt{2\pi}} \exp\left(-\left(\frac{4x^2 + y^2}{8}\right)\right)\left[\exp(i\kappa x) - \exp\left(-\frac{\kappa^2}{2}\right)\right], \tag{5.43}$$

where κ is a constant whose value depends on assumptions about the bandwidth, but is approximately equal to 3.5. Scaling and translating this "mother wavelet" produces a collection of filters which (in the same way that the Frei–Chen basis set did) represents how much of an image neighborhood resembles the feature, and from which the image can be completely reconstructed.

5A.5 Vocabulary

You should know the meaning of the following terms.

```
Canny edge detector
Nonmaximum suppression
Wavelet
```

Assignment 5.A1

```
At high levels of scale, only              objects are
visible. (Fill in the blank.)
```

Assignment 5.A2

What does the following expression estimate? $(f \otimes h_1)^2 + (f \otimes h_2)^2 = E$, where the kernel h_1 is defined by

0.05	0.08	0.05
−0.136	−0.225	−0.136
0.05	0.08	0.05

,

and h_2 is created by the transpose.

Choose the best answer:

(a) the first derivative with respect to x (where x is the horizontal direction)
(b) the second derivative with respect to x
(c) the Laplacian
(d) the second derivative with respect to y
(e) the quadratic variation.

Assignment 5.A3

You are to use the idea of differentiating a Gaussian to derive a kernel. What variance does a one-dimensional (zero mean) Gaussian need to have to have the property that the extrema of its first derivative occur at $x = \pm 1$?

Assignment 5.A4

What is the advantage of the quadratic variation as compared to the Laplacian?

Assignment 5.A5

Let $E = (f - Hg)^T (f - Hg)$. Using the equivalence between the kernel form of a linear operator and the matrix form, write an expression for E using kernel notation.

Assignment 5.A6

The purpose of this assignment is to walk you through the construction of an image pyramid, so that you will more fully understand the potential utility of this data structure, and can use it in a coding and transmission application. Along the way, you will pick up some additional understanding of the general area of image coding. Coding is not a principal learning objective of this course, but in your research career, you are sure to encounter people who deal with coding all the time. It will prove to be useful to have a grasp of some of the more basic concepts.

(1) Locate the image "asterix512.ifs," and verify that it is indeed 512 × 512 (use ifs info). If it is not 512 × 512, then first write a program which will pad it out to that size.

(2) Write a **program** which will take a two-dimensional $n \times n$ image and produce an image, of the same data type, which is $n/2 \times n/2$. The program name should be **ShrinkByTwo** and called by

ShrinkByTwo inimg outimg

The program should NOT simply take every other pixel. Instead, each pixel of the output image should be constructed by averaging the corresponding four pixels in the input image. Note the requirement that the output image be of the same type as the input. That is easily accomplished by using **ifscreate** with the first argument being the character string defining the type of the input. For example ifscreate (in->ifsdt, len, IFS_CR_ALL,0). Use your program to create asterix256, asterix128, asterix64, and asterix32. Don't bother going below a 32 × 32 image. Turn in your program and a printout of your images.

(3) Write a **subroutine** to zoom an image. It should have the following calling convention:

```
ZoomByTwo(inimg,outimg)
IFSIMG inimg, outimg;
```

The calling program is responsible for creating, reading, etc. of the images. The subroutine simply fills outimg with a zoomed version of inimg. Use any algorithm you wish to fill in the missing pixels. (We recommend that the missing pixels be some average of the input pixels.)

Before we can proceed, we need to consider a "pyramid coder." When you ran ShrinkByTwo, the set of images you produced is the pyramid representation of asterix512. In a pyramid coder, the objective is to use the pyramid representation to transmit as little information as possible over the channel. Here is the idea: First, transmit all of asterix32. Then, both the transmitter and receiver run ZoomByTwo to create a zoomed version of asterix32, something like

```
ZoomByTwo(a32, a64prime)
```

When we created asterix32 from asterix64, we threw away some information, and we cannot easily get it back, so a64prime will not be identical to asterix64. If, however, ZoomByTwo (which in the image coding literature is called the *predictor*) is pretty good, the difference between a64prime and asterix64 will be small (small in value that is; it is still 64×64). Therefore, compute diff64, the difference between a64prime and asterix64. If the predictor were perfect, the resultant difference would be a 64×64 image of all zeros, which could be coded in some clever way (by run-length encoding for example), and transmitted with very few bits. Let us now transmit diff64 to the receiver. By simply adding diff64 to the version of a64prime generated at the receiver, we can correct the errors made by the predictor, and we now have a correct version of asterix64 at the receiver, but all we transmitted was diff64. Clever, aren't we? Now, from asterix64, we play the same game and get asterix128 created by transmitting diff128, *et cetera*, *ad nauseam*. Now, the assignment:

(4) Create the images described above, diff64, diff128, diff256, diff512. Measure the approximate number of bits required to transmit each of them. To make that measurement: Compute the standard deviation of, say, diff64. Take the log base 2 of that standard deviation, and that is the average number of bits per pixel required to code that image. Assume you were to transmit asterix512 directly. That would require $512 \times 512 \times 8$ bits (assuming the image is 8 bit – you better verify). Now, compare with the performance of your pyramid coder by adding up all the bits/pixel you found for each of the diff images you transmitted. Did your coder work well? Discuss this in your report.

References

[5.1] V. Anh, J. Shi, and H. Tsai, "Scaling Theorems for Zero Crossings of Bandlimited Signals," *IEEE Transactions on Pattern Analysis and Machine Intelligence*, **18**(3). 1996.

[5.2] J. Babaud, A. Witkin, M. Baudin, and R. Duda, "Uniqueness of the Gaussian Kernel for Scale-space Filtering," *IEEE Transactions on Pattern Analysis and Machine Intelligence*, **8**(1), 1986.

[5.3] G. Bilbro and W. Snyder, "Optimization of Functions with Many Minima," *IEEE Transactions on Systems, Man, and Cybernetics*, **21**(4), July/August, 1991.

[5.4] K. Boyer and S. Sarkar, "On the Localization Performance Measure and Optimal Edge Detection," *IEEE Transactions on Pattern Analysis and Machine Intelligence*, **16**(1), 1994.

[5.5] P. Burt and E. Adelson, "The Laplacian Pyramid as a Compact Image Code," *Computer Vision, Graphics, and Image Processing*, **16**, pp. 20–51, 1981.

[5.6] J. Canny, "A Computational Approach to Edge Detection," *IEEE Transactions on Pattern Analysis and Machine Intelligence*, **8**(6), 1986.

[5.7] W. Chojnacki, M. Brooks, A. van der Hengel, and D. Gawley, "On the Fitting of Surfaces to Data with Covariances," *IEEE Transactions on Pattern Analysis and Machine Intelligence*, **22**(11), 2000.

[5.8] J. Crowley, *A Representation for Visual Information*, Ph.D. Thesis, Carnegie-Mellon University, 1981.

[5.9] J. Daugman, "Two-dimensional Spectral Analysis of Cortical Receptive Fields," *Vision Research*, **20**, pp. 847–856, 1980.

[5.10] J. Daugman, "Uncertainly Relation for Resolution in Space, Spatial Frequency, and Orientation Optimized by Two-dimensional Visual Cortical Filters," *Journal of the Optical Society of America*, **2**(7), 1985.

[5.11] W. Deng and S. Iyengar, "A New Probabilistic Scheme and Its Application to Edge Detection," *IEEE Transactions on Pattern Analysis and Machine Intelligence*, **18**(4), 1996.

[5.12] W. Frei and C. Chen, "Fast Boundary Detection: A Generalization and a New Algorithm," *IEEE Transactions on Computers*, **26**(2), 1977.

[5.13] D. Hubel and T. Wiesel, "Receptive Fields, Binocular Interaction, and Functional Architecture in the Cat's Visual Cortex," *Journal of Physiology (London)*, **160**, pp. 106–154, 1962.

[5.14] D. Hubel and T. Wiesel, 'Functional Architecture of Macaque Monkey Visual Cortex," *Proceedings of the Royal Society of London, B*, **198**, pp. 1–59, 1977.

[5.15] L. Iverson and S. Zucker, "Logical/Linear Operators for Image Curves," *IEEE Transactions on Pattern Analysis and Machine Intelligence*, **17**(10), 1995.

[5.16] P. Jackway and M. Deriche, "Scale-space Properties of the Multiscale Morphological Dilation–Erosion," *IEEE Transactions on Pattern Analysis and Machine Intelligence*, **18**(1), 1996.

[5.17] J. Jones and L. Palmer, "An Evaluation of the Two-dimensional Gabor Filter Model of Simple Receptive Fields in the Cat Striate Cortex," *Journal of Neurophysiology*, **58**, pp. 1233–1258, 1987.

[5.18] E. Joseph and T. Pavlidis, "Bar Code Waveform Recognition using Peak Locations," *IEEE Transactions on Pattern Analysis and Machine Intelligence*, **16**(6), 1994.

[5.19] M. Kelly, in *Machine Intelligence*, volume 6, University of Edinburgh Press, 1971.

[5.20] M. Kisworo, S. Venkatesh, and G. West, "Modeling Edges at Subpixel Accuracy using the Local Energy Approach," *IEEE Transactions on Pattern Analysis and Machine Intelligence*, **16**(4), 1994.

[5.21] A. Klinger, "Pattern and Search Statistics," in *Optimizing Methods in Statistics*, New York, Academic Press, 1971.

[5.22] P. Kube and P. Perona, "Scale-space Properties of Quadratic Feature Detectors," *IEEE Transactions on Pattern Analysis and Machine Intelligence*, **18**(10), 1996.

[5.23] T. Lee, "Image Representation Using 2-D Gabor Wavelets," *IEEE Transactions on Pattern Analysis and Machine Intelligence*, **18**(10), 1996.

[5.24] Y. Leung, J. Zhang, and Z. Xu, "Clustering by Scale-space Filtering," *IEEE Transactions on Pattern Analysis and Machine Intelligence*, **22**(12), 2000.

[5.25] T. Lindeberg, "Scale-space for Discrete Signals," *IEEE Transactions on Pattern Analysis and Machine Intelligence*, **12**(3), 1990.

[5.26] T. Lindeberg, "Scale-space Theory, A Basic Tool for Analysing Structures at Different Scales," *Journal of Applied Statistics*, **21**(2), 1994.

[5.27] D. Marr and E. Hildreth, "Theory of Edge Detection," *Proceedings of the Royal Society of London, B*, **207**, pp. 187–217, 1980.

[5.28] D. Marr and T. Poggio, "A Computational Theory of Human Stereo Vision," *Proceedings of the Royal Society of London, B*, **204**, pp. 301–328, 1979.

[5.29] R. Nelson, "Finding Line Segments by Stick Growing," *IEEE Transactions on Pattern Analysis and Machine Intelligence*, **16**(5), 1994.

[5.30] E. Pauwels, L. Van Gool, P. Fiddelaers, and T. Moons, "An Extended Class of Scale-invariant and Recursive Scale Space Filters," *IEEE Transactions on Pattern Analysis and Machine Intelligence*, **17**(7), 1995.

[5.31] P. Perona, "Deformable Kernels for Early Vision," *IEEE Transactions on Pattern Analysis and Machine Intelligence*, **17**(5), 1995.

[5.32] P. Perona and J. Malik, "Scale-space and Edge Detection using Anisotropic Diffusion", *IEEE Transactions on Pattern Analysis and Machine Intelligence*, **12**(7), pp. 629–639, 1990.

[5.33] W. Pratt, *Digital Image Processing*, Chichester, John Wiley and Sons, 1978.

[5.34] H. Tagare and R. deFigueiredo, "Reply to 'On the Localization Performance Measure and Optimal Edge Detection'," *IEEE Transactions on Pattern Analysis and Machine Intelligence*, **16**(1), 1994.

[5.35] A. Taratorin and S. Sideman, "Constrained Regularized Differentiation," *IEEE Transactions on Pattern Analysis and Machine Intelligence*, **16**(1), 1994.

[5.36] F. van der Heijden, "Edge and Line Feature Extraction Based on Covariance Models," *IEEE Transactions on Pattern Analysis and Machine Intelligence*, **17**(1), 1995.

[5.37] M. Van Horn, W. Snyder, and D. Herrington, "A Radial Filtering Scheme Applied to Intracoronary Ultrasound Images," *Computers in Cardiology*, September, 1993.

[5.38] P. Verbeek and L. van Vliet, "On the Location Error of Curved Edges in Low-pass Filtered 2-D and 3-D Images," *IEEE Transactions on Pattern Analysis and Machine Intelligence*, **16**(7), 1994.

[5.39] I. Weiss, "High-order Differentiation Filters that Work," *IEEE Transactions on Pattern Analysis and Machine Intelligence*, **16**(7), 1994.

[5.40] M. Werman and Z. Geyzel, "Fitting a Second Degree Curve in the Presence of Error," *IEEE Transactions on Pattern Analysis and Machine Intelligence*, **17**(2), 1995.

[5.41] R. Young, "The Gaussian Derivative Model for Spatial Vision: I. Retinal Mechanisms," *Spatial Vision*, **2**, pp. 273–293, 1987.

6 Image relaxation: Restoration and feature extraction

To change, and to change for the better are two different things

German proverb

In this chapter, we move toward developing techniques which remove noise and degradations so that features can be derived more cleanly for segmentation. The techniques of *a posteriori* image restoration and iterative image feature extraction are described and compared. While image restoration methods remove degradations from an image [6.3], image feature extraction methods extract features such as edges from noisy images. Both are shown to perform the same basic operation: image relaxation. In the advanced topics section, image feature extraction methods, known as *graduated nonconvexity* (GNC) and *variable conductance diffusion* (VCD), are compared with a restoration/feature extraction method known as *mean field annealing* (MFA). This equivalence shows the relationship between energy minimization methods and spatial analysis methods and between their respective parameters of temperature and scale. The chapter concludes by discussing the general philosophy of extracting features from images.

6.1 Relaxation

The term "relaxation" was originally used to describe a collection of iterative numerical techniques for solving simultaneous nonlinear equations (see [6.18] for a review). The term was extended to a set of iterative classification methods by Rosenfeld and Kak [6.64] because of their similarity. Here, we provide a general definition of the term which will encompass these methods as well as those more recent techniques which are the emphasis of this discussion.

A relaxation process is an iteration.

Definition

A *relaxation* process is a multistep algorithm with the property that (1) the output of a single step is of the same form as the input, so that the algorithm may be applied iteratively, and (2) it converges to a bounded result. Some researchers also require that the operation on any element (any pixel, in our application) be dependent only on the state of the pixels in some well defined, finite "neighborhood" of that element. We will see that all the algorithms discussed here are relaxation processes, according to these criteria.

6.2 Restoration

In an image restoration problem, we assume that an ideal image, *f*, has been corrupted to create the measured image, g. The usual model for the corruption is a distortion operation, denoted by *D*, followed by the addition of random noise

$$g = D(f) + n, \tag{6.1}$$

where $g = [g_1, \ldots, g_N]^{\mathrm{T}}$ and g_i denotes the ith pixel in a column vector representation of the image $g.f$ and n are similarly defined. The restoration problem, then, is the problem of finding a best estimate of *f* given the measurement, g, some knowledge of the distortion (often called "blur"), and the statistics of the noise.

Restoration is often referred to as an *inverse problem*. That is, we have a process (in this case blur) which takes an input and produces an output. We can only measure the output, and we wish to infer the input.

6.2.1 Inverse problems and ill-posedness

A problem $g = D(f)$ is said [6.32] to be *well-posed* if

- for each *f*, a solution, g, exists
- the solution is unique
- the solution g continuously depends on the data *f*.

Definition of an ill-posed problem.

If these three conditions do not all hold, the problem is said to be "ill-posed." Ill-posedness is normally caused by the ill-conditioning of the problem. Conditioning of a mathematical problem is measured by the sensitivity of output to changes in input. For a well-conditioned problem, a small change of input does not affect the output much; while for an ill-conditioned problem, a small change of input can change the output a great deal.

Condition number is the measurement of the conditioning of a problem. Generally, it is defined as Eq. (6.2). The larger the condition number, the more ill-conditioned the problem is:

$$\text{condition number} \approx \frac{\text{change in output}}{\text{change in input}}. \tag{6.2}$$

The conditioning of a linear system $Ax = b$ is determined by the condition number of matrix A. The relative condition number K is defined as Eq. (6.3),

$$K = \|A\| \|A^{-1}\| \tag{6.3}$$

where $\|.\|$ usually indicates the 2-norm. K is in the range of $[1, \infty)$. When $K \gg 1$, the linear system is ill-conditioned.

Table 6.1. *The system of resulting linear equations.*

$ga =$		$0.5b+$		$0.5d$					
$gb =$	$0.33a+$		$0.33c+$		$0.33e$				
$gc =$		$0.5b+$				$0.5f$			
$gd =$	$0.33a+$				$0.33e+$		$.33g$		
$ge =$		$0.25b+$		$0.25d+$		$0.25f+$		$0.25h$	
$gf =$			$0.33c+$		$0.33e+$				$0.33i$
$gg =$				$0.5d+$				$0.5h$	
$gh =$					$0.33e+$		$0.33g+$		$0.33i$
$gi =$						$0.5f+$		$0.5h$	

Denote $G = [ga\ gb\ gc\ gd\ ge\ gf\ gg\ gh\ gi]^{\mathrm{T}}$, $F = [a\ b\ c\ d\ e\ f\ g\ h\ i]^{\mathrm{T}}$.

If the blur process turns out to be linear and space-invariant, we call the process of inverting it *deconvolution*.

In Eq. (6.1), suppose we know how the blur process works. Then it would seem that we should be able to undo it. We will see why that might not be the case.

As an example, look at a very simple image, just about the simplest image we can come up with, a 3×3 image, and give each pixel a name, using the letters $a, \ldots, i$. Now, suppose this image is blurred by a linear blur process in which each pixel is replaced by the average of its neighbors (suppose the 4-neighbor definition is used). In the case of edge or corner pixels, fewer pixels are in the neighborhood. This new, blurred image has values $ga, \ldots, gi$, and these are related to the original values by the system of linear equations shown in Table 6.1.

a	b	c
d	e	f
g	h	i

$$H = \begin{bmatrix} 0 & 0.5 & 0 & 0.5 & 0 & 0 & 0 & 0 & 0 \\ 0.33 & 0 & 0.33 & 0 & 0.33 & 0 & 0 & 0 & 0 \\ 0 & 0.5 & 0 & 0 & 0 & 0.5 & 0 & 0 & 0 \\ 0.33 & 0 & 0 & 0 & 0.33 & 0 & 0.33 & 0 & 0 \\ 0 & 0.25 & 0 & 0.25 & 0 & 0.25 & 0 & 0.25 & 0 \\ 0 & 0 & 0.33 & 0 & 0.33 & 0 & 0 & 0 & 0.33 \\ 0 & 0 & 0 & 0.5 & 0 & 0 & 0 & 0.5 & 0 \\ 0 & 0 & 0 & 0 & 0.33 & 0 & 0.33 & 0 & 0.33 \\ 0 & 0 & 0 & 0 & 0 & 0.5 & 0 & 0.5 & 0 \end{bmatrix}$$

Then, we may represent the process of blur by $G = HF$ and solve for the values before the blur using $F = H^{-1}G$. It appears that this should be simple. If we have a model for the distortion process (the model in this case is the matrix H), all we need do is invert it, and multiply. Let us look to see why that might not work too well. First, calculate the inverse of H numerically. Whoops! Our matrix inverter program tells us that the matrix is singular. Guess that will not work. Is it a bad choice of blur operations?

It is hard, it turns out, to contrive a numerical example which is not singular. It is possible, of course, but difficult. Here is the real key: *Even if the distortion matrix turns out to be nonsingular, the problem is still likely to be "ill-conditioned."*

That is, (let's review) we measure $ga, \ldots, gi$, and use matrix multiplication to determine $a, \ldots, i$. If H is not singular, then in a perfect world it should work. As engineers know, however, in fact there is always noise, so instead of measuring ga, we actually measure $ga + \varepsilon$, where ε is some perturbation due to noise. If the system, that is, the distortion matrix, is ill-conditioned (and it is), then this small change in ga may produce large differences in the estimates of $a, \ldots, i$. For this reason, simple matrix inversion does not work, even if the system is linear.

Another, perhaps simpler example of ill-conditioning [6.36] is as follows: Consider the linear system described by a blur A, and unknown image f, and a measurement g, where

$$g = Af$$
$$A = \begin{bmatrix} 1 & 1 \\ 1 & 1.01 \end{bmatrix} \quad f = \begin{bmatrix} f_1 \\ f_2 \end{bmatrix} \quad g = \begin{bmatrix} 1 \\ 1 \end{bmatrix}. \tag{6.4}$$

You decide whether this change is "trivially small."

The condition number for matrix A is 402.0075 which is much larger than 1. This system has solution $f_1 = 1,\ f_2 = 0$. Now, suppose the measurement, g, is corrupted by noise, producing $g = [1\ 1.01]^{\mathrm{T}}$. Then, the solution is $f_1 = 0,\ f_2 = 1$. A trivially small change in the measured data caused a dramatic change in the solution.

There are many ways to approach these ill-posed restoration problems. They all share a common structure: the *regularization theory*. Generally speaking, any regularization method tries to analyze a related *well-posed* problem whose solution approximates the original ill-posed problem [6.57].

The first approach one might think of is to produce an image estimate which has the minimum expected mean square error. That is, find the unknown image f which minimizes

$$E = \sum_i (g_i - (f_i \otimes h))^2, \tag{6.5}$$

Not only is this version also ill-posed, it is the same problem! (Recall the way to write application of kernel operators by using a matrix?)

where the sum is over all the pixels in the image, and the distortion is represented by application of a kernel operator corresponding to a blur h. Simply minimizing E does not work, as the problem is still ill-conditioned. Making some assumptions about the noise can give us a bit better performance. If the distortion is linear, space-invariant, and the noise is stationary, the Wiener filter gives the optimal solution according to this criterion. (See [6.28] for a tutorial presentation.)

6.3 The MAP approach

In this section, we introduce a bit of mathematics we need before we can move any further.

6.3.1 Bayes' rule

Bayes' rule concerns three probability functions: the *a priori* probability density, $p(f)$, the conditional probability density, $p(g|f)$, and the *a posteriori* conditional probability density, $p(f|g)$.

We define $p(f)$ to represent the *a priori* probability density that some particular picture f occurs. (If we consider the brightness values as continuous, then we need to use a probability density rather than a probability. Which we use has no impact on the resulting form of the derivations given below.) That is, it is the probability of picture f occurring before any measurements are made. As a discrete example of the *a priori* probability, suppose we have a factory which manufactures flanges and gaskets, but makes ten times as many flanges as gaskets. Flanges and gaskets may come down the conveyor at random times. But because of our *a priori* knowledge that the plant manufactures ten times as many flanges as gaskets we know that we are much more likely to see a flange than a gasket if we choose to look at the conveyor at some random time. Thus the *a priori* possibility that our camera sees a flange is 0.9 and the *a priori* probability of gaskets is 0.1.

We define $p(g|f)$ to represent the *conditional probability density* that image g is measured, given that the measured image is known to be some corruption of image f. The probability density function may be characterized in several possible ways. One is by simply tabulating the number of times a particular value occurs for each possible value of the variable, in this case, length. Such a tabulation is referred to as a *histogram* of the variables. Properly normalized, a histogram can be a useful representation of a probability density function. Unfortunately, it is difficult to represent the density function of images as a histogram. One may also describe a density function in a parametric way using some analytic function (e.g., the Gaussian).

We define $p(f|g)$ to represent the *a posteriori* conditional probability (density) that the observed image g is really a corruption of image f. $p(f|g)$ is what we are looking for. We will use it as our decision rule or, more correctly, as our *discrimination function*. Our decision rule will then be as follows.

A MAP algorithm seeks the image which maximizes the probability AFTER the measurement.

For a measurement g made on an unknown image, compute $p(f_i|g)$ for each possible f_i. Then decide that the unknown is the image f_i for which $p(f_i|g)$ is greater than $p(f_j|g)$ for all $i \neq j$. When we make classification decision based on $p(f_i|g)$, we are using a **maximum *a posteriori* (MAP) image processing algorithm**.

We can relate the three probability functions just defined by using *Bayes' rule*:

$$p(f|g) = \frac{p(g|f)p(f)}{something} \tag{6.6}$$

$$something = p(g). \tag{6.7}$$

In Eq. (6.6) we used "something" to represent the denominator for the conditional probability density. We used the word "something" to call attention to the fact that this number represents the probability density of that value of g occurring, independent of the original, uncorrupted image. Since this number is independent of f, and is the same for all possible fs, it therefore does not provide us any help in distinguishing which class is most likely. Instead, it is a normalization constant which we use to ensure that the number $p(f|g)$ has the desirable properties of a probability; that is, it lies between 0 and 1 and sums to 1 when summed over all possible images (that is, the observed object belongs to at least one of the classes which we are considering).

We wish to maximize the *a posteriori* probability density $p(f|g)$ of the unknown correct image f given measured image g. We will be relating the probabilities of the entire image to the probabilities of individual pixels. Using Bayes' rule, we have the proportionality

Watch for the notation change here! Now, the subscript on f refers to a pixel.

$$p(f_i|g_i) \propto p(g_i|f_i)p(f_i) \tag{6.8}$$

where i is the pixel index $f = [f_1, f_2, \ldots, f_N]^T$.

For the purposes of this discussion, one can ignore the normalizing constant of proportionality and maximize the right-hand side of Eq. (6.8).

For now we assume that there is no distortion to the image except noise, and we assume that the noise is statistically independent from one pixel to the next. Then we can write

$$p(g|f) = \prod_i p(g_i|f_i). \tag{6.9}$$

Since the only difference between the measurement g_i and the true but unknown image pixel f_i is the noise, and if we assume a Gaussian noise model, we can replace the conditional density of the individual pixels with the density of the noise, producing

$$p(g|f) = \prod_i \frac{1}{\sqrt{2\pi}\sigma} \exp\left(-\frac{n_i^2}{2\sigma^2}\right). \tag{6.10}$$

Moving the product inside the exponential allows us to write [6.8, 6.27, 6.37, 6.38, 6.80]:

$$p(g|f) = \left(\frac{1}{\sqrt{2\pi}\sigma}\right)^N \exp\left(\frac{-\sum_i (f_i - g_i)^2}{2\sigma^2}\right). \tag{6.11}$$

A number of researchers have noted that an image is appropriately represented by a two- or three-dimensional Markov field [6.5, 6.33, 6.47, 6.56] which allows [6.6, 6.27, 6.45] the representation of the *a priori* probability of a given pixel value by a Gibbs distribution.

$$p(f_i) \propto \exp\left(\frac{-\sum_{j\in\aleph_i} V_{ij}}{T}\right). \tag{6.12}$$

The sum is taken over $\aleph_i$, the neighborhood of pixel i. Recall from Chapter 4 that the *aura* of a set of pixels A relative to a set B is the collection of all points in B that are neighbors of pixels in A, where the concept of "neighbor" is given by a problem-specific definition. Here, the concept is similar, except that we consider only the neighborhood of a single pixel rather than the neighborhood of a set. Like the definition of aura, the definition of neighborhood is allowed to be problem-dependent, and in this sense two pixels do not have to be adjacent or even necessarily "close" in the image to be considered neighbors. In essentially all practical applications, however, the neighbors of a particular pixel are those pixels which are adjacent. T is an adjustable width parameter, and the Vs are *potential functions* which are, in general, functions of the pixels in the neighborhood.

The way we formulate the prior probability for the entire image is to once again write a product

$$p(f) = \prod_i p(f_i). \tag{6.13}$$

Check this out! The answer which maximized the conditional probability is the same as the answer which minimized the squared error! This happens because the noise is additive Gaussian.

Forming the product of Eqs. (6.11) and (6.12) as indicated in Eq. (6.8) and eliminating the constant[1] term, we then take natural logarithms and change the sign thereby converting the problem from maximizing a probability to minimizing an objective function.

$$H(f, g) = \left(\sum_i \frac{(f_i - g_i)^2}{2\sigma^2}\right) + \sum_i \sum_{j\in\aleph_i} V_{ij}. \tag{6.14}$$

We will refer to the first, conditional, term of Eq. (6.14) as the "noise" term [6.27] and to the second as the "prior." This gives the following form:

$$H(f, g) = H_{\mathrm{n}}(f, g) + H_{\mathrm{p}}(f). \tag{6.15}$$

Examination of the noise term reveals that it is minimized by $f = g$, which implies that the restoration, f, should resemble the measurement g.

We may choose the prior (regularization) term to emphasize whatever property we think the image should have, i.e., having only a few known brightness values [6.70];

1 These terms do not affect the location of the minimum. The σ which remains in Eq. (6.14) allows us to weight the relative importance of the noise and prior terms.

having a brightness which varies smoothly, except at boundaries [6.9, 6.16]; or the most common: having brightness which is constant in local areas and discontinuous at boundaries [6.16, 6.38].

6.3.2 Digression: The problem with inverse problems

We are performing an operation that some (but not we) refer to as "denoising."

Remember what we said about ill-posed problems.

The problem we have described to this point is an inverse problem. That is, we know the distortion process (it is just noise so far, but we could add blur), and we have the result of that process operating on an unknown image. Estimating that unknown image is the inverse problem. Such problems are most often approached by choosing a function (like the function of Eq. (6.14)) to be minimized, which is the sum of two terms, one dependent on the measured data and one on the unknown image. The image may be represented, as we do here, as a collection of point values which must be solved for, or as a function (cf. [6.82]). In the case that f is, conceptually at least, a function, then f has values between the sample points. In such a case, we may think of the process of solving for f as *interpolation*.

6.3.3 An objective function for edge-preserving smoothing

In section 6.3.1, we briefly discussed the prior term which is chosen to emphasize the property of the unknown image. The most common property used is that the pixel brightness should be constant in local areas and discontinuous at boundaries [6.16, 6.38]. In order to minimize the objective function of Eq. (6.14), the potential function V is formulated as a penalty function.

Let Λ_i represent some scalar measure of the brightness variation about pixel i. We are about to develop the penalty function that will follow a recommendation of Besag [6.6] "to encourage smooth variation." It says that V_i "should be strictly increasing" in the absolute value of Λ_i and if "occasional abrupt changes" are expected, it should "quickly reach a maximum." Let's see if we can understand why he makes this recommendation.

The quadratic form $V_i = b\Lambda_i^2$ with $b > 0$ is attractive because it is simple. It also penalizes noise in a nonlinear way. That is, Λ_i is some measure of the strength of the gradient at pixel i, so that the bigger the gradient, the bigger V_i. The minimization process chooses solutions – images – which result in SMALL values of V_i. Thus, we are most likely to find a solution which does not have abrupt changes; no abrupt changes means no sharp edges. This solution results in blurred edges that are unacceptable in many applications. It is appropriate only for "smooth variation in the true scene" [6.6]. The key to preserving sharp edges is simple: Penalize brightness variations due to noise – the bigger the variation, the bigger the penalty – but at some point, do not penalize any more (so edges get no *additional* penalty), as illustrated in Fig. 6.1. Unfortunately (see section 6A.1), this function is not differentiable, and we therefore cannot do gradient descent with it.

An image which minimizes the quadratic form will be blurry.

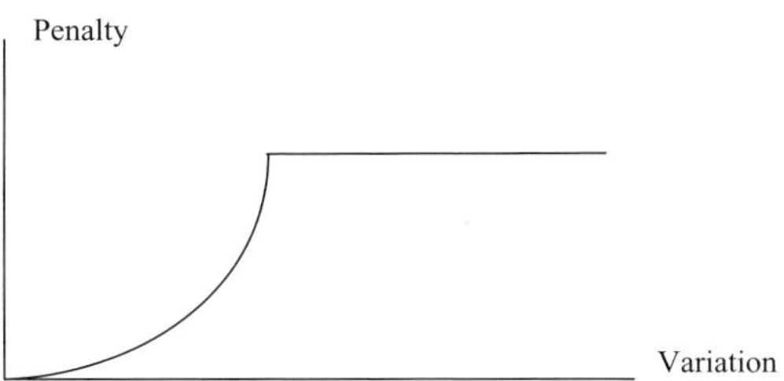

Fig. 6.1. The penalty should be stronger for higher noise. Assuming local variations in brightness are due to noise, a larger variation is penalized more. However, local brightness variations can also be due to edges, which should not be penalized (otherwise, they will be blurred). Therefore we want our penalty function to have an upper limit.

Instead, we will choose an inverted Gaussian and write:

$$H_{\mathrm{p}}(f) = -\frac{b}{\sqrt{2\pi}\tau}\left(\sum_i \exp\left(-\frac{\Lambda_i^2}{2\tau^2}\right)\right). \tag{6.16}$$

The $\sqrt{2\pi}$ does not mean anything at all!

In Eq. (6.16), the constants are irrelevant. We include them only so that it looks like a Gaussian. τ is a soft threshold: It represents *a priori* knowledge of the roughness of the surface. The form for Λ makes this knowledge explicit. It is hoped that the spatial derivatives Λ_i will become small almost everywhere as the algorithm proceeds. This concept will be explored in more detail in the next section.

Combining Eqs. (6.15) and (6.16), we have an objective function which, if minimized, will result in a restored image which resembles the data (in the least squares sense) while at the same time consisting of regions of uniform brightness separated by abrupt edges. We could use mean field annealing (MFA) to minimize this objective function.

6.4 Mean field annealing

Mean field annealing (MFA) is a technique for finding a good minimum of complex functions which typically have many minima. The mean field approximation in statistical mechanics allows a continuous representation of the energy states of a collection of particles. In the same sense, MFA approximates the stochastic algorithm called "simulated annealing" (SA). SA has been shown to converge in probability to the global minimum, even for nonconvex problems [6.27]. Because SA also takes an unacceptably long time to converge, a number of techniques have been derived to accomplish speedups [6.43]; MFA is one of those techniques.

Since its introduction in 1989 [6.8, 6.11], this technique has found applications in restoration of locally homogeneous [6.37, 6.38] and locally smooth [6.8, 6.9] images; in image segmentation [6.69, 6.70]; in motion analysis [6.1]; and in sensor fusion

[6.7]. MFA is based on a combination of the concepts of simulated annealing [6.48] and the mean field approximation of statistical mechanics [6.17]. The earlier work in MFA followed the lead of researchers in simulated annealing, and relied on an analogy to statistical mechanics to justify the technique. However, it was later [6.13] demonstrated that such an analogy, although insightful and correct, is not necessary, and that MFA may be derived purely from information-theoretic considerations.

Use of the term "MFA" can be confusing, since the acronym represents first and foremost, a technique for deriving algorithms. Secondarily, however, the term has come to represent a class of image restoration algorithms. These algorithms are derived by posing an image restoration problem as a minimization problem, and then following a particular methodology (the mean field approximation) to derive the minimization method. The resulting method will combine gradient descent and "annealing," that is, monotonic variation of a control parameter (known as "temperature") through the duration of the algorithm.

Another popular technique for image noise removal, known as "graduated non-convexity" (GNC) [6.16], bears many similarities to MFA in that GNC uses a descent method and reduces a control parameter. The similarities are so strong that GNC may in fact be derived using MFA as a basis [6.12].

Much of the work done to date with MFA has used the MFA methodology to derive restoration-like algorithms. However, MFA may be considered an image feature-extraction method and is equivalent to another relaxation method known as "image diffusion."

The student is referred to the above references for formal derivations of MFA. Here, we present what we have found to be a more pedagogically attractive explanation of the same concept, showing MFA to be a particular type of *continuation method* [6.2].

A continuation method is an algorithm which implements a *homotopy*, that is, a continuous deformation of one (hyper) surface into another. In the problem described by Eq. (6.14), we use MFA to distort a convex N-surface into one which has (typically) many local minima. For purposes of illustration, let us consider the simplest possible version of Eq. (6.15) which still captures the essence of pixel interactions:

Did you observe that the 2 is missing from the denominator? We said the constants don't really matter, but stay on your toes anyway!

$$H(f_1, f_2) = (f_1 - g_1)^2 + (f_2 - g_2)^2 - \frac{1}{\tau}\exp\left(-\frac{(f_1 - f_2)^2}{\tau^2}\right). \tag{6.17}$$

Here we have a (not very interesting) image consisting of only two pixels, f_1 and f_2, which have been corrupted by noise to result in measured pixel values g_1 and g_2. The prior term chosen in this case encourages solutions in which $f_1 = f_2$. The principal result of the MFA derivations, when applied to a function of the type described by Eq. (6.17), is the replacement of τ by $\tau + T$, where T is a parameter (called "temperature" in the literature) which is initialized to a "large" value, and gradually

Beware of notation! Here, we are doing "gradient descent" to minimize a function of f, so the pixel values are what we are differentiating WRT. We also saw different "gradients" in the last chapter. Those were derivatives of f with respect to x and y.

reduced to approach zero. Performing these substitutions in Eq. (6.17) we have a new objective function

$$H_T(f_1, f_2) = (f_1 - g_1)^2 + (f_2 - g_2)^2 - \frac{1}{\tau + T} \exp\left(-\frac{(f_1 - f_2)^2}{(\tau + T)^2}\right). \tag{6.18}$$

We denote the result of the MFA derivation (and make the dependence on T explicit) by referring to this as H_T. Differentiating the H_T of Eq. (6.18) with respect to the two elements of the vector f results in

$$\frac{\partial}{\partial f} H_T(f_1, f_2) = \begin{bmatrix} 2(f_1 - g_1) + \dfrac{1}{\tau + T} \exp\left(-\dfrac{(f_1 - f_2)^2}{(\tau + T)^2}\right)\left(\dfrac{2(f_1 - f_2)}{(\tau + T)^2}\right) \\ 2(f_2 - g_2) + \dfrac{1}{\tau + T} \exp\left(-\dfrac{(f_1 - f_2)^2}{(\tau + T)^2}\right)\left(\dfrac{2(f_2 - f_1)}{(\tau + T)^2}\right) \end{bmatrix}. \tag{6.19}$$

The MFA approach is as follows.

(1) Set $T = T_{\text{initial}}$ (a problem-dependent parameter).
(2) Using Eq. (6.19), perform gradient descent or some other minimization technique, and find the f which minimizes H_T.
(3) Reduce T.
(4) If $T > T_{\text{final}}$, go to (2).

The simplest version of gradient descent applicable here is the iteration

$$\begin{bmatrix} f_1 \\ f_2 \end{bmatrix}^{k+1} \Leftarrow \begin{bmatrix} f_1 \\ f_2 \end{bmatrix}^{k} - \alpha \frac{\partial}{\partial f} H_T\left(f_1^k, f_2^k\right) \tag{6.20}$$

for some small scalar constant α. To see how this works, consider the case of large T: Making T large results in

$$\min_f H_T(f_1, f_2) = \min_f \left((f_1 - g_1)^2 + (f_2 - g_2)^2\right) \quad \text{as} \quad (T \to \infty) \tag{6.21}$$

which is a convex function (a paraboloid) minimized by $f_i = g_i$. Now, we have an initial starting point for the gradient descent. We may now make T slightly smaller, repeat the descent step, find a new solution, and iterate again, using smaller and smaller T until T becomes insignificant in comparison to τ.

Even though we have simplified the problem to only two pixels, the two-dimensional space of f_1, f_2 described by Eq. (6.18) is still difficult to visualize, and we simplify still further to a one-dimensional problem to capture the essence of the homotopy. In Fig. 6.2, we illustrate a one-dimensional function:

$$(f - k)^2 - \frac{1}{\tau + T} \exp\left(-\frac{(f - l)^2}{(\tau + T)^2}\right)$$

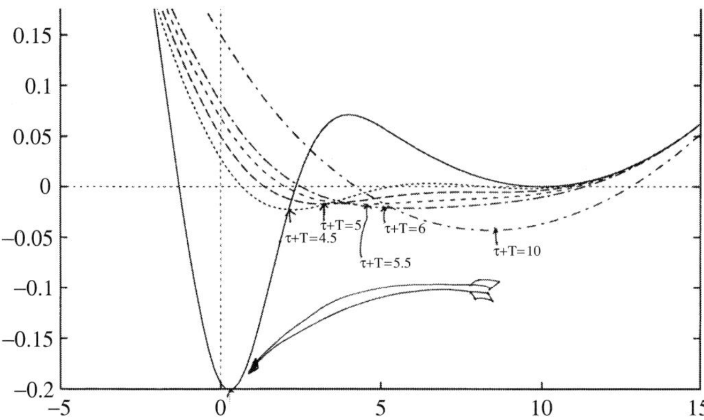

Fig. 6.2. Continuous deformation of a function which is initially convex to find the global minimum of a nonconvex function.

with scalar constants k and l. At high values of $T(T + \tau = 10)$, the curve is completely convex, and as T is reduced, it assumes its true form. At each iteration, the minimum is tracked, as indicated by the arrows, concluding in the global minimum.

6.4.1 Selecting a prior term

A good exam question would be "why is the noise term called the noise term?" or "what assumptions were involved in deriving the noise term from conditional probabilities of noise?"

Let us review the meaning of the terms in the MFA energy function. First, consider the noise term, and write the noise term in a more general way:

$$\sum_i ((D(f))_i - g_i)^2, \tag{6.22}$$

where $(D(f))_i$ denotes some distortion of image f in the vicinity of pixel i. Finding the image f which minimizes this term produces the image which, when distorted, most closely (in the sum-squared error sense) resembles the measurement g. Now, let's look at the prior term.

The term "fidelity" refers to how well the reconstruction resembles the measurement; minimizing these terms preserves fidelity to the measurement.

Writing the prior in a slightly more general way

$$-\frac{1}{\tau}\left[\sum_i \exp\left(-\left(\frac{(R(f))_i^2}{\tau^2}\right)\right)\right] \tag{6.23}$$

where the term $(R(f))_i$ denotes some function of the (unknown) image f at pixel i, and τ incorporates the function of $\tau + T$ in Eq. (6.18). What kind of image minimizes this term? Let's look at what it means.

Most errors are made in implementing this algorithm by dropping minus signs!

First, observe the minus sign in front of the prior term. With that sign there, to minimize the function, we should find the image which causes the exponential to be maximized. What kind of image maximizes the exponential? Now look at the argument of the exponential. Observe the minus sign, and observe that both terms are squared, and therefore always positive. Thus, the argument of the exponential is always negative. What negative argument causes an exponential to be maximal? Answer, zero (well, minus zero actually). Thus, to maximize the exponential, choose any image *f* which causes *R*(*f*) to be zero.

What do we conclude from this? For any function *R*(*f*), the *f* which causes *R*(*f*) to be zero is the *f* which the prior term will seek. This observation gives us a lot of freedom of design. We can choose the function *R*(*f*) to produce the type of solution we seek. Let's now look at some examples.

EXAMPLE

Piecewise-constant images

Consider this form for the prior

$$R^2(f) = \left(\frac{\partial f}{\partial x}\right)^2 + \left(\frac{\partial f}{\partial y}\right)^2. \tag{6.24}$$

In order for this term to be zero, both partial derivatives must be zero. The only type of surface which satisfies this condition is one which does not vary in either direction – flat, but not flat everywhere. To see why the solution is piecewise-constant rather than completely constant, you need to recognize that the total function being minimized is the sum of both the prior and the noise terms. The prior term seeks a constant solution, but the noise term seeks a solution which is faithful to the measurement. The optimal solution to this problem is a solution which is flat in segments, as illustrated in one dimension in Fig. 6.3. The function *R*(*f*) is nonzero only for the points where *f* undergoes an abrupt edge. To see more clearly what this produces, consider the extension to continuous functions. If x is continuous, then the summation in Eq. (6.23) becomes an integral. The argument of the integral is nonzero at only a small, finite number of points (referred to as a set of measure zero), which is insignificant compared to the rest of the integral.

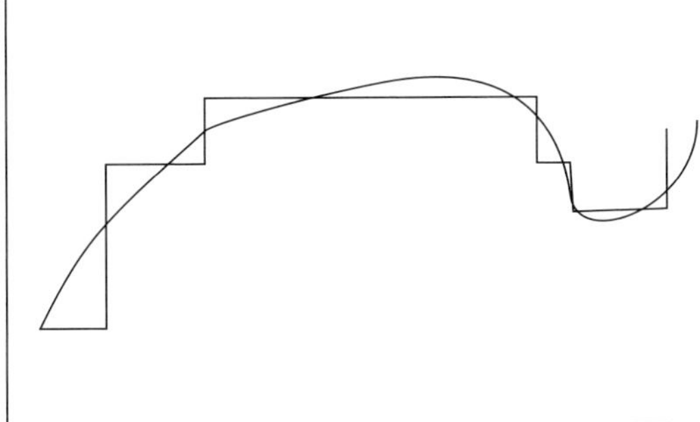

Fig. 6.3. A piecewise-constant solution to the fitting of a surface. The solution has a derivative equal to zero at almost every point. Nonzero derivatives exist only at the points where steps exist.

This is the *quadratic variation*. The Laplacian also involves second derivatives. Ask yourself how it differs from the quadratic variation.

EXAMPLE

Piecewise-planar images

Let's try another example and see what that does. Consider

$$R^2(f) = \left(\frac{\partial^2 f}{\partial x^2}\right)^2 + \left(\frac{\partial^2 f}{\partial y^2}\right)^2 + 2\left(\frac{\partial^2 f}{\partial x \partial y}\right)^2 . \tag{6.25}$$

What does this do? What kind of function has a zero for all its second derivatives? Answer, a plane. Thus, using this form for *R*(*f*) will produce an image which is planar, but still maintains fidelity to the data – a piecewise-planar image. Another alternative operator which is also based on the second derivative is the Laplacian,

$$\frac{\partial^2 f}{\partial x^2} + \frac{\partial^2 f}{\partial y^2} .$$

We saw both of these back in Chapter 5.

You might ask your instructor, "Breaking a brightness image into piecewise-linear segments is the same as assuming the actual surfaces are planar, right?" To which you would probably get "Yes, except for variations in lighting, reflectance, and albedo." Ignoring that, you charge on, saying "But real surfaces aren't all planar." The answer is twofold: First is the trivial and useless observation that all surfaces are planar, you just have to examine a sufficiently small region. More seriously, whether breaking an image into planar patches makes sense depends on the application. For example [6.14, 6.74], one could do a piecewise-constant segmentation to remove noise and then treat each patch as a plane and get improved estimates of optic flow [6.41] or stereo [6.72] that way. Some of the underlying theory of planar approximations to images may be found in [6.62].

Interestingly, Yi and Chelberg [6.83] observe that second-order priors like this one require a great deal more computation than first-order priors, and that first-order priors can be made approximately invariant. However, in our own experiments, we have not found that second-order priors impose such a severe computational penalty, and they do provide more flexibility in reconstruction.

A one-dimensional example of such a solution is shown in Fig. 6.4. The idea of modeling an image as piecewise-planar has recently received additional support from some work by Elder and co-workers [6.22, 6.23, 6.24] which suggests that "the edge representation of an image is, to a good approximation, invertible." They accomplish this remarkable result by assuming that except at edges, an image satisfies the Laplace equation $\nabla^2 f(x, y) = 0$.

In conclusion, you can choose any function for the argument of the exponential that you wish, as long as the image you want is produced by setting the argument to zero. Some more general properties of prior models have been stated by Li [6.51] as a function of the local image gradient Λ: A prior should (1) be continuous in its first derivative; (2) be even ($h(\Lambda) = h(-\Lambda)$); (3) be positive $h(\Lambda) > 0$; and (4) have

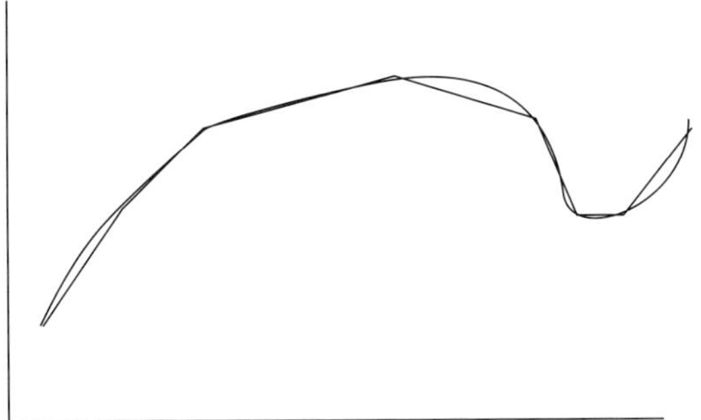

Fig. 6.4. A piecewise-linear solution to the data of Fig. 6.3. Clearly, the piecewise-linear solution preserves fidelity to the data more accurately than the piecewise-constant.

a negative derivative for positive argument $h'(\Lambda) < 0$, for $\Lambda > 0$; go to a constant in the limit $\lim_{\Lambda \to \infty} |\Lambda h(\Lambda)| = C$.

6.4.2 Annealing: Avoiding local minima

MFA gives us one more capability that distinguishes it from other MAP methods, the ability to avoid most local minima.

The magic is all in the use of the parameter τ in Eq. (6.23). Begin with "large" τ. Do gradient descent. As the descent iterations proceed, "slowly" reduce τ. When τ reaches an "appropriate" minimum, stop. Now, let's see what the words in quotes in the previous sentences mean.

This is the definition of the Kroneker delta. The Dirac delta is infinity when its argument goes to zero and zero otherwise, and integrates to one.

We wish to develop a function which is minimized when the number of pixels which are equal to their neighbors is maximal. One way to count such pixels is to use the delta function, which is one if the difference between two pixels is zero:

$$\delta(f_i - f_j) = \begin{cases} 1 & \text{if} \quad f_i = f_j \\ 0 & \text{otherwise} \end{cases} \tag{6.26}$$

so the number of times a pixel is equal to its neighbor is

$$\sum_i \sum_{j \in \aleph_i} \delta(f_i - f_j). \tag{6.27}$$

We want a function which is minimal when this double summation is maximal, so we simply introduce a minus sign

$$H_{\mathrm{p}} = -\sum_i \sum_{j \in \aleph_i} \delta(f_i - f_j). \tag{6.28}$$

The delta function is not convenient, since it is not differentiable, and we will want to use gradient descent to solve this problem. But there is another problem with this formulation: If the image is continuously valued (or even if it is represented in floating point), what does it mean for f_i to equal to f_j? How close should they be before they are considered equal? How about $|f_i - f_i| < 0.01$? Is this small enough? How about 0.001? OK? Do you accept that? So we insist that two points which differ by more than 0.001 contribute to the error. What about two points that differ by 0.000 999? That pair does not contribute at all. Does that make sense?

What we have generated is very similar to the problem of describing the probability that a measure has a particular value. The probability, for example, of being exactly 6.000 000 (for an arbitrary number of zeros) is zero, and we therefore resort to a different representation for the concept of likelihood, and we invent the probability density. In a similar way, in this problem, we pursue the same philosophy, instead of using the Kroneker delta, we replace the delta function with a continuous, differentiable function, which represents the same intuition,

$$H_{\mathrm{p}} = -\sum_i \sum_{j \in \aleph_i} \frac{1}{\sqrt{2\pi}\tau} \exp\left(-\frac{(f_i - f_j)^2}{2\tau^2}\right). \tag{6.29}$$

This form also allows the concept of annealing: Start with large τ and reduce τ until it approaches zero. The square root of a constant is not particularly meaningful, but it does serve to ensure that the function remains bounded in the appropriate way.

The details of why this process of annealing avoids local minima are described elsewhere [6.8, 6.11, 6.12], but result from the analogy to simulated annealing [6.27].

Initial value of τ

We will choose an initial value of f to be g. Consider what happens if τ_{initial} is large. A large τ will cause the argument of the exponential to be close to zero, and the value of the exponential itself to be approximately one. But that value is, itself, divided by τ, so that when τ is large, the value of the prior is on the order of $1/\tau$; and if τ is a big number, then the prior is insignificant relative to the noise term. We can ensure that τ is initially "large" by choosing τ to be larger than, say, twice the average value of the numerator,

$$\tau_{\mathrm{initial}} = 2\langle R(f)\rangle. \tag{6.30}$$

Decreasing τ

MFA is based upon the mathematics of simulated annealing. One can show that in simulated annealing, a global minimum can be achieved by following a logarithmic annealing schedule like

$$\tau^K = \frac{1}{\ln K} \tag{6.31}$$

where K is the iteration number. This schedule decreases τ extremely slowly; so slowly as to be impractical. Instead, one could choose a schedule like

$$\tau^K = 0.99\tau^{K-1} \tag{6.32}$$

which has been shown to work satisfactorily in many applications, and reduces τ much faster than the logarithmic schedule.

6.4.3 How to differentiate a function containing a kernel operator

In previous discussions, we excluded the effect of blur and assumed noise is the only source of distortion. In this section, we extend the results we derived above and include blur in the formation of the measured image g, such that the noise term can be written as Eq. (6.33). More important, we illustrate how to differentiate a function containing a kernel operator (or a blur operator)

$$\sum_i ((f \otimes h)_i - g_i)^2. \tag{6.33}$$

Taking a one-dimensional example, assume f_i is a pixel from the original (unknown) image, g_i is a pixel from the measured image, and h is the horizontal blur kernel with a finite kernel size 5, as shown in Fig. 6.5. We now explain the derivative of the noise term, Eq. (6.33), in gradient descent in detail. First write out all the terms involving a pixel (f_4) at which a measurement (g_4) was made in the noise term H_n above

These are the only terms in the sum of Eq. (6.33) involving f_4.

$$\begin{aligned} E_4 &= ((f \otimes h)_2 - g_2)^2 + ((f \otimes h)_3 - g_3)^2 + ((f \otimes h)_4 - g_4)^2 \\ &\quad + ((f \otimes h)_5 - g_5)^2 + ((f \otimes h)_6 - g_6)^2 \\ &= (f_0 h_{-2} + f_1 h_{-1} + f_2 h_0 + f_3 h_1 + f_4 h_2 - g_2)^2 \\ &\quad + (f_1 h_{-2} + f_2 h_{-1} + f_3 h_0 + f_4 h_1 + f_5 h_2 - g_3)^2 \\ &\quad + (f_2 h_{-2} + f_3 h_{-1} + f_4 h_0 + f_5 h_1 + f_6 h_2 - g_4)^2 \\ &\quad + (f_3 h_{-2} + f_4 h_{-1} + f_5 h_0 + f_6 h_1 + f_7 h_2 - g_5)^2 \\ &\quad + (f_4 h_{-2} + f_5 h_{-1} + f_6 h_0 + f_7 h_1 + f_8 h_2 - g_6)^2 \end{aligned} \tag{6.34}$$

where $(f \otimes h)_i$ denotes the application of kernel h to image f with the origin of h (in this case, the center) located at pixel f_i. The derivative of H_n with respect to pixel f_4 can then be derived as Eq. (6.35), and further generalized as Eq. (6.36),

$$\begin{aligned} \frac{\partial H_\text{n}}{\partial f_4} &= 2((f \otimes h)_2 - g_2)h_2 + 2((f \otimes h)_3 - g_3)h_1 + 2((f \otimes h)_4 - g_4)h_0 \\ &\quad + 2((f \otimes h)_5 - g_5)h_{-1} + 2((f \otimes h)_6 - g_6)h_{-2} \end{aligned} \tag{6.35}$$

$$\frac{\partial H_\text{n}}{\partial f_4} = (((f \otimes h) - g) \otimes h_\text{rev})_4 \tag{6.36}$$

where $h_\text{rev} = h_2, h_1, h_0, h_{-1}, h_{-2}$, and $(f \otimes h - g)$ is computed at all points. The

...	f_2	f_3	f_4	f_5	f_6	...
	h_{-2}	h_{-1}	h_0	h_1	h_2	

Fig. 6.5. A one-dimensional image and a one-dimensional kernel with five pixels.

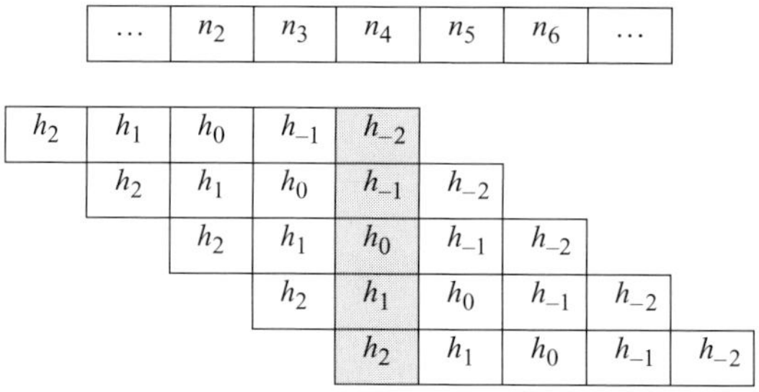

	…	n_2	n_3	n_4	n_5	n_6	…	
h_2	h_1	h_0	h_{-1}	h_{-2}				
	h_2	h_1	h_0	h_{-1}	h_{-2}			
		h_2	h_1	h_0	h_{-1}	h_{-2}		
			h_2	h_1	h_0	h_{-1}	h_{-2}	
				h_2	h_1	h_0	h_{-1}	h_{-2}

Fig. 6.6. The reverse kernel in the derivation of the noise term.

application of h_{rev} is illustrated in Fig. 6.6 more clearly where we assume $n_i = ((f \otimes h) - g)_i$.

The application of the reverse kernel with a two-dimensional image follows the same rule as that in the one-dimensional case. Equation (6.37) shows a 3×3 kernel function (h) and the corresponding reverse kernel h_{rev}.

$$h = \begin{bmatrix} h_{-1,-1} & h_{-1,0} & h_{-1,1} \\ h_{0,-1} & h_{0,0} & h_{0,1} \\ h_{1,-1} & h_{1,0} & h_{1,1} \end{bmatrix}, \quad h_{\text{rev}} = \begin{bmatrix} h_{1,1} & h_{1,0} & h_{1,-1} \\ h_{0,1} & h_{0,0} & h_{0,-1} \\ h_{-1,1} & h_{-1,0} & h_{-1,-1} \end{bmatrix}. \tag{6.37}$$

Before we discuss the prior term, let's conclude the general form of differentiation when a kernel function is involved. Let $R(f \otimes h)$ be some differentiable function. The derivative with respect to f is

$$\frac{\partial}{\partial f} R(f \otimes h) = (R'(f \otimes h)) \otimes h_{\text{rev}} \tag{6.38}$$

where $R'(t) = \frac{\partial}{\partial t} R(t)$.

In [6.63], differentiation of a function containing a kernel operator is used in a particularly interesting case, where we only have measurements at every other pixel. Thus, this problem may be viewed as not only image estimation, but interpolation as well [6.78]. The experimental results and performance evaluation of MFA missing data estimation algorithm are demonstrated in detail in [6.63].

Besides the noise term, the prior term can also contain a kernel operator. Recall that the prior energy function models the neighborhood operation and thus can be represented as a kernel operation with the kernel selected dependent on the property of the image. We illustrate the gradient using the following prior:

$$H_{\text{p}} = -\sum_i \exp(-(f \otimes r)^2) \tag{6.39}$$

$$\frac{\partial H_{\text{p}}}{\partial f} = [2(f \otimes r) \exp(-(f \otimes r)^2)] \otimes r_{\text{rev}}. \tag{6.40}$$

Recall that the derivative of the prior $\partial H_{\text{p}}/\partial f$ is itself an image, and that image is derived by applying r to f, multiplying (pixel-by-pixel) with the exponential to produce another image, and applying the reverse of r to that image.

6.4.4 Practical considerations: Edge-preserving smoothing

Collecting all we have learned in this chapter so far into two equations, we choose an objective function which preserves fidelity (the restoration resembles the measurement) but which is piecewise-linear.

$$H(f) = \sum_i \frac{(f_i - g_i)^2}{2\sigma^2} - \frac{\beta}{\tau} \sum_i \exp\left(-\frac{(R(f))_i^2}{\tau^2}\right) \tag{6.41}$$

where $(R(f))_i$ will be the quadratic variation of Eq. (6.25) at pixel i. Of course, in specific applications, different priors may be indicated. To perform gradient descent, we must find the derivative with respect to f. This problem becomes complicated when we recognize that the numerator of the argument of the exponential varies in both x and y. We have two choices.

- We could recognize that R is a sum of three terms, that the exponential of a sum is the product of the exponentials, and use the product rule of derivatives to construct a rather complicated expression for the derivative.
- We could say "instead of putting the summation in the argument of the exponential, let's just add the exponentials."

Of course, these are not equivalent expressions. However, minimizing either gets at the same idea – a piecewise-linear image. Since the second is simpler to implement, being engineers, we choose the second option. We know how to do the derivatives, so we end up with the following algorithm.

The derivative of the noise term is trivial: On each iteration, simply change pixel i by $dnoise_i = (f_i - g_i)/\sigma^2$.

To determine the derivative of the prior requires a bit more work: according to Eq. (6.40), the derivative of the prior term is

$$\frac{\beta}{\tau}\left(\left(\frac{f \otimes \Delta}{\tau^2}\right) \exp\left(-\frac{(f \otimes \Delta)^2}{2\tau^2}\right)\right) \otimes \Delta_{\text{rev}}.$$

Define three kernel operators which we will use to estimate the three partial second derivatives in the quadratic variation

$$\Delta_{xx} = \frac{1}{\sqrt{6}}\begin{bmatrix} 0 & 0 & 0 \\ 1 & -2 & 1 \\ 0 & 0 & 0 \end{bmatrix}, \quad \Delta_{yy} = \frac{1}{\sqrt{6}}\begin{bmatrix} 0 & 1 & 0 \\ 0 & -2 & 0 \\ 0 & 1 & 0 \end{bmatrix}, \quad \text{and}$$

$$\Delta_{xy} = 2\begin{bmatrix} -0.25 & 0 & 0.25 \\ 0 & 0 & 0 \\ 0.25 & 0 & -0.25 \end{bmatrix}.$$

Notice that these kernels are symmetric. Therefore $\Delta = \Delta_{\text{rev}}$.

Compute three images, where the ith pixels of those images are

$$r_{ixx} = (\Delta_{xx} \otimes f)_i, \quad r_{iyy} = (\Delta_{yy} \otimes f)_i, \quad \text{and} \quad r_{ixy} = (\Delta_{xy} \otimes f)_i.$$

Create the image s_{xx} whose elements are

$$\frac{1}{\tau^3} r_{ixx} \exp\left(-\frac{r_{ixx}^2}{2\tau^2}\right),$$

and similarly create s_{yy} and s_{xy}.

For the purposes of gradient descent, the change in pixel i from the prior term is then

$$dprior_i = \beta((\Delta_{xx} \otimes s_{xx})_i + (\Delta_{yy} \otimes s_{yy})_i + (\Delta_{xy} \otimes s_{xy})_i).$$

The gradient descent rule says to change each element of f using $f_i \Leftarrow f_i - \alpha d_i$, where $d_i = dnoise_i + dprior_i$.

The learning coefficient α should be $\alpha = \gamma\sigma\sqrt{\tau}/RMS(d_i)$, where γ is a small dimensionless number, like 0.04; $RMS(d)$ is the root mean square norm of the gradient d; σ can be determined as the variance of the noise in the image (note that this is NOT a good estimate in synthetic images). We observe that in this form, α changes every iteration.

The coefficient β is on the order of σ, and choosing $\beta = \sigma$ is usually adequate.

Implementing this algorithm and annealing over a couple of orders of magnitude of τ should give reductions of noise similar to those illustrated in Figs. 6.10–6.13.

6.5 Conclusion

Consistency is easy to see in some of the applications of image optimization. For example, in a paper by one of the authors [6.7], use is made of consistency in fusing a (noisy) range image with a (noisy) brightness image of the same object. Since both are images of the same object, both the range image, which depends on the geometry, and the brightness image, which depends on the reflectivity (and therefore on the geometry) must be consistent.

Optimization methods are so pervasive in this chapter that the chapter title could almost be "image optimization." We set up an objective function, a function of the measured image and the (unknown) true image and find the (unknown) image which minimizes the objective function. We introduce two terms, a noise term – which depends on the measurement – and a prior term – which depends only on the true image. We then find the "true" image by finding the image which minimizes the objective function. A variety of minimization techniques could be used. In this

Gradient descent with annealing.

chapter, we use gradient descent with annealing, but other, more sophisticated and faster techniques, such as conjugate gradient, could be used.

6.6 Vocabulary

You should know the meanings of the following terms.

Anisotropic diffusion (see 6A.2)
Annealing
Bayes' rule
GNC (see 6A.1)
Inverse problem
MAP algorithm
Relaxation
Restoration

Assignment 6.1

Equation (6.34) illustrates the partial derivative of an expression involving a kernel, by expanding the kernel into a sum. Use this approach to prove that Eq. (6.40) can be derived from Eq. (6.39). Do your proof using a one-dimensional problem, and use a kernel which is 3×1 (denote the elements of the kernel as h_{-1}, h_0, and h_1).

Assignment 6.2

Implement Eq. (6.65) on the image angio.ifs, or some other image which your instructor selects. Experiment with various run times and parameter settings.

Assignment 6.3

In Eq. (6.25), the quadratic variation is presented as a prior term. A very similar prior would be the Laplacian. What is the difference? That is, are there image features which would minimize the Laplacian and not minimize the quadratic variation? Vice versa?

Assignment 6.4

Which of the following expressions represents a Laplacian?

(a) $\frac{\partial^2 f}{\partial x^2} + \frac{\partial^2 f}{\partial y^2}$ (c) $\left(\frac{\partial^2 f}{\partial x^2}\right)^2 + \left(\frac{\partial^2}{\partial y^2}\right)^2$ (e) $\sqrt{\left(\frac{\partial^2 f}{\partial x^2}\right)^2 + \left(\frac{\partial^2}{\partial y^2}\right)^2}$

(b) $\frac{\partial f}{\partial x} + \frac{\partial f}{\partial y}$ (d) $\left(\frac{\partial f}{\partial x}\right)^2 + \left(\frac{\partial f}{\partial x}\right)^2$ (f) $\sqrt{\left(\frac{\partial f}{\partial x}\right)^2 + \left(\frac{\partial f}{\partial x}\right)^2}$

Assignment 6.5

One form of the diffusion equation is written $df/dt = h_x \otimes (c(h_x \otimes f)) + h_y \otimes (c(h_y \otimes f))$ where h_x and h_y estimate the first derivatives in the x and y directions, respectively. This suggests that four kernels must be applied to compute this result. Simple algebra, however, suggests that this could be rewritten as $df/dt = c(h_{xx} \otimes f + h_{yy} \otimes f)$, which requires application of only two kernels. Is this simplification of the algorithm correct? If not, explain why not, or under what conditions it would be true.

Assignment 6.6

Consider the following image Hamiltonian

$$H(f) = \left(\sum_i \left(\frac{f_i - g_i}{\sigma^2}\right)^2\right) - \sum_i \frac{1}{\tau} \exp\left(-\frac{(h \otimes f)^2}{\tau^2}\right) \equiv H_n(f) + H_p(f)$$

where $\otimes$ denotes application of a kernel operator, the pixels in the image are lexicographically indexed by i, and the kernel h is

$$\begin{bmatrix} -1 & 2 & -1 \\ -2 & 4 & -2 \\ -1 & 2 & -1 \end{bmatrix}.$$

Let $G_p(f_k)$ denote the partial derivative of H_p with respect to pixel k. $G_p(f_k) = \partial/\partial f_k H_p(f)$. Write an expression for $G_p(f_k)$. Use kernel notation.

Assignment 6.7

Continuing problem Assignment 6.6, you are to consider ONLY the prior term. Write an equation which describes the change in image brightness at pixel k as one iteration of a simple gradient descent algorithm. Denote the gradient by $G_{\mathrm{p}}(f_k)$ and use that in this answer.

Assignment 6.8

Continuing problem Assignment 6.7, expand this differential equation (assume the brightness varies only in the x direction) by substituting the form for $G_{\mathrm{p}}(f_k)$ which you derived. Is this a type of diffusion equation? Discuss. (Hint: Replace the application of kernels with appropriate derivatives.)

Assignment 6.9

In a diffusion problem, you are asked to diffuse a VECTOR quantity, instead of the brightness which you did in your project. Replace the terms in the diffusion equation with the appropriate vector quantities, and write the new differential equation. (Hint: You may find the algebra easier if you denote the vector as $[a,b]^{\mathrm{T}}$.)

Assignment 6.10

The time that the diffusion runs is somehow related to blur. This is why some people refer to diffusions of this type as "scale space." Discuss this use of terminology.

Topic 6A Alternative and equivalent algorithms

6A.1 GNC: An alternative algorithm for noise removal

Just as the MFA approach described in the previous section minimizes an objective function in order to find an image with sharp edges, graduated nonconvexity (GNC) does the same, but uses an objective function which treats the presence of edges explicitly.

We consider the case in which our *a priori* knowledge states that the image is uniform in brightness, except for step discontinuities. Blake and Zisserman [6.16] refer to this case as the "weak membrane," and a similar MFA instance is referred to [6.12] as "piecewise-uniform." The similarities can be seen both in the objective function (compare Fig. 6.7 and Fig. 6.9) and in the restorations (Figs. 6.10–6.13). There exist other formulations [6.12] which

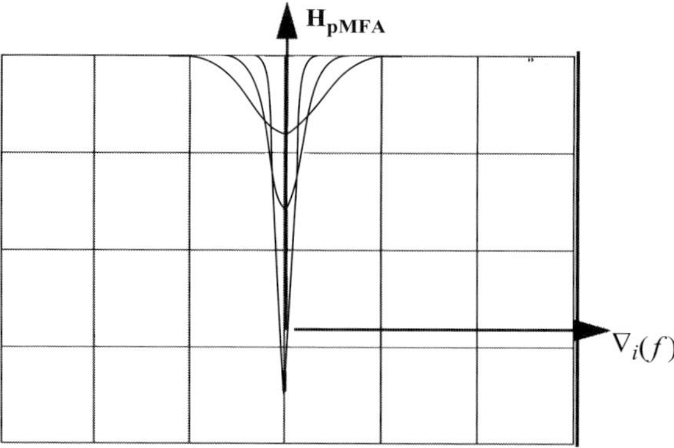

Fig. 6.7. Prior energy for MFA, for various values of *T*. Smaller *T* results in sharper peaks.

pose the MFA problem in a manner even more similar to GNC, a similarity first noted by Geiger and Girosi [6.25]. In the "weak membrane" application of GNC, the minimization problem is

$$min_{f,l} H_{\mathrm{GNC}} \tag{6.42}$$

where

$$H_{\mathrm{GNC}} = H_{\mathrm{n}} + S + P, \quad S = \lambda^2 \sum_i |\nabla_i(f)|^2 (1 - l_i), \quad \text{and} \quad P = \alpha \sum_i l_i \tag{6.43}$$

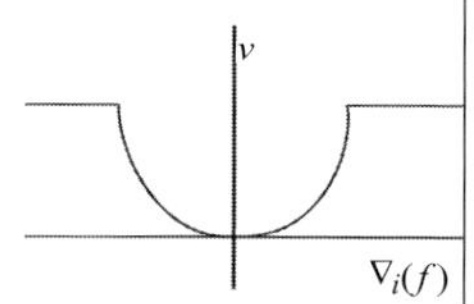

Fig. 6.8. Prior energy of the GNC algorithm.

and the notation $\Delta_i(f)$ is interpreted to mean "the gradient of the image at point i." Here, the $l_i \in \{0, 1\}$ denotes a discontinuity in f at the ith pixel. That is, if $l_1 = 1$, the pixel at point i is interpreted as an edge point. Similarly, f_i will denote the brightness of the ith pixel. It has been shown [6.16] that minimizing H_{GNC} can be reduced to the following problem, which involves only continuous variables

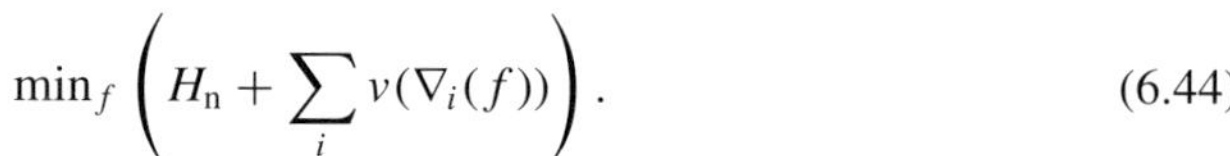

$$\min_f \left(H_{\mathrm{n}} + \sum_i v(\nabla_i(f)) \right). \tag{6.44}$$

In Eqs. (6.43) and (6.44), the $|\nabla(.)|$ represents any operator which returns a scalar measure of the local "edginess" of the image such as $(\partial f/\partial x)^2 + (\partial f/\partial y)^2$, and the v function of Eq. (6.44) is the "clipped parabola" illustrated in Fig. 6.8.

Note: This "clipped parabola" has the same general shape as the inverted Gaussian we mentioned in Eq. (6.16), for the same reasons: We want to penalize noise (so the bigger the noise, the bigger the penalty) but we don't want to penalize edges (so at some point, we stop making the penalty any larger).

The minimization problem posed by Eq. (6.44) is unsolvable by techniques such as gradient descent, since the function defined by Eq. (6.44) is in general not convex. That is, it may possess many minima. Instead, GNC approximates v with the piecewise-smooth function

$$v^*(t) = \begin{cases} \lambda^2 t^2 & \text{if } (|t| < q) \\ \alpha - c^*(|t| - r)^2/2 & \text{if } (q \le |t| < r) \\ \alpha & \text{if } (|t| \ge r) \end{cases} \tag{6.45}$$

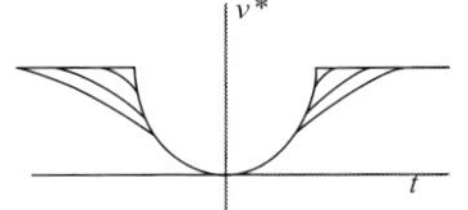

Fig. 6.9. Smoothed approximations to the energy of Fig. 6.8. Smaller values of *p* result in approximations which are closer to the desired prior. The *t* used here (in GNC) is equivalent to the edge strength ∇of MFA.

where c^* is a scalar constant, $c = c^*/p$,

$$r^2 = \alpha\left(\frac{2}{c^*} + \frac{1}{\lambda^2}\right), \quad \text{and} \quad q = \frac{\alpha}{\lambda^2 r}. \tag{6.46}$$

Equations (6.45) and (6.46) then define the algorithm. Reducing the parameter p from 1 to 0 steadily changes v^* until it becomes precisely equal to v. This produces a family of prior energies, illustrated in Fig. 6.9.

The process of gradually reducing p begins by minimizing a function which is convex, and therefore has a unique minimum. Then, from that minimum, the local minimum is tracked continuously as p is reduced from 1 to 0.

6A.2 Variable conductance diffusion

Variable conductance diffusion (VCD) [6.31, 6.59, 6.62] is a powerful method of image feature extraction in which blurring is allowed to occur except at edges. The term "edge" may be liberally interpreted to mean any image feature which is of interest. For example, Whitaker [6.79] operates on the gradient of the image rather than the image itself and allows smoothing to occur except where ridges (sharp changes of gradient direction) occur. Such an operation is not a restoration by any means, since most of the information in the original image is lost. It is, however, a very robust way to extract a central axis of an object in a gray scale image.

Relation of spatial derivatives (RHS) to temporal derivatives (LHS).

VCD operates by simulating the diffusion equation

$$\frac{\partial f_i}{\partial t} = \nabla \cdot (c_i \cdot \nabla_i f) \tag{6.47}$$

where t is time, and $\nabla_i f$ denotes the spatial gradient of f at pixel i. The diffusion equation models the flow of some quantity (the most commonly used example is heat) through a material with conductivity (e.g., thermal conductivity) c.

If c_i is constant, independent of the pixel number i, ($c_i = c$) then the partial differential Eq. (6.47) has a solution which is the same as convolution with a Gaussian, in which the variance of that Gaussian depends on c and on the time over which the diffusion is run. Specifically, let f, a function of space and time, be described by a specific partial differential equation (PDE). If it is possible to write f in the following form:

$$f(x,t) = \int_{-\infty}^{\infty} G(x,x',t) f(x',0)\, dx', \tag{6.48}$$

then we say that $G(x, x', t)$ is the *Green's function* of the PDE. The special case of isotropic diffusion may be stated formally in one dimension:

Theorem

The Gaussian is the Green's function of the PDE

$$\frac{\partial f}{\partial t} = c \frac{\partial^2 f}{\partial x^2}. \tag{6.49}$$

Proof

This is accomplished by writing the Gaussian as

$$G(x, x', t) = \frac{1}{\sigma} \exp\left(-\frac{(x - x')^2}{2\sigma^2}\right),$$

where σ will turn out to be a function of time (we have omitted the $1/\sqrt{2\pi}$ since it occurs on both sides of the PDE and cancels out).

Substitute Eq. (6.48) into Eq. (6.49), producing on the left-hand side (LHS) the partial with respect to t of an integral in which σ is a function of t. Taking this partial, make the LHS equal to

$$\left[\frac{(x - x')^2}{\sigma^4} - \frac{1}{\sigma^2}\right] \frac{\partial \sigma}{\partial t}. \tag{6.50}$$

Similarly, we can take the second partial derivative with respect to x to create the right-hand side (RHS):

$$\frac{c}{\sigma}\left[\frac{(x - x')^2}{\sigma^4} - \frac{1}{\sigma^2}\right]. \tag{6.51}$$

Equating Eqs. (6.50) and (6.51) produces the equation

$$\frac{\partial \sigma}{\partial t} = \frac{c}{\sigma} \tag{6.52}$$

whose solution is

$$\sigma^2 = 2ct. \text{ QED.} \tag{6.53}$$

In the case of VCD, the conductance becomes a function of the spatial coordinates, in this instance parameterized by i. In particular it becomes a property of the local image intensities themselves. The conductance c_i is usefully seen as a factor by which space is locally compressed.

To smooth, except at edges, we let c_i be small if i is an edge pixel, i.e., if a selected image property is locally nonuniform. If c_i is small (in the heat transfer analogy), little heat flows (space is stretched), and in the image, little smoothing occurs. If, on the other hand, c_i is large, then much smoothing is allowed in the vicinity of pixel i (space is compressed). VCD then, just as the forms of MFA and GNC discussed, implements an operation which after repetition produces a nearly piecewise uniform result.

6A.3 Edge-oriented anisotropic diffusion

As we observed in Eq. (6.48), the Gaussian is the Green's function of the diffusion equation. That is, a diffusion process running on an image produces the same result as convolution with a Gaussian, where the variance of the Gaussian depends on how long the diffusion has been running. The constant-conductance diffusion equation is

$$f_t = c(f_{xx} + f_{yy}). \tag{6.54}$$

If there is an edge in the image, we want to remove noise on both sides of the edge, but not blur the edge. Therefore it makes sense to diffuse in a direction tangent to the edge. The normal and tangent vectors to an edge in a two-dimensional image are given by

$$N = \frac{[f_x \; f_y]^{\mathrm{T}}}{\sqrt{f_x^2 + f_y^2}} \quad \text{and} \quad T = \frac{[-f_y \; f_x]^{\mathrm{T}}}{\sqrt{f_x^2 + f_y^2}}.$$

Consider now, the second partial derivatives of f taken in the N and T directions: f_{NN} and f_{TT}.

Since the Laplacian is rotation-invariant, we can write the diffusion PDE (Eq. (6.54)) in the new coordinates by $f_t = c(f_{NN} + f_{TT})$.

One may derive the following relationships between the partials

$$\begin{aligned} f_{NN} &= \left(f_x^2 f_{xx} + 2 f_x f_y f_{xy} + f_y^2 f_{yy}\right)/\left(f_x^2 + f_y^2\right) \\ f_{TT} &= \left(f_y^2 f_{xx} - 2 f_x f_y f_{xy} + f_x^2 f_{yy}\right)/\left(f_x^2 + f_y^2\right). \end{aligned} \tag{6.55}$$

Substituting this form into Eq. (6.54) and subtracting the normal flow, we end up with a PDE which smooths along edges without smoothing across edges:

$$f_t = \left(f_y^2 f_{xx} - 2 f_x f_y f_{xy} + f_x^2 f_{yy}\right)/\left(f_x^2 + f_y^2\right), \tag{6.56}$$

which we denote as anisotropic diffusion.

6A.4 A common description of image relaxation operators

6A.4.1 MFA and GNC

In both cases, we have an energy function which increasingly penalizes the presence of gradients in the image. In the GNC case the prior retains its original shape, and the "annealing" process (that is, the lowering of p) results in successively closer fits to the predetermined shape of the prior. In the MFA case, the shape of the prior itself changes, retaining a constant area, but becoming narrower as T is reduced. With this observation, it should not be surprising that piecewise-constant MFA and the weak membrane of GNC achieve the same result.

Fig. 6.10. Original image.

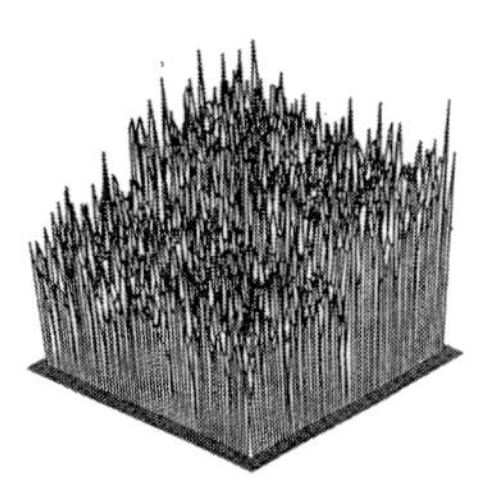

Fig. 6.11. Corrupted image.

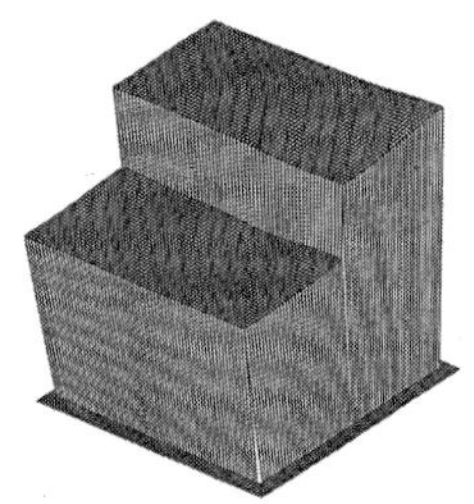

Fig. 6.12. MFA restoration.

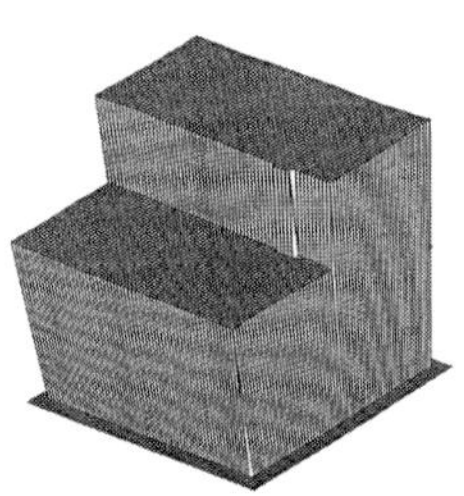

Fig. 6.13. GNC restoration.

The equivalence of the techniques is demonstrated in the next section and proven formally in [6.12], to which we refer the reader for in-depth analysis. The following experiments are also described in that paper and are reprinted here to assist the reader in understanding the action of the algorithms.

The two approaches were each used to restore the same image with various signal-to-noise ratios (SNRs). On each application of MFA and GNC to a noisy image, the respective parameters were varied to achieve the best possible image restoration. Several hundred runs with distinct parameter values were completed for each algorithm. We found that for each noisy image there existed some parameter set for each algorithm such that the restored images were of comparable quality.

The resulting image quality achieved is depicted above with the original image (Fig. 6.10), the image corrupted with SNR $= 2$ (Fig. 6.11), the MFA restored image (Fig. 6.12), and the GNC restored image (Fig. 6.13).

Coding of the GNC algorithm can be found in [6.16], which performs descent using successive over-relaxation (SOR). By using an implementation of MFA which also uses SOR, we found the execution times of MFA to be roughly ten times faster than GNC for high noise cases (SNR < 3). For cleaner images, SNR ≥ 4, the GNC execution times were faster.

6A.4.2 MFA and VCD – equivalent algorithms

Before we can complete this comparison, we need to elaborate on the nature of a spatial derivative. A derivative of brightness with respect to distance in an image could be written

$$\frac{\partial f}{\partial x} = \lim_{\Delta x \to 0} \frac{f(x + \Delta x) - f(x)}{\Delta x}. \tag{6.57}$$

In a sampled image (as all digital images are), however, the limiting process makes no sense, for as ter Haar Romeny *et al.* point out [6.73], one cannot take differences at scales smaller than a pixel. Instead, one must estimate the derivative at a point by operations performed on some *neighborhood* of that point. The topic of estimating derivatives is an old one, and we will not examine it further except to point out that most analyses have concluded that such estimates (for higher order derivatives as well) are generally computed by a kernel operator in one dimension and by the Euclidean norm of an array of n operators in n dimensions. For the purposes of this derivation, we consider only derivatives in the x direction. We will generalize this in a few paragraphs. Thus, we may rewrite the prior term as

$$H_{\mathrm{p}}(f) = \frac{-b}{\sqrt{2\pi}T} \sum_i \exp\left(-\frac{(f \otimes r)_i^2}{2T^2}\right). \tag{6.58}$$

By the notation $(f \otimes r)_i$ we mean the result of applying a kernel r to the image f at point i. The kernel may be chosen to emphasize problem-dependent image characteristics. This general form has been used to remove noise from piecewise-constant [6.38] and piecewise-linear [6.8, 6.9] images.

In the following derivation, we consider only the prior term.

To perform gradient descent we will need the derivative, so we calculate

$$\frac{\partial H_{\mathrm{p}}}{\partial f_i} = \frac{b}{\sqrt{2\pi}T} \left[\left(\frac{(f \otimes r)}{T^2} \exp\left(-\frac{(f \otimes r)^2}{2T^2}\right)\right) \otimes r_{\mathrm{rev}} \right]_i \tag{6.59}$$

where r_{rev} denotes the mirror image of the kernel r.

Students: Pay attention! The magic is here.

Now, let us assume that the variation Λ_i, to be made small almost everywhere, is the magnitude of the gradient of the image. It is well-known [6.53] that using the derivative of a Gaussian as a kernel is an excellent way to estimate image derivatives in the presence of noise. Denoting by G_x the derivative with respect to x of a Gaussian, we could substitute $G_x, G_y = \nabla G$ for r in Eq. (6.59); writing ∇f for $f \otimes \nabla G$ results in

$$\frac{\partial H_{\mathrm{p}}}{\partial f} = -\kappa(\nabla((\nabla f) \exp(-(\nabla f)^2))). \tag{6.60}$$

And here.

In the equation above, we have lumped the constants together into κ and set the annealing control parameter, T, to 1 for clarity. We have then made use of the observation that for centered first derivative kernels, $f \otimes h = -(f \otimes h_{\mathrm{rev}})$. We will discuss the impact of T in the next section.

Finally, we consider the use of $\frac{\partial}{\partial f} H(f)$ in a gradient descent algorithm. In the simplest implementations of gradient descent, f is updated by (compare with Eq. (6.20))

$$f_i^{k+1} = f_i^k - \alpha \frac{\partial H}{\partial f_i} \tag{6.61}$$

where f^k denotes the value of f at iteration k, and where α is some small constant (or, in more sophisticated algorithms, a function of the Hessian of H). Rewriting Eq. (6.61),

$$\frac{\partial H}{\partial f_i} = \frac{f_i^{k+1} - f_i^k}{\alpha} \tag{6.62}$$

we note that the LHS of Eq. (6.62) then represents a change in f between iterations k and $k+1$, and in fact bears a strong resemblance to the form of the derivative of f. We make this similarity explicit by defining that iteration k is calculated at time t and iteration $k+1$ calculated at time $t + \Delta t$. (In similar contexts, t is sometimes known as the "evolution parameter.") Since t is an artificially introduced parameter, without physically meaningful units, we may scale it by any convenient proportionality constant, and we have

$$\frac{\partial H}{\partial f_i} = \frac{f_i(t + \Delta t) - f_i(t)}{\Delta t} \equiv \frac{\partial f_i}{\partial t}, \tag{6.63}$$

where we have allowed the constant α to be renamed Δt so the expression looks like a derivative with respect to time. Substituting this (re)definition into Eq. (6.60), we have our final result: By simply changing notation and making explicit the time between iterations, the derivative of the MFA prior term may be rewritten as

$$\frac{\partial f_i}{\partial t} = -\kappa(\nabla((\nabla f)\ \exp(-(\nabla_T f)^2))), \tag{6.64}$$

where $\nabla_T f$ denotes the scaling of the gradient magnitude by T.

Writing the diffusion equation

$$\frac{\partial f_i}{\partial t} = -\kappa(\nabla(c\nabla f)) \tag{6.65}$$

in which the conductivity, c, is replaced by the exponential, we observe that Eq. (6.64) is precisely the form of the diffusion equation used in (VCD) [6.31, 6.59, 6.62]. The constant κ simply incorporates into one place all the constants that are involved.

By Eq. (6.65), we have shown the equivalence of MFA and VCD, provided that MFA is performed without using the noise term, Eq. (6.11). This equivalence provides for the union of two schools of thought concerning image analysis: The first (optimization) school considers the properties that an image *ought* to have. It then sets up an optimization problem whose solution is the desired image. One might also call this the *restoration* school. The second (process) school is more concerned with determining the appropriate spatial analysis to apply. Adaptive filtering, diffusion, template matching, etc. are more concerned with the process itself than with what that process does to some hypothetical "energy function" of the image. The result of this section demonstrates that these two schools are not just philosophically equivalent, at least for this particular form of edge-preserving smoothing, they are *precisely* equivalent.

The above equivalence considered only the prior term of the MFA objective function. Addition of the noise term converts an image feature extraction algorithm into a constrained restoration algorithm.

Nordström [6.59] also observed a similarity between diffusion techniques and regularization (optimization) methods. He notes that "the anisotropic diffusion method (which Whitaker [6.79] calls VCD) does not seek an optimal solution of any kind." This is *not quite true*: "The developer of the technique did not *intend* for it to be used as a minimization technique" is possibly a better way to say it. Nordström then proceeds to "unify from the original outlook quite different methods of regularization and anisotropic diffusion." He proceeds quite elegantly and precisely to define a cost function whose behavior is an anisotropic diffusion, in a similar manner to the derivation presented here. Nordström also argues for the necessity of adding a "stabilizing cost" to "restrict the space of possible estimated image functions." At this point the reader should not be surprised to learn that the form of the stabilized cost is

$$\sum_i (f_i - g_i)^2 \tag{6.66}$$

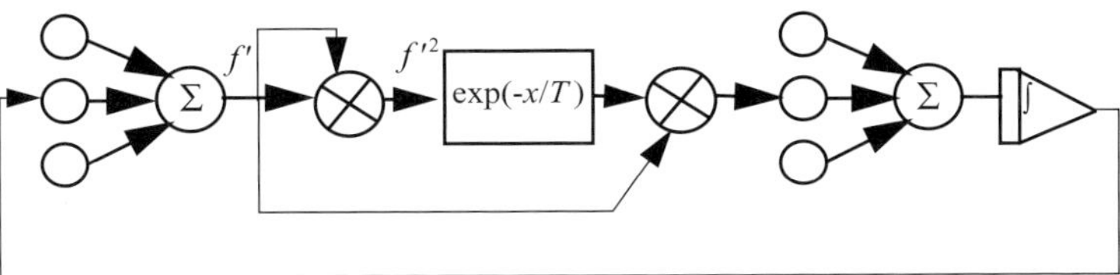

Fig. 6.14. It is straightforward to configure a locally connected, recurrent network which performs the image relaxation networks described here.

which we showed in Eq. (6.11) is a measure of the effect of Gaussian noise in a blur-free imaging system. Thus we see that *biased anisotropic diffusion* (BAD) [6.59] may be thought of as a maximum *a posteriori* restoration of an image. This observation now permits researchers in VCD/BAD to consider use of different stabilizing costs, if they have additional information concerning the noise generation process.

6A.5 Relationship to neural networks

The relationship between a Hopfield neural network and an optimization problem is well-known. Given an objective of the "Ising" type, it is straightforward to find a recurrent neural network whose stable states are minima of the objective functions (see [6.39] as well as [6.40]).

The most straightforward forms of this type of recurrent nets deal with binary-valued variables, with one neuron representing each variable. In this usage, a "neuron" is a sum-of-products operator which produces a weighted sum of its inputs. That sum is then passed through a monotonic nonlinearity – usually a limiter of some sort, e.g., a sigmoid. Following this definition, we may describe their operations represented by Eq. (6.59) as a two-layer network, in which each layer is only locally connected as shown in Fig. 6.14 (see also [6.13]).

The local connectivity is important from an implementation point of view, since that aspect makes construction of parallel, real-time hardware very feasible.

6A.6 Conclusion

MFA combines annealing with gradient descent.

An equivalence exists between the problems of image optimization and diffusion. Others have made similar observations. In addition to Nordström [6.59], Geiger and Yuille came to a similar conclusion [6.26] for an energy function requiring explicit line processes. Since it has been shown [6.38] that the line processes are not required, their result may now be interpreted more generally. In an image optimization problem (in particular, restoration), one defines a criterion function which is to be minimized, and applies some minimization scheme to find a global (or at least a good) minimum. Thus, an image restoration problem may be considered as a combination of goals: (1) to preserve the information in the original image, and thereby produce a resulting image which resembles that original (or the result of an operation on it) in some way; and (2) to produce an image which possesses certain properties, such as local smoothness except at boundaries. If one abandons the first goal, then restoration problems become iconic feature extraction problems. Wu and Doerschuk [6.81] have developed an attractive extension to this work.

Finally, recall that in [6.19], we demonstrated that MFA operations could be calculated by a two-layer, locally connected, recurrent neural network. From this paper one may then conclude that GNC and VCD are likewise implementable by straightforward neural networks.

These results lead us to conjecture the following guiding principles for the design of feature extraction algorithms.

- Relaxation is a central concept. A relaxation algorithm possesses the following properties. (1) It must be iterative. That is, the output of one pass of the algorithm is the same format as the input, so that the algorithm may be applied to its own output. (2) It must converge.[2]
- The relaxation algorithm needs to be local in nature. That is, at any time, changes to any pixel depend only on the local neighborhood of that pixel. Compliance with this guideline allows global interactions to occur smoothly over time (iteration number) and space, and provides a theoretical basis, via the Gibbs/Markov field equivalence for analyzing the operations.
- The equivalence between diffusion and optimization is helpful in understanding the performance of both forms of algorithms. For those designing diffusion algorithms, seeing them as an optimizing relaxation is helpful: Since all relaxation algorithms minimize *something*, it is useful to look at the integral of the diffusion step (although this process is often intractable) to see what properties are actually being minimized in the development/application of an ad hoc technique. For those designing optimization methods, seeing the result as a spatial deformation according to the local degree of nonuniformity followed by an averaging can help to understand the spatiotemporal effects of the optimization.
- In designing feature extraction algorithms, it is helpful to consider the algorithm as a restoration, even if no residual is explicitly minimized, for then, the precise effect of the algorithm on the image may be better comprehended.
- Scale variation in spatial analysis algorithms and temperature control in annealing algorithms are closely related. The power that has been found for both of them is related.
- Finally, the nonlinear operations (exponentials) indicated in all the algorithms mentioned are absolutely essential to success. The Kolmogorov theorem [6.49] demonstrates sufficiency by stating that a linear operation followed by the appropriate nonlinearity allows computation of arbitrary mapping. Here, we claim that such nonlinearity is not only sufficient but necessary. This fact probably contributes most significantly to the recent success of a number of neural network applications.

Bibliography

[6.1] I. Abdelqader, S. Rajala, W. Snyder, and G. Bilbro, "Energy Minimization Approach to Motion Estimation using Mean Field Annealing," *Signal Processing*, July 1992.

[6.2] E. Allgower and K. Georg, *Numerical Continuation Methods*, Berlin, Springer-Verlag, 1990.

[2] Some authors also include a third requirement, "localness," to the definition of a relaxation. We prefer to include this in a separate bullet point.

[6.3] H. Andrews and B. Hunt, *Digital Image Restoration*, Englewood Cliffs, NJ, Prentice-Hall, 1977.

[6.4] D. Baker and J. Aggarwal, "Geometry Guided Segmentation of Outdoor Scenes," *SPIE Applications of Artificial Intelligence*, **VI**, pp. 576–583, 1988.

[6.5] J. Besag, "Spatial Interaction and the Statistical Analysis of Lattice Systems," *Journal of the Royal Statistical Society, B*, **36**, pp. 192–326, 1974.

[6.6] J. Besag, "On the Statistical Analysis of Dirty Pictures," *Journal of the Royal Statistical Society, B*, **48** (3), 1986.

[6.7] G. Bilbro and W. Snyder, "Fusion of Range and Luminance Data," *IEEE Symposium on Intelligent Control*, Arlington, August, 1988.

[6.8] G. Bilbro and W. Snyder, "Range Image Restoration using Mean Field Annealing," In *Advances in Neural Network Information Processing Systems*, San Mateo, CA, Morgan-Kaufmann, 1989.

[6.9] G. Bilbro and W. Snyder, "Mean Field Annealing, an Application to Image Noise Removal," *Journal of Neural Network Computing*, Fall, 1990.

[6.10] G. Bilbro and W. Snyder, "Optimization of Functions with Many Minima," *IEEE Transactions on Systems, Man, and Cybernetics*, **21**(4), July/August, 1991.

[6.11] G. Bilbro, R. Mann, T. Miller, W. Snyder, D. Van den Bout and M. White, "Optimization by Mean Field Annealing," In *Advances in Neural Information Processing Systems*, San Mateo, CA, Morgan-Kauffman, 1989.

[6.12] G. Bilbro, W. Snyder, S. Garnier, and J. Gault, "Mean Field Annealing: a Formalism for Constructing GNC-like Algorithms," *IEEE Transactions on Neural Networks,* **3**(1) pp. 131–138, 1992.

[6.13] G. Bilbro, W. Snyder, and R. Mann, "Mean Field Approximation Minimizes Relative Entropy," *Journal of the Optical Society of America, A*, **8**(2), February 1991.

[6.14] M. Black and A. Jepson, "Estimating Optical Flow in Segmented Images Using Variable-order Parametric Models with Local Deformations," *IEEE Transactions on Pattern Analysis and Machine Intelligence*, **18**(10), 1996.

[6.15] A. Blake, "Comparison of the Efficiency of Deterministic and Stochastic Algorithms for Visual Reconstruction," *IEEE Transactions on Pattern Analysis and Machine Intelligence*, **1**(1), 1989.

[6.16] A. Blake and A. Zisserman, *Visual Reconstruction*, Cambridge, MA, MIT Press, 1987.

[6.17] E. Brezin, D. LeGuillon, and J. Zinn-Justin, "Field Theoretical Approaches to Critical Phenomena," *Phase Transitions and Critical Phenomena*, volume 6, eds. C. Domb and M. Green, New York, Academic Press, 1976.

[6.18] R. Burden, J. Faires, and A. Reynolds, *Numerical Analysis,* Boston, Prindle, 1981.

[6.19] H. Chang and M. Fitzpatrick, "Geometrical Image Transformation to Compensate for MRI Distortions," *SPIE Medical Imaging IV*, **1233**, pp. 116–127, February, 1990.

[6.20] H. Derin and H. Elliot, "Modeling and Segmentation of Noisy and Textured Images using Gibbs Random Fields," *IEEE Transactions on Pattern Analysis and Machine Intelligence*, **9**, pp. 39–55, 1987.

[6.21] H. Ehricke, "Problems and Approaches for Tissue Segmentation in 3D-MR Imaging," *SPIE Medical Imaging IV: Image Processing*, **1233**, pp. 128–137, February, 1990.

[6.22] J. Elder, "Are Edges Incomplete?" *International Journal of Computer Vision*, **34**(2), 1999.

[6.23] J. Elder and R. Goldberg, "Image Editing in the Contour Domain," *IEEE Transactions on Pattern Analysis and Machine Intelligence*, **23**(3), 2001.

[6.24] J. Elder and S. Zucker, "Scale Space Localization, Blur, and Contour-based Image Coding," *IEEE Conference on Computer Vision and Pattern Recognition*, San Francisco, CA, 1996.

[6.25] D. Geiger and F. Girosi, "Parallel and Deterministic Algorithms for MRFS: Surface Reconstruction and Integration," *AI Memo*, No 1114, Cambridge, MA, MIT, 1989.

[6.26] D. Geiger and A. Yuille, "A Common Framework for Image Segmentation by Energy Functions and Nonlinear Diffusion," *MIT AI Lab Report*, Cambridge, MA, 1989.

[6.27] D. Geman and S. Geman, "Stochastic Relaxation, Gibbs Distributions, and the Bayesian Restoration of Images," *IEEE Transactions on Pattern Analysis and Machine Intelligence*, **6**(6), November, 1984.

[6.28] R. Gonzalez and P. Wintz, *Digital Image Processing*, 2nd edn, Reading, MA, Addison-Wesley, 1987.

[6.29] A. Gray, J. Kay, and D. Titterington, "An Empirical Study of the Simulation of Various Models used for Images," *IEEE Transactions on Pattern Analysis and Machine Intelligence*, **16**(5), 1994.

[6.30] B. Groshong, G. Bilbro, and W. Snyder, "Restoration of Eddy Current Images by Constrained Gradient Descent," *Journal of Nondestructive Evaluation*, December, 1991.

[6.31] S. Grossberg, "Neural Dynamics of Brightness Perception: Features, Boundaries, Diffusion, and Resonance," *Perception and Psychophysics*, **36**(5), pp. 428–456, 1984.

[6.32] J. Hadamard, *Lectures on the Cauchy Problem in Linear Partial Differential Equations*, New Haven, CT, Yale University Press, 1923.

[6.33] J. Hammersley and P. Clifford, "Markov Field on Finite Graphs and Lattices," unpublished.

[6.34] F. Hansen and H. Elliot, "Image Segmentation using Simple Markov Field Models," *Computer Graphics and Image Processing*, **20**, pp. 101–132, 1982.

[6.35] R. Haralick and G. Shapiro, "Image Segmentation Techniques," *Computer Vision, Graphics, and Image Processing*, **29**, pp. 100–132, 1985.

[6.36] E. Hensel, *Inverse Theory and Applications for Engineers*, Englewood Cliffs, NJ, Prentice-Hall, 1991.

[6.37] H. Hiriyannaiah, *Signal Reconstruction using Mean Field Annealing.* Ph.D. Thesis, North Carolina State University, Raleigh, NC, 1990.

[6.38] H. Hiriyannaiah, G. Bilbro, W. Snyder, and R. Mann, "Restoration of Locally Homogeneous Images using Mean Field Annealing," *Journal of the Optical Society of America A*, **6**, pp. 1901–1912, December, 1989.

[6.39] J. Hopfield, "Neurons with Graded Response Have Collective Computational Properties Like Those of Two-state Neurons," *Proceedings of the National Academy of Science* USA, **81**, pp. 3058–3092.

[6.40] J. Hopfield, and D. Tank, "Neural Computations of Decisions in Optimization Problems," *Biological Cybernetics*, **52**, pp. 141–152, 1985.

[6.41] M. Irani, B. Rousso, and S. Peleg, "Recovery of Ego-motion Using Region Alignment," *IEEE Transactions on Pattern Analysis and Machine Intelligence*, **19**(3), 1997.

[6.42] A. Kak and M. Slaney, *Principles of Computerized Tomographic Imaging*, New York, IEEE Press, 1988.

[6.43] S. Kapoor, P. Mundkur, and U. Desai, "Depth and Image Recovery using a MRF Model," *IEEE Transactions on Pattern Analysis and Machine Intelligence*, 16(11), 1994.

[6.44] I. Kapouleas and C. Kulikowski, "A Model-based System for Interpretation of MR Human Brain Scans," *Proceedings of the SPIE, Medical Imaging* II, vol. 914, February, 1988.

[6.45] R. Kashyap and R. Chellappa, "Estimation and Choice of Neighbors in Spatial-interaction Model of Images," *IEEE Transactions on Information Theory*, **29**, pp. 60–72, January, 1983.

[6.46] M. Kelly, In *Machine Intelligence*, vol 6, Edinburgh, University of Edinburgh Press, 1971.

[6.47] R. Kindermann and J. Snell, *Markov Random Fields and Their Applications*, Providence, RI, American Mathematical Society, 1980.

[6.48] S. Kirkpatrick, Gelatt C, and Vecchi M, "Optimization by Simulated Annealing," *Science*, **220**, pp. 671–668, 1983.

[6.49] A. Kolmogorov, "On the Representation of Continuous Functions of One Variable by Superposition of Continuous Functions of One Variable and Addition," *AMS Translation*, **2**, pp. 55–59, 1957.

[6.50] R. Lee, and R. Leahy, "Multi-spectral Tissue Classification of MR Images Using Sensor Fusion Approaches," *SPIE Medical Imaging IV: Image Processing*, **1233**, pp. 149–157, February, 1990.

[6.51] S. Li, "On Discontinuity-adaptive Smoothness Priors in Computer Vision," *IEEE Transactions on Pattern Analysis and Machine Intelligence*, **17**(6), 1995.

[6.52] R. Malik and T. Whangbo, "Angle Densities and Recognition of 3D Objects," *IEEE Transactions on Pattern Analysis and Machine Intelligence*, **19**(1), 1997.

[6.53] D. Marr, *Vision*, San Francisco, CA, Freeman, 1982.

[6.54] J. Marroquin, *Probabilistic Solution to Inverse Problems*, Doctoral Dissertation, MIT, 1985.

[6.55] P. Morris, in *Nuclear Magnetic Resonance Imaging in Medicine and Biology*, Oxford, Clarendon Press, 1986.

[6.56] J. Moussouris, "Gibbs and Markov Systems with Constraints," *Journal of Statistical Physics*, (10), pp. 11–33, 1974.

[6.57] M. Nashed, "Aspects of Generalized Inverses in Analysis and Regularization," in *Generalized Inverses and Applications*, ed. by M. Nashed, New York, Academic Press, 1976.

[6.58] T. Nelson, "Propagation Characteristics of a Fractal Network: Applications to the His-Purkinje Conduction System," *SPIE Medical Imaging IV: Image Processing*, **1233**, pp. 23–32, February, 1990.

[6.59] N. Nordström, "Biased Anisotropic Diffusion – A Unified Regularization and Diffusion Approach to Edge Detection," *Image and Vision Computing*, **8**(4), pp. 318–327, 1990.

[6.60] T. Pavlidis, *Structural Pattern Recognition*, Berlin, Springer-Verlag, 1977.

[6.61] A. Pentland, "Interpolation using Wavelet Bases," *IEEE Transactions on Pattern Analysis and Machine Intelligence*, **16**(4), 1994.

[6.62] P. Perona and J. Malik, "Scale-space and Edge Detection using Anisotropic Diffusion," *IEEE Transactions on Pattern Analysis and Machine Intelligence*, **12**, pp. 629–639, July, 1990.

[6.63] H. Qi, *A High-resolution, Large-area Digital Imaging System*, Ph.D. Thesis, North Carolina State University, 1999.

[6.64] A. Rosenfeld and A. Kak, *Digital Picture Processing,* 2nd edn, Vol. 2, New York, Academic Press, 1982.

[6.65] Samet, *The Design and Analysis of Spacial Data Structures*, Reading, MA, Addison-Wesley, 1989.

[6.66] P. Santago, K. Link, W. Snyder, J. Worley, S. Rajala, and Y. Han, "Restoration of Cardiac Magnetic Resonance Images," *Symposium on Computer Based Medical Systems*, Chapel Hill, NC, June 3–6, 1990.

[6.67] S. Shemlon and S. Dunn, "Rule-based Interpretation with Models of Expected Structure," *SPIE Medical Imaging IV*, **1233**, pp. 33–44, February, 1990.

[6.68] W. Snyder, G. Bilbro, A. Logenthiran, and S. Rajala, "Optimal Thresholding – A New Approach," *Pattern Recognition Letters*, **11**(12), December, 1990.

[6.69] W. Snyder, P. Santago, A. Logenthiran, K. Link, G. Bilbro, and S. Rajala, "Segmentation of Magnetic Resonance Images using Mean Field Annealing," *XII International Conference on Information Processing in Medical Imaging*, Kent, England, July 7–11, 1991.

[6.70] W. Snyder, A. Logenthiran, P. Santago, K. Link, G. Bilbro, and S. Rajala, "Segmentation of Magnetic Resonance Images using Mean Field Annealing," *Image and Vision Computing*, **10**(6), pp. 361–368, 1992.

[6.71] C. Soukoulis, K. Levin, and G. Grest, "Irreversibility and Metastability in Spin-glasses. I. Ising Model," *Physical Review B*, **28**(3), pp. 1495–1509, 1983.

[6.72] B. Super and W. Klarquist, "Patch-based Stereo in a General Binocular Viewing Geometry," *IEEE Transactions on Pattern Analysis and Machine Intelligence*, **19**(3), 1997.

[6.73] B. ter Haar Romeny, L. Florack, J. Koenderink, and M. Viergever, "Scale Space: Its Natural Operators and Differential Invariants," *XII International Conference on Information Processing in Medical Imaging*, Kent, England, July 7–11, 1991.

[6.74] P. Torr, R. Szeliski, and P. Anandan, "An Integrated Bayesian Approach to Layer Extraction from Image Sequences," *IEEE Transactions on Pattern Analysis and Machine Intelligence*, **23**(3), 2001.

[6.75] J. van Laarhoven and E. Aarts, *Simulated Annealing: Theory and Applications*, Norwell, MA, Reidel, 1988.

[6.76] M. Vannier, *et al.*, "Multispectral Analysis of Magnetic Resonance Images," *Radiology*, **154**, pp. 221–224, January, 1985.

[6.77] D. Van den Bout and T. Miller, "Graph Partitioning using Annealed Neural Networks," *IEEE Transactions on Neural Networks*, **1**(2), pp. 192–203, 1990.

[6.78] C. Wang, W. Snyder, and G. Bilbro, "Optimal Interpolation of Images," *Neural Networks for Computing Conference*, Snowbird, UT, April, 1995.

[6.79] R. Whitaker, "Geometry-limited Diffusion in the Characterization of Geometric Patches in Images," TR91-039, Dept. of Computer Science, UNC, Chapel Hill, NC, 1991.

[6.80] G. Wolberg and T. Pavlidis, "Restoration of Binary Images Using Stochastic Relaxation With Annealing," *Pattern Recognition Letters*, **3**(6), pp. 375–388, December, 1985.

[6.81] C. Wu and P. Doerschuk, "Cluster Expansions for the Deterministic Computation of Bayesian Estimators Based on Markov Random Fields," *IEEE Transactions on Pattern Analysis and Machine Intelligence*, **17**(3), 1975.

[6.82] M. Yaou and W. Chang, "Fast Surface Interpolation Using Multiresolution Wavelet Transform," *IEEE Transactions on Pattern Analysis and Machine Intelligence*, **16**(7), 1994.

[6.83] J. Yi and D. Chelberg, "Discontinuity-preserving and Viewpoint Invariant Reconstruction of Visible Surfaces Using a First-order Regularization," *IEEE Transactions on Pattern Analysis and Machine Intelligence*, **17**(6), 1995.

7 Mathematical morphology

A man's discourse is like to a rich Persian carpet, the beautiful figures and patterns of which can be shown only by spreading and extending it out; when it is contracted and folded up, they are obscured and lost

Plutarch

The suffix "-ology" means "study of-," so obviously, "morphology" is the study of morphs; answering critical questions like: "How come they only come out at night, and then fly toward the light?" and "Why is it that bug zappers only toast the harmless critters, leaving the 'skeeters alone?" and – HOLD IT! That's MORPH-ology, the study of SHAPE, not moths! Try again . . .

7.1 Binary morphology

We begin by considering ONLY BINARY images. That's important, remember it! Only binary! We will discuss a couple of operators first. Then, once you understand how they work, we'll explain how they are used in binary images. As an extension to binary morphology, we also describe gray scale morphology operations and the corresponding operators.

7.1.1 Dilation

First, the intuitive definition: The dilation of a (BINARY) image is that same image with all the foreground regions made just a little bit bigger.

Now, formally: We consider two images, f_A and f_B, and let A and B be sets of ordered pairs, consisting of the coordinates of each foreground pixel in f_A and f_B, respectively.

Consider one pixel in f_B, and its corresponding element (ordered pair) of B, call that element $b \in B$. Create a new set by adding the ordered pair b to EVERY ordered pair in A. Let's look at a tiny example.

For this image, $A = \{(2, 8), (3, 6), (4, 4), (5, 6), (6, 4), (7, 6), (8, 8)\}$. Adding the pair $(-1, 1)$ results in the set $A_{(-1,1)} = \{(1, 9), (2, 7), (3, 5), (4, 7), (5, 5), (6, 7), (7, 9)\}$. The corresponding image is also shown in Fig. 7.1, and we hope you

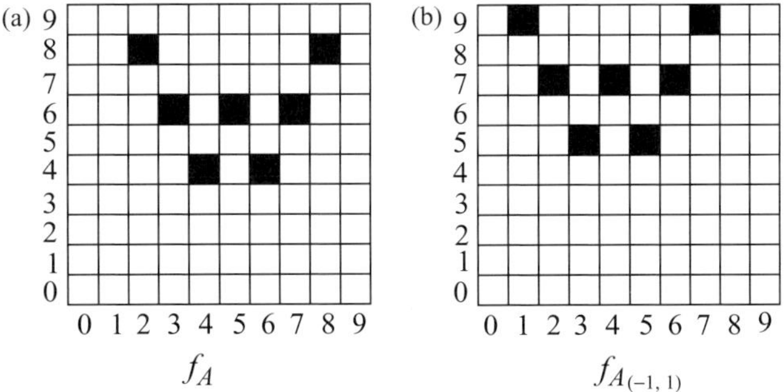

Fig. 7.1. **Example of dilation. (a) The original binary image. (b) The binary image dilated by $B = \{(-1, 1)\}$.**

observed that $A_{(-1,1)}$ is nothing more than a translation of A. With this concept firmly understood, think about what would happen if you constructed a SET of translations of A, one for each pair in B. We denote that as $\{A_b, b \in B\}$, that is, b is one of the ordered pairs in B.

Formally, we define the DILATION of A by B as $A \oplus B = \{a + b | (a \in A, b \in B)\}$, which is the same as the union of all those translations of A,

$$A \oplus B = \bigcup_{b \in B} A_b \tag{7.1}$$

and we use the same symbol to denote the dilation of images: $f_A \oplus f_B$. Here is another example.

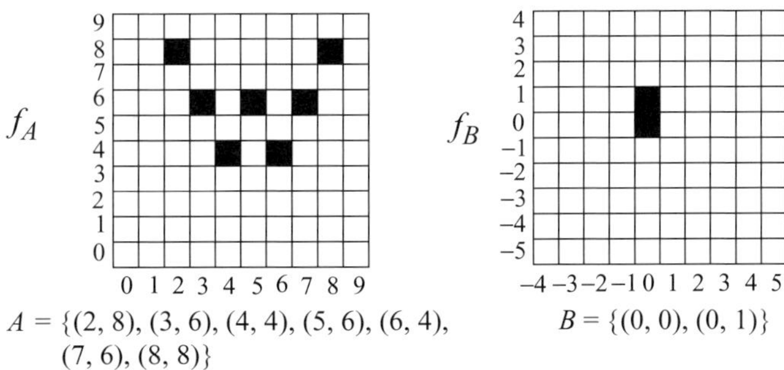

Based on the definition,

$$A \oplus B = A_{(0,0)} \cup A_{(0,1)} \tag{7.2}$$

$$= \{(2, 8), (2, 9), (3, 6), (3, 7), (4, 4), (4, 5), (5, 6), (5, 7), (6, 4), (6, 5), (7, 6), (7, 7), (8, 8), (8, 9)\} \tag{7.3}$$

and

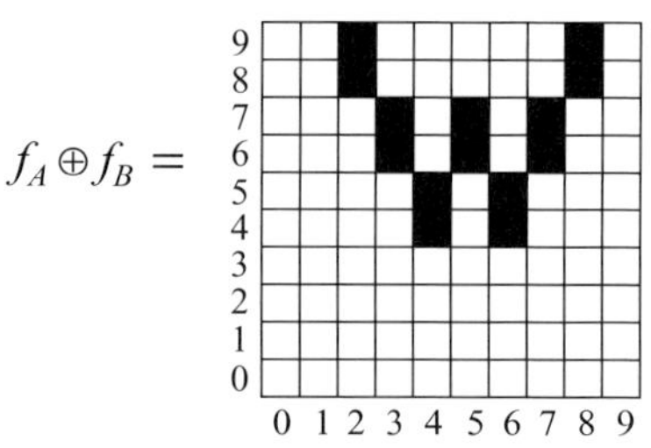

It's time to start getting used to set notation.

Denote by #*A* the number of elements in set *A*. In this example, $\#A = 7$, and $\#(A \oplus B) = 14$. This happened to be true, only coincidentally, because there was no overlap between $A(0, 0)$ and $A(0, 1)$, or said another way: $A_{(0,0)} \cap A_{(0,1)} = \emptyset$.

For a more general problem, this will not be the case. In general,

$$\#(A \oplus B) \leq \#A \cdot \#B. \tag{7.4}$$

To go further, we need to define some notation: If x is an ordered pair, then (1) the *translation*[1] of a set A by x is A_x, (2) the *reflection* of A is $\tilde{A} = \{(-\mathrm{x}, -\mathrm{y}|(x, y) \in A)\}$ and (3) the complement of set A is A_c. An example of reflection is

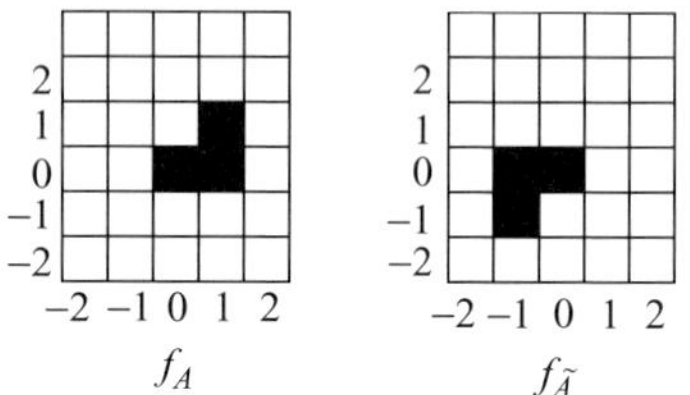

In this example, $A = \{(0, 0), (1, 0), (1, 1)\}$ and the reflection of A is $\{(0, 0), (-1, 0), (-1, -1)\}$.

7.1.2 Erosion

Now, we define the (sort of) inverse of dilation, erosion,

$$A \ominus B = \{a | (a + b) \in A \text{ for every } (a \in A, b \in B)\} \tag{7.5}$$

which can be written in terms of translations by

$$A \ominus B = \bigcap_{b \in \tilde{B}} A_b. \tag{7.6}$$

[1] We do not have to be limited to a 2-space. As long as x and A are drawn from the same space, it works. More generally, if A and B are sets in a space ε, and $x \in \varepsilon$, the translation of A by x is denoted $A_x = \{y|$ for some $a \in A, a = a + x\}$.

Notice two things: The second set, B, is reflected, and the intersection symbol is used. Let's do an example.

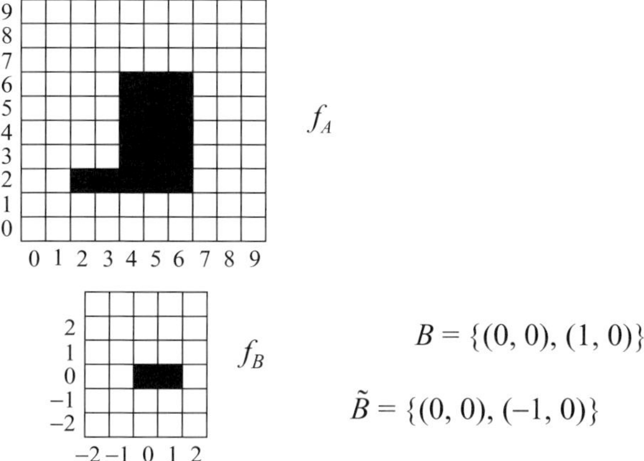

Rather than the tedious job of listing all the 17 elements of A, just draw

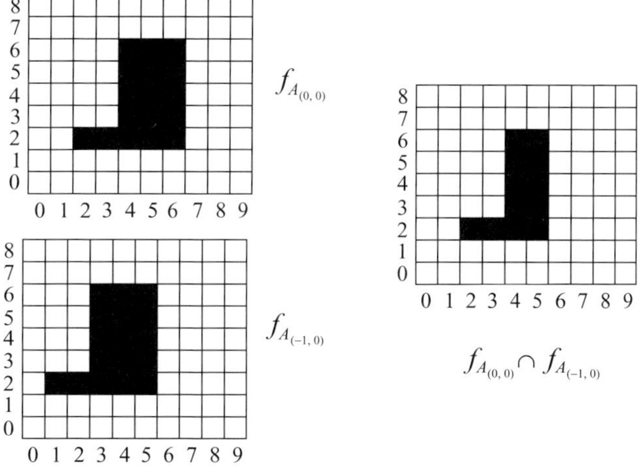

So now we have dilation and erosion defined. You will observe that usually (for all practical purposes) one of the images is "small" compared to the other; that is, in the example above

$$\#A \gg \#B. \tag{7.7}$$

When this is the case, we refer to the smaller image, f_B, as the "structuring element" (s.e.).

7.1.3 Properties of dilation and erosion

- *Commutative property*. Dilation is commutative, which means we can dilate images in any order without changing the result

$$A \oplus B = B \oplus A. \tag{7.8}$$

- *Associative property*. Dilation is also associative, which means we can group images in a dilation in any way we want without changing the result

$$(A \oplus B) \oplus C = A \oplus (B \oplus C). \tag{7.9}$$

- *Distributive property*. This property says that in an expression where we need to dilate an image with the union of two images, we can dilate first and then take the union. In other words, the dilation can be distributed over all the terms inside the parentheses

$$A \oplus (B \cup C) = (A \oplus B) \cup (A \oplus C). \tag{7.10}$$

- *Increasing property*. If $A \subseteq B$, then

$$A \oplus K \subseteq B \oplus K, \tag{7.11}$$

for any s.e. K. When this property holds, we say the operator is "increasing."

An example proof: Dilation is increasing

Let the set A consist of elements A_i. $A = \{A_1, A_2, \ldots, A_n\}$, and let B be similarly denoted. Furthermore, suppose $B \subseteq A$. Now, suppose both A and B are dilated by the same s.e., K. Take a single element of K, say K_1, and dilate each element of A by K_1, $A \oplus K_1 = \{A_1 + K_1, A_2 + K_1, \ldots, A_n + K_1\}$, and similarly dilate B. Since every element of B was also an element of A, every element of $B \oplus K_i$ is in $A \oplus K_i$. Since this observation is true for an arbitrary element of K, it is true for all elements of K. Now consider the union of the results of applying two elements of the s.e. K to A: $A_{12} = (A \oplus K_1) \cup (A \oplus K_2)$. Since $B \oplus K_1 \subseteq A \oplus K_1$ and $B \oplus K_2 \subseteq A \oplus K_2$, we know from set theory that if $R_1 \subseteq S_1$ and $R_2 \subseteq S_2$ then $R_1 \cup R_2 \subseteq S_1 \cup S_2$ and we are done.

- *Extensive and anti-extensive properties*. If we say an operator is "extensive," we mean that applying this operator to a set, A, produces an answer which contains A. If the s.e. contains the origin (that is, element (0, 0)), dilation is extensive:

$$A \oplus K \supseteq A. \tag{7.12}$$

As you might guess, erosion has some extensive properties as well: That is, erosion is "anti-extensive," if $o \in B$, then $A \ominus B \subseteq A$, where o indicates the element at the origin.

- *Duality.* Duality is similar to DeMorgan's laws: It relates set complement, dilation, and erosion.

$$\begin{aligned}(A \ominus B)^c &= A^c \oplus \tilde{B} \\ (A \oplus B)^c &= A^c \ominus \tilde{B}\end{aligned} \tag{7.13}$$

where the superscript c denotes set complement.

- *Other properties of erosion.*

$$A \ominus (B \oplus C) = (A \ominus B) \ominus C \tag{7.14}$$

$$(A \cup B) \ominus C \supseteq (A \ominus C) \cup (B \ominus C) \tag{7.15}$$

$$A \ominus (B \cap C) \supseteq (A \ominus B) \cup (A \ominus C) \tag{7.16}$$

$$A \ominus (B \cup C) = (A \ominus B) \cap (A \ominus C). \tag{7.17}$$

Erosion and dilation are not inverses of each other.

A cautionary note: You cannot do cancellation with morphological operators as you might suspect. For example, if $A = B \ominus C$, we could dilate both sides by C to get $A \oplus C = (B \ominus C) \oplus C$ and if dilation and erosion were true inverses, the RHS would be just B. However, the RHS is in fact the opening of B by C and not simply B.

7.1.4 Opening and closing

The OPENING of f_A by an s.e. f_B is written as

$$f_A o f_B = (f_A \ominus f_B) \oplus f_B \tag{7.18}$$

and, as you might guess, the CLOSING of f_A by an s.e. f_B is written as

$$f_A \bullet f_B = (f_A \oplus f_B) \ominus f_B. \tag{7.19}$$

An application

So what is the purpose of all this? Let's do an example: Inspection of printed circuit (PC) boards. Here's a picture of a PC board with two traces on it shorted together by a hair which was stuck to the board when it went through the wave solder machine. We will use opening to identify the short.

First, erode the image using a small s.e. We choose an s.e. which is smaller than the features of interest (the traces), but larger than the defect. The erosion then looks like this:

Now, we dilate back using the same s.e.

and surprise, surprise, the defect is gone. For inspection purposes, one could now subtract the original from the opened, and the difference image would contain only the defect. Furthermore these operations can be done in hardware, blindingly fast.

Another way to think of opening

The opening of f_A by f_K selects precisely those points in f_A that "match" f_K in the following sense: Take the structuring element f_K and place it so that *every* foreground pixel in f_K covers a foreground pixel in f_A. If you can find a position where the s.e. can be located to make this true, then every pixel of f_A which is covered is in the opening. Now, move f_K around (by translations only), and find every such position – every place the s.e. can be placed where it is totally inside the image. The set of covered pixels determines the opening of f_A by f_K. Here's an example of opening in 1D, viewed from this perspective (from Haralick and Shapiro [7.12]). Let A be the open interval

$$A = (3.1, 7.4) \cup (11.5, 11.6) \cup (18.9, 19.8) \tag{7.20}$$

and open it with the s.e. $K = (-1, 1)$. Then

$$AoK = (3.1, 7.4). \tag{7.21}$$

This example illustrates first of all that morphological concepts can be extended to a continuous domain. (For the time being, remember, however, that this is continuous in resolution, not in brightness value; still binary. We will fix that soon.) Second, it illustrates the fact that opening preserves exactly the geometry of objects which are "big enough," and totally erases smaller objects. In this sense, opening resembles the functioning of the median filter, where each pixel is replaced by the median of its neighbors.

7.1.5 Properties of opening and closing

Some properties of opening and closing are listed below (some you should be prepared to figure out).

- Duality: $(AoK)^c = A^c \bullet \tilde{K}$.
  Proof of duality. Notice how this proof is done. We expect students to do proofs this carefully.
  1. $(AoK)^c = [(A \ominus K) \oplus K]^c$ definition of opening
  2. $= (A \ominus K)^c \ominus \tilde{K}$ complement of dilation
  3. $= (A^c \oplus K) \ominus \tilde{K}$ complement of erosion
  4. $= A^c \bullet \tilde{K}$ definition of closing.

Can you prove this? Is it even true?

- Idempotency: Opening and closing are idempotent. That is, repetitions of the same operation have no further effect:

$$AoK = (AoK)oK$$
$$(A \bullet K) = (A \bullet K) \bullet K .$$

- Closing is extensive: $A \bullet K \supseteq A$.
- Opening is anti-extensive: $AoK \subseteq A$.
- Images dilated by f_K remain invariant under opening by f_K. That is, $f_A \oplus f_K = (f_A \oplus f_K) o f_K$.
  Proof
  1. $A \bullet K \supseteq A$ since closing is extensive
  2. $(A \bullet K) \oplus K \supseteq A \oplus K$ dilation is increasing
  3. $((A \oplus K) \ominus K) \oplus K \supseteq A \oplus K$ definition of closing
  4. $(A \oplus K)oK \supseteq A \oplus K$ definition of opening
  5. for any B, $BoK \subseteq B$ opening is anti-extensive
  6. therefore, $(A \oplus K)oK \subseteq A \oplus K$ substitution of $A \oplus K$ for B
  7. $(A \oplus K)oK = A \oplus K$ since $A \oplus K$ is both greater or equal to and less or equal to $(A \oplus K)oK$, so only equality can be true.

Computing opening and closing quickly

Any translation-invariant increasing operation (Ψ) such as opening or closing can be computed by a union operation of the form

$$\Psi(A) = \bigcup_i A \ominus K_i \tag{7.22}$$

for some set of structuring elements $K = \{K_1, K_2, \ldots\}$, where the set K is said to be a *basis* set for this operation. The erosions could all be done in parallel and the union performed very quickly using lookup table methods. More details are available in [7.17].

7.2 Gray-scale morphology

Up to now, we have assumed that the images with which we are dealing are binary. Now, we relax that requirement and allow f_A to take on continuous values in a finite range $fmin_A \leq f_A \leq fmax_A$. With this extension, we no longer have a simple definition for such operations as dilation, since the union operator is not defined. To define gray-scale morphological operations, we need first to define a new concept, the "umbra."

The **umbra**, $U(f_A)$ of a two-dimensional gray-scale image f_A is the set of all ordered triples, (x, y, U), which satisfy $0 < U \leq f_A(x, y)$. If we think of f_A as being continuously valued, then $U(f_A)$ is an infinite set. To make morphological operations feasible, we assume f_A is quantized to M values,

$$\#U(f_A) \leq M \cdot \#A \tag{7.23}$$

To illustrate the concept of umbra, let f_A be one dimensional. (We illustrate a one-dimensional function here as a two-dimensional function in which one dimension is always zero. That way, you have an example that is easily extended to two dimensions):

$$A = \{(0, 0), (1, 0), (2, 0), (3, 0), (4, 0), (5, 0), (6, 0)\}$$

and the pixel value at the corresponding coordinate is

$$f_A(x, 0) = [1,\ 2,\ 3,\ 1,\ 2,\ 3,\ 3].$$

Notice the new notation: Since f_A takes on various values, depending on which element of A is being considered, we use functional notation. Drawing f_A, we have Fig. 7.2.

In Fig. 7.2, the heavy black line represents f_A, and the umbra is the shaded area under f_A. Following this figure, the heavy black line is the TOP of the umbra. We could write the umbra as a set of ordered triples:

$$U(f_A) = \{(0, 0, 1), (1, 0, 1), (1, 0, 2), (2, 0, 1), (2, 0, 2), (2, 0, 3), (3, 0, 1),$$
$$(4, 0, 1), (4, 0, 2), (5, 0, 1), (5, 0, 2), (5, 0, 3), (6, 0, 1), (6, 0, 2), (6, 0, 3)\}.$$

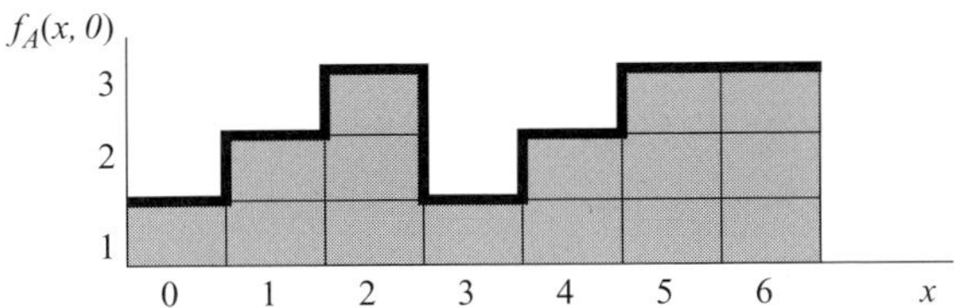

Fig. 7.2. An illustration of umbra.

Here's the trick: Although the gray-level image is no longer binary (and therefore not representable by set membership) *the umbra does have those properties.*

We may therefore define the dilation of a gray-scale image f_A by a gray-scale s.e., f_B as

$$f_A(x, y) \oplus f_B(x, y) = \mathrm{TOP}(U(f_A) \oplus U(f_B)) \tag{7.24}$$

and erosion is similarly defined. Furthermore gray-scale opening and closing can be defined in terms of gray-scale dilation and erosion.

Generalizing this concept to two-dimensional images, the umbra becomes three dimensional, a set of triples

$$U(f(x, y)) = \{(x, y, z) | (z \leq f(x, y))\}. \tag{7.25}$$

Then, gray-scale dilation and erosion may be written compactly as

$$g(x, y) \oplus h(x, y) = \{(x, y, z) | (z \leq \max(g(x - x_1, y - y_1) + h(x_1, y_1)))\} \quad \forall (x_1, y_1)$$
$$g(x, y) \ominus h(x, y) = \{(x, y, z) | (z \leq \min(g(x - x_1, y - y_1) - h(x_1, y_1)))\} \quad \forall (x_1, y_1)$$

for $(x_1, y_1) \in \Omega \subset Z \times Z$, where Ω denotes the set of possible pixel locations, assumed here to be positive and integer.

7.3 The distance transform

A very important application of morphological operations is to derive the distance transform (DT). The distance transform may be defined in several ways. We will give one definition in section 7A.4, but we will present another, simpler one here. This is an iconic representation of an image, in which each pixel in the DT contains the distance of the corresponding pixel in the input image from some feature. Most often, the feature is an edge. In this section, we consider the set of points ∂R, denoting the boundary of region R. Fig. 7.3 illustrates the DT of the interior of a region.

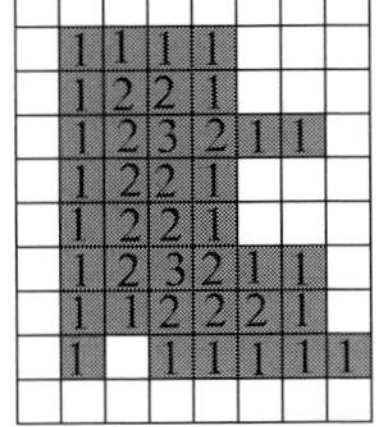

Fig. 7.3. Pixels interior to the region are shaded. The boundary is assumed to lie just outside this region. The DT of the region, computed using a 4-connected definition, is shown.

Formally, the DT is described by

$$DT(x) = \min_{y \in \partial R} \|x - y\| \tag{7.26}$$

where x and y are 2-vectors of coordinates. This transform is a solution to the differential equation

$$\|\nabla DT(x)\| = 1 \tag{7.27}$$

when the initial conditions are $DT(x) = 0$ for $x \in \partial R$.

One may compute the DT using conventional morphological operations: Let us suppose we want the distance from the outside edge of an object. Repetitively erode the image, using some "appropriate" structuring element. Each time a pixel disappears, record the iteration at which that pixel vanished. Store the iteration number in the corresponding pixel as the DT. Pretty simple, isn't it? That definition does not quite give the Euclidian distance. Huang and Mitchell [7.15] show how to obtain the Euclidian distance transform using gray-scale morphology, and Breu *et al.* [7.6] show how to compute the Euclidian DT in linear time. One interesting question in this kind of iterative erosion is "where do the points go?" It turns out for a structuring element which is strictly convex, the points go in the direction of the normal to the boundary of the region being eroded [7.43].

7.3.1 Using a mask to compute the DT

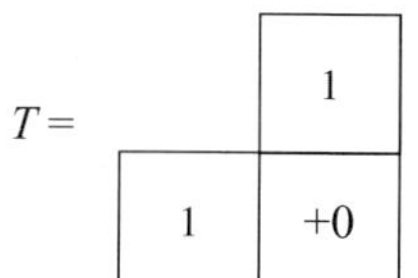

Fig. 7.4. Mask used to calculate the 4-connected DT. The origin of the mask is denoted by a plus sign.

Computing the distance transform may be accomplished by iterative application of a mask, such as the one illustrated in Fig. 7.4.

At iteration m, the distance transform is updated using the equation:

$$D^m(x, y) = \min_{k,l \in T}(D^{m-1}(x + k, y + l) + T(k, l)). \tag{7.28}$$

A more detailed explanation follows. First, the distance transform, $D(x, y)$, is initialized to a special symbol denoting "infinity" at every nonedge point, $D^0(x, y) = \infty, \forall(x, y) \notin \partial R$, and to zero at every edge point $D^0(x, y) = 0, \forall(x, y) \in \partial R$. Then, application of the mask starts at the upper left corner of the image, placing the origin of the mask over the (1, 1) pixel of the image, and applying Eq. (7.28) to calculate a new value for the DT at (1, 1). In the example of Fig. 7.5, the DT is illustrated with infinities denoted by blank squares, and edges denoted by zeros. The mask of Fig. 7.4 is applied in the shaded area. The application of Eq. (7.28) produces $\min(1 + 0, 1 + \infty)$ for the distance transform value in the pixel indicated by the arrow.

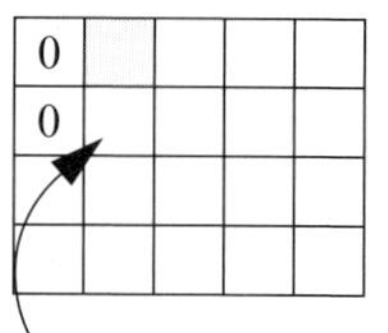

Fig. 7.5. Calculation of the distance transform value at one particular point. The result is 1.

After one pass, top-to-bottom, left-to-right, the mask is reversed (in both directions), and applied again, bottom-to-top, right-to-left.

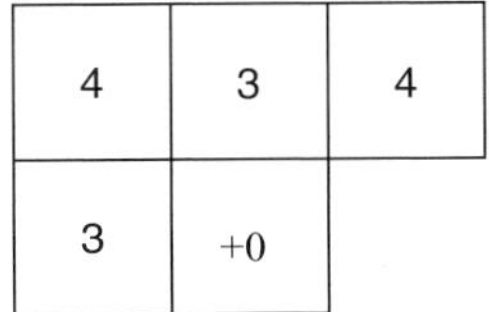

Fig. 7.6. A mask whose application produces the chamfer map.

This process is repeated at each pixel until all pixels in the image have been processed, and then iterated until all pixels in the DT are marked with a finite value. Masks other than that of Fig. 7.4 produce other variations of the DT. In particular, Fig. 7.6 produces the chamfer map. If divided by three, the chamfer map produces a DT which is not a bad approximation to the Euclidian distance.

7.3.2 The Voronoi diagram

In later sections of this book, we will be concerned with the junctions between regions, and the relationships of adjacent regions. We will occasionally need to consider relationships between regions which do not actually touch. For this, the concept of the Voronoi diagram will be useful. We introduce it here because of the similarities it shares with the distance transform.

Consider the image illustrated in Fig. 7.7. In that image, several regions are indicated in gray, inside the white circle. For any region, i, the Voronoi domain of that region is the set of points such that those points are closer to points in that region than to points in any other region:

$$V_i = \{x | d(x, P_i) < d(x, P_j), \forall (j \neq i)\} \tag{7.29}$$

where P_i denotes any point in region i.

The set of points equidistant from two of those regions is referred to as the Voronoi diagram, and is illustrated by the dark lines in the figure. This is equivalent to saying

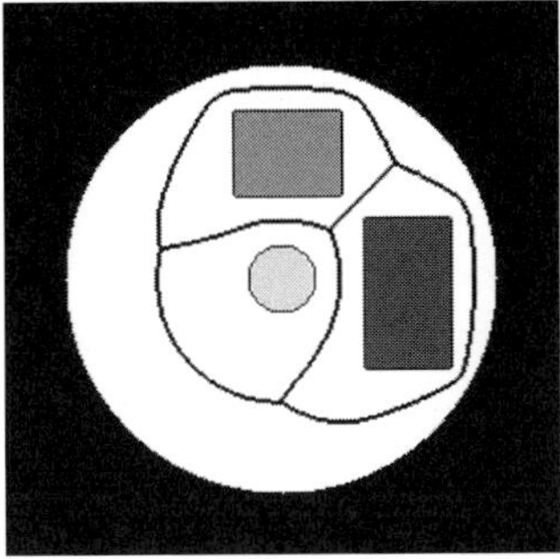

Fig. 7.7. Several regions and the resulting Voronoi diagram.

that points in the Voronoi diagram do not belong to the Voronoi domain of any region.

7.4 Conclusion

In this chapter, we have looked at a particular approach to processing the shape of regions. Morphological operators are particularly useful in binary images, but may be applied to gray-scale images as well. Unlike most chapters in this book, we did not make explicit use of either optimization methods or consistency.

7.5 Vocabulary

You should know the meanings of the following terms.

Closing
Dilation
Distance transform
Erosion
Extensive
Increasing
Opening
Umbra
Voronoi diagram

Assignment 7.1

In section 7.1.3, we stated that dilation is commutative because addition is commutative. Erosion also involves addition, but one of the two images is reversed. Is erosion commutative? Prove or disprove it.

Assignment 7.2

In section 7.3, a mask is given and it is stated that application of this mask produces a distance transform which is "not a bad approximation" to the Euclidian distance from the interior point to the nearest edge point. How bad is it? Contrive an example where the value produced by the application of the mask is different from the Euclidian distance to the nearest edge point.

Assignment 7.3

Consider a region with an area of 500 pixels with 120 pixels on the boundary. You need to find the distance transform from each pixel in the interior to the boundary, using the Euclidian distance. (Note: Pixels ON the boundary are not considered IN the region — at least in this problem.) What is the computational complexity? Note: You may come up with some algorithm more clever than that used to produce any of the answers below, so if your algorithm does not produce one of these answers, explain it.

(a) 60 000
(b) 120 000
(c) 45 600
(d) 91 200

Assignment 7.4

Trick question: For Problem Assignment 7.3, how many square roots must you calculate to determine this distance transform? Remember, this is the Euclidian distance.

Assignment 7.5

Prove (or disprove) that binary images eroded by a kernel, K, remain invariant under closing by K. That is, prove that $A \Theta K = (A \Theta K) \bullet K$.

Assignment 7.6

Show that dilation is not necessarily extensive if the origin is not in the s.e.

Assignment 7.7

Prove that dilation is increasing.

Assignment 7.8

Let C be the class of binary images that have only **one** dark pixel. For a particular image, let that pixel be located at (i_0, j_0).

Using erosion and dilation by kernels that have $\{(0, 0)\}$ as an element, devise an operator, that is, a set of erosions and dilations, and structuring elements (you may need only one s.e. or you may need more than one), which, when applied to an element of C, would output an image with the dark pixel shifted to (i_0+2, j_0+1) (disregard boundaries).

Assignment 7.9

Which of the following statements is correct? (You should be able to reason through this, without doing a proof.)

$$\text{(a)}\quad (A \ominus B) \ominus C = A \ominus (B \ominus C) \tag{7.30}$$

$$\text{(b)}\quad (A \ominus B) \ominus C = A \ominus (B \oplus C) \tag{7.31}$$

Assignment 7.10

Use the thresholded images you created in Assignment 5.5 and Assignment 5.6. Choose a suitable structuring element, and apply opening to remove the noise.

Topic 7A Morphology

7A.1 Computing erosion and dilation efficiently

Equation (7.14) (or Eq. (7.31)), as we hope you determined from Assignment 7.9, is correct. It says that erosion by a large s.e., say K, can be broken down into two sequential erosions, first by B and then by C, provided we are able to find B and C such that $K = B \oplus C$. This is sometimes referred to as the "chain rule" for erosion. It has substantial impact on hardware implementations.

Suppose we have custom hardware that can compute erosion by a 3×3 s.e. at video rates, but in some application we need to erode by a particular 4×4. The chain rule says that if we can (somehow) find two 3×3 s.e.s such that their dilation is the 4×4 we want, we can get the same result by passing the input image through our special hardware twice. But how to find B and C? To give an indication of how this is done, let us consider a very simple decomposition, into a set of s.e.s each of which contains only two elements, the origin and one other point. That is, we wish to find $H_1, H_2, \ldots, H_N$ such that $A \oplus H = (\ldots[(A \oplus H_1) \oplus H_2]\ldots)$, and $H_i = (0, p_i)$. The ps are found [7.12] using the following approach: Search among pairs of points in H for a pair p_1 and p_2 such that H is invariant under opening by the difference of the two points. $H = Ho\{0, p_1 - p_2\}$. If two such points can be found, then $H_i = (0, p_1 - p_2)$, and we reduce H by $H' = H \ominus H_i$. The process is repeated recursively. If no such pair of points can be found, then a search is made of quadruples of points p_1, p_2, p_3, p_4, to see if $H = Ho\{p_1 - p_2, p_3 - p_4\}$, and so on.

Matheron [7.21] proved that any of a large class of morphological operations can be computed as a union of erosions, or by using duality, as an intersection of dilations. Choosing the set of "basis sets" so that a given operation by a given structuring element may be calculated in this way has been the subject of considerable research [7.18, 7.20].

7A.1.1 Decomposition into 3 × 3 structuring elements

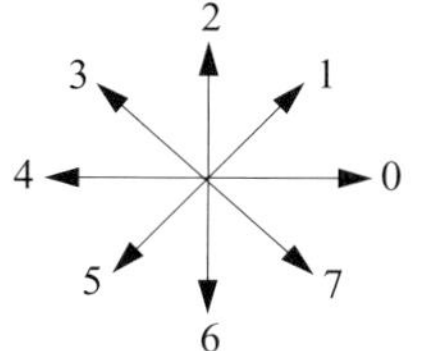

Fig. 7.8. The eight directions in which one might step in going from one pixel to another around a boundary.

In the following, we illustrate the technique of Park and Chin [7.23], using their examples. The algorithm is presented here. The reader is referred to the original papers for the theory. The example given in this section is from [7.23].

In order to understand this method for decomposing structuring elements, we need knowledge of the chain code from section 9.5. In a chain code, we represent a counterclockwise walk around a region by a sequence of numbers, all between 0 and 7, designating the direction of each step (Fig. 7.8). In this work, the concept of a chain code is slightly generalized to include not only single pixel steps in one of the eight cardinal directions, but also to include specific concave boundary segments.

The structuring element will be decomposed into a set of 3 × 3 elements. First, observe that there are only 28 distinct concave boundary segments which fit into a 3 × 3 area. These are listed and named in Fig. 7.9.

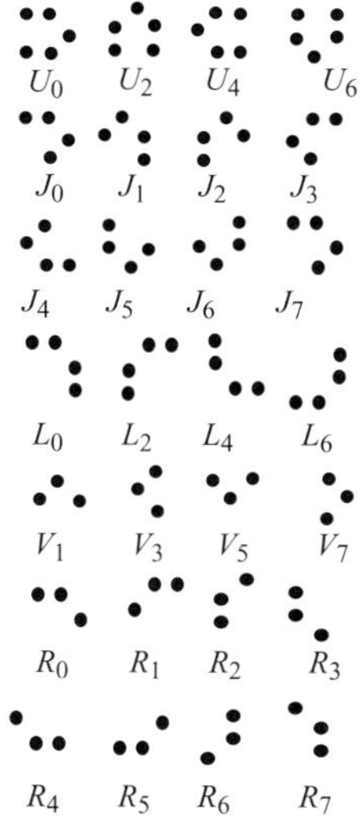

Fig. 7.9. Possible concave boundaries in 3 × 3 images. The number denotes the direction of the first chain code.

We now define a rather general class of structuring elements which are simply connected (no holes) and whose boundaries have a form given by a regular expression involving the concave segments and the chain code directions. For example, the regular expression $L_0 2^2 4^2$ denotes the curve created by starting with the L_0 segment, followed by two steps in the "2" direction (the superscript denotes repetition) followed by two steps in the "4" direction. This is illustrated in Fig. 7.10.

The class of structuring elements which can be decomposed using this algorithm is the set of all simply connected structuring elements whose boundary can be written in the form (the order of the subscripts is important to the definition)

$$U_0^{S_{U0}} J_0^{S_{J0}} L_0^{S_{L0}} R_0^{S_{R0}} 0^{S_0} J_1^{S_{J1}} V_1^{S_{V1}} R_1^{S_{R1}} 1^{S_1} \ldots J_7^{S_{J7}} V_7^{S_{V7}} R_7^{S_{R7}} 7^{S_7} \tag{7.32}$$

where any of the superscripts may be zero. For example, $V_1 1^2 2 R_4^2 4 6^3 R_7$ is a member of this set, but $J_7 1 J_2$ is not.

Definition

An image A is a *factor* of an image S if and only if it is possible to write S as the dilation of A, that is, $S = A \oplus B$. A factor A is a *prime factor* of S if and only if A cannot be factored into anything other than itself and single-pixel images. In Table 7.1 are listed all the prime factors which start with R_0. The prime factors are not required to be in the form of Eq. (7.32). In Table 7.2, we present other prime factors, listing only their chain code representation.

Now we present the approach to decomposition of the structuring element using an example. We will decompose the structuring element illustrated in Fig. 7.11, whose chain code is $S = L_0 0^3 1 2^4 R_4 4^3 R_6 6^2$. The concave portions of this boundary are denoted by $v_1 = L_0$, $v_2 = R_4$, $v_3 = R_6$. The convex portions by $d_1 = 0$, $d_2 = 1$, $d_3 = 2$, $d_4 = 4$, $d_5 = 6$.

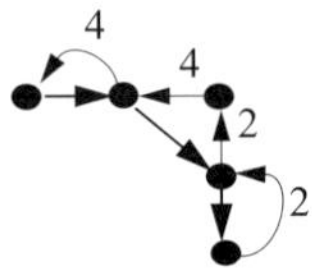

Fig. 7.10. Illustration of how the sequence $L_0 2^2 4^2$ describes a figure. The L_0 is drawn in bold.

Table 7.1. *Prime factors starting with R_0.*

$R_0 2^2 45$	$R_0 2^2 V_5 6$	$R_0 2^2 R_5$	$R_0 235$	$R_0 2 R_4 6$	$R_0 24^2$	$R_0 V_3 4^2 6$	$R_0 V_3 45$
	$R_0 V_3 V_5 6$	$R_0 V_3 R_5$	$R_0 R_3 5$	$R_0 R_3 46$	$R_0 3^2 6$	$R_0 34$	

Table 7.2. *Other prime factors starting with U_0, J_0, I_0, V_1, R_0 and R_1. To obtain V_i, J_i, etc. for $i = 2, 4, 6$, increase each chain code by i. To obtain $i = 3, 5, 7$, increase each chain code by $(i - 1)$.*

U_0	$U_0 0^2 2^2 4^2$	$U_0 0^2 234$	$U_0 0124^2$	$U_0 0134$	
J_0	$J_0 02^2 4^2$	$J_0 0234$	$J_0 124^2$	$J_0 134$	
J_1	$J_1 2^2 4^2 6$	$J_1 2^2 45$	$J_1 2346$	$J_1 235$	
L_0	$L_0 2^2 4^2$	$L_0 234$			
V_1	$V_1 2^2 4^2 V_7$	$V_1 2^2 4 R_6$	$V_1 2^2 V_5 6^2$	$V_1 2^2 V_5 V_7$	$V_1 2^2 R_5 6$
V_1	$V_1 234 V_7$	$V_1 23 R_6$	$V_1 2 R_4 6^2$	$V_1 2 R_4 V_7$	$V_1 24^2 6$
V_1	$V_1 V_3 4^2 6^2$	$V_1 V_3 4^2 V_7$	$V_1 V_3 456$	$V_1 V_3 4 R_6$	$V_1 V_3 V_5 6^2$
V_1	$V_1 V_3 R_5 6$	$V_1 V_3 5^2$	$V_1 R_3 46^2$	$V_1 R_3 4 V_7$	$V_1 R_3 56$
V_1	$V_1 2^2 5^2$	$V_1 245$	$V_1 V_3 V_5 V_7$	$V_1 R_3 R_6$	
V_1	$V_1 3^2 6^2$	$V_1 3^2 V_7$	$V_1 346$	$V_1 35$	
R_0	$R_0 2^2 45$	$R_0 2^2 V_5 6$	$R_0 2^2 R_5$	$R_0 235$	$R_0 2 R_4 6$
R_0	$R_0 242$	$R_0 V_3 4^2 6$	$R_0 V_3 45$	$R_0 V_3 V_5 6$	$R_0 V_3 R_5$
R_0	$R_0 R 35$	$R_0 R 346$	$R_0 3^2 6$	$R_0 34$	
R_1	$R_1 24^2 V_7$	$R_1 24 R_6$	$R_1 2 V_5 6^2$	$R_1 2 V_5 V_7$	$R_1 2 R_5 6$
R_1	$R_1 25^2$	$R_1 346^2$	$R_1 34 V_7$	$R_1 356$	$R_1 3 R_6$
R_1	$R_1 R_4 6^2$	$R_1 R_4 V_7$	$R_1 4^2 6$	$R_1 45$	

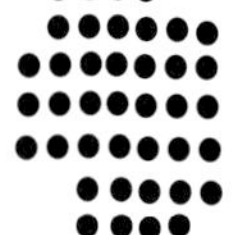

Fig. 7.11. A structuring element to be decomposed into 3×3 elements.

First, we identify all the prime factors involving L_0, R_4, and R_6, which are compatible with this image. These are illustrated in Fig. 7.12. To understand more clearly how these are compatible with the image, consider the segment $R_4 6^2 01$. Observe that this matches the R_4 segment which is at the upper right of Fig. 7.11.

The next step in the process is to construct a matrix Θ, where Θ_{ij} represents the number of times v_i occurs in A_j. In this example,

$$\Theta = \begin{bmatrix} 1 & 0 & 0 & 0 & 0 \\ 0 & 1 & 1 & 0 & 0 \\ 0 & 0 & 0 & 1 & 1 \end{bmatrix},$$

$L_0 2^2 4^2$	$R_4 6^2 01$ $(R_0 2^2 45)$	$R_4 60^2$ $(R_0 24^2)$	$R_6 02^2$ $(R_0 24^2)$	$R_6 12$ $(R_0 34)$
A_1	A_2	A_3	A_4	A_5

Fig. 7.12. The prime factors which match segment of the boundary of Fig. 7.11. The chain codes in parentheses indicate what the boundary would be if rotated to R_0 equivalents.

where we can see that R_4 occurs once in A_2 and A_3, but not at all in A_1, A_4, or A_5. Next, we construct a matrix Ω, where Ω_{ij} counts the number of times d_i occurs in A_j. Here,

$$\Omega = \begin{bmatrix} 0 & 1 & 2 & 1 & 0 \\ 0 & 1 & 0 & 0 & 1 \\ 2 & 0 & 0 & 2 & 1 \\ 2 & 0 & 0 & 0 & 0 \\ 0 & 2 & 1 & 0 & 0 \end{bmatrix}.$$

Two vectors are defined, Y, representing the number of times v_i occurs in the original boundary and Z, the number of times d_i occurs in the original boundary. In this example, $Y = [1\ \ 1\ \ 1]^{\mathrm{T}}$ and $Z = [3\ \ 1\ \ 4\ \ 3\ \ 1]^{\mathrm{T}}$. A vector X which satisfies

$$\begin{aligned} \Theta X &= Y \\ \Omega X &\leq Z \end{aligned} \tag{7.33}$$

is a solution for the decomposition. In this example, $X = [10110]^{\mathrm{T}}$ satisfies these two equations. Note that $\Theta X = [30421]^{\mathrm{T}}$ which is less than or equal to Z, in the sense that each element of ΘX is less than or equal to the corresponding element of Z. Thus, we can decompose the boundary by dilating by A_1 once, by A_2 zero times, A_3 once, A_4 once, and A_5 zero times, or $S = A_1 \oplus A_3 \oplus A_4 \oplus B$.

All that remains is to determine a kernel B. This is accomplished by looking at the difference between ΘX and Z. $Z - \Theta X = [01011]^{\mathrm{T}}$. Thus, we need a kernel whose boundary is described by the sequence d_2, d_4, d_5, each repeated once. This sequence is 146, as in Fig. 7.13. Thus, we have a sequence of structuring elements, each 3×3, whose sequential application produces the same result as the kernel in Fig. 7.11.

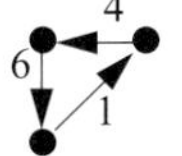

Fig. 7.13. The s.e. described by the sequence 146.

7A.2 Morphological sampling theorem

Here, we are going to be talking about sampling, much as Shannon did in his famous sampling theorem, except that we will start with an image which has been sampled already, and represented on a digital grid, and consider (sub)sampling to a smaller grid.

First, define a sampling grid. A sampling grid, as we use the term here, is just an image, with a foreground point at every point at which we will want to sample our image. Any old grid will do, except that it must satisfy

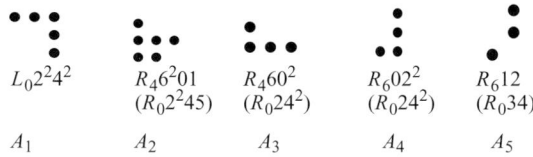

$$S \oplus S = S \tag{7.34}$$

and

$$S = \tilde{S}. \tag{7.35}$$

Equation (7.34) represents an example of the property that this particular S is "closed under dilation." A convenient grid to use is every third point, as illustrated in Fig. 7.14.

So now, sampling means to read and remember the image values at all the pixels where the grid is black. Now suppose K is some s.e. We propose a rather simple reconstruction algorithm. Here is the idea: Let's sample our image using the sampling grid specified. Then, we ask the question: Under what conditions will the dilation of the sampled image by the s.e. K, be the original image? Florencio and Schafer [7.7] have shown that the following properties are required: First, the sampling grid itself, when dilated by the s.e., must be the entire space:

You might want to verify that this grid satisfies the properties mentioned above.

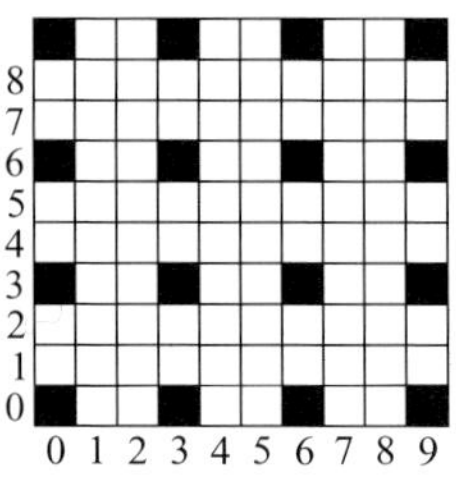

Fig. 7.14. An image which represents sampling at every third point.

$$S \oplus K = \xi, \tag{7.36}$$

the translations of K by all the points in S form a partition of the space:

$$\forall((x, y) \in S, x \neq y), K_x \cap K_y = \emptyset, \tag{7.37}$$

and K must contain the origin.

For the sampling grid shown above, here's an s.e. which satisfies these three conditions:

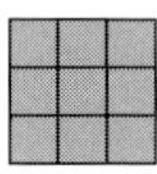

Actually, this is not very interesting – just the center pixel (the origin) and its eight neighbors. But it does satisfy the three conditions.

If these properties hold for S and K, then the theorem is as follows. Let F be some image, let P be the set of images F_i satisfying $F_i = A \oplus K$ for some $A \subseteq S$, and let Q be the set of images F_j satisfying $F_j = F_j o K = F_j \bullet K$. Then:

Part A The samples of $F \in P$ at the points in S are necessary and sufficient to perfectly reconstruct F.

Part B The samples of $F \in Q$ at the points in S are necessary to reconstruct F with bounded error and sufficient to reconstruct F with an error at most r (K) where r (K) is the radius of the smallest circle containing K.

There is a lot to talk about regarding this theorem: First, what does part A mean? Can you figure it out from the set notation? You should be able to, but you may say to yourself "if this means what I think it means, it's trivial." Well, what it means is this: If F can be generated by taking some collection of the points in S, and dilating them with kernel K, then all you have to remember is which points in S were needed. Yes, it is kind of trivial, isn't it? But there are some profound implications, which we will get to when we talk about the Nyquist rate.

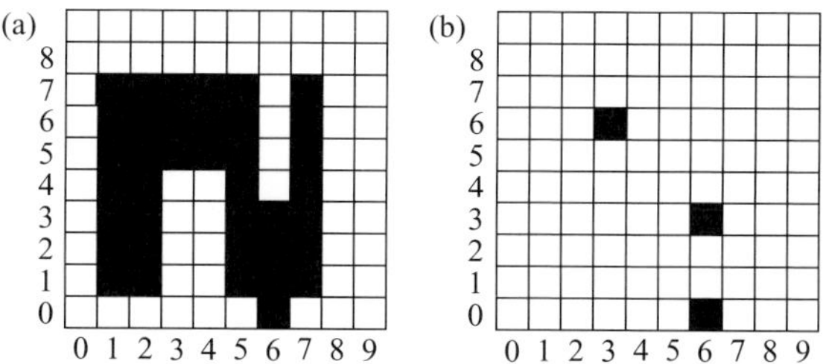

Fig. 7.15. (a) Original image. (b) Result of sampling the original image with *S* of Fig. 7.14.

Now to understand part B: The Haussdorf distance is a measure of the difference between two (sub)images. Given two sets of points $S = \{s_1, s_2, \ldots s_n\}$ and $T = \{t_1, t_2, \ldots t_m\}$, the Haussdorf distance is defined by

$$d_H(S, T) = \max(h(S, T), h(T, S)) \tag{7.38}$$

where $h(S, T) = \max_{s \in S} \min_{t \in T} \|s - t\|$. That is, the Haussdorf distance is the largest of the minimum distances between elements of one set to the elements of the other set.

Part B says that if F is closed under opening by K, (that's quite an expression, don't you think? "Closed under opening"), that is, if F does not change when it's opened by K and closed under closing by K, then the samples of F at the points of S are sufficient to represent F almost exactly. By "almost exactly" we mean the set of sample points, when dilated by K, is very nearly equal to the original F.

Now let's do that example: Fig. 7.15 represents an original image, F_A and what we get when we sample it with S.

Since we know that we could only get data on rows 0, 3, 6, or 9 and similarly for columns, we could toss out those rows and columns in our reduced resolution version, and get a smaller image. So there, on the left of Fig. 7.16, we have the subsampled, smaller version of the original image. Now let's zoom it back, using dilation: Place K down at each of the sampling points and we get the reconstruction of Fig. 7.16. Hmmm. doesn't look much like the original, unsampled image, does it? (Whatever happened to "exact reconstruction"?) And here's a tough question. This theorem claims to say that one can reconstruct a signal by sampling every third point. Doesn't this contradict Shannon? Doesn't Shannon say we have

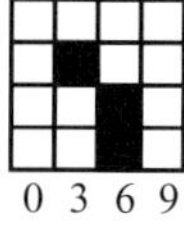

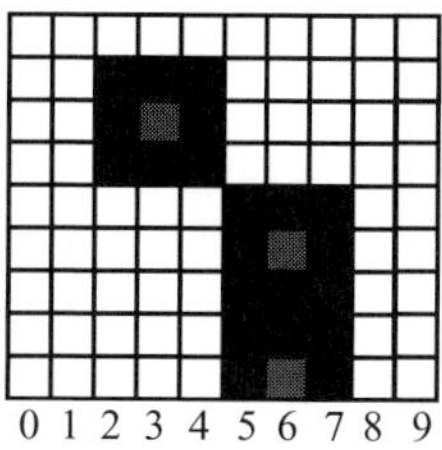

Fig. 7.16. Subsampled image and reconstruction by dilation. Students: Note the original image in Fig. 7.15(a) – does that image belong to *P* or *Q*?

to sample every other point anyway? (Actually, Shannon's theorem was defined for analog signals. Students: How DOES the Shannon theorem apply to this case? What IS the Nyquist rate?) The sampling theorem does not say we can reconstruct any signal, it says something about frequencies, doesn't it? (Say yes.) In fact, the sampling theorem basically says that a sampling grid with a particular frequency (which is the same as the grid spacing) cannot be used to store information which changes more than half the grid spacing. In morphological sampling, the theorem is complicated by the fact that we not only can choose a grid, we can also choose a kernel. The restrictions are given by part B of the theorem. Unless the image is one which has already been "prefiltered" by K, K cannot restore it. Haralick [7.12, p. 252] says it this way: "The morphological sampling theorem cannot produce a reconstruction whose positional accuracy is better than the radius of the circumscribing disk of the reconstruction structuring element K."

7A.3 Choosing a structuring element

Generally, we choose structuring elements which are small and symmetric, with the origin in the center. Schonfeld [7.34] provides some suggestions, but generally, these are the only guidelines.

7A.4 Closing gaps in edges and surfaces[2]

The algorithm presented in this section is presented in a bit more detail than usual in this book. A graduate student in the machine vision discipline needs to learn how to write a good journal paper, and this chapter is written to illustrate the components of such a paper. It includes an introduction with references to relevant literature, a body that describes and explains an algorithm, and a results section that illustrates performance of the algorithm in comparison to other published results. The student should not only read this chapter for its technical content, but should pay attention to style and organization.

7A.4.1 Introduction

In this section, we apply the concepts of the distance transform, connected components (see Chapter 8), and binary morphology-like operations to solve the problem of closing gaps in edges (see [7.16] for alternative strategies). In two dimensions, an edge is a curve, and in three dimensions, a surface. As we have seen, any edge operator is certain to occasionally fail, resulting in both extraneous edges and gaps in edges. If an edge has gaps in it, then connected component labeling routines will fail, labeling interior and exterior points the same. Thus, we must develop techniques which correct such edge detection errors. We will relate these techniques to morphological operators. In [7.10], we developed a technique called distance-transform-based closing, and have implemented this technique in both two and three dimensions. This new technique is compared with binary morphology (mask erosion techniques [7.1]) and iterative parallel thinning [7.44], a 3D parallel thinning technique [7.42]

2 Several of the figures in this chapter are from a paper written by one of the authors [7.36].

and with 3D morphological techniques. In every case, the technique reported here produces superior performance by better preserving the shape of sharp corners.

An aside: Connected component labeling

The intended use of this book is for students to read the introductory portions of each chapter, and then come back and read the advanced topics sections (like this one). In the introductory portion of Chapter 8, an operation called "connected component labeling (CCL)" is explained. You need to use the concept of connected components right now. If you haven't read that section yet, please read and understand section 8.3 first, and come back to this section.

Problem definition

We are given an edge image. Due to noise, blur, or other error, the edge/surface resulting from edge detection in a real image will occasionally have "gaps" – areas in which the edge detector response is not sufficiently strong to make a positive determination. (See Pratt [7.25, Section 17] for an excellent discussion of edge detector errors.) Such gaps may be closed by various types of morphological operators.The algorithm described here is denoted "DT-driven closing."

The word "closing" refers to closing gaps in edges, not the morphological closing operator.

7A.4.2 The distance transform

We have already seen that the distance transform, $D(x, y)$ gives some measure of the distance from point x, y to the nearest edge point [7.28, 7.29, 7.31, 7.32]. We use a particular variant of the distance transform called the "chamfer map," denoted $C(x, y)$ and refer to the values of this map as the "DT distance."

Similarly, we may extend the concept of the distance transform to three-dimensional space. The transform $D(x, y, z)$ is parameterized identically to $g(x, y, z)$ and contains the DT distance from point $\langle x, y, z\rangle$ to the closest edge.

The distance transform may be used to assist in the measurement of the properties of regions: At every point, one knows the maximum size kernel which may be applied without crossing an edge [7.35].

The k-neighbors of a point $\langle x, y\rangle$ are defined in either two or three dimensions as:

$$\begin{aligned}\aleph_k(x, y) &= \{\langle x_{k1}, y_{k1}\rangle, \langle x_{k2}, y_{k2}\rangle, \ldots, \langle x_{km}, y_{km}\rangle\}\\ &= \{\langle x_{k\lambda}, y_{k\lambda}\rangle\} : \text{DT distance } (\langle x_{k\lambda}, y_{k\lambda}\rangle, \langle x, y\rangle) = k.\end{aligned}$$

Here $m = 8k, k \geq 1$ in 2-space and $m = (2k + 1)^3 - (2k - 1)^3$ in 3-space.

We referred to the distance transform resulting from application of any masks using Eq. (7.28) as "chamfer [7.2] maps" in our earlier work [7.10], but in other literature [7.4, 7.5, 7.31], the term "chamfer" is restricted to the use of the second of the masks in Fig. 7.17. Either definition will suffice for the edge closing method described in the next subsection.

1	1	1
1	0	1
1	1	1

4	3	4
3	0	3
4	3	4

Fig. 7.17. Two distance definitions which may be used to compute distance transforms. The one on the right produces the chamfer map, closely approximating the Euclidian distance.

2	2	2	3	3	3	3	3	3	3
1	1	2	2	2	2	2	2	2	2
	1	1	1	1	2	1	1	1	1
1				1	2	1			
1	1	1	1	1	2	1	1	1	1
2	2	2	2	2	2	2	2	2	2

Fig. 7.18. Distance transform near a gap in an edge.

To construct a complete DT, Eq. (7.28) is iterated until no more changes occur between iterations. This is the approach followed by Bister *et al.* [7.3]. For the application discussed here, however, we assume some *a priori* knowledge of maximum gap size. This allows us to define a value $K_{\max}$ reflecting this knowledge. $K_{\max}$ will represent a distance so great that any pixel this far away could not possibly be part of the edge. Normally, we use values $K_{\max} \leq 4$, since this allows gaps of six pixels to be bridged. Fig. 7.18 illustrates a distance transform near a gap in an edge.

Generation of the three-dimensional distance transform proceeds in an analogous manner. Observe that the computational complexity of distance transform generation is proportional to the number of edge points in the image and to the size of $K_{\max}$. Larger values of $K_{\max}$ increase the size of the kernel $\aleph_k$ substantially in three dimensions.

7A.4.3 Segmentation using connected components

Pixels whose DT distance from an edge is less than $K_{\max}$ are labeled as "TBA" (to be assigned), meaning, they could possibly belong to an edge. The remaining pixels are segmented into disjoint regions. The straightforward way to accomplish this segmentation is to use connected component analysis on the pixels which are not labeled TBA. However, one could use more sophisticated cooperative processes such as those described in Chapter 8 and in [7.14]. Somehow we produce a label image like the label images we will discuss in Chapter 8, in which $\mathrm{L}(x, y) = j$ is interpreted as "the pixel at $\langle x, y\rangle$ belongs to region *j*." Pixels in L close to an edge are labeled as potentially part of the edge:

$$\{\mathrm{DT}(x, y) < K_{\max}\} \rightarrow \{\mathrm{L}(x, y) = \mathrm{TBA}\}.$$

Bister *et al.* [7.3] identify local maxima in the DT, and each maximum results in a potential region. They note, however, "Since the distance transform is sensitive to noise producing irregularities in the borders of the regions, one cavity (region) can contain many local maxima very close to one another. In order to eliminate these spurious maxima, a filter merges the maxima for which the sum of the heights is much larger than the geometrical distance between them." We conjecture that the performance of this filter is equivalent to our choice of K_{max}. Instead of searching for a maximum, we use connected components (Chapter 8). Both techniques identify an area within a region which robustly characterizes that region.

7A.4.4 Relabeling unassigned points

The last step in the algorithm relabels the TBA points, using the distance transform $DT(x, y)$ and the label image $L(x, y)$ to create a new label image $L'(x, y)$. Working inward from those points where $DT(x, y) \geq K_{max}$ (the set of points K_{max} or farther from an edge point), each point is relabeled by assigning it the label of the most appropriate neighbor. This "erosion" of the questionable pixels is performed iteratively. At each iteration, i, only the TBA pixels in the label image $L(x, y)$ that correspond to pixels such that $DT(x, y) = K_{max} - i$ are relabeled in each pass.

TBA pixels are relabeled only when there is strong evidence (defined in the next section) indicating to which region the pixel should be assigned. If the reassignment is uncertain, then the pixel is left TBA and the decision is postponed until the next pass. When all of the TBA pixels in $L(x, y)$ corresponding to the k-valued pixels of the distance transform have been reassigned to valid image regions or when no further change takes place in an iteration, k is decremented and the next set of TBA points is considered for relabeling. The relabeling is complete when all of the edge pixels ($k = 0$) have been assigned to valid image regions. When single-pixel resolution between regions is required, more sophisticated region normalization techniques may be used [7.8].

A voxel is a three-dimensional pixel.

The key to closing gaps lies in the strategy used to select the most appropriate region to which to reassign the currently considered pixel $L(x, y)$. To accomplish this in two dimensions, we examine the pixel's eight surrounding neighbors. In three dimensions, we examine the voxel's 26 neighbors. We refer to this as finding the pixel's or voxel's "best neighbor."

The surrounding pixels may come from one of three possible classes, depending on how they are labeled at the current iteration:

(1) A label corresponding to an object or a region of an object in the label image $L(x, y)$.
(2) A label corresponding to the image background.
(3) The TBA label.

We relabel the current pixel to one of the first two classes by simply counting how many neighboring pixels come from that class and selecting the largest as the "best" class. This strategy will only fail if all the neighbors are TBA, or constraints (see next paragraph) we have placed on the relabeling make it undesirable to reassign the pixel to the apparent best neighbor.

One constraint on the relabeling algorithm is the selection of the background region as the best neighbor. Neighbors belonging to the background are considered in a more restricted fashion than those belonging to regions. For the background value to be selected as the best neighbor, the background pixel must be face-connected to the pixel under consideration. This avoids the undesirable result of occasionally finding isolated background pixels inside a closed boundary. (See [7.30] for more discussion on this connectivity paradox.)

When $k = 0$, the TBA pixels we seek to relabel are either true edges or noise pixels. An edge in an image, by definition, represents an interface between an object region and some other region in that image. The other region may be either another object region or the background. We choose to relabel an edge pixel as part of an image region regardless of the outcome of the enumeration. Only in the case that the edge pixel is completely surrounded by background, do we choose background as the best neighbor.

Relabeling algorithm – three-dimensional data

```
/* in this function example, we have omitted several details like definitions of
variables. Still, this example captures the essence of the algorithm */
void relabel()
{
  /* loop over all the voxels */
  for(frame = 0; frame < numberframes; frame++)
  for(row = 0; row < numberrows; row++)
  for(col = 0; col < numbercols; col++) {
    if(DT[frame][row][col] >= Kmax)
      Lprime[frame][row][col] = L[frame][row][col];
    else if(DT[frame][row][col] == 0)
      Lprime[frame][row][col] = EDGE;
    else Lprime[frame][row][col] = TBA;
    }
  for(k = Kmax-1; k > 0;k--){
    changing = TRUE;
    while(changing == TRUE) {
      changing = FALSE;
      for(frame = 0; frame < numberframes; frame++)
      for(row = 0; row < numberrows; row++)
      for(col = 0; col < numbercols; col++){
        if(((L[frame][row][col] == TBA)
          || (L[frame][row][col] == EDGE))
          &&(DT[frame][row][col] >= k)) {
          Lprime[frame][row][col] =
  Best26Neighbor(frame,row,col);
          changing = TRUE;
          }
  else Lprime[frame][row][col] =
  L[frame][row][col];
      }/* end for loop on frame,row,col*/
      copyarray(Ltemp, Lprime); /*copy entire array Lprime into Ltemp*/
      copyarray(Lprime, L);
```

```
    copyarray(L, Ltemp);
    }/* end while */
  }/* end for k */
}/* end relabel */
/*==============================================================*/
/* in this function, p and n are data structures containing the frame, row, and col*/
/* coordinates of a voxel*/
int Best26Neighbor(p)
  {
  while ((n = neighbor(p)) != NULL)
    {
    if (L(n) != EDGE)
      {
      if(L(n) != BACKGROUND) Card[L(n)]++;
      else if(faceconnected(n,p)) Card[BACKGROUND]++;
      }
    }
    if ((maximum(Card) == BACKGROUND) && (L(n) == EDGE))
      return NextMax(Card);
      else return maximum(Card);
} /* end Best26Neighbor */
```

In this algorithm, "Card" is simply an array which keeps a count (cardinality) of the number of times a particular label is adjacent to the voxel of interest.

7A.4.5 Examples

In this section, we will provide the technique described above with competing morphological strategies.

Closely spaced regions

The output of an edge detection process is illustrated in Fig. 7.19. Note that the typical gaps in the edge are larger than the spacing between the regions. A distance transform is applied to the edge image. The distance transform and the result of connected component labeling are shown in Fig 7.20 and Fig. 7.21 shows the segmentation resulting from the DT-based closing method.

Other two-dimensional images

A binary image was processed containing several cutlery objects (knives, spoons, and forks). Figs. 7.22–7.24 show the results obtained for this input, which are presented in the order: (a) the original image; (b) the DT itself; (c) the "thickened edge image", which consists of all pixels such that $C(x, y) < K_{\max}$ are initially relabeled as TBA; (d) labeled image produced from thickened edge image; (e) relabeled image based on DT and labeled image.

Note the similarity between the thickened edge image and the morphological dilation of the edge.

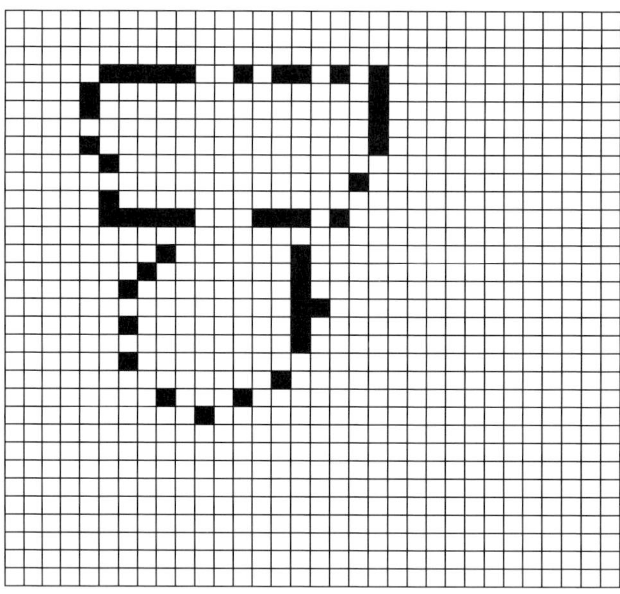

Fig. 7.19. Two regions with large gaps in their boundaries. From [7.36]. Used with permission.

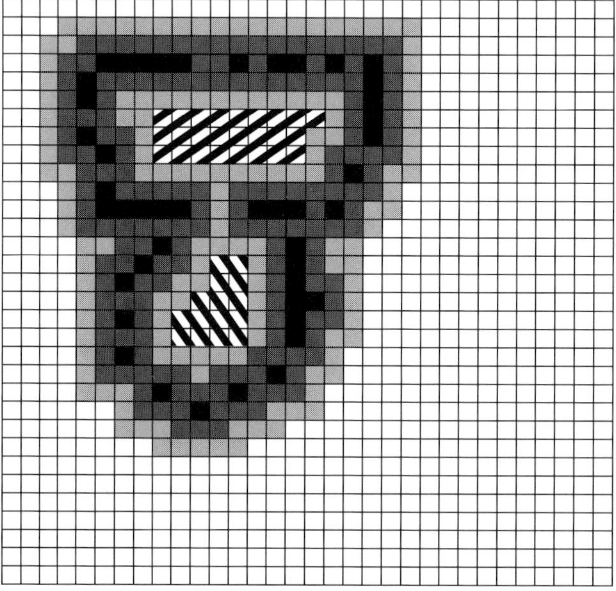

Fig. 7.20. Distance transform and result of connected component labeling of Fig. 7.19. From [7.36]. Used with permission.

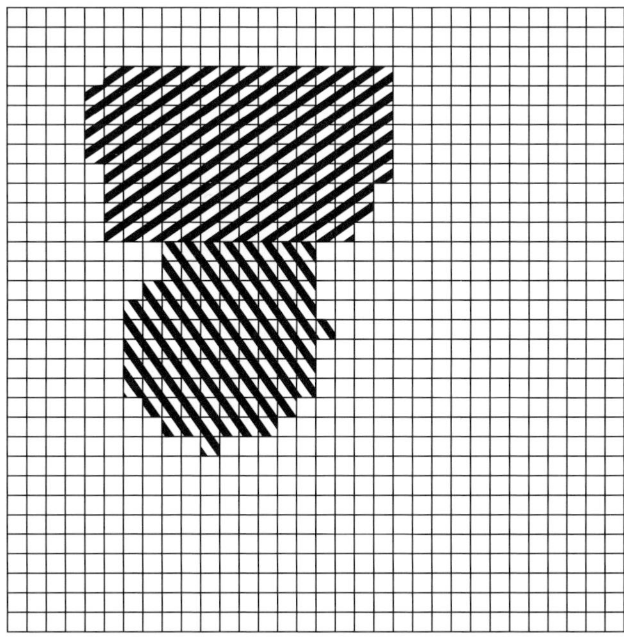

Fig. 7.21. Segmentation resulting from applying DT-driven closing. Note that the regions are correctly partitioned, with a precision accurate to the individual pixel. From [7.36]. Used with permission.

7A.4.6 Comparison with published two-dimensional methods

Thinning differs from skeletonization in that thinning preserves connectivity: A boundary which is intact when wide will still be intact after thinning.

To evaluate the performance of DT-driven edge closing in two-dimensional images, we compared it with two other types of binary thinning techniques – iterative erosion and iterative thinning. As input to these algorithms we used the bit-mapped "cutlery" image shown in Fig. 7.22 after it was dilated by a 5×5 kernel to close all gaps. The basic idea is this: Use dilation to close the gaps in the boundary. Then use thinning to reduce the "fat" boundary to a thin one. The results of these comparisons are described in the following sections.

Note that although both thinning and skeleton perform similar operations, thinning preserves the connectivity of the edge, while skeleton does not. A skeleton may be defined in terms of morphological operations as follows.

The kth order homothetic of an s.e., A, denoted kA, is the result of applying the operator of interest – in this case dilation – to A. That is, dilating $\{0\}$ by A, and then dilating the result by A again, until k such dilations have been done.

A conventional morphological skeleton of an image X is formed by decomposing an image into a number of skeleton subsets, S_i where

$$S_i = (X \ominus iA) \backslash [(X \ominus iA)o(A)] \tag{7.39}$$

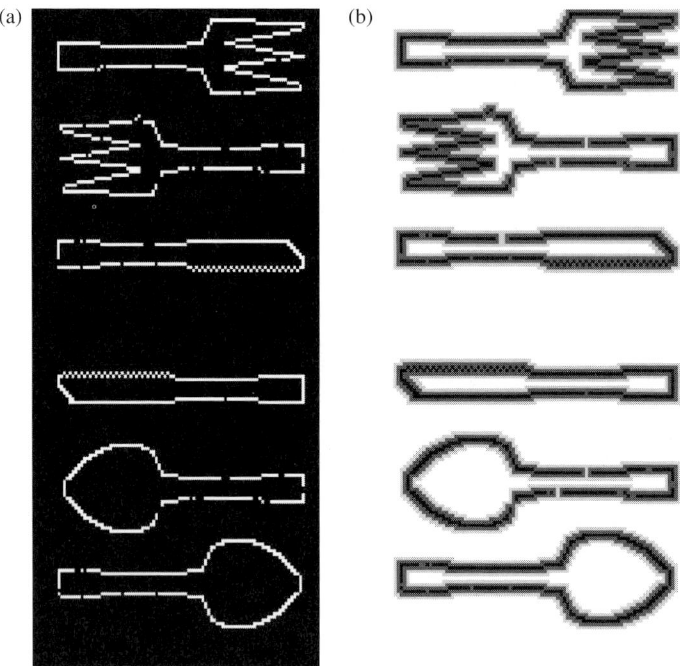

Fig. 7.22. (a) Original cutlery image, containing gaps in edges. (b) Distance transform of original cutlery image. From [7.36]. Used with permission.

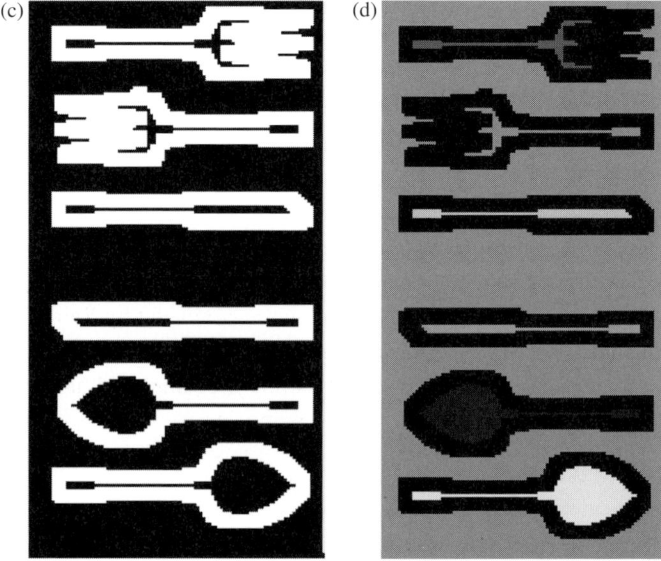

Fig. 7.23. (c) Thickened edge image. (d) Label image based on dilated edges. From [7.36]. Used with permission.

(e)

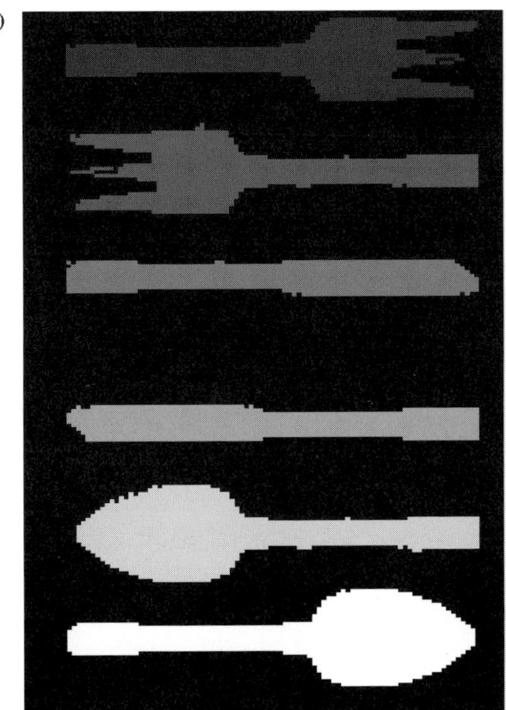

Fig. 7.24. (e) Relabeled cutlery image. From [7.36]. Used with permission.

where $X \backslash Y$ denotes all the elements of X which are not in Y. The skeleton is then constructed by

$$Skeleton = \bigcup_{i=0}^{N} S_i \oplus iA. \tag{7.40}$$

The subsets contain information about size, orientation, and connectivity. A minimal skeleton has the property of being able to exactly reconstruct the original image, but it does not necessarily preserve path or surface connectivity [7.20]. An alternative to the morphological skeleton, not considered in this book, is morphological shape decomposition (MSD) [7.24]. MSD and the morphological skeleton are compared by Reinhardt and Higgins [7.27] who conclude that MSD performs slightly better. While morphological skeletonization can be used for many applications (such as image coding) and has been widely studied, it is not directly comparable to 2D or 3D thinning, in which connectivity is preserved. Since, in this application of edge/surface gap closing, we insist on preservation of connectivity, we consider below only techniques which possess this property.

Morphological 2D thinning (Arcelli, Cordella, and Levialdi)

Arcelli *et al.* [7.1] use a sequence of masks to implement thinning. The original image is eroded by each of eight 3 × 3 masks in sequence and the resulting image is used as input

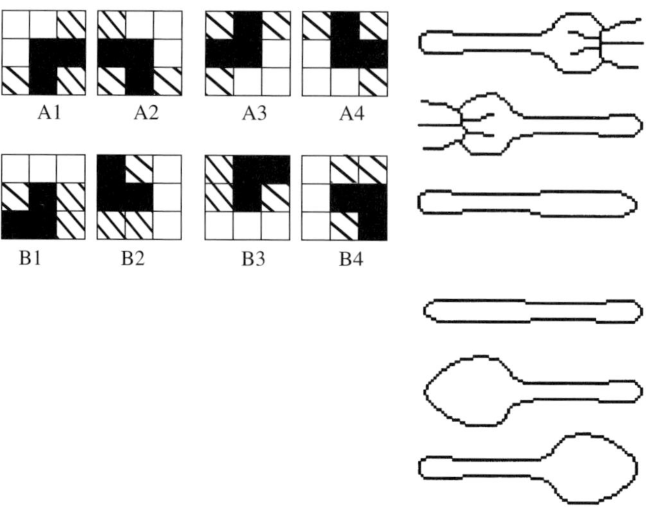

Fig. 7.25. 3 × 3 masks (left) and result of erosion thinning on the cutlery image (right). From [7.36]. Used with permission.

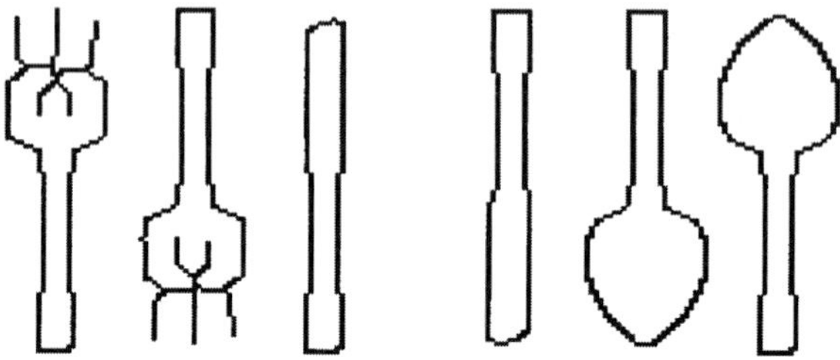

Fig. 7.26. Results of iterative thinning. From [7.36]. Used with permission.

to the next iteration until all possible pixels have been eroded. In each mask shown in Fig. 7.25 [7.1, 7.26], the positions marked in black denote edge pixels, white denotes background, and hashed are pixels which do not become involved in the computation. A mask is said to "match" an image at coordinates (k, l) if, when the center of the mask is registered to pixel (k, l), then all image pixels covered by the mask are edge or background as denoted by the corresponding mask pixel. If a mask matches an image at a particular edge point, then the edge at that point in the image may be reset to background. The masks are applied in the following order: A1, B1, A2, B2, A3, B3, A4, B4. The result of thinning Fig. 7.23 using this algorithm is shown in Fig. 7.25. Note how this technique distorts the shape of sharp vertices like the fork tines.

Iterative 2D thinning (Zhang and Suen)

Zhang and Suen's algorithm for thinning binary images [7.44] iteratively passes over the image, deciding if contour points can be deleted. A contour point is an edge pixel with at least one 8-neighbor that is a background pixel. Each iteration contains two passes and at each iteration, the decision to remove a pixel is based on the number of edge neighbors of a pixel, the number of 0–1 transitions in a sequence around the pixel, and on two sets of background neighbor configurations. The result of iteratively thinning the dilated cutlery image (Fig. 7.23) is shown in Fig. 7.26. It is interesting to note the similarity with the technique of Arcelli *et al.*

7A.4.7 Three-dimensional images

DT-driven closing was applied to a three-dimensional image of a box with broken interior partitions; Figs. 7.27–7.29 show the results. The order of the images is the same as shown for the 2D results. Note that even though the gaps are large in x, y, they are successfully bridged, and the edges are still sharp in the final relabeled image.

The three-dimensional algorithm was also tested on a synthetic ellipsoid (Fig. 7.30) which was deliberately undersampled to produce large gaps. The results of DT-driven closing with a K_{max} value of 3 are shown in Fig. 7.30.

Fig. 7.27. (a) Original broken box image, frame 20. (b) Distance transform of broken box.

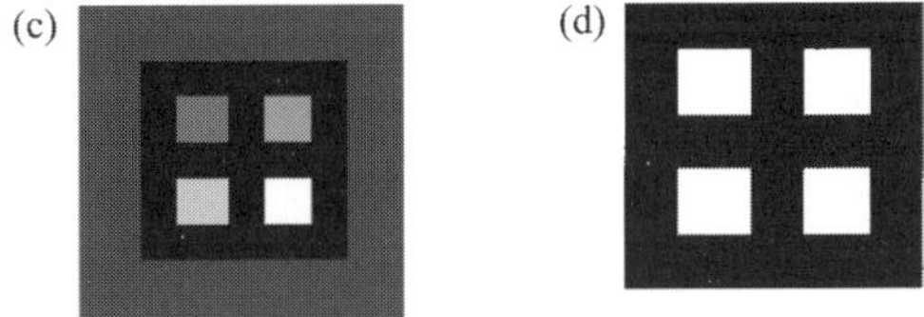

Fig. 7.28. (c) Thickened edge image. (d) Label image based on dilated edges.

(e)

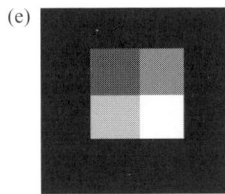

Fig. 7.29. (e) Relabeled broken box image.

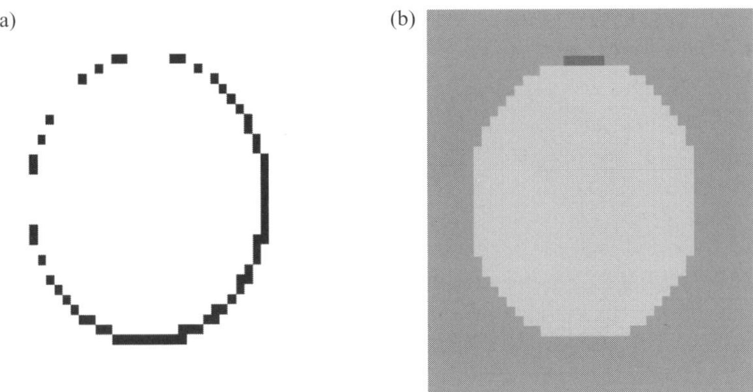

Fig. 7.30. Ellipsoid with gaps (a) and relabeled ellipsoid (b).

7A.4.8 Comparison with published three-dimensional method

Other research in three-dimensional thinning and skeletonization [7.11, 7.19, 7.41, 7.42] is primarily based on computing the Euler [7.13] connectivity number, N, for each three-dimensional solid where $N = V - E + F$ and V, E, and F denote the numbers of vertices, edges and faces of the object, respectively. For a solid which does not contain tunnels or cavities, $N = 2$. Each tunnel or hole penetrating an object decreases N by two; each cavity in the solid increases N by two. The connectivity number must be preserved during thinning to maintain the topology of the original object [7.26, 7.33].

3D thinning (Tsao and Fu)
DT-driven thinning was compared to a topological three-dimensional thinning algorithm described by Tsao and Fu [7.42] which preserves both path and surface connectivity. If surface connectivity is maintained, then path connectivity or number of components is also maintained. This algorithm examines the center voxel in a $3 \times 3 \times 3$ cube and classifies it as a border point in each of six subiterations, N(orth), S(outh), E(ast), W(est), U(p), or D(own), if the corresponding N, S, E, W, U or D neighbor is zero. If the deletion of a voxel does not change the connectivity of the remaining voxels in the $3 \times 3 \times 3$ window and if its removal does not change the connectivity in two "checking planes," the voxel can be deleted. The result of thinning the thickened edge image in Fig. 7.28(c) by Tsao and Fu's method is shown in Fig. 7.31.

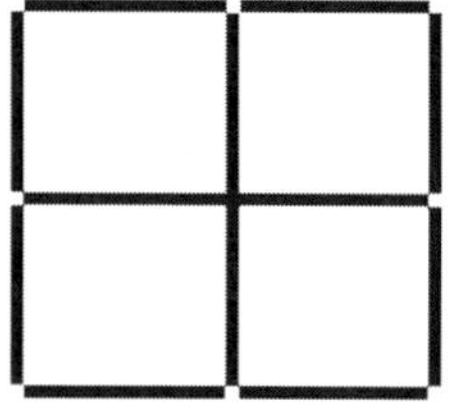

Fig. 7.31. Tsao and Fu thinning of dilated box.

The ellipsoid in Fig. 7.30 was also dilated using a $5 \times 5 \times 5$ kernel to close the gaps (Fig. 7.32(a)) and thinned using Tsao and Fu's algorithm (Fig. 7.32(b)). The extra lines at the top result from quantization effects in adjacent planes.

7A.4.9 Preserving geometry

Probably the most significant aspect of the performance of DT-driven closing is its ability to preserve the geometry of surfaces, particularly near vertices. Its two-dimensional performance is particularly well demonstrated in Fig. 7.24. To demonstrate how it processes vertex geometry in three dimensions, we synthesized a cone and extracted the surface with a three-dimensional edge detector.

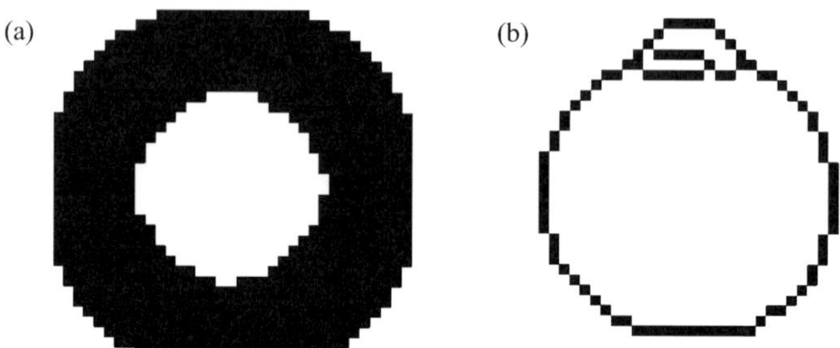

Fig. 7.32. (a) Dilated ellipsoid and (b) Tsao and Fu thinned ellipsoid. From [7.36]. Used with permission.

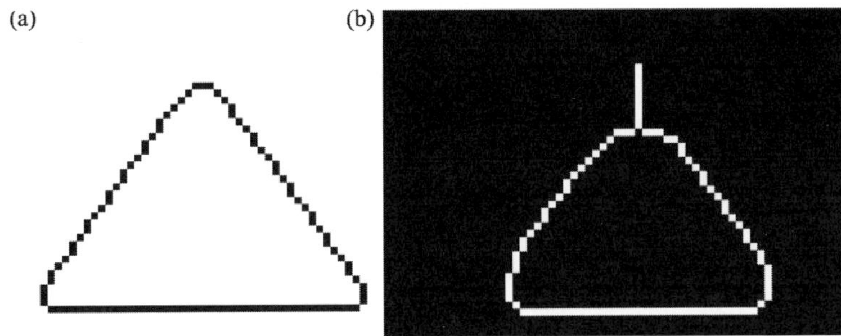

Fig. 7.33. A cross section through the apex of a cone which has been processed by DT-based closing (a) and by Tsao–Fu thinning (b). From [7.36]. Used with permission.

Gaps in the surface were then closed with DT-driven closing. The same cone surface was dilated using conventional dilation to close gaps, and then thinned using the Tsao–Fu algorithm. The results are shown in Fig. 7.33. Since the process of dilation replaces the edge information, the Tsao–Fu thinning algorithm has no memory of the original surface when it thins the dilated image. Of course, neither technique processes the geometry perfectly. However, since DT-driven closing retains, via the DT, a "memory" of the original, undilated, geometry, this technique is better able to restore that geometry after gaps are closed. See [7.38] for more details and computation speed.

7A.4.10 Why this section was included

We included this rather long section on DT-driven edge closing for several reasons: First, it is important for the student to see that the implementation of any real technique requires that you know a bit about more than just one technique. You had to use knowledge of connected components, edge detection, morphological dilation, etc. Second, since this book is directed primarily at graduate students, we wanted you to see how to write a journal paper. This chapter is similar in many ways to (and includes figures from) [7.38]. Note the format: Introduce the problem, describe the algorithm, cite the literature, and most important, compare the new technique with existing techniques. Follow these simple rules, and you too can publish!

7A.5 Vocabulary

You should know the meanings of the following terms.

Chamfer map
Prime factor
Sampling

Assignment **7.A1**

In section 7A.1.1 an example decomposition of a structuring element was given. Prove that this decomposition gives the same result as the original when used to dilate the image {(0,0)}.

Assignment 7.A2

What is the major difference between the output of a thinning algorithm and the maxima of the distance transform? Choose the best answer from the following.

(a) A thinning algorithm preserves connectivity. The maxima of the DT are not necessarily connected.
(b) The maxima of the DT are unique, thinning algorithms do not produce unique results.
(c) The DT preserves connectivity. Thinning algorithms do not.
(d) Thinning algorithms produce the intensity axis of symmetry. The DT does not.

Assignment 7.A3

Use the distance transform to compute the medial axis of an image. The name of the image will be given in class.

Bibliography

[7.1] C. Arcelli, L. Cordella, and S. Levialdi, "Parallel Thinning of Binary Pictures," *Electronics Letters*, **11**, pp. 148–149, 1975.

[7.2] H.G. Barrow, "Parametric Correspondence and Chamfer Matching," *Proceedings of the 5th International Joint Conference on Artificial Intelligence*, August, 1977, pp. 659–663.

[7.3] M. Bister, Y. Taeymans, and J. Cornelis, "Automated Segmentation of Cardiac MR Images," *Computers in Cardiology*, Washington, DC, IEEE Computer Society Press, pp. 215–218, 1989.

[7.4] G. Borgefors, "Distance Transformations in Arbitrary Dimensions," *Computer Graphics, Vision, and Image Processing*, **27**, pp. 321–345, 1984.

[7.5] G. Borgefors, "Distance Transformations in Digital Images," *Computer Graphics, Vision, and Image Processing*, **34**, pp. 344–371, 1986.

[7.6] H. Breu, J. Gil, D. Kirkpatrick, and M. Werman, "Linear Time Euclidian Distance Transform Algorithms," *IEEE Transactions on Pattern Analysis and Machine Intelligence*, **17**(5), 1995.

[7.7] D. Florencio and R. Schafer, "Homotopy and Critical Morphological Sampling," *Proceedings of the SPIE*, **2308**, June, 1994.

[7.8] J.D. Foley, A. vanDam, S.K. Feiner, and J.F. Hughes, *Computer Graphics: Principles and Practice*, Reading, MA, Addison-Wesley, pp. 91–99, 1990.

[7.9] W. Gong and G. Bertrand, "A Simple Parallel 3D Thinning Algorithm," 10th *International Conference on Pattern Recognition*, June, 1990.

[7.10] B. R. Groshong, and W. E. Snyder, "Using Chamfer Maps to Segment Images," Technical Report CCSP-WP-86/11, Center for Communications and Signal Processing, North Carolina State University, Raleigh, NC, USA, December, 1986.

[7.11] K. Hafford and K. Preston Jr., "Three-dimensional Skeletonization of Elongated Solids," *Computer Vision, Graphics, and Image Processing*, **27**, pp. 78–91, 1984.

[7.12] R. Haralick and L. Shapiro, *Computer and Robot Vision*, Volume 1, New York, Addison-Wesley, 1992.

[7.13] D. Hilbert and S. Cohn-Vossen, *Geometry and the Imagination*, New York, Chelsea, 1952.

[7.14] H. Hiriyannaiah, G. Bilbro, and W. Snyder, "Restoration of Locally Homogeneous Images using Mean Field Annealing," *Journal of the Optical Society of America, A*, **6**(12), pp. 1901–1912, 1989.

[7.15] C. Huang and O. Mitchell, "A Euclidian Distance Transform using Grayscale Morphology Decomposition," *IEEE Transactions on Pattern Analysis and Machine Intelligence*, **16**(4), 1994.

[7.16] X. Jiang, "An Adaptive Contour Closure Algorithm and its Experimental Evaluation," *IEEE Transactions on Pattern Analysis and Machine Intelligence*, **22**(11), 2000.

[7.17] R. Jones and I. Svalbe, "Morphological Filtering as Template Matching," *IEEE Transactions on Pattern Analysis and Machine Intelligence*, **16**(4), 1994.

[7.18] R. Jones and I. Svalbe, "Algorithms for the Decomposition of Gray-scale Morphological Operations," *IEEE Transactions on Pattern Analysis and Machine Intelligence*, **16**(6), 1994.

[7.19] S. Lobregt, P. Verbeek, and F. Groen, "Three-Dimensional Skeletonization: Principle and Algorithm," *IEEE Transactions on Pattern Analysis and Machine Intelligence*, **2**(1), pp. 75–77, 1980.

[7.20] P. Maragos and R. Schafer, "Morphological Skeleton Representation and Coding of Binary Images," *IEEE Transactions on Acoustics, Speech, and Signal Processing*, **34**, pp. 1228–1244, 1986.

[7.21] G. Matheron, *Random Sets and Integral Geometry*, New York, Wiley, 1975.

[7.22] H. Park and R. Chin, "Optimal Decomposition of Convex Morphological Structuring Elements for 4-connected Parallel Array Processors," *IEEE Transactions on Pattern Analysis and Machine Intelligence*, **16**(3), 1994.

[7.23] H. Park and R. Chin, "Decomposition of Arbitrarily Shaped Morphological Structuring Elements," *IEEE Transactions on Pattern Analysis and Machine Intelligence*, **17**(1), 1995.

[7.24] I. Pitas and A. Venetsanopoulos, "Morphological Shape Decomposition," *IEEE Transactions on Pattern Analysis and Machine Intelligence*, **12**(1), 1990.

[7.25] Pratt, W. K., *Digital Image Processing*, New York, Wiley, 1978.

[7.26] K. Preston, and M. Duff, *Modern Cellular Automata Theory and Applications*, New York, Plenum Press, 1984.

[7.27] J. Reinhardt and W. Higgins, "Comparison Between the Morphological Skeleton and Morphological Shape Decomposition," *IEEE Transactions on Pattern Analysis and Machine Intelligence*, **18**(9), 1996.

[7.28] A. Rosenfeld, "Distance Functions on Digital Images," *Pattern Recognition*, **1**, pp. 33–61, 1968.

[7.29] A. Rosenfeld, *Picture Processing by Computer*, New York, Academic Press, 1969.

[7.30] A. Rosenfeld, "Connectivity in Digital Pictures," *Journal of the Association for Computing Machinery*, **17**(1), pp. 146–160, 1970.

[7.31] A. Rosenfeld and A. Kak, *Digital Picture Processing*, Volume 2, New York, Academic Press, 1982.

[7.32] A. Rosenfeld and J. Pfaltz, "Sequential Operations in Digital Picture Processing," *Journal of the Association for Computing Machinery*, **13**, pp. 471–494, 1968.

[7.33] P. Saha and B. Chaudhuri, "Detection of 3-D Simple Points for Topology Preserving Transformations with Application to Thinning," *IEEE Transactions on Pattern Analysis and Machine Intelligence*, **16**(10), 1994.

[7.34] D. Schonfeld, "Optimal Structuring Elements for the Morphological Pattern Restoration of Binary Images," *IEEE Transactions on Pattern Analysis and Machine Intelligence*, **16**(6), 1994.

[7.35] W. Snyder and G. Bilbro, "Segmentation of Three Dimensional Range Images," CCSP-TR-84/7, North Carolina State University, November, 1984.

[7.36] W. Snyder and A. Cowart, "An Iterative Approach to Region Growing Using Associative Memories," *IEEE Transactions on Pattern Analysis and Machine Intelligence*, **5**(6), pp. 638–651, 1985.

[7.37] W. Snyder and C. Savage, "Content-Addressable Read-Write Memories for Image Analysis," *IEEE Transactions on Computers*, **31**(10), pp. 963–968, 1982.

[7.38] W. Snyder, M. Hsiao, K. Boone, T. Hudacko, and B. Groshong, "Closing Gaps in Edges and Surfaces," *Image and Vision Computing*, October, 1992.

[7.39] J. Srihari, K. Uhurpa, and M. Yau, "Understanding the Bin of Parts," IEEE Conference on Decision Control, pp. 44–49, December, 1979.

[7.40] H. Tagare, F. Vos, C. Jaffe, and J. Duncan, "Arrangement: A Spatial Relation Between Parts for Evaluating Similarity of Tomographic Section," *IEEE Transactions on Pattern Analysis and Machine Intelligence*, **17**(9), 1995.

[7.41] J. Toriwaki and Y. Tatsuhiro, "Topological Properties and Topology- preserving Transformation of a Three-dimensional Binary Picture," *Proceedings of the Sixth International Pattern Recognition Conference*, pp. 414–419, Munich, 1982.

[7.42] Y. Tsao and K. Fu, "A Parallel Thinning Algorithm for 3-D Pictures," *Computer Graphics and Image Processing*, **17**, pp. 315–331, 1981.

[7.43] R. van den Boomgaard and A. Smeulders, "The Morphological Structure of Images: The Differential Equations of Morphological Scale-space," *IEEE Transactions on Pattern Analysis and Machine Intelligence*, **16**(11), 1994.

[7.44] T. Zhang and C. Suen, "A Fast Parallel Algorithm for Thinning Digital Patterns," *Communications of the Association for Computing Machinery*, **27**(3), pp. 236–239, 1984.

8 Segmentation

Galia est omnes divisa in partes tres

Julius Caesar

Segmentation is the process of separating objects from background. It is the building block for all the subsequent processes like shape analysis, object recognition, etc. In this chapter, we first discuss several popular segmentation algorithms, including threshold-based, region-based (or connected component analysis), edge-based, and surface-based. We also describe some recently developed segmentation algorithms in the topics section.

8.1 Segmentation: Partitioning an image

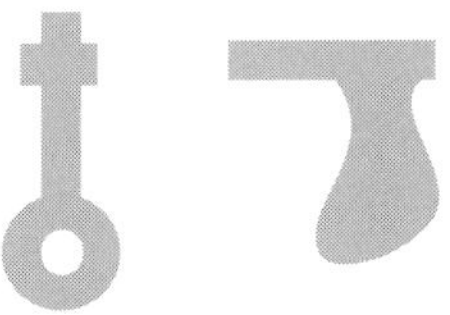

Fig. 8.1. An image with two foreground regions.

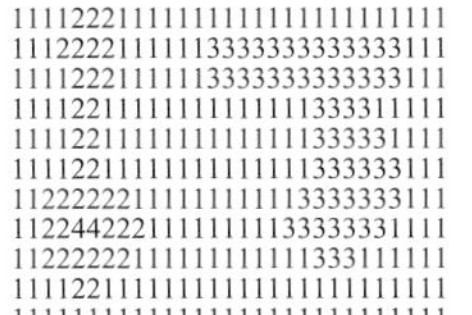

Fig. 8.2. A segmentation and labeling of the image in Fig. 8.1.

In many machine vision applications, the set of possible objects in the scene is quite limited. For example, if the camera is viewing a conveyer, there may be only one type of part which appears, and the vision task could be to determine the position and orientation of the part. In other applications, the part being viewed may be one of a small set of possible parts, and the objective is to both locate and identify each part.[??] Finally, the camera may be used to inspect parts for quality control.

In this section, we will assume that the parts are fairly simple and can be characterized by their two-dimensional projections, as provided by a single camera view. Furthermore, we will assume that the shape is adequate to characterize the objects. That is, color or variation in brightness is not required. We will first consider dividing the picture into connected regions.

A *segmentation* of a picture is a partitioning into connected regions, where each region is homogeneous in some sense and is identified by a unique label. For example, in Fig. 8.2 (a "label image"), region 1 is identified as the background. Although region 4 is really background as well, it is labeled as a separate region since it is not connected to region 1.

The term "homogeneous" deserves some discussion. It could mean all the pixels are the same brightness, but that criterion is too strong for most practical applications. It could mean that all pixels are close to some representative (mean) brightness.

Stated more formally [8.80], a region is homogeneous if the brightness values are consistent with having been generated by a particular probability distribution (see also the analysis by Ng and Lee [8.44]). In the case of range imagery [8.35], where we (might) have an equation which describes the surface, we could say a region is homogeneous if it can be described by the combination of that equation and some probabilistic deformation. For example, if all the points in a region of a range image lay in the same plane except for deviations whose distance from the plane may be described by a particular Gaussian distribution, we might say this region is homogeneous.

There are several ways to perform segmentation. Threshold-based techniques are guaranteed to form closed regions, for they simply assign all pixels above (or below, depending on the problem) a specified threshold to be in the same region. Edge-based techniques assume that regions are separated by neighborhoods where the edge strength is high. Region-based methods start with elemental (e.g., homogeneous) regions and split or merge them. Then, there are a variety of hybrid methods, including watershed [8.5] techniques. A watershed method generally operates on the gradient of the image; segmentation consists of flooding the image with (by analogy) water, in which region boundaries (areas of high edge strength) are erected to prevent water from different seed points from mixing. Traditional "region growing" methods are really variations on watershed methods [8.1].

Before we can go much further in our discussion of issues and techniques in segmentation, you need to understand some of the interesting and unexpected things that happen to geometry and topology when you deal with digital images. Remember the connectivity paradox from section 4.5? We discovered that an object could have a closed boundary but still have an inside and outside which were connected? As another example, consider the problem of finding the perimeter of a region, or even the length of a line from a sampled version of that line. That problem has immediate applications in segmentation, and yet it is not obvious how to estimate it [8.31]. Keep in mind that these sorts of things can happen, because we are going to talk a lot about connectivity in this chapter.

8.2 Segmentation by thresholding

In applications where specific gray values of regions are not important, we can segment a picture into "objects" and "background" by simply choosing a threshold in brightness. We define any region whose brightness is above the threshold as *object* and all below the threshold as *background*.

There are several different ways to choose thresholds, ranging from the trivially simple to the very sophisticated. As the sophistication of the technique increases, performance improves but at the cost of increased computational complexity.

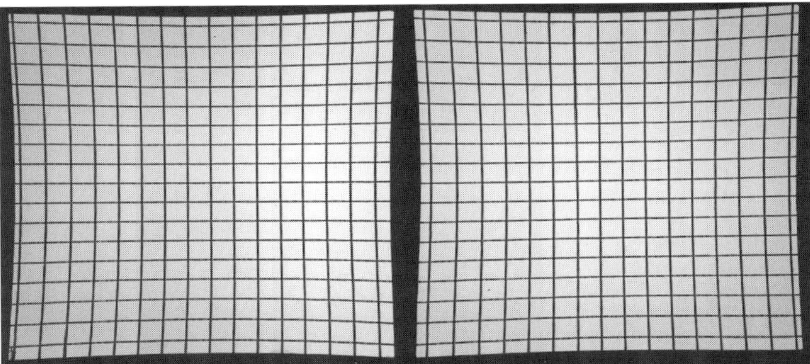

Fig. 8.3. Two detectors forming an image of a rectilinear grid. The light source is uniform, however, the images exhibit both radiometric (brightness) distortion (the left image is brighter in the right center and the right image is brighter in the middle) and geometric distortion (straight lines are distorted in a characteristic pincushion form).

Probably the most important factor to note is the *local nature* of thresholding. That is, a single threshold is almost never appropriate for an entire scene. It is nearly always the local contrast between object and background that contains the relevant information. Since camera sensitivity drops off from the center of the picture to the edges due to parabolic distortion and/or vignetting as shown in Fig. 8.3, it is often useless to attempt to establish a global threshold. A dramatic example of this effect can be seen in an image of a rectilinear grid, in which the "uniform" white varies significantly over the surface.

Effects such as parabolic distortion and vignetting are quite predictable and easy to correct. In fact, off-the-shelf hardware is available for just such applications. It is more difficult, however, to predict and correct effects of nonuniform ambient illumination, such as sunlight through a window, which changes radically over the day.

Since a single threshold cannot provide sufficient performance, we must choose local thresholds. The most common approach is called *block thresholding*, in which the picture is partitioned into rectangular blocks and different thresholds are used on each block. Typical block sizes are 32×32 or 64×64 for 512×512 images. The block is first analyzed and a threshold is chosen; then that block of the image is thresholded using the results of the analysis. In more sophisticated (but slower) versions of block thresholding, the block is analyzed and the threshold computed. Then that threshold is applied only to the single pixel at the center of the block. The block is then moved over one pixel, and the process is repeated.

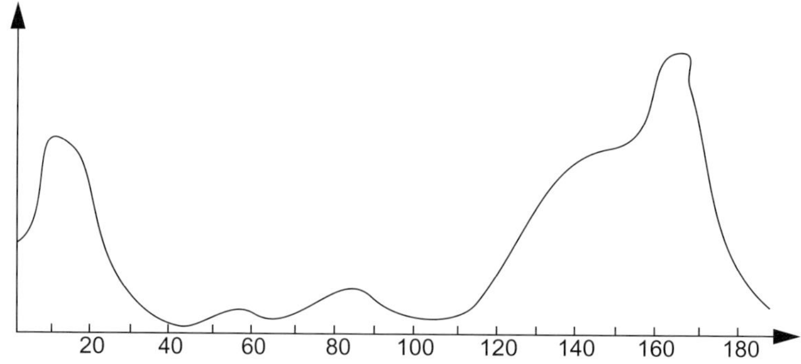

Fig. 8.4. A histogram of a bimodal image, with many bright pixels (intensity around 169) and many dark pixels (intensity around 11).

8.2.1 Choosing a threshold

The simplest strategy for choosing a threshold is to average the intensity over the block and choose $i_{\text{avg}} + \Delta i$ as the threshold, where Δi is some small increment, such as 5 out of 256 gray levels. Such a simple thresholding scheme can have surprisingly good results.

However, when the simpler schemes fail, one is forced to move to more sophisticated techniques, such as thresholding based on histogram analysis. Before we describe this technique, we will first define a histogram.

The *histogram h*(i) of an image $f(x, y)$ is a function of the permissible intensity values. In a typical imaging system, intensity takes on values between 0X00 (black) and 0XFF (white). A graph that shows, for each gray level, the number of times that level occurs in the image is called the histogram of an image. Illustrated in Fig. 8.4 is a histogram of black parts on a white conveyor.

In Fig. 8.4 we note two distinct peaks, one at gray level 11, almost pure black, and one at gray level 169, bright white. With the exception of noise pixels, every point in the image belongs to one of these regions. A good threshold, then, is anywhere between the two peaks.

Histograms are seldom as "nice" as the one in Fig. 8.4 and some additional processing is generally needed ([8.14, 8.51, 8.70] explain and experimentally compare several such methods). In the following section, we describe a more sophisticated technique for finding the optimal threshold.

Fitting a sum of Gaussians

In [8.7], a technique was developed which finds the global minimum of a function of several variables, even for functions which have more than one minimum. That technique, known as *tree annealing* (TA) may be applied to the problem of histogram analysis and thresholding in the following way [8.62].

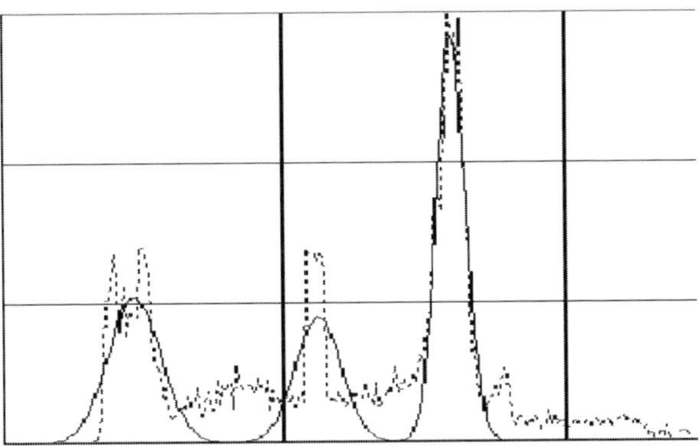

Fig. 8.5. MSE fit of the sum of three Gaussians to a histogram.

Given a bimodal histogram, $h(f)$ of an image, where f represents a brightness value, a standard technique [8.14] to find the best threshold of that image is to fit the histogram with the sum of two Gaussians:

$$h(x) = \frac{A_1}{\sqrt{2\pi}\sigma_1} \exp\left[-\frac{(x-\mu_1)^2}{2\sigma_1^2}\right] + \frac{A_2}{\sqrt{2\pi}\sigma_2} \exp\left[-\frac{(x-\mu_2)^2}{2\sigma_2^2}\right]. \tag{8.1}$$

If $h(f)$ is properly normalized, one may adjust the usual normalization of the two component Gaussians so that each sums to unity on the 256 discrete gray levels (rather than integrating to unity on the continuous interval), and thereby admit the additional constraint that $A_1 + A_2 = 1$. Use of this constraint reduces the number of parameters to be estimated from six to five; however, we have determined experimentally that TA actually solves the problem more accurately by using the six variables, without readjusting the normalization on each iteration. Conventional descent often terminates at a suboptimal local minimum for this two-Gaussian problem and is even less reliable for a three-Gaussian problem. TA deals easily with either. The result of fitting a sum of three Gaussians to the histogram of an image is shown in Fig. 8.5.

Whatever algorithm we use, the philosophy of histogram-based thresholding is the same: Find peaks in the histogram and choose the threshold to be between them.

In many industrial environments, the lighting may be extremely well controlled. With such control, the best thresholds will be constant over time and may be chosen interactively during system set up. However, in general, different thresholds are used in different areas of the picture.

8.3 Connected component analysis

Let us assume, for now, that a good threshold has been chosen and that our picture has been partitioned into regions of pure black and pure white, as shown in Fig. 8.1.

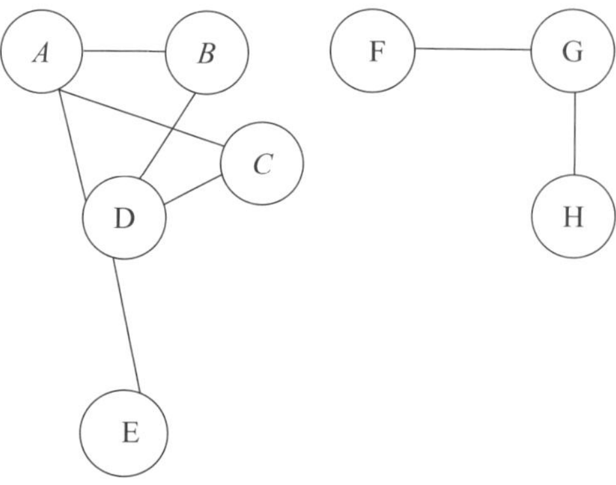

Fig. 8.6. A graph with two connected components.

The production of a segmented picture such as Fig. 8.2 requires an analysis of connectedness. That is, a pixel is in region i if it is above threshold and is adjacent to a pixel in region i. Since regions may curve and fork, the analysis cannot be as simple as starting at the top and marking connected pixels going down. Instead, a more sophisticated technique is needed.

Fig. 8.7. An image containing two connected components.

The term "connected component" comes from graph theory. Consider the graph in Fig. 8.6.

This graph has eight vertices, eight edges, and two connected components. That is, there is a path along edges from A to D, or B to E, or F to H, etc. But there is no path from A to F. In image analysis, we **think of** the foreground pixels as vertices in a graph, and the "adjacent to" property as determining the edges in the graph. With this definition, we see that the image in Fig. 8.7 contains two connected components. Graphs will be revisited in more detail in Chapter 12.

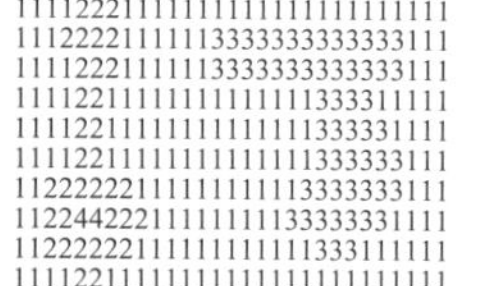

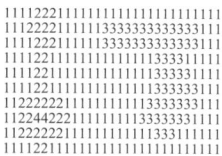

```
1111222111111111111111111111
1112222111113333333333333111
1111222111113333333333333111
1111221111111111111333311111
1111221111111111111333331111
1111221111111111111333333111
1122222221111111113333333111
1122442221111111133333331111
1122222221111111111333111111
1111221111111111111111111111
```

Fig. 8.8. The label image corresponding to Fig. 8.7.

In order to make use of the concept of connected component labeling (CCL), we must also define a "label image" which is an iconic representation, isomorphic to the original image, in which each pixel contains the number of the component to which it belongs.

One algorithm which produces a label image is known as "region growing." It utilizes a label memory corresponding to the frame buffer just as Fig. 8.8 corresponds to Fig. 8.7. In this description, we will refer to "black" pixels as object and "white" as background.

Initially, each cell in the label memory L is set to zero. We will refer to the picture memory as f. Thus the labeling operation can be written as $L(x, y) \leftarrow N$ for some label number N.

8.3.1 Recursive region growing algorithm

This algorithm implements region growing by using a pushdown stack on which to temporarily keep the coordinates of pixels in the region.

(1) Find an unlabeled black pixel; that is, $L(x, y) = 0$. Choose a new label number for this region, call it N. If all pixels have been labeled, stop.
(2) $L(x, y) \leftarrow N$.
(3) If $f(x-1, y)$ is black and $L(x-1, y) = 0$, push the coordinate pair $(x-1, y)$ onto the stack.
If $f(x+1, y)$ is black and $L(x+1, y) = 0$, push $(x+1, y)$ onto the stack.
If $f(x, y-1)$ is black and $L(x, y-1) = 0$, push $(x, y-1)$ onto the stack.
If $f(x, y+1)$ is black and $L(x, y+1) = 0$, push $(x, y+1)$ onto the stack.
(4) Choose a new (x, y) by popping the stack.
(5) If the stack is empty, go to 1, else go to 2.

This labeling operation results in a set of connected regions, each assigned a unique label number. To find the region to which any given pixel belongs, the computer has only to interrogate the corresponding location in the L memory and read the region number.

EXAMPLE

Applying region growing

Fig. 8.9 shows a 4×7 array of pixels. Assume the initial value of $\langle x, y\rangle$ is $\langle 2, 4\rangle$. Apply algorithm "grow" and show the contents of the stack and L each time step (3) is executed. Let the initial value of N be 1.

Solution

Pass 1. Immediately after execution of step (3). The algorithm has examined pixel $\langle 2, 4\rangle$, examined its 4-neighbors, and detected only one 4-neighbor in the foreground, the pixel at $\langle 3, 4\rangle$. Thus, the coordinates of that pixel are placed on

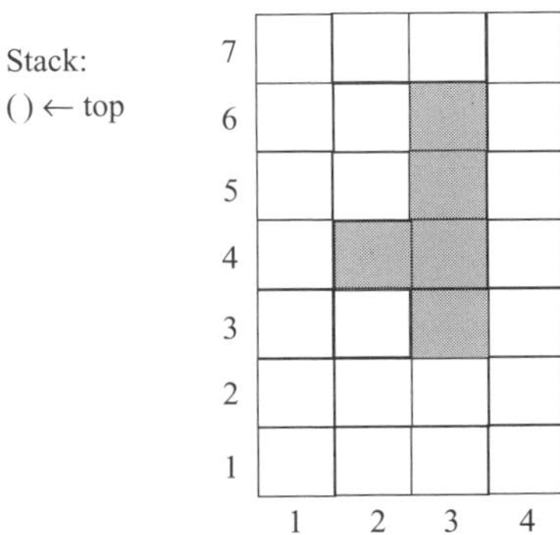

Fig. 8.9. An example image after thresholding.

the stack.

Stack:

$$\boxed{\langle 3, 4\rangle} \leftarrow \text{top}$$

$$L = \begin{array}{c} 7 \\ 6 \\ 5 \\ 4 \\ 3 \\ 2 \\ 1 \\ \\ \end{array} \begin{array}{c} \left[\begin{array}{cccc} 0 & 0 & 0 & 0 \\ 0 & 0 & 0 & 0 \\ 0 & 0 & 0 & 0 \\ 0 & 1 & 0 & 0 \\ 0 & 0 & 0 & 0 \\ 0 & 0 & 0 & 0 \\ 0 & 0 & 0 & 0 \end{array}\right] \\ \begin{array}{cccc} 1 & 2 & 3 & 4 \end{array} \end{array}$$

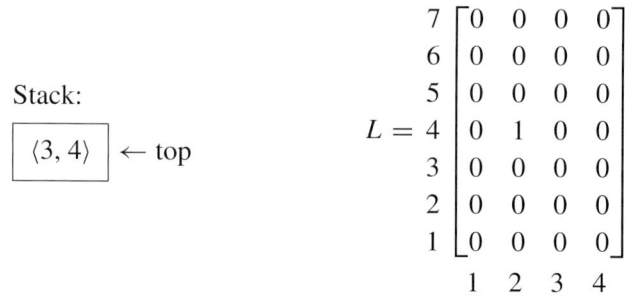

Pass 2. The pixel at $\langle 3, 4\rangle$ was removed from the top of the stack and marked with a 1 in the L image, its neighbors were examined and two 4-neighbors were found, the pixels at $\langle 3, 3\rangle$ and at $\langle 3, 5\rangle$; both were put on the stack.

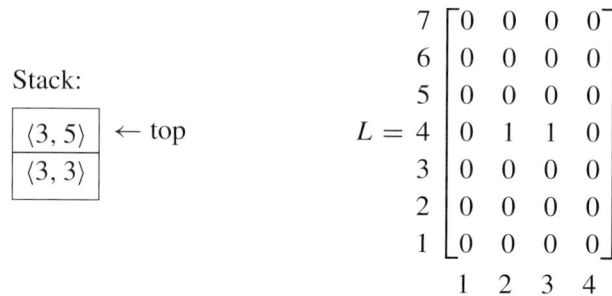

Stack:

$$\begin{array}{|c|l} \hline \langle 3, 5\rangle & \leftarrow \text{top} \\ \hline \langle 3, 3\rangle & \\ \hline \end{array}$$

$$L = \begin{array}{c} 7 \\ 6 \\ 5 \\ 4 \\ 3 \\ 2 \\ 1 \\ \\ \end{array} \begin{array}{c} \left[\begin{array}{cccc} 0 & 0 & 0 & 0 \\ 0 & 0 & 0 & 0 \\ 0 & 0 & 0 & 0 \\ 0 & 1 & 1 & 0 \\ 0 & 0 & 0 & 0 \\ 0 & 0 & 0 & 0 \\ 0 & 0 & 0 & 0 \end{array}\right] \\ \begin{array}{cccc} 1 & 2 & 3 & 4 \end{array} \end{array}$$

Pass 3. The top of the stack contained $\langle 3, 5\rangle$. That pixel was removed from the stack and marked with a 1 in the L image. All its neighbors were examined and one 4-neighbor was found, the pixel at $\langle 3, 6\rangle$. Thus the coordinates of this pixel are put on the stack.

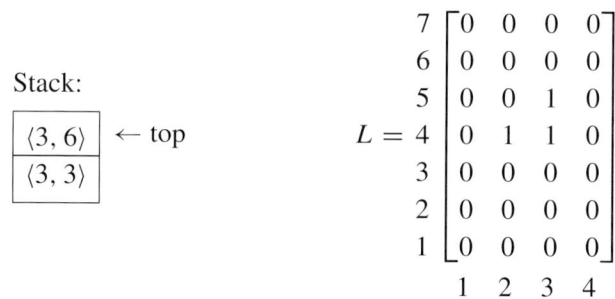

Stack:

$$\begin{array}{|c|l} \hline \langle 3, 6\rangle & \leftarrow \text{top} \\ \hline \langle 3, 3\rangle & \\ \hline \end{array}$$

$$L = \begin{array}{c} 7 \\ 6 \\ 5 \\ 4 \\ 3 \\ 2 \\ 1 \\ \\ \end{array} \begin{array}{c} \left[\begin{array}{cccc} 0 & 0 & 0 & 0 \\ 0 & 0 & 0 & 0 \\ 0 & 0 & 1 & 0 \\ 0 & 1 & 1 & 0 \\ 0 & 0 & 0 & 0 \\ 0 & 0 & 0 & 0 \\ 0 & 0 & 0 & 0 \end{array}\right] \\ \begin{array}{cccc} 1 & 2 & 3 & 4 \end{array} \end{array}$$

Pass 4. The stack was "popped" again, this time removing the $\langle 3, 6\rangle$ and marked with a 1 in the L image. That pixel was examined and determined to have no 4-neighbors which had not already been labeled.

Stack:

| ⟨3, 3⟩ | ← top |

$$L = \begin{matrix} 7 \\ 6 \\ 5 \\ 4 \\ 3 \\ 2 \\ 1 \end{matrix} \begin{bmatrix} 0 & 0 & 0 & 0 \\ 0 & 0 & 1 & 0 \\ 0 & 0 & 1 & 0 \\ 0 & 1 & 1 & 0 \\ 0 & 0 & 0 & 0 \\ 0 & 0 & 0 & 0 \\ 0 & 0 & 0 & 0 \end{bmatrix}$$
$$\quad\quad 1 \quad 2 \quad 3 \quad 4$$

Pass 5. The stack was popped again, removing the ⟨3, 3⟩ and marked with a 1 in the L image. That pixel was examined and determined to have no 4-neighbors which had not already been labeled.

Stack :
() ← top

$$L = \begin{matrix} 7 \\ 6 \\ 5 \\ 4 \\ 3 \\ 2 \\ 1 \end{matrix} \begin{bmatrix} 0 & 0 & 0 & 0 \\ 0 & 0 & 1 & 0 \\ 0 & 0 & 1 & 0 \\ 0 & 1 & 1 & 0 \\ 0 & 0 & 1 & 0 \\ 0 & 0 & 0 & 0 \\ 0 & 0 & 0 & 0 \end{bmatrix}$$
$$\quad\quad 1 \quad 2 \quad 3 \quad 4$$

Pass 6. The stack was popped again, producing a return value of "stack empty" and the algorithm is complete since all black pixels had been labeled.

This region growing algorithm is just one of several strategies for performing *connected component analysis*. Other strategies exist which are faster than the one described, including some that run at raster-scan rates [8.6]. We will now consider one such technique.

8.3.2 An iterative approach to connected component analysis

Since the region growing technique always results in closed regions, this technique is often preferable to other techniques which are based on edge detection or line fitting. Numerous variations and applications of the basic region growing technique have been proposed in the past [8.10, 8.16]. Although region growing has proved to be an integral part of scene analysis, its use can quickly become computationally prohibitive, particularly for high-resolution images. This has prompted the consideration of alternative, faster, more hardware-specific methods of region partitioning.

In this section we present an alternative algorithm to recursive region growing. This algorithm is functionally identical to recursive region growing in that it returns a set of labeled pixels, meeting the adjacency and similarity criteria.

While this algorithm is functionally identical to traditional region growing, it is fundamentally different in concept and potential implementation.

The objective in developing this algorithm was to find a way to achieve the results of region growing, but to do so "on the fly" with a single pass over the data. This result was achieved by using the concept of a content-addressable memory. This memory may be a physical piece of hardware or a lookup-table-driven access method in simulation software.

This algorithm is based on the concept of equivalence relationships between the pixels of an image. Equivalency is defined as follows. Two pixels a and b are defined to be equivalent (denoted $R(a, b)$) if they belong to the same region of an image. This relationship can be shown to be reflexive ($R(a, a)$), symmetric ($R(a, b) \Rightarrow R(b, a)$), and transitive ($R(a, b)$ AND $R(b, c) \Rightarrow R(a, c)$); which makes it an "equivalence relation."

The transitive property enables all pixels in a region to be determined by considering only local adjacency properties. In this algorithm, each pixel will be compared to each adjacent pixel in a left-to-right, top-to-bottom raster-scan fashion. The assignment of a region label to a pixel results from this comparison operation. Pixels in a simple binary image are labeled in raster-scan order. The region labeling proceeds in a straightforward manner: First look to the left, consider that your neighbor and use its label. Then look up to the pixel above, and consider that pixel your neighbor. If both are labeled and the labels are different, you have a problem. Fig. 8.10 demonstrates the situation that can arise as result of this comparison when the equivalence relation $R(1, 2)$ is discovered at the pixel designated by the question mark.

```
111  22
111  22
111  22
111  22
111 11?
```

Fig. 8.10. Ambiguity in label assignment.

The system proposed in [8.60] employs hardware to assign region labels to pixels and to maintain a table of equivalence relationships. Fig. 8.12 shows that this hardware resides between the image memory and a host computer. Functionally, this hardware is transparent to the host computer. For the example of Fig. 8.10, all pixels will be perceived by the host computer as belonging to region 1 (the lower numbered region label takes precedence in the equivalence relationship here).

The operation of the hardware is described in the flowchart of Fig. 8.11. In order to understand this flowchart, the following notation is introduced.

- $f(x, y)$ is the gray-scale value of the (x, y) pixel in the image memory.
- $(x, y)_i$ is the ith adjacent neighbor of the (x, y) pixel.
- $f_i(x, y)$ is the gray-scale value of the ith adjacent neighbor of the (x, y) pixel.
- $L(x, y)$ is the region label corresponding to the (x, y) pixel in the image memory.
- $L_i(x, y)$ is the region label number corresponding to the ith adjacent neighbor of the (x, y) pixel.
- $K(i)$ is the contents or the ith element in the equivalence memory. This memory is a content-addressable memory.

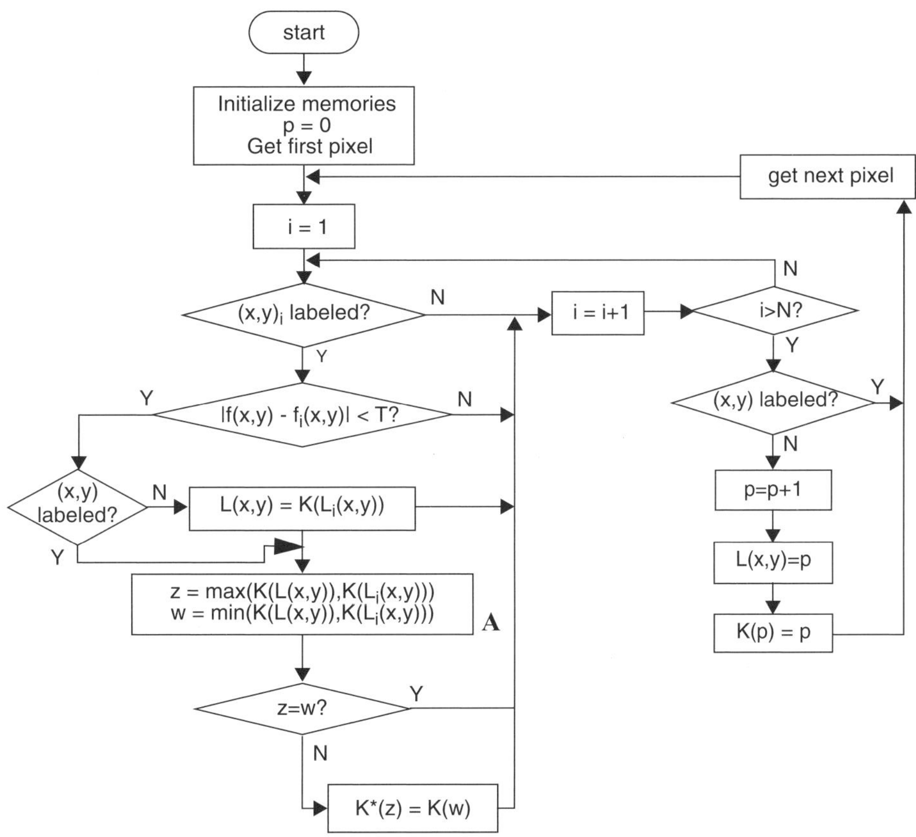

Fig. 8.11. Flowchart of the algorithm (from [8.60]).

- $K^*(i) = K(j)$ implies the sequence:
  - read $K(j)$
  - search K using i as a search key (i.e., determine all l such that $K(l) = i$)
  - write $K(j)$ to all positive responders of the search.
- T is a threshold.
- p is the highest numbered region label (initially 1).
- N is the number of pixels in a neighborhood (four or eight connectivity).

The algorithm is illustrated in Fig. 8.13 for an arbitrary region adapted from Milgram *et al.* [8.41]. The reader should note that the relation $|f(x, y) - f_i(x, y)| < T$ tests if two pixels are similar. There are other similarity measures which could be used, including local first- and/or second-order statistics. If two pixels meet this criterion and they are adjacent, then they are in the same region. By definition, if two pixels are in the same region, then $R(a, b)$ holds. That is,
$\{\text{ADJACENT}(\langle x, y\rangle, \langle x', y'\rangle) \wedge |f(x, y) - f_i(x, y)| < T\} \leftrightarrow R(\langle x, y\rangle, \langle x', y'\rangle).$

The transitive property of R cannot be used to infer that $R(\langle x, y\rangle, \langle x', y'\rangle) \rightarrow |f(x, y) - f_i(x, y)| < T$ without also considering the adjacency property.

As the region partitioning proceeds in real time (i.e., synchronously with the raster scan), two activities must be performed. First, the L memory must be loaded with the region label number of each pixel under consideration, and second, the memory must be updated with all equivalence relationships discovered. For example, if region 4 is actually identical to region 2, then both $K(2)$ and $K(4)$ will contain 2 (the lower numbered region label takes precedence). Hence, when the host computer interrogates pixel (x, y) of the L memory, the interface/processor interprets $L(x, y)$ in terms of the memory and returns $K(L(x, y))$ to the computer. For example, pixel $(11, 11)$ in Fig. 8.13 would be returned as belonging to region 1 since $K(L(11, 11)) = K(4) = 1$.

The difficulty generally encountered in this type of procedure is the problem of chaining. That is, if $R\,(2, 4)$ and $R\,(3, 4)$ have been determined, then $R(2, 3)$ must also be deduced. However, to require that the machine search out all such possibilities after the image has been processed defeats the original objective of performing region partitioning during scan time. Chaining is avoided in this algorithm by ensuring that $K(2) = K(3) = K(4) = 2$. However, this merely transfers the chaining problem to the scanning and labeling process.

Block A of the algorithm flowchart shown in Fig. 8.11 resolves the chaining problem. Whenever an equivalence relationship is detected, all locations in the K memory containing the larger region label number are loaded with the smaller region label number. While the execution of this step in real time may not be feasible for conventional random access memories, it is within the capability of the content-addressable memories discussed in the next section.

An architecture for implementation

The architecture proposed to implement the algorithm of this section is shown in Fig. 8.12. This hardware is intended as a special-purpose processor for an existing computer-based image processing system. The architecture contains four major components: Image memory (f), Region label memory (L), Equivalence memory (K), and an Interface/processor. The gray scale values of the image reside in the image memory. Typically, the f memory would contain 512×512 bytes. The region labels assigned to individual pixels are contained in the region label memory. However, the contents of the L memory also include all intermediate region labels for which equivalence labels were determined. The contents of the L memory are interpreted in terms of the contents of the equivalence memory by straightforward table lookup hardware. The size of the L memory is directly related to the bit length required to represent the region label (including intermediate region labels) associated with each pixel.

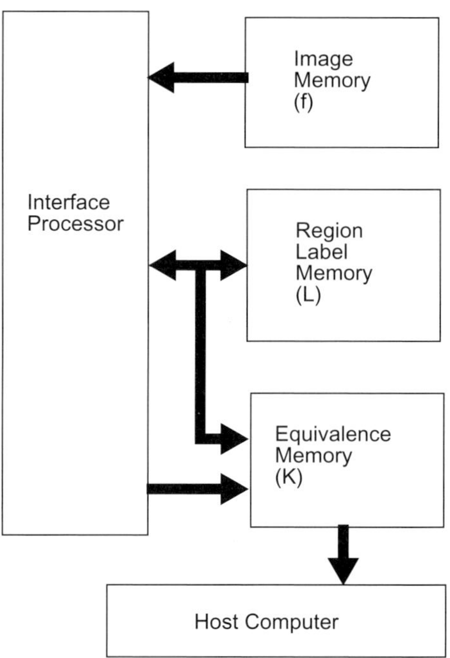

Fig. 8.12. Architecture of a region labeling system.

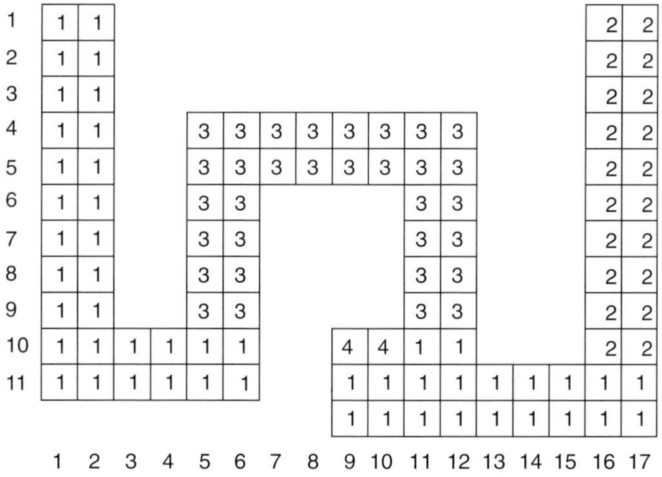

	1	2	3	4	5	6	7	8	9	10	11	12	13	14	15	16	17
1	1	1														2	2
2	1	1														2	2
3	1	1														2	2
4	1	1			3	3	3	3	3	3	3	3				2	2
5	1	1			3	3	3	3	3	3	3	3				2	2
6	1	1			3	3					3	3				2	2
7	1	1			3	3					3	3				2	2
8	1	1			3	3					3	3				2	2
9	1	1			3	3					3	3				2	2
10	1	1	1	1	1	1			4	4	1	1				2	2
11	1	1	1	1	1	1			1	1	1	1	1	1	1	1	1
									1	1	1	1	1	1	1	1	1

Fig. 8.13. An example of a difficult labeling problem (from [8.60]).

Fig. 8.13 illustrates a difficult labeling problem, and Table 8.1 illustrates the labeling process for this example.

The *f* memory and the *L* memory are both conventional random access memories. However, the equivalence memory has two modes of operation. It may be used on a conventional RAM where the address input corresponds to the region table

Table 8.1. *The contents of equivalence memory as the labeling algorithm progresses.*

Pixel		*K memory address*			
x	*y*	1	2	3	4
1	1	1			
16	1	1	2		
5	4	1	2	3	
5	10	1	2	1	
9	10	1	2	1	4
11	10	1	2	1	1
16	11	1	1	1	1

and data output is the equivalent table. In associative memory mode it is used to update that table. In this mode, two activities occur in synchronism with a two-phase clock.

- Phase 1: All memory cells whose contents match the contents of the data bus set their corresponding enable flip-flops (see Fig. 8.14).
- Phase 2: All memory cells whose enable flip-flops are set read the contents of the data bus.

This operation effectively updates the equivalence table in parallel during the scan.

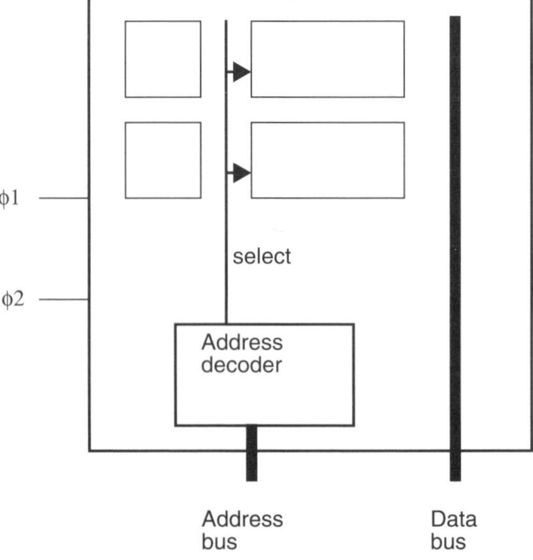

Fig. 8.14. Organization of the *K* memory.

It is also possible to update the table by a search algorithm at the end of the scan. However, doing the updating in parallel during the scan tremendously reduces the number of equivalences (since the lowest number is always used), thus reducing the number of bits needed for the K memory. A content-addressable memory has the property that memory cells can be accessed or loaded by their contents [8.45, 8.69, 8.77].

The parameter most crucial to the development of a satisfactory memory, and hence a system capable of operating in real time, is the memory size. A near real-time system will result if the memory size necessitates a compromise in the access speed. Issues of memory size are discussed further in [8.60] where simulation involving real images is presented.

The final component of the architecture is the interface/processor. The primary purpose of this unit is to execute the algorithm described in this section and flowcharted in Fig. 8.11. Additionally, it must be capable of (1) processing the video signal input into gray-scale values for storage in the memory, and (2) interpreting the L memory in terms of the K memory.

Simulation

The algorithm was applied to a 512×512 image of text data that was thresholded prior to segmentation. Two parameters were of interest: (1) the number of elemental regions (those whose labels are stored in L), since this affects the word width of L and K and the length of K; and (2) the number of regions perceived by the algorithm since this determines the amount of further processing which the host computer must perform before useful information can be gleaned from the image.

The results of one simulation are summarized below.

- *Four-neighbor connectivity*. 912 elemental regions, 138 perceived regions.
- *Eight-neighbor connectivity*. 883 elemental regions, 109 perceived regions.

These results indicate that a $512 \times 512 \times 10$ bit L memory and a 1024×10 bit memory would be required for this image.

In this section we have addressed the issue of performing image analysis operations in real time on television-scanned data. We have shown that it is possible to design hardware which can perform the operation of region growing in this way. The concept of using equivalence relations to partition an input set is fundamental to the algorithm. Furthermore, the use of content-addressable read/write memories facilitates the implementation of such equivalence relation processing in real time.

These concepts were developed by considering potential hardware structures; however, nothing about the algorithm prevents its implementation on a digital computer. The program described here was written to simulate the effectiveness of this approach. Since then, we have used it to label regions in an image segmentation. Its speed of operation, even in simulation, is superior to our earlier region grower, for identical performance.

8.3.3 An alternative to the label image

In this chapter, we primarily use the label image as the means for identifying which pixels belong to which regions. Other authors [8.73] have used a tree-based approach. That is, a graph is established, and all the children of a given parent are assumed to lie in the same region. Thus, levels of the graph correspond to levels of scale. That is, a "parent node" is at a higher (more blurred) level of scale than the children of that node. A child node is defined as being related to a parent if (1) it is geometrically close to the parent and (2) it is similar in intensity. More details are available in [8.73].

8.4 Segmentation of curves

Sometimes, we already have the boundaries of regions, and are interested in describing the boundary in some way which is appropriate for us to be able to characterize either the entire boundary or individual segments. While there are many ways to approach this problem [8.17, 8.42, 8.52], almost all of them involve identifying distinguished points [8.15, 8.26, 8.55, 8.79] along the boundary, and then characterizing the curves between these "salient" points. Obviously, the definition of saliency [8.18, 8.23, 8.48] is critical to the performance of the algorithm.

First, recognize that a curve is a one-dimensional function, which is simply bent in 2-space. That is, a curve can be parameterized using a single parameter, which is usually arc length. However, the arc length parameterization is not invariant to affine transforms [9.1]. For this reason, Rivlin and Weiss [8.49] develop invariants of curves without using a parameter. The x–y coordinates of a point on the curve can then be written as one-dimensional functions of the arc length, $\psi(s) = [x(s), y(s)]$, that is, how far along the curve we have traveled. One can construct any smooth curve up to a rigid body motion if the curvature as a function of arc length is known [8.76]. Of course, we never deal with a smooth, arc-length parameterized curve, we only have sampled versions of such curves, and the relationship between true arc length and the number of pixels through which the curve passes is not as simple as one might think [8.31]. In fact, even for noise-free curves, digitization introduces errors [8.76]. Interestingly, if one *adds* noise, curve positions (at least for straight lines) can be estimated with increased accuracy [8.37].

The *speed* of a curve at a point s is

$$\dot{\psi}(s) = \sqrt{\left(\frac{\partial x}{\partial s}\right)^2 + \left(\frac{\partial y}{\partial s}\right)^2}.$$

The *outward normal direction* to a curve at point s is

$$n_{\psi}(s) = \frac{\left(\frac{\partial}{\partial s}y(s) - \frac{\partial}{\partial s}x(s)\right)}{\psi(s)}.$$

Suppose the curve is closed, then the concepts of INSIDE and OUTSIDE make sense. Given a point in the plane $x = [x_i, y_i]$ which is not on the curve, let ψ_x represent the closest point on the curve to x (at this point, the arc length is defined to be s_x). Then we say x is INSIDE the curve if $[x - \psi_x]n_{\psi}(s_x) \leq 0$ and OUTSIDE otherwise.

There is a way [8.54] to perform curve evolution (see section 9.8) such that the enclosed area remains constant.

You do not necessarily have to find salient points. Chen *et al.* [8.13] simply apply an orientation-selective filter at all possible orientations. If two segments have sufficiently different orientations, the filter response will have multiple peaks, and the position of the peaks can be used to identify the segments. This approach seems to work well for images consisting of straight lines with X or T intersections (see Chapter 10 for a discussion of the types of intersections).

Rosen and West [8.50] propose a slightly different strategy for finding salient points. They fit the sequence of data points with whatever function seems appropriate (ellipses or straight lines). The data point which fits most poorly becomes the salient point. The curve is then divided into two segments, and the fit is repeated recursively on each segment.

8.5 Active contours (snakes)

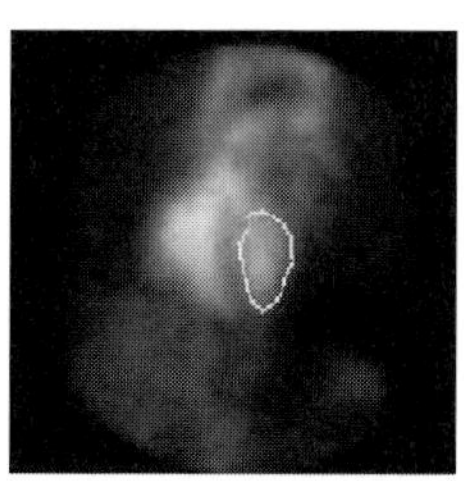

Fig. 8.15. The left ventricle imaged using nuclear medicine.

The concept of active contours was originally developed to address the fact that any edge detection algorithm will fail on some images, because in certain areas of the image, the edge simply does not exist. For example, Fig. 8.15 illustrates a human heart imaged using nuclear medicine. A radioactive drug is introduced into the circulation, and an image is made which reflects the radiation at each point. The brightness at a point is a measure of the integral in a direction perpendicular to the imaging plane of the amount of blood in the area subtended by that pixel. The volume of blood within the ventricle can thus be calculated by summing the brightness over the area of the ventricle. Of course, this requires an accurate segmentation of the ventricle boundary, a problem made difficult by the fact that there is essentially no contrast in the upper left corner of the ventricle. This occurs because radiation from other sources behind the ventricle (superior and inferior vena cava, etc.) contributes to blur out the contrast. Thus, a technique is required which can bridge these rather large gaps – gaps which are really too large to be bridged using the closing techniques of Chapter 7.

The snake is a closed boundary which moves, retaining its connectivity as it moves.

Following the active contour philosophy, a contour is first initialized, either by a user or automatically. The boundary then moves, and moves until many/most of the contour points align with image edge points. An animation of the contour as it searches for boundary points reminds the viewer of the movements of a snake, hence these boundaries are often referred to as "snakes."

Two philosophies are followed in deriving snake algorithms: Energy minimization and partial differential equations (PDEs).

8.5.1 The energy minimization philosophy

The boundary moves to reduce an energy: $E = E_{\mathrm{I}} + E_{\mathrm{E}}$, where the internal energy (E_{I}) characterizes the curve itself, and the external energy (E_{E}) characterizes the image at the points where the snake is currently located.

Internal energy measures the degree of bending along the curve, how big the curve is, etc.

The term with the β coefficient should remind you of something.

The exact form of the internal energy is application-dependent, but the following form is commonly used.

$$E_{\mathrm{I}} = \sum \alpha \|X_i - X_j\| + \beta \|X_{i-1} - 2X_i + X_{i+1}\|,$$

where $X_i = [x_i \; y_i]^{\mathrm{T}}$ is the snake point. Minimizing the first term produces a curve where the snake points are close together. The curve which minimizes the second term will have little bending. A negative aspect of the first term is that it is minimized by a snake which shrinks to a single point. Because of this, many applications also introduce an "expansion" term which causes the entire curve to grow larger.

The external energy measures edginess of the region through which the boundary passes. Again, there are many functions which may be used for this. Our favorite is

$$E_{\mathrm{E}} = \sum \exp(-\|\nabla f(X_i)\|).$$

For two dimensions, the minimization problem is solvable using dynamic programming [9.19]. However, rather than using local edginess at the boundary, one could use the difference in average contrast between outside and inside [9.55], which is only meaningful if the outside is relatively homogeneous.

Observation: This problem fits perfectly into the MAP philosophy, and simulated annealing (SA) can be used [9.78]. However, the search neighborhood is problematic. That is, as we have discussed, SA is guaranteed to find the state which globally minimizes a set of states. However, the set of states must be sampled in order for SA to work. In [9.78], an existing contour is used as a starting point, and at each iteration, the only states sampled are contours within one pixel of the current contour, and from that set a minimum is chosen. The resultant contour is the best contour of the set sampled, but not necessarily from the entire region of interest.

It is also important that the forms chosen for the energy functions be invariant to scale, translation, and rotation. One way to accomplish this is to use two snakes, suitably weighted, one outside the hypothesized boundary and contracting, and one initiated inside and expanding [9.25].

8.5.2 The PDE philosophy

The movement of a contour C can be described by $dC_i/dt = s(X_i)N(X_i)$, where $N(X_i)$ is the normal vector at the ith contour point, C_i is a point on the contour, and s the speed. The notation on the LHS simply describes the motion of the ith boundary point. The RHS says the boundary moves in the normal direction. "Movement" of point C_i actually consists of changing the x, y coordinates of that point.

The speed, s, is the product of two functions

$$s(x, y) = s_{\mathrm{I}}(x, y)s_{\mathrm{E}}(x, y) \tag{8.2}$$

where $s_{\mathrm{I}}(x, y) = \pm 1 - \varepsilon\kappa(x, y)$ and $s_{\mathrm{E}}(x, y) = 1/(1 + \Delta(x, y))$. $\Delta(x, y)$ is a measure of the "edginess" in the image at point x, y, and $\kappa(x, y)$ measures the curvature of the contour at x, y.

Manhaeghe *et al.* obtain a snake-like result using Kohonen maps [9.47]; see also [9.88]. One advantage of this approach is that the computations are local. One can simply look at a point on the present boundary, and consider where that point could be moved. Choose one candidate location and determine if moving to that point increases or decreases the energy (if you are using the energy minimization method).

Considering only the movement of boundary points, however, introduces some problems. The first is the difficulty in accurately computing the curvature from the boundary points. As you know, any derivative-based operator is super-sensitive to noise. Since the curvature involves a second derivative, it is even worse. Another problem is that there is no really effective way to allow for the possibility that the boundary might divide into separate components. The following level-set approach resolves those differences.

Remember the distance transform? From Chapter 7, the distance transform resulted in a function $DT(x, y)$ which was equal to zero on boundary pixels and got larger as one went away from the boundary. Now consider a new version of the distance transform, which is exactly the same OUTSIDE the contour of interest. (Remember, the contour is closed, and so the concepts of INSIDE and OUTSIDE make sense.) Inside the contour, this new function (which we will refer to as the *metric function*) is the negative of the distance transform.

This is the key point to understanding level sets. You are modifying EVERY point, outside and inside the contour.

$$\psi(x, y) = \begin{cases} DT(x, y) & \text{if } (x, y) \text{ is outside the contour} \\ -DT(x, y) & \text{if } (x, y) \text{ is inside the contour.} \end{cases} \tag{8.3}$$

It is important to note that for points on the contour, the metric function is equal to zero. The set of points where $\psi(x, y) = C$ is called the C-level set of ψ, and we are particularly interested in the zero-level set.

Now we will modify the metric function ψ. For every point (x, y) we compute a new value of $\psi(x, y)$. There are several ways we could modify those points, and we will mention some of them below, but remember, the contour of interest is still the set of points where the (modified) metric function takes on a value of zero. We initialized it to be the distance transform, but from here on, you should no longer think of it as a distance transform (although it will retain some of those characteristics). Instead, just think of it as another "brightness," a function of x and y.

What is the gradient of the metric function? You knew how to compute the gradient when it was brightness. It is no different. Thinking about a level set as an isophote, you knew that the gradient is normal to the isophote, so, given the gradient vector, how did you get the normal? Do you remember how to calculate the gradient?

$$\boldsymbol{G}(x, y) = \left[\frac{\partial}{\partial x}\psi(x, y), \frac{\partial}{\partial y}\psi(x, y)\right]^{\mathrm{T}}, \tag{8.4}$$

then the normal is just the normalized (naturally) version of the gradient.

$$\boldsymbol{n}(x, y) = \frac{\boldsymbol{G}(x, y)}{|\boldsymbol{G}(x, y)|}. \tag{8.5}$$

So we can relate the normal to the contour to the gradient of the metric function. We can also relate [9.46] the movement of the contour in the normal direction to a function (which is called the speed function), describing how rapidly the metric changes, producing a differential form for the change in ψ:

$$\frac{\partial}{\partial t}\psi(x, y) = s(\kappa)|\nabla\psi(x, y)|, \tag{8.6}$$

where s is a problem-dependent speed function, with curvature and image edginess as parameters, similar to that defined in Eq. (8.2).

We could modify the metric function in a variety of ways. For example, we could use a form which looks like a gradient descent

Now precisely what is the difference between equations (8.7) and (8.8)?

$$\psi^{n+1}(x, y) = \psi^{n}(x, y) - \alpha s(x, y)|\nabla\psi(x, y)|. \tag{8.7}$$

Or a form which looks like a differential equation

$$\frac{\psi^{n+1}(x, y) - \psi^{n}(x, y)}{\Delta t} = s(x, y)|\nabla\psi(x, y)|, \tag{8.8}$$

where s involves something about the brightness variations in the image and also involves the curvature (in 2D) of the isophote at x, y. Of course, if you insist on using the 2D curvature of the zero level set, you need to relate that to the function ψ, which fortunately is not too hard. Since the normal vector has already been worked out, and the curvature can be related to changes in normal direction, it is possible to show that:

To be consistent with the literature (and so it will fit on one line), we are using the subscript notation for partial derivatives here.

$$\kappa = \frac{\psi_{xx}\psi_x^2 - 2\psi_x\psi_y\psi_{xy} + \psi_{yy}\psi_y^2}{\left(\psi_x^2 + \psi_y^2\right)^{3/2}}, \tag{8.9}$$

where the functional notation has been dropped.

Over the course of the algorithm, the metric function evolves following a rule like Eq. (8.7). As it evolves, it will have zero values at different points, and those points will define the evolution of the contour.

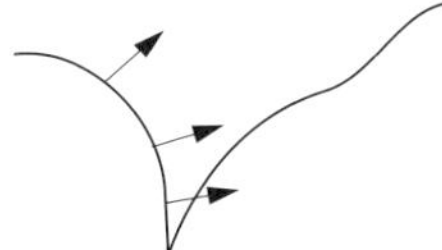

Fig. 8.16. A contour with a very sharp crease in it. A unit step in the normal direction near the crease moves the new contour inside the old.

One interesting detail which one must consider when implementing an algorithm like this is the possibility that the contour might cross itself. For example, consider the contour segment illustrated in Fig. 8.16. The current contour contains a sharp concavity. Some typical normal vectors are illustrated. A unit step in the direction of the normal at the lowest point would place the new contour point inside the contour. One approach to dealing with this is a simple heuristic which states that points which were labeled inside can never again be considered as outside.

The idea of using level sets for adaptive contours was first proposed by Sethian [9.66, 9.67]. Malladi *et al.* [9.46] extended this by observing that there are advantages to considering only a set of points near the current contour. Taubin and Ronfard [9.84] use the concept of a level set implicitly in fitting piecewise-linear curves. Kimmel *et al.* [9.40] demonstrate that level sets may be used for other things, such as finding the shortest path on a surface.

Not all algorithms which use a deformable contour philosophy follow the strategies described in section 8.5. For example, Lai and Chin [8.33] describe a variation which treats the contour points as a sequence of random variables, which may therefore be described by a Markov process, and optimized using MAP strategies. Space does not permit a discussion of those algorithms here, but the reader may find adequate direction in the sources cited in the bibliography at the end of this chapter.

8.6 Segmentation of surfaces

In range images, we have (typically) numerous surfaces. There are generally two philosophies which can be followed. First, one may simply seek surfaces which do not bend too quickly. Following that philosophy produces algorithms which seek smooth solutions, and segment regions along lines of high surface curvature. One example of this philosophy was discussed in Chapter 6 where we describe an algorithm which removes noise while seeking the best piecewise-linear fit to the data points. Such a fit is equivalent to fitting a surface with a set of planes. Points where planes meet produce either "roof" edges or "step" edges, depending on viewpoint. If an annealing algorithm is used such as MFA, good segmentations to more general surfaces can be produced simply by not running the algorithm all the way to a truly planar solution [8.6]. A second philosophical approach to range image segmentation is to assume some equation for the surface, e.g., a quadric (a general second-order surface, defined in section 8.6.1). Then, all points which satisfy that equation and which are adjacent belong to the same surface. This philosophy mixes the problems of segmentation and fitting, for we do not know which points to use to estimate the parameters of the surface until we have some sort of segmentation [8.7, 8.59]. In the next section, we look at these two philosophies in a bit more detail.

8.6.1 Describing surfaces

Implicit and explicit forms of equations were defined in Chapter 4.

In general, we must fit a surface to the data. Taubin *et al.* [8.68] have looked carefully at the question of fitting surfaces to data and observe first that polynomial surfaces are very attractive, but such a polynomial should be of even order. The implicit form is clearly much more attractive, but is much more difficult to fit. Consider for example the second-order forms mentioned in Chapter 4. An explicit representation might be

$$z = ax^2 + by^2 + cxy + dx + ey + f \tag{8.10}$$

and an implicit form is

$$ax^2 + by^2 + cz^2 + dxy + exz + fyz + gx + hy + iz + j = 0. \tag{8.11}$$

The expression of Eq. (8.11) is called a *quadric*, and it is a general form which describes all second-order surfaces (cones, spheres, planes, ellipsoids, etc.). In Chapter 5 you learned how to fit an explicit function to data, by minimizing the squared error. Unfortunately, the explicit form does not allow the possibility of higher order terms in z. You could solve Eq. (8.11) for z, using the quadratic form, and then have an explicit form, but now you have a square root on the right-hand side, and lose the ability to use linear methods for solving for the vector of coefficients.

We can use the implicit form by first defining $f(x, y, z) \equiv ax^2 + by^2 + cz^2 + dxy + exz + fyz + gx + hy + iz + j$ and making the following observation: If the point $[x_i, y_i, z_i]^T$ is on the surface described by the parameter vector $[a, b, c, d, e, f, g, h, i, j]^T$, then $f(x_i, y_i, z_i)$ should be exactly zero. We define a *level set* of a function as the collection of points $[x_i, y_i, z_i]^T$ such that $f(x_i, y_i, z_i) = L$ for some scalar constant L. Thus, we?? can find the coefficients by minimizing $E = \sum_i (f(x_i, y_i, z_i))^2$ (also known as the *algebraic distance* from the point (x_i, y_i, z_i) to the surface). In some cases, this gives good results, but it is not really what we want; we really should be minimizing $\sum_i d([x_i, y_i, z_i], f(x, y, z))$, where d is some distance metric, for example the Euclidean distance from the point to the surface (this is known as the *geometric distance* [8.66] to the surface). Again, this turns out to be algebraically intractable. (To implement this see [8.67] and [17.37] for important details.) Although methods based on the algebraic distance work relatively well most of the time, they can certainly fail. Whatever distance measure we use, it should have the following properties [17.37]: (1) The measure should be zero whenever the true (Euclidean, geometric) distance is zero (the algebraic distance does this); (2) at the sample points, the derivatives with respect to the parameters are the same for the true distance and the measure.

Of course, whatever representation you choose to use (polynomials are popular), there is always a desire for a representation that is invariant to affine transforms [8.27].

8.6.2 Fitting ellipses and ellipsoids

Although this section discusses fitting surfaces, we have introduced the concept of algebraic distance, and this seems to be an appropriate place to talk about the simpler case of fitting ellipses to curve data, as well as the three-dimensional extension to fitting ellipsoids. An ellipse is described by the general equation for a conic section:

$$ax^2 + bxy + cy^2 + dx + ey + f = 0. \tag{8.12}$$

This implicit form describes not only ellipses, but lines, hyperbolae, parabolae, and circles. To guarantee that the resulting curve is an ellipse we must also ensure that it satisfies

$$b^2 - 4ac < 0. \tag{8.13}$$

Satisfying this constraint produces an optimization problem which is nonlinear. If instead, we find the coefficients a–f which minimize

$$\sum_i \left(ax_i^2 + bx_i y_i + cy_i^2 + dx_i + ey_i + f\right)^2, \tag{8.14}$$

we get a solution which tends to fit areas of low curvature with hyperbolic arcs rather than with ellipses. Similar difficulties occur when attempting to fit ellipsoids to range data. See Wang *et al.* [8.74], Rosen and West [8.50], and Fitzgibbon *et al.* [8.19] for more details.

In performing such fits, it is important to know when a point is simply an "outlier," that is, it has been corrupted by substantial noise, but actually belongs to the surface under consideration, or when it really belongs to another, possibly occluding, surface. Darrell and Pentland [13.9] examined this question in some detail and demonstrated that "M-estimates" lead to excellent segmentations. Cabrera and Meer [8.11] remove the bias from fits of an ellipse using an iterative algorithm called "bootstrapping."

How you fit a function to data also depends on the nature of the noise or corruption to the data. If the noise is additive, zero-mean Gaussian (which is what we almost always assume) then the minimum vertical distance (MMSE) or the minimal normal distance (which we called eigenvector line fitting) methods work well. If the noise is not Gaussian, then other methods are more appropriate. For example, nuclear medicine images are corrupted primarily by counting (Poisson) noise. Such a noise differs from Gaussian in two important ways – it is never negative, and it is signal dependent. Well away from zero, Poisson noise may be reasonably modeled by additive Gaussian with a variance equal to the signal. Other sensors produce other types of noise. Stewart [8.64] considers the case of inliers and outliers, but assumes that the bad data are randomly distributed over the dynamic range of the sensor. That is, the noise is not additive.

Given one segmentation, should you merge two adjacent regions? If they are adjacent and satisfy the same equation to within some noise measure, they should be merged [8.8, 8.29, 8.34, 8.56]. Other relevant papers on fitting surfaces include [8.4, 8.75, 8.78].

One is also faced with the issue of what surface measurements to use as a basis for segmentation. Curvature would appear to be particularly attractive since the measurement curvature is invariant to viewpoint. However, "curvature estimates are very sensitive to quantization noise" [8.71].

8.7 Evaluating the quality of a segmentation

As you have concluded by now, there are many algorithms and variations on algorithms for segmentation. But which is the best? Who knows? We need an algorithm to evaluate the quality of segmentations. But which is the best such evaluation algorithm? We need an algorithm which evaluates the quality of evaluation algorithms which ... (help!).

There are several approaches to evaluating segmentation quality. Since one result of a segmentation is edges, you could indirectly infer segmentation quality by measuring edge positions. Pratt [5.33] provides one such algorithm.

Bilbro and Snyder [8.6] first remove noise, then fit the resulting surface. They only consider the quality of the noise removal, which can be tested very simply: Subtract the segmented, cleaned image from the original. What you SHOULD see is just the noise. If your noise-removal algorithm produces an image containing features, then it removed something other than noise.

It is really difficult to evaluate the quality of segmentation of a brightness image, since different human observers will come to different conclusions as to what the correct answer is. With range images, however, it is somewhat easier to determine "truth" since surfaces can physically be measured.

Hoover *et al.* [8.22] propose the following formalism for comparing the quality of a machine-segmented (MS) image using a human-segmented ground-truth (GT) image as the "gold standard." Let M and G denote the MS and GT images respectively; let $M_i(i = 1, \ldots, m)$ denote a region in M; and let $G_j(j = 1, \ldots, m)$ denote a region in G. $|R|$ will denote the number of pixels in region R. Let O_{ij} be the number of pixels which belong to both region i in the MS image and region j in the GT image. Finally, let T be a threshold, $0.5 < T \leq 1.0$.

There are five different segmentation results, defined as follows.

(1) A correct classification occurs when $O_{ij} \geq T|M|$ and $O_{ij} \geq T|G|$.
(2) An *oversegmentation* occurs when a region in the GT image is broken into several regions in the MS image. Formally, given one region in GT, i, and several regions in MS, $j_1, j_2, \ldots, j_n$, if
    (a) at least 100T percent of the pixels in each region of MS actually belong to region i. $(O_{i,j_l} \geq T|j_l|, \forall l)$, and
    (b) at least 100T percent of the pixels which actually belong to region i are marked as belonging to the union of regions $j_1, j_2, \ldots j_n$, $(\sum_{l=1}^{n} O_{i,j_l} \geq T|i|)$.
(3) *Undersegmentation* occurs when pixels in distinct regions in the GT image are identified as belonging to the same region in the MS image. The definition is isomorphic to the definition of oversegmentation with the two images reversed.
(4) A *missed classification* occurs when a region in the GT image is neither correctly segmented, nor is part of an oversegmentation or an undersegmentation.
(5) A *noise classification* is identical to a missed classification except that the region belongs to the MS image.

Two correct segmentations can be further compared in the case of range imagery by computing the normal vectors to the GT region and the corresponding MS region, and finding the absolute value of the angle between these two vectors.

With these definitions, we can evaluate the quality of a segmentation by counting correct or erroneous segmentations, and measuring total angle error. By plotting these measures vs. T and comparing these plots, a measure of segmenter performance may be determined.

Hoover *et al.* [8.22] use this approach to do a thorough evaluation of the quality of four different range image segmentation algorithms.

8.8 Conclusion

In this chapter, we used the concepts of consistency to identify the components of a region. In the first example we studied, if all pixels were the same brightness, they were defined to be in the same region. In the example of section 8.6.1, all the points satisfying the same surface equation were defined to fall in the same region.

Minimum squared error.

In section 8.2.1 we used an optimization method, minimum squared error, to find the best threshold. In section 8.5, we obtained a closed boundary using the philosophy of active contours, by specifying a problem-specific objective function and finding the boundary which minimizes that function. Any appropriate minimization technique could be used. In section 8A.5, we will once again see function minimization (based on gradient descent with annealing) used in a maximum *a posteriori* method which will find the picture which minimizes a particular objective function, in this case resulting in a segmentation.

Gradient descent.

8.9 Vocabulary

You should know the meanings of the following terms.

Active contour
Algebraic distance
Connected component
Explicit form
Geometric distance
Histogram
Homogeneous
Implicit form
Label image
Normal direction
Oversegmentation
Quadric
Region growing
Salient point
Segmentation
Snake
Speed of a curve
Thresholding
Undersegmentation

Assignment 8.1

Section 2.3.5 of the text by Haralick and Shapiro [4.18] describes a labeling algorithm similar in some ways to that presented in this section. Write a

Table 8.2. *Lookup table.*

Location	0	1	2	3	4	5	6	7	8	9
Contents	0	0	2	2	4	4	2	4	2	2

report which compares and contrasts these two techniques. Consider: (1) ease of implementation on a uniprocessor (simplicity of code); (2) speed on a uniprocessor; and (3) potential for parallel implementation.

Assignment 8.2

In the process of a connected component labeling scheme, we find at one point that the lookup table appears as shown in Table 8.2.

Now, we discover the following equivalence: $9 = 7$. On the blank row in Table 8.2, show the contents of the lookup table after resolving the equivalence.

Topic 8A Segmentation

In this chapter, we have up to this point considered partitioning of images into areas which differ in some way in their brightness or range. The algorithms we have presented are applicable, however to any feature which characterizes a pixel or area around a pixel. For this reason, we mention here some other measures which may be used, including texture, color, and motion. In this section, we also mention other approaches to segmentation, including segmentation based on edges.

8A.1 Texture segmentation

In section 4A.2.2, texture is discussed. If one is able to quantify the concept of texture – to assign two different numbers to two different textures in order to distinguish them – then texture can be used as a feature in a segmentation algorithm. Instead of defining ADJACENT as similar brightnesses, one simply defines them as having similar textures. One could also combine color and texture [8.46, 8.65].

Several researchers have observed that there are really two fundamentally different types of textures, those that are in some sense "deterministic," and those that are in some sense "random," and which can in principle be modeled by Markov random fields [8.2]. Liu and Picard [8.38] and others [8.20, 8.21, 8.58] observe that the peaks in the Fourier transform give hints to how separate these texture characteristics are.

Table 8.3. *Halvings which occur to compute the fractal dimension.*

n	0	1	2	n
ε	1	1/2	1/4	$1/2^n$
M_ε	1	2	4	2^n

Table 8.4. *Fractal dimension.*

n	0	1	2	n
ε	1	1/2	1/4	$1/2^n$
M_ε	1	4	16	2^{2n}

8A.1.1 Fractal dimension

The fractal dimension measures the self-similarity of a shape to itself when measured on a different scale. In that way it provides a measure of the spatial distribution of the points within the foreground (which presumably are the object of interest). The utility of the fractal dimension may be seen by considering a set S of points in a 2-space and defining the fractal dimension of S as

$$\dim(S) = \lim_{\varepsilon \to 0} \frac{\log M_\varepsilon}{\log(1/\varepsilon)} \tag{8.15}$$

where M_ε is the minimum number of $\varepsilon \times \varepsilon$ boxes required to cover S. Let's look at the fractal dimension of some example objects. We begin with a single point. Obviously, a point may be covered by precisely one square box, independent of the size of the box. Therefore, in Eq. (8.15), M_ε is always equal to one, and the limit of the denominator is infinity. Therefore we find the fractal dimension of a point is simply zero. Now let us consider a straight line of unit length. Clearly, such a line may be covered by a 1×1 box. However, it may also be covered by two boxes, each $1/2 \times 1/2$, or four boxes, each $1/4 \times 1/4$. Each time the size of the box is halved, the number of boxes required doubles. We could tabulate this process in terms of a parameter n, equal to the number of halvings that have been done (Table 8.3).

From Table 8.3, we may evaluate dim(S) from

$$\dim(S) = \lim_{n \to \infty} \frac{\log 2^n}{\log(1/(1/2))^n} = 1. \tag{8.16}$$

Finally, let S be a square area, and without loss of generality, let its sides be of length one. We could cover this with a single 1×1 box, or by four $1/2 \times 1/2$ boxes, or by 16 $1/4 \times 1/4$ boxes, etc. An argument similar to the one for the straight line results in Table 8.4.

From Table 8.4, the fractal dimension is

$$\dim(S) = \lim_{n \to \infty} \frac{\log 2^{2n}}{\log(1/(1/2))^n} = \lim_{n \to \infty} \frac{\log 2^{2n}}{\log 2^n} = \lim_{n \to \infty} \frac{2n}{n} = 2. \tag{8.17}$$

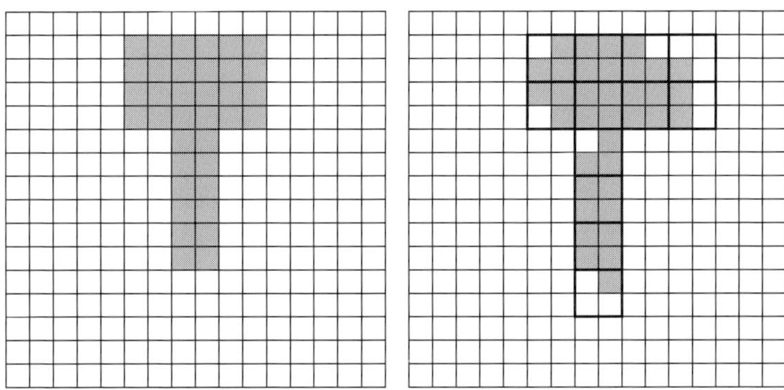

Fig. 8.17. The figure on the left has a fractal dimension of 2.0, the figure on the right has a fractal dimension of 1.58. In the right-hand figure, the covering 2×2 squares are shown with dark lines.

So, we end up with an intuitively appealing result: A point is a zero-dimensional thing, a line is one dimensional, and a square is two dimensional. At least, the result agrees with intuition for these very simple shapes.

Given an image, we must give some thought to how to apply Eq. (8.15) to the discrete pixel domain, since we obviously cannot take a limiting box size less than $\varepsilon = 1$. Solving Eq. (8.15) for M_ε we find for small values of ε

$$\log M_\varepsilon = \left(\log\left(\frac{1}{\varepsilon}\right)\right)\dim(S) \tag{8.18}$$

which simplifies to

$$\log M_\varepsilon = -(\log\varepsilon)\dim(S). \tag{8.19}$$

Thus, for small ε, the log of M_ε is linearly related to the log of ε, with the slope being the dimension of S. We may therefore estimate the dimension of S by considering the two smallest values we have for ε, 1 and 2, and computing that slope, using

$$\dim(S) \approx -\frac{\log M_1 - \log M_2}{\log 1 - \log 2} = \frac{\log M_1 - \log M_2}{\log 2} \tag{8.20}$$

producing a simple algorithm: Find M_1, the number of 1×1 boxes required to cover the object; this is simply the area of the object measured in pixels. Find M_2, the number of 2×2 boxes required to cover the same object, and take the difference of the logs.

Consider the following example. The image on the left of Fig. 8.17 has a foreground region which has an area of 36, and which can be covered by 9 2×2 squares. Thus, its fractal dimension is $(\log 36 - \log 9)/\log 2 = 2$. The figure on the right has the same area, but requires 12 2×2 squares to cover it, and therefore has a fractal dimension of $(\log 36 - \log 12)/\log 2 = 1.58$.

The concept of fractal dimension can be extended to gray-scale images, and gray-scale image features extracted using this measure [8.12].

8A.2 Segmentation of images using edges

One way to segment is to use a connected components algorithm, but to assume there are certain points which cannot be connected to anything. These might be, for example, edge points [8.75].

The usual problem with using edges for segmentation is that edge detectors, as you well know, are both prone to producing extra edges and to losing portions of edges, producing gaps. Jacobs [8.24] has an interesting approach to this problem. He first defines an "acceptable region" in terms of its edges by requiring that the region be convex and that the set of edges bounding the region be mostly real measured rather than inferred. That is, between any set of points, we could hypothesize that edges exist. However, if a particular set of points is really connected, with only a few gaps, and those gaps are small in totality compared to the perimeter of the region, we might believe such a region is real, that is, a salient group. Jacobs comes up with some very clever heuristics to reduce this combinatorial search of possible regions into something fairly manageable.

The Hough transform, described in Chapter 11, will provide another means for identifying segments of edges, even if some parts are missing.

8A.3 Motion segmentation

If one could identify all the connected pixels which have the same motion characteristics, one could apply the connected component methods we have already discussed. Although some papers emphasize motion segmentation – Patras *et al.* [8.47] use watershed methods and a multistage segmentation method – it is critical to any motion segmentation algorithm that it be based on an effective method for detecting and representing motion.

The problem of characterizing image motion has been the object of a great deal of research, and is discussed in much more detail in section 9A.3.

8A.4 Color segmentation

Just as texture variation can produce segmentations, so can color variations. Variations on color segmentation include posing an optimization problem and minimizing an objective function. The image which produces the minimum of the objective function is then the segmentation. Liu and Yang [8.36] use simulated annealing to find a good color-based segmentation. Clustering, which is described briefly in Chapter 15, also gives an approach [8.72].

8A.5 Segmentation using MAP methods

In 1991, Snyder *et al.* [8.63] made the observation that one could convert a restoration problem into a classification problem in a very straightforward way if one had knowledge of the mean and, optionally, the variance, of the brightness of each class [8.39, 8.63]. In 2000, the same conclusion was reached using variational methods [8.53]. In order to "classify" an image surface into different classes, that is, different segments, we can use the same MAP methods as we did in solving the image restoration problems.

In order to make use of the MAP methods, all one need do is modify the prior term to include the expected brightness. For example, suppose we have prior knowledge that an image should be smooth except for step discontinuities, and in addition, each pixel is allowed to only have one of three brightnesses (e.g. csf, white matter, and gray matter). The following prior term is maximal when the brightness at pixel i is identical to its neighbors or when it has the value k_1 or the value k_2, or the value k_3.

$$H(f) = \sum_i \sum_{j \in \aleph_i} \exp(-(f_i - f_j)^2 (f_i - k_1)^2 (f_i - k_2)^2 (f_i - k_3)^2). \tag{8.21}$$

For further reading, Zhu and Yuille [8.80] show that several similar techniques may be combined. Further, they illustrate relationships in a variety of image representations.

8A.6 Human segmentation

One of the aspects of segmentation which, regrettably, we do not have time to address in this book is the issue of how humans do segmentation. That is, what is the "correct" segmentation? For example, Koenderink and van Doorn [8.28] observed that humans tend to perceive the projection of three-dimensional objects as collections of ellipses. See [8.57] for an overview.

Bibliography

[8.1] R. Adams and L. Bischof, "Seeded Region Growing," *IEEE Transactions on Pattern Analysis and Machine Intelligence*, **16**(6), 1994.

[8.2] P. Andrey and P. Tarroux, "Unsupervised Segmentation of Markov Random Field Modeled Textured Images using Selectionist Relaxation," *IEEE Transactions on Pattern Analysis and Machine Intelligence*, **20**(3), 1998.

[8.3] K.E. Batcher, "STARAN Parallel Processor System Hardware," *Proceedings of AFIPS National Computer Conference*, vol 43, pp. 405–410, 1974.

[8.4] J. Berkmann and T. Caelli, "Computation of Surface Geometry and Segmentation using Covariance Techniques," *IEEE Transactions on Pattern Analysis and Machine Intelligence*, **16**(11), 1994.

[8.5] S. Beucher, "Watersheds of Functions and Picture Segmentation," *IEEE International Conference on Acoustics, Speech and Signal Processing*, Paris, May, 1982.

[8.6] G. Bilbro and W. Snyder, "Range Image Restoration using Mean Field Annealing," In *Advances in Neural Network Information Processing Systems*, San Mateo, CA, Morgan-Kaufmann, 1989.

[8.7] G. Bilbro and W. Snyder, "Optimization of Functions with Many Minima," *IEEE Transactions on Systems, Man, and Cybernetics*, **21**(4), July/August, 1991.

[8.8] G. Blais and M. Levine, "Registering Multiview Range Data to Create 3D Computer Objects," *IEEE Transactions on Pattern Analysis and Machine Intelligence*, **17**(8), 1995.

[8.9] K. Boyer, M. Mirza, and G. Ganguly, "The Robust Sequential Estimator: A General Approach and its Application to Surface Organization in Range Data," *IEEE Transactions on Pattern Analysis and Machine Intelligence*, **16**(10), 1994.

[8.10] C.R. Brice and C.L. Fennema, "Scene Analysis using Regions," *Artificial Intelligence*, **1**, pp. 205–226, Fall, 1970.

[8.11] J. Cabrera and P. Meer, "Unbiased Estimation of Ellipses by Bootstrapping," *IEEE Transactions on Pattern Analysis and Machine Intelligence*, **18**(7), 1996.

[8.12] B. Chaudhuri and N. Sarkar, "Texture Segmentation using Fractal Dimension," *IEEE Transactions on Pattern Analysis and Machine Intelligence*, **17**(1), 1995.

[8.13] J. Chen, Y. Sato, and S. Tamura, "Orientation Space Filtering for Multiple Orientation Line Segmentation," *IEEE Transactions on Pattern Analysis and Machine Intelligence*, **22**(5), 2000.

[8.14] C. Chow and T. Keneko, "Automatic Detection of the Left Ventricle from Cine-angiograms," *Computers and Biomedical Research*, **5**, pp. 388–410, 1972.

[8.15] T. Davis, "Fast Decomposition of Digital Curves into Polygons using the Haar Transform," *IEEE Transactions on Pattern Analysis and Machine Intelligence*, **21**(8), August, 1999.

[8.16] R.O. Duda and P.E. Hart, *Pattern Classification and Scene Analysis*, New York, Wiley, 1973.

[8.17] M. Fishler and R. Bolles, "Perceptual Organization and Curve Partitioning," *IEEE Transactions on Pattern Analysis and Machine Intelligence*, **8**(1), 1986.

[8.18] M. Fishler and H. Wolf, "Locating Perceptually Salient Points on Planar Curves," *IEEE Transactions on Pattern Analysis and Machine Intelligence*, **16**(2), 1994.

[8.19] A. Fitzgibbon, M. Pilu, and R. Fisher, "Direct Least Square Fitting of Ellipses," *IEEE Transactions on Pattern Analysis and Machine Intelligence*, **21**(5), May, 1999.

[8.20] J. Francos, "Orthogonal Decomposition of 2-D Random Fields and Their Applications in 2-D Spectral Estimation," *Signal Processing and its Applications*, ed. N.K. Bose and C.R. Rao, Amsterdam, North Holland, 1993.

[8.21] J. Francos, Z. Meiri, and B. Porat, "A Unified Texture Model Based on a 2-D Wold Like Decomposition," *IEEE Transactions on Signal Processing*, **41**, pp. 2665–2678, August, 1993.

[8.22] A. Hoover, G. Jean-Baptiste, X. Jiang, P. Flynn, H. Bunke, D. Goldgof, K. Bowyer, D. Eggbert, A. Fitzgibbon, and R. Fisher, "An Experimental Comparison of Range Image Segmentation Algorithms," *IEEE Transactions on Pattern Analysis and Machine Intelligence*, **18**(7), 1996.

[8.23] L. Itti, C. Koch, and E. Niebur, "A Model of Saliency-based Visual Attention for Rapid Scene Analysis," *IEEE Transactions on Pattern Analysis and Machine Intelligence*, **20**(11), 1998.

[8.24] D. Jacobs, "Robust and Efficient Detection of Salient Convex Groups," *IEEE Transactions on Pattern Analysis and Machine Intelligence*, **18**(1), 1996.

[8.25] K. Kanatani, "Statistical Bias of Conic Fitting and Renormalization", *IEEE Transactions on Pattern Analysis and Machine Intelligence*, **16**(3), 1994.

[8.26] N. Katzir, M. Lindenbaum, and M. Porat, "Curve Segmentation under Partial Occlusion," *IEEE Transactions on Pattern Analysis and Machine Intelligence*, **16**(5), 1994.

[8.27] D. Keren, "Using Symbolic Computation to Find Algebraic Invariants," *IEEE Transactions on Pattern Analysis and Machine Intelligence*, **16**(11), 1994.

[8.28] J. Koenderink and A. Van Doorn, "The Shape of Smooth Objects and the Way Contours End," *Perception*, **11**, pp. 129–137, 1982.

[8.29] K. Köster and M. Spann, "MIR: An Approach to Robust Clustering – Application to Range Image Segmentation," *IEEE Transactions on Pattern Analysis and Machine Intelligence*, **22**(5), 2000.

[8.30] L. Krakauer, "Computer Analysis of Visual Properties of Curved Objects," Project MAC TR-82, 1971.

[8.31] S. Kulkarni, S. Mitter, T. Richardson, and J. Tsitsiklis, "Local vs. Global Computation of Length of Digitized Curves," *IEEE Transactions on Pattern Analysis and Machine Intelligence*, **16**(7), 1994.

[8.32] S. Kumar, S. Han, D. Goldgof, and K. Bowyer, "On Recovering Hyperquadrics from Range Data," *IEEE Transactions on Pattern Analysis and Machine Intelligence*, **17**(11), 1995.

[8.33] K. Lai and R. Chin, "Deformable Contours: Modeling and Extraction," *IEEE Transactions on Pattern Analysis and Machine Intelligence*, **17**(11), pp. 1084–1090, 1995.

[8.34] S. LaValle and S. Hutchinson, "A Bayesian Segmentation Methodology for Parametric Image Models," *IEEE Transactions on Pattern Analysis and Machine Intelligence*, **17**(2), 1995.

[8.35] K. Lee, P. Meer, and R. Park, "Robust Adaptive Segmentation of Range Images," *IEEE Transactions on Pattern Analysis and Machine Intelligence*, **20**(2), 1998.

[8.36] J. Liu and Y. Yang, "Multiresolution Color Image Segmentation," *IEEE Transactions on Pattern Analysis and Machine Intelligence*, **16**(7), 1994.

[8.37] X. Liu and R. Ehrich, "Subpixel Edge Location in Binary Images using Dithering," *IEEE Transactions on Pattern Analysis and Machine Intelligence*, **17**(6), 1995.

[8.38] F. Liu and R. Picard, "Periodicity, Directionality, and Randomness: Wold Features for Image Modeling and Retrieval," *IEEE Transactions on Pattern Analysis and Machine Intelligence*, **18**(7), 1996.

[8.39] A. Logenthiran, W. Snyder, and P. Santago, "MAP Segmentation of Magnetic Resonance Images using Mean Field Annealing," *SPIE Symposium on Electronic Imaging, Science and Technology*, February, 1991.

[8.40] A. Matheny and D. Goldgof, "The Use of Three- and Four-dimensional Surface Harmonics for Rigid and Nonrigid Shape Recovery and Representation," *IEEE Transactions on Pattern Analysis and Machine Intelligence*, **17**(10), 1995.

[8.41] D.L. Milgram, A. Rosenfeld, T. Willett, and G. Tisdale, "Algorithms and Hardware Technology for Image Recognition," *Final Report to U.S. Army Night Vision Lab*, March 31, 1978.

[8.42] F. Mokhtarian and A. Mackworth, "Scale-based Description and Recognition of Planar Curves and Two-dimensional Shapes," *IEEE Transactions on Pattern Analysis and Machine Intelligence*, **8**(1), 1986.

[8.43] K. Mori, M. Kidode, H. Shinoda, and H. Asada, "Design of Local Parallel Pattern Processor for Image Processing," *Proceedings AFIPS National Computer Conference*, vol 47, pp. 1025–1031 June, 1978.

[8.44] W. Ng and C. Lee, "Comment on Using the Uniformity Measure for Performance Measure in Image Segmentation," *IEEE Transactions on Pattern Analysis and Machine Intelligence*, **18**(9), 1996.

[8.45] B. Parhami, "Associative Memories and Processors: An Overview and Selected Bibliography," *Proceedings of the IEEE*, **61**, pp. 772–730, June, 1973.

[8.46] D. Panjwani and G. Healey, "Markov Random Field Models for Unsupervised Segmentation of Textured Color Images," *IEEE Transactions on Pattern Analysis and Machine Intelligence*, **17**(10), 1995.

[8.47] I. Patras, E. Hendriks, and R. Lagendijk, "Video Segmentation by MAP Labeling of Watershed Segments," *IEEE Transactions on Pattern Analysis and Machine Intelligence*, **23**(3), 2001.

[8.48] A. Pikaz and I. Dinstein, "Using Simple Decomposition for Smoothing and Feature Point Detection of Noisy Digital Curves," *IEEE Transactions on Pattern Analysis and Machine Intelligence*, **16**(8), 1994.

[8.49] E. Rivlin and I. Weiss, "Local Invariants for Recognition," *IEEE Transactions on Pattern Analysis and Machine Intelligence*, **17**(3), 1995.

[8.50] P. Rosen and G. West, "Nonparametric Segmentation of Curves into Various Representations," *IEEE Transactions on Pattern Analysis and Machine Intelligence*, **17**(12), 1995.

[8.51] A. Rosenfeld and A. Kak, *Digital Picture Processing*, 2nd edition, New York, Academic Press, 1997.

[8.52] G. Roth and M. Levine, "Geometric Primitive Extraction using a Genetic Algorithm," *IEEE Transactions on Pattern Analysis and Machine Intelligence*, **16**(9), 1994.

[8.53] C. Samson, L. Blanc-Fèraud, G. Aubert, and J. Zerubia, "A Variational Model for Image Classification and Restoration," *IEEE Transactions on Pattern Analysis and Machine Intelligence*, **22**(5), 2000.

[8.54] G. Sapiro and A. Tannenbaum, "Area and Length Preserving Geometric Invariant Scale Spaces," *IEEE Transactions on Pattern Analysis and Machine Intelligence*, **17**(1), 1995.

[8.55] H. Sheu and W. Hu, "Multiprimitive Segmentation of Planar Curves – a Two-level Breakpoint Classification and Tuning Approach," *IEEE Transactions on Pattern Analysis and Machine Intelligence*, **21**(8), 1999.

[8.56] H. Shum, K. Ikeuchi, and R. Reddy, "Principal Component Analysis with Missing Data and Its Application to Polyhedral Object Modeling," *IEEE Transactions on Pattern Analysis and Machine Intelligence*, **17**(9), 1995.

[8.57] K. Siddiqi and B. Kimia, "Parts of Visual Form: Computational Aspects," *IEEE Transactions on Pattern Analysis and Machine Intelligence*, **17**(3), 1995.

[8.58] R. Sriram, J. Francos, and W. Pearlman, "Texture Coding Using a Wold Decomposition Model," *Proc. International Conference on Pattern Recognition*, Jerusalem, October, 1994.

[8.59] W. Snyder and G. Bilbro, "Segmentation of Range Images," *International Conference on Robotics and Automation*, St. Louis, March, 1985.

[8.60] W. Snyder and A. Cowart, "An Iterative Approach to Region Growing," *IEEE Transactions on Pattern Analysis and Machine Intelligence*, **5**(3), 1983.

[8.61] W.E. Snyder and C.D. Savage, "Content-Addressable Read-Write Memories for Image Analysis," *IEEE Transactions on Computers*, **31**(10), pp. 963–967, 1982.

[8.62] W. Snyder, G. Bilbro, A. Logenthiran, and S. Rajala, "Optimal Thresholding, A New Approach," *Pattern Recognition Letters*, **11**(11), 1990.

[8.63] W. Snyder, P. Santago, A. Logenthiran, K. Link, G. Bilbro, and S. Rajala, "Segmentation of Magnetic Resonance Images using Mean Field Annealing," *XII International Conference on Information Processing in Medical Imaging*, Kent, England, July 7–11, 1991.

[8.64] C. Stewart, "MINIPRAN: A New Robust Estimator for Computer Vision," *IEEE Transactions on Pattern Analysis and Machine Intelligence*, **17**(10), 1995.

[8.65] P. Suen and G. Healey, "The Analysis and Recognition of Real-world Textures in Three Dimensions," *IEEE Transactions on Pattern Analysis and Machine Intelligence*, **22**(5), 2000.

[8.66] S. Sullivan, L. Sandford, and J. Ponce, "Using Geometric Distance Fits for 3-D Object Modeling and Recognition," *IEEE Transactions on Pattern Analysis and Machine Intelligence*, **16**(12), 1994.

[8.67] G. Taubin, "Nonplanar Curve and Surface Estimation in 3-space," *IEEE Robotics and Automation Conference*, Philadelphia, May, 1988.

[8.68] G. Taubin, F. Cukierman, S. Sullivan, J. Ponce, and D. Kriegman, "Parameterized Families of Polynomials for Bounded Algebraic Curve and Surface Fitting," *IEEE Transactions on Pattern Analysis and Machine Intelligence*, **16**(3), 1994.

[8.69] K.J. Thurber, *Large-Scale Computer Architecture: Parallel and Associative Processors*, Rochelle Park, NJ, Hayden, 1976.

[8.70] O. Trier and A. Jain, "Goal-directed Evaluation of Binarization Methods," *IEEE Transactions on Pattern Analysis and Machine Intelligence*, **17**(12), 1995.

[8.71] E. Trucco and R. Fisher, "Experiments in Curvature-based Segmentation of Range Data," *IEEE Transactions on Pattern Analysis and Machine Intelligence*, **17**(2), 1995.

[8.72] T. Uchiyama and M. Arbib, "Color Image Segmentation Using Competitive Learning," *IEEE Transactions on Pattern Analysis and Machine Intelligence*, **16**(12), 1994.

[8.73] K. Vincken, A. Koster, and M. Viergever, "Probabilistic Multiscale Image Segmentation," *IEEE Transactions on Pattern Analysis and Machine Intelligence*, **19**(2), 1997.

[8.74] R. Wang, A. Hanson, and E. Riseman, "Fast Extraction of Ellipses," *Ninth International Conference on Pattern Recognition*, Rome, 1988.

[8.75] M. Wani and B. Batchelor, "Edge-region-based Segmentation of Range Images," *IEEE Transactions on Pattern Analysis and Machine Intelligence*, **16**(3), 1994.

[8.76] M. Worring and A. Smeulders, "Digitized Circular Arcs: Characterization and Parameter Estimation," *IEEE Transactions on Pattern Analysis and Machine Intelligence*, **17**(6), 1995.

[8.77] S. Yau and H. Jung, "Associative Processor Architecture – A Survey," *Computer Surveys*, 9, pp. 3–26, March, 1977.

[8.78] X. Yu, T. Bui, and A. Kryzak, "Robust Estimation for Range Image Segmentation and Reconstruction," *IEEE Transactions on Pattern Analysis and Machine Intelligence*, **16**(5), 1994.

[8.79] P. Zhu and P. Chirlian, "On Critical Point Detection of Digital Shapes," *IEEE Transactions on Pattern Analysis and Machine Intelligence*, **17**(8), 1995.

[8.80] S. Zhu and A. Yuille, "Region Competition: Unifying Snakes, Region Growing, and Bayes/MDL for Multiband Image Segmentation," *IEEE Transactions on Pattern Analysis and Machine Intelligence*, **18**(9), 1996.

9 Shape

Space tells matter how to move, and matter tells space to get bent

Douglas Adams

In this chapter, we assume a successful segmentation, and explore the question of characterization of the resulting regions. We begin by considering two-dimensional regions which are denoted by each pixel in the region having value 1 and all background pixels having value 0. We assume only one region is processed at a time, since in studying connected component labeling, we learned how to realize these assumptions.

In the process of generating the segmented version of a picture, the computer performs a region growing operation that acts on each pixel in the region. In so doing, the computer can easily keep track of the area. Area is one of many features that can help us to distinguish one type of object from another. For example, the image of a connecting rod typically occupies more area (more black pixels) than does the image of a valve. Thus, by measuring the area of a region, we may discern the type of object.

9.1 Linear transformations

One topic we will want to consider in this chapter is invariance to various types of linear transformations on regions. That is, consider all the pixels in a region, and write the x, y coordinates of each pixel as a 2-vector, and operate on that set of vectors. The first transformations of interest are the orthogonal transformations such as

$$R_z = \begin{bmatrix} \cos\theta & -\sin\theta \\ \sin\theta & \cos\theta \end{bmatrix},$$

which operate on the coordinates of the pixels in a region to produce new coordinate pairs. For example

$$\begin{bmatrix} x' \\ y' \end{bmatrix} = R_z \begin{bmatrix} x \\ y \end{bmatrix},$$

where R_z is as defined above, represents a rotation about the z axis. Given a region s, we can easily construct a matrix, which we denote by S, in which each column contains the x, y coordinates of a pixel in the region. For example, suppose the region $s = \{(1, 2), (3, 4), (1, 3), (2, 3)\}$, then the corresponding coordinate matrix S is

$$S = \begin{bmatrix} 1 & 3 & 1 & 2 \\ 2 & 4 & 3 & 3 \end{bmatrix}.$$

We can apply an orthogonal transformation such as a rotation to the entire region by matrix multiplication

$$S' = R_z S = \begin{bmatrix} \cos\theta & -\sin\theta \\ \sin\theta & \cos\theta \end{bmatrix} \begin{bmatrix} 1 & 3 & 1 & 2 \\ 2 & 4 & 3 & 3 \end{bmatrix}.$$

That works wonderfully well for rotations, but how can we include translation in this formalism? To accomplish that desire, we augment the rotation matrices by adding a row and a column, all zeros except for a 1 in the lower right corner. With this new definition,

$$R_z = \begin{bmatrix} \cos\theta & -\sin\theta & 0 \\ \sin\theta & \cos\theta & 0 \\ 0 & 0 & 1 \end{bmatrix}.$$

We also augment the definition of a point by adding a 1 in the third position, so the coordinate pair (x, y) in this new notation becomes

$$\begin{bmatrix} x \\ y \\ 1 \end{bmatrix}.$$

Now, we can combine translation and rotation into a single matrix representation (called a homogeneous transformation matrix). We accomplish this by changing the third column to include the translation. For example, to rotate a point about the origin by an amount θ, and then translate it by an amount dx in the x direction and dy in the y direction, we perform the matrix multiplication

$$x' = \begin{bmatrix} \cos\theta & -\sin\theta & dx \\ \sin\theta & \cos\theta & dy \\ 0 & 0 & 1 \end{bmatrix} \begin{bmatrix} x \\ y \\ 1 \end{bmatrix}. \tag{9.1}$$

Thus, it is possible to represent rotation in the viewing plane (about the z axis) and translation in that plane by a single matrix multiplication. All the transformations mentioned above are elements of a class of transformations called "similarity transformations." Similarity transformations are characterized by the fact that they may move an object around, but they do not change its shape.

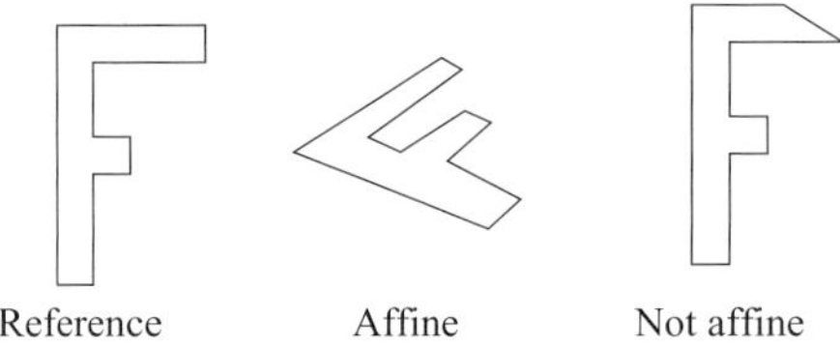

Fig. 9.1. Affine transformations can scale the coordinate axes. If the axes are scaled differently, the image undergoes a shear distortion.

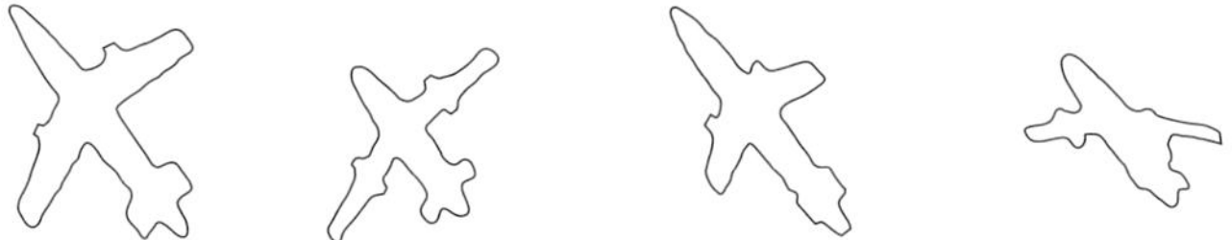

Fig. 9.2. Aircraft images which are affine transformations of each other (from [9.1]).

But what can we do, if anything, to represent rotations out of the camera plane? To answer this, we need to define an *affine transformation*. An affine transformation of the 2-vector $\boldsymbol{x} = [x, y]^{\mathrm{T}}$ produces the 2-vector $\boldsymbol{x}' = [x', y']^{\mathrm{T}}$, where

$$\boldsymbol{x}' = A\boldsymbol{x} + \boldsymbol{b}, \tag{9.2}$$

and $\boldsymbol{b}$ is likewise a 2-vector. This looks just like the similarity transformations mentioned above, except that we do not require the matrix A be orthonormal, only non-singular. An affine transformation may distort the shape of a region. For example, shear may result from an affine transformation, as illustrated in Fig. 9.1.

As you probably recognize, rotation out of the field of view of a planar object is equivalent to an affine transformation of that object. This gives us a (very limited) way to think about rotation out of the field of view. If an object is nearly planar, and the rotation out of the field of view is small, that is, nothing gets occluded, we can represent this 3D motion as a 2D affine transformation. For example, Fig. 9.2 shows some images of an airplane which are all affine transformations of each other.

The matrix which implements an affine transform (after correction for translation) can be decomposed into its constituents [9.62] – rotation, zoom, and shear:

$$\begin{bmatrix} a_{11} & a_{12} \\ a_{21} & a_{22} \end{bmatrix} \begin{bmatrix} x \\ y \end{bmatrix} = \begin{bmatrix} \cos\theta & \sin\theta \\ -\sin\theta & \cos\theta \end{bmatrix} \begin{bmatrix} \alpha & 0 \\ 0 & \delta \end{bmatrix} \begin{bmatrix} 1 & \beta \\ 0 & 1 \end{bmatrix} \begin{bmatrix} x \\ y \end{bmatrix}. \tag{9.3}$$

Now, what does one do with these concepts of transformations? One can correct transformations and align objects together through inverse transformations to assist in shape analysis. For example, one can correct for translation by shifting so the centroid is at the origin. Correcting for rotation involves rotating the image until the principal axes of the object are in alignment with the coordinate axes.

Finding the principal axes is accomplished by a linear transformation which transforms the covariance matrix of the object (or its boundary) into the unit matrix. This transformation is related to the whitening transformation and the K–L transform. Unfortunately, once such a transformation has been done, Euclidian distances between points in the region have been changed [9.92].

In the previous paragraph, the word "distance" occurs. Although one usually thinks that "distance" means "Euclidian distance," in this book we will use this word many times and in many forms. For that reason, we should be just a bit more rigorous about exactly what the concept means. The Euclidian distance (for example) is a type of measure called a "metric." Any metric $d(a, b)$ has the following properties:

Properties of a metric.

- $d(a, a) = 0 \qquad \forall a$
- $d(a, b) = d(b, a) \qquad \forall (a, b)$
- $d(a, b) + d(b, c) \geq d(a, c) \qquad \forall (a, b, c)$

We will have several opportunities to examine metrics in later chapters.

9.2 Transformation methods based on the covariance matrix

Consider the distribution of points illustrated in Fig. 9.3. Each point may be exactly characterized by its location, the ordered pair, (x_1, x_2), but neither x_1 nor x_2, by itself, is adequate to describe the point.

Now consider Fig. 9.4, which shows two new axes, y_1 and y_2. Again, the ordered pair $\langle y_1, y_2 \rangle$ is adequate to exactly describe the point, but y_2 is (compared to y_1) very nearly zero, most of the time. Thus, we would lose very little if we simply discarded y_2 and used the scalar, y_1 to describe each point. Our objective in this section will be to learn how to determine y_1 and y_2 in an optimal manner.

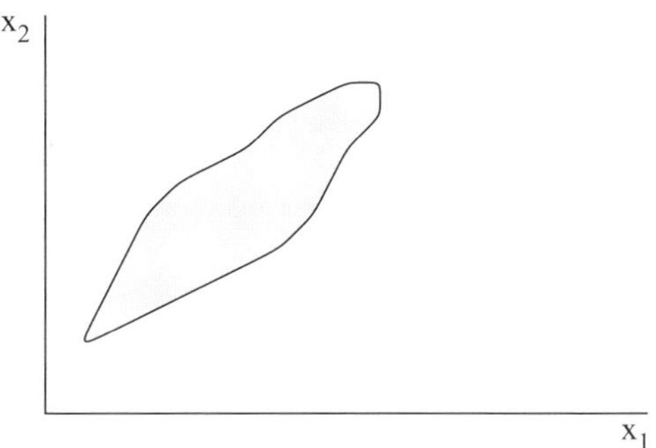

Fig. 9.3. A region which is approximately an ellipsoid.

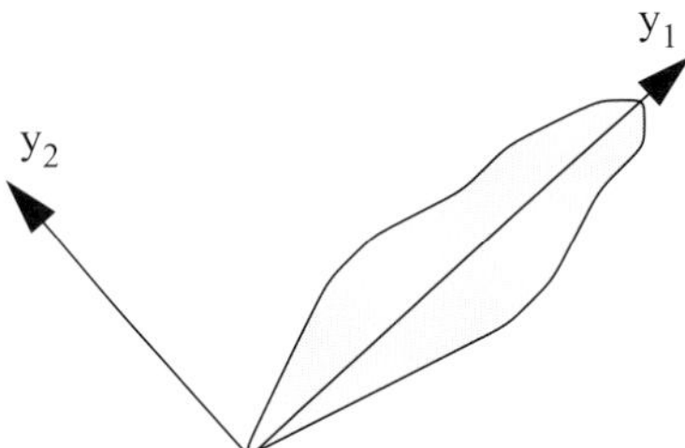

Fig. 9.4. A new set of coordinates, derived by a rotation of the original coordinates, in which one coordinate well represents the data.

9.2.1 Derivation of the K–L expansion

Let $\boldsymbol{x}$ be a d-dimensional random vector. We will describe $\boldsymbol{x}$ in terms of a set of *basis vectors*. That is, represent $\boldsymbol{x}$ by

$$\boldsymbol{x} = \sum_{i=1}^{d} y_i \boldsymbol{b}_i. \tag{9.4}$$

Here, the vectors $\boldsymbol{b}_i$ are deterministic (and may in general, be specified in advance). Since any random vector $\boldsymbol{x}$ may be expressed in terms of the same d vectors, $\boldsymbol{b}_i (i = 1, \ldots, d)$, we say the vectors $\boldsymbol{b}_i$ *span the space* containing $\boldsymbol{x}$, and refer to them as a *basis set* for $\boldsymbol{x}$. To make further use of the basis set, we will require:[1]

(1) The $\boldsymbol{b}_i$ vectors are linearly independent;
(2) The $\boldsymbol{b}_i$ vectors are orthonormal, i.e.,

$$\boldsymbol{b}_i^{\mathrm{T}} \boldsymbol{b}_j = \begin{cases} 1 & (i = j) \\ 0 & (i \neq j) \end{cases}. \tag{9.5}$$

Under these conditions, the y_i may be found by *projections*, where the projection operation is defined by

$$y_i = \boldsymbol{b}_i^{\mathrm{T}} \boldsymbol{x} (i = 1, \ldots, d) \tag{9.6}$$

and we let

$$\boldsymbol{y} = [y_1, \ldots, y_d]^{\mathrm{T}}. \tag{9.7}$$

Here, we say the number y_i results from projecting $\boldsymbol{x}$ onto the basis vector $\boldsymbol{b}_i$.

Suppose we wish to ignore all but $m (m < d)$ components of $\boldsymbol{y}$ (which we will call the *principal components*) and yet we wish to still represent $\boldsymbol{x}$, although with some error. We will thus calculate (by projection onto the basis vectors) the first m

1 To be a basis, they do not have to be orthonormal, just not parallel, but here, we will require orthonormality.

elements of $\boldsymbol{y}$, and replace the others with constants, forming the estimate

$$\hat{\boldsymbol{x}} = \sum_{i=1}^{m} y_i \boldsymbol{b}_i + \sum_{i=m+1}^{d} \alpha_i \boldsymbol{b}_i . \tag{9.8}$$

The error which we have introduced by using some arbitrary constants, the alphas of Eq. (9.8), rather than the elements of $\boldsymbol{y}$, is given by

$$\begin{aligned} \Delta \boldsymbol{x} &= \boldsymbol{x} - \hat{\boldsymbol{x}} \\ &= \boldsymbol{x} - \left[\sum_{i=1}^{m} y_i \boldsymbol{b}_i + \sum_{i=m+1}^{d} \alpha_i \boldsymbol{b}_i \right] \\ &= \sum_{i=m+1}^{d} [y_i - \alpha_i] \boldsymbol{b}_i . \end{aligned} \tag{9.9}$$

If we think of $\boldsymbol{x}$ and therefore $\Delta \boldsymbol{x}$ as random vectors, we can use the expected magnitude of $\Delta \boldsymbol{x}$ to quantify how well our representation works.

$$\begin{aligned} \varepsilon^2(m) &= E\left\{ \sum_{i=m+1}^{d} \sum_{j=m+1}^{d} (y_i - \alpha_i) \boldsymbol{b}_i^{\mathrm{T}} (y_j - \alpha_j) \boldsymbol{b}_j \right\} \\ &= E\left\{ \sum \sum (y_i - \alpha_i)(y_j - \alpha_j) \boldsymbol{b}_i^{\mathrm{T}} \boldsymbol{b}_j \right\} \end{aligned} \tag{9.10}$$

noting that y_i is a scalar, and recalling Eq. (9.5), we have

$$\varepsilon^2(m) = \sum_{i=m+1}^{d} E\left\{ (y_i - \alpha_i)^2 \right\}. \tag{9.11}$$

To find the optimal choice for α_i, we minimize ε^2 with respect to α_i:

$$\frac{\partial \varepsilon^2}{\partial \alpha_i} = \frac{\partial}{\partial \alpha_i} E\left\{ (y_i - \alpha_i)^2 \right\} = -2(E\{y_i\} - \alpha_i) = 0, \tag{9.12}$$

resulting in

$$\alpha_i = E\{y_i\} = \boldsymbol{b}_i^{\mathrm{T}} E\{\boldsymbol{x}\}. \tag{9.13}$$

So, we should replace those elements of $\boldsymbol{y}$ which we do not measure by their expected values – mathematically and intuitively appealing.

Substituting Eq. (9.13) into Eq. (9.11), we have

$$\varepsilon^2(m) = \sum_{i=m+1}^{d} E\left[(y_i - E\{y_i\})^2 \right]. \tag{9.14}$$

Substituting Eq. (9.6) into Eq. (9.14):

$$\varepsilon^2(m) = \sum_{i=m+1}^{d} E\left[\left(\boldsymbol{b}_i^{\mathrm{T}} \boldsymbol{x} - E\left\{ \boldsymbol{b}_i^{\mathrm{T}} \boldsymbol{x} \right\} \right)^2 \right]$$

$$= \sum_i E\left[\left(b_i^{\mathrm{T}} x - E\left\{b_i^{\mathrm{T}} x\right\}\right)(x^{\mathrm{T}} b_i - E\{x^{\mathrm{T}} b_i\})\right] \tag{9.15}$$
$$= \sum_i E\left[b_i^{\mathrm{T}}(x - E\{x\})(x^{\mathrm{T}} - E\{x^{\mathrm{T}}\})b_i\right].$$

Equation (9.15) can be reformatted into

$$\varepsilon^2(m) = \sum_i b_i^{\mathrm{T}} E[(x - E\{x\})(x - E\{x\})^{\mathrm{T}}] b_i \tag{9.16}$$

and we now recognize the term between the ***b***s in Eq. (9.16) as the covariance of ***x***:

$$\varepsilon^2(m) = \sum_{i=m+1}^{d} b_i^{\mathrm{T}} K_x b_i. \tag{9.17}$$

It can be shown (and we will show it below, for the special case of fitting a straight line) that the choice of vector b_i which minimizes Eq. (9.17) also satisfies

$$K_x b_i = \lambda_i b_i. \tag{9.18}$$

That is, the optimal basis vectors are the eigenvectors of K_x.

The covariance of ***y*** can be easily related to K_x

Proof is a homework.

$$K_y = E[(y - E\{y\})(y - E\{y\})^{\mathrm{T}}] \tag{9.19}$$
$$= B^{\mathrm{T}} K_x B \tag{9.20}$$

where the matrix B has columns made from the basis vectors, $b_1, b_2, \ldots, b_d$.

Furthermore, in the case that the columns of B are the eigenvectors of K_x, then B will be the transformation which diagonalizes K_x, resulting in

$$K_y = \begin{bmatrix} \lambda_1 & 0 & \cdots & 0 \\ 0 & \lambda_2 & \cdots & 0 \\ \cdots & \cdots & \cdots & \cdots \\ 0 & 0 & \cdots & \lambda_d \end{bmatrix}. \tag{9.21}$$

Substituting Eq. (9.21) into Eq. (9.17), we find

$$\varepsilon^2(m) = \sum_{i=m+1}^{d} b_i^{\mathrm{T}} \lambda_i b_i. \tag{9.22}$$

Since λ_i is scalar,

$$\varepsilon^2(m) = \sum_{i=m+1}^{d} \lambda_i b_i^{\mathrm{T}} b_i \tag{9.23}$$

and, again remembering the orthonormal conditions on b_i,

$$\varepsilon^2(m) = \sum_{i=m+1}^{d} \lambda_i. \tag{9.24}$$

Thus, we represent a d-dimensional measurement, $\boldsymbol{x}$, by an m-dimensional vector, $\boldsymbol{y}$, where $m < d$,

$$y_i = \boldsymbol{b}_i^{\mathrm{T}} \boldsymbol{x}, \tag{9.25}$$

and the $\boldsymbol{b}_i$s are the eigenvectors of the covariance of $\boldsymbol{x}$.

This expansion of a random vector in terms of the eigenvectors of the covariance matrix is referred to as the *Karhunen–Loève* expansion, or the "K–L expansion."

9.2.2 Properties of the K–L expansion

Without loss of generality, we will sort the eigenvectors $\boldsymbol{b}_i$ in terms of their eigenvalues. That is, assign subscripts to the eigenvalues such that

$$\lambda_1 \geq \lambda_2 \geq \lambda_3 \geq \ldots \lambda_d. \tag{9.26}$$

Then, we refer to $\boldsymbol{b}_1$, corresponding to λ_1, as the "major eigenvector."

Interpretation as a hyperellipsoid

Representing data as an ellipsoid will be useful when we discuss aspect ratio.

If we think of the distribution of points, $\boldsymbol{x}$, as represented by a hyperellipsoid, the major axis of that ellipse will pass through the center of gravity of the data, and will be in the direction of the eigenvector corresponding to the largest eigenvalue of K_x. This is illustrated in Fig. 9.5. Thus, the K–L transform fits an ellipse to two-dimensional data, an ellipsoid to three-dimensional data, and a hyperellipsoid to higher dimensional data.

Use in straight line fitting

Consider the set of instances of the random vector $\boldsymbol{x}$.

$$\{\boldsymbol{x}_i\}(i = 1, \ldots n). \tag{9.27}$$

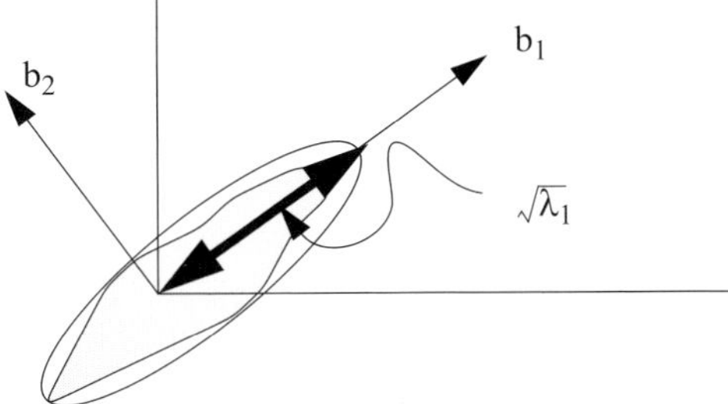

Fig. 9.5. A covariance matrix may be thought of as representing a hyperellipsoid, oriented in the directions of the eigenvectors, and with extent in those directions equal to the square root of the corresponding eigenvalues.

We wish to find the straight line which best fits this set of data. Move the origin to the center of gravity of the set. Then, characterize the (currently unknown) best fitting line by its unit normal vector, $\boldsymbol{n}$. Then, for each point $\boldsymbol{x}_i$, the perpendicular distance from $\boldsymbol{x}_i$ to the best fitting line will be equal to the projection of that point onto $\boldsymbol{n}$. Denote this distance by $d_i(\boldsymbol{n})$:

$$d_i^2(\boldsymbol{n}) = (\boldsymbol{n}^{\mathrm{T}}\boldsymbol{x}_i)^2. \tag{9.28}$$

To find the best fitting straight line, we minimize the sum of the squared perpendicular distances

$$\varepsilon^2 = \sum_{i=1}^{n} d_i^2(\boldsymbol{n}) = \left(\sum_{i=1}^{n}(\boldsymbol{n}^{\mathrm{T}}\boldsymbol{x}_i)^2\right) = \sum_{i=1}^{n}(\boldsymbol{n}^{\mathrm{T}}\boldsymbol{x}_i)(\boldsymbol{x}_i^{\mathrm{T}}\boldsymbol{n}) = \boldsymbol{n}^{\mathrm{T}}\left[\sum_{i=1}^{n}\boldsymbol{x}_i\boldsymbol{x}_i^{\mathrm{T}}\right]\boldsymbol{n} \tag{9.29}$$

which we wish to minimize, subject to the constraint that $\boldsymbol{n}$ is a unit vector,

$$\boldsymbol{n}^{\mathrm{T}}\boldsymbol{n} = 1. \tag{9.30}$$

We perform the constrained minimization using a Lagrange multiplier and minimizing

$$\boldsymbol{n}^{\mathrm{T}}\left(\sum_i \boldsymbol{x}_i\boldsymbol{x}_i^{\mathrm{T}}\right)\boldsymbol{n} - \lambda(\boldsymbol{n}^{\mathrm{T}}\boldsymbol{n} - 1). \tag{9.31}$$

Defining $S = \sum_i \boldsymbol{x}_i\boldsymbol{x}_i^{\mathrm{T}}$, we take

$$\frac{\partial}{\partial \boldsymbol{n}}(\boldsymbol{n}^{\mathrm{T}}S\boldsymbol{n} - \lambda(\boldsymbol{n}^{\mathrm{T}}\boldsymbol{n} - 1)). \tag{9.32}$$

Differentiating the quadratic form $\boldsymbol{n}^{\mathrm{T}}S\boldsymbol{n}$, we get $2S\boldsymbol{n}$, and setting the derivative to zero results in

$$2S\boldsymbol{n} - 2\lambda\boldsymbol{n} = 0, \tag{9.33}$$

which is the same eigenvalue problem mentioned earlier. Thus we may state:

The best fitting straight line passes through the mean of the set of data points, and will lie in the direction corresponding to the major eigenvector of the covariance of the set.

We have now seen two different ways to find the straight line which best fits data: The method of least squares described in section 5.3, if applied to fitting a line rather than a plane, minimizes the vertical distance from the data points to the line. The method described in this section minimizes the perpendicular distance described by Eq. (9.29). Other methods also exist. For example, [9.53] finds piecewise representations which preserve moments up to an arbitrarily specified order.

Fitting functions to data occurs in many contexts. For example, O'Gorman [9.54] has looked at fitting not only straight edges, but points, straight lines, and regions with straight edges. By so doing, subpixel precision can be obtained.

In the following, we will consider a few of the many simple features which may be used to characterize the shape of regions (for additional features, see also [9.2]).

9.3 Simple features

In this section, we describe several simple features which can be used to describe the shape of a patch – the output of the segmentation process. Many of these features can be computed as part of the segmentation process itself. For example, since the connected component labeling program must touch every pixel in the region, it can easily keep track of the area.

The following is a list of simple features which are likewise simple to calculate.

- *Average gray value*. In the case of black and white "silhouette" pictures, this is simple to compute.
- *Maximum gray value*. Is straightforward to compute.
- *Minimum gray value*. Is straightforward to compute.
- *Area (A)*. A count of all pixels in the region.
- *Perimeter (P)*. Several different definitions exist. Probably the simplest is a count of all pixels in the region that are adjacent to a pixel not in the region.
- *Diameter (D)*. The diameter describes the maximum chord – the distance between those two points on the boundary of the region whose mutual distance is maximum [9.68, 9.71]. We will discuss computation of this parameter in the next section.
- *Thinness (also called compactness)*[2] *(T)*. Two definitions for compactness exist: $T_a = (P^2/A) - 4\pi$ measures the ratio of the squared perimeter to the area; $T_b = D/A$ measures the ratio of the diameter to the area. Fig. 9.6 compares these two measurements on example regions.
- *Center of gravity (CG)*. The x and y coordinates of the center of gravity may be written as

  $$m_x = \frac{1}{N}\sum x, \quad m_y = \frac{1}{N}\sum y$$

  for all N points in a region. However, we prefer the vector form

  $$\boldsymbol{m} = \frac{1}{N}\sum_{i=1}^{N}\begin{bmatrix} x_i \\ y_i \end{bmatrix}. \tag{9.34}$$

2 Some authors [9.69] prefer not to confuse the mathematical definition of compactness with this definition, and thus refer to this measure as the *isoperimetric measure*.

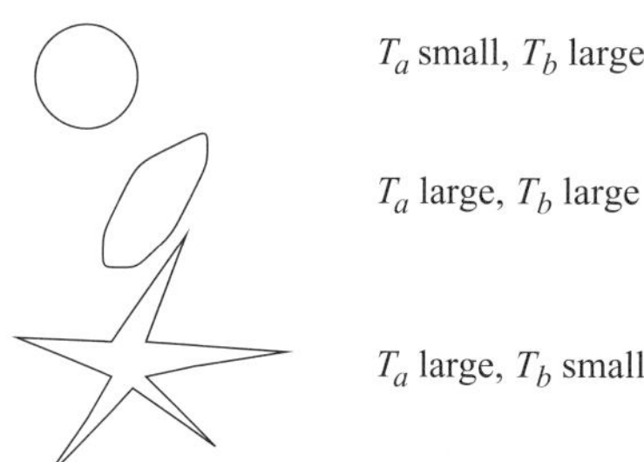

Fig. 9.6. Results of applying two different compactness measures to various regions. Since a circle has the minimum perimeter for a given area, it minimizes T_a. A starfish on the other hand, has a large perimeter for the same area.

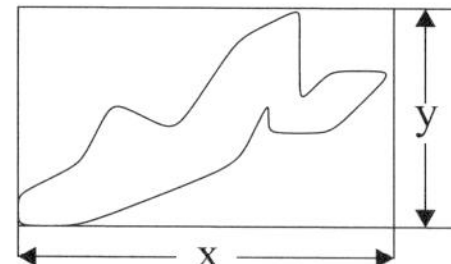

Fig. 9.7. y/x is the aspect ratio using one definition, with horizontal and vertical sides to the bounding rectangle.

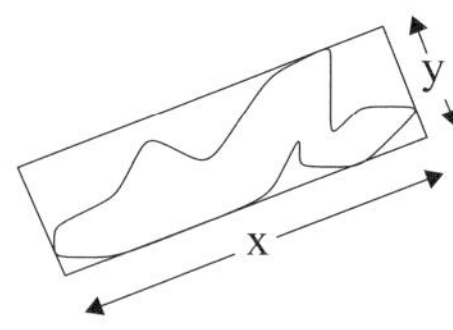

Fig. 9.8. y/x is the minimum aspect ratio.

The methods of section 9.2.2 give a simple way to estimate the aspect ratio.

- *X–Y aspect ratio.* See Fig. 9.7. The aspect ratio is the length/width ratio of the bounding rectangle of the region. This is simple to compute.
- *Minimum aspect ratio.* See Fig. 9.8. Again, a length/width, but much more computation is required to find the minimum such rectangle.

  The minimum aspect ratio can be a difficult calculation, since it requires a search for extremal points. A very good approximation can be obtained if we think of a region as represented by an ellipse-shaped distribution of points. In this case, as we discussed in Fig. 9.5, the eigenvalues of the covariance of points are measures of the distribution of points along orthogonal axes – major and minor. The ratio of those eigenvalues is quite a good approximations of the minimum aspect ratio.
- *Number of holes.* One feature that is very descriptive and reasonably easy to compute is the number of holes in a region.
- *Triangle similarity.* Consider three points on the boundary of a region, P_1, P_2, P_3, let $d(P_i, P_j)$ denote the Euclidian distance between two of those points, and let $S = d(P_1, P_2) + d(P_2, P_3) + d(P_3, P_1)$ be the perimeter of that triangle. The 2-vector

$$\left[\frac{d(P_1, P_2)}{S}, \frac{d(P_2, P_3)}{S}\right], \tag{9.35}$$

  simply the ratio of side lengths to perimeter, is invariant to rotation, translation, and zoom [9.6].
- *Symmetry.* In two dimensions, a region is said to be mirror-symmetric if it is invariant under a reflection about a line. That line is referred to as the *axis of symmetry.* A region is said to have rotational symmetry of order n if it is invariant under a rotation of $2\pi/n$ about some point, usually the center of gravity of the region. There are two challenges in determining the symmetry of regions. The first is simply determining the axis. The other is to answer the question: "How symmetric is it?" Most papers prior to 1995 which analyzed symmetry of regions

in machine vision applications treated symmetry as a predicate: Either the region is symmetric or it is not. Zabrodsky *et al.* [9.97] present a measure called the symmetry distance which quantifies *how* symmetric a region is.

9.3.1 Computing the diameter

The diameter, mentioned previously, is a robust measure of shape, when taken with other suitable measures, and when calculated appropriately. For this reason, we devote a section to the process of actually performing that calculation.

Consider for a moment, the following scenario: "I" (who happen to be simulating a digital computer) am looking for a region in an image. I am only willing to accept regions which have a particular appearance – namely, such regions should be long, thin, and oriented in a particular direction. How can I quantify such vague terms as "long," "thin," and what precisely is the orientation of a region? After all, a region is simply a list of points which was the output from a region growing routine. There are, of course, a large number of methods described in the literature for describing shape of regions. One approach to solving such problems makes use of the location of the "extremes" or diameter of the region; that is, those two points, A and B, whose mutual distance $d(A, B)$ is greater than or equal to the distance between any other two points in the region.

This calculation can find applications when it is necessary to quantify the "shape" of a region in an image, as might occur in classification of parts on an assembly line, optical character recognition, etc., or in a robotic application which requires that the orientation of a region be determined in order to pick up an object.

Suppose we have a region in a two-dimensional image plane described by a set of points, R. The problem then is to find two points, $A, B \in R$, such that $\forall p_1, p_2 \in R, d(A, B) \geq d(p_1, p_2)$. If the set R is small (say, 10–20 points) a simple comparison of every point with every other point is the most straightforward approach. However, when R becomes larger, the number of comparisons needed grows as $n(n-1)/2$ or $O(n^2)$.

In this section, we mention a technique for solving this problem. It uses eigenvalue analysis to find the "best" (in a generalized least squares sense) estimate of the major axis of the region, and defines the extrema as being the two points furthest out in the direction of that axis. A second approach, which is applicable to nonconvex regions, is described in section 9A.1. This technique is heuristic and relatively complicated to implement, but is fast and guaranteed to converge.

The principal axis approach

In this method we first find the major axis of the region and those two points on the boundary of the region which are closest to that axis. In this approach, minor deviations, such as the spur shown in Fig. 9.9, will be ignored, even though they may actually contain one of the extreme points.

Assignment: Is this algorithm new, or is it repeated somewhere else in the text?

The first step in the process is calculation of the major axis. This is performed by a minimization-of-squared-error technique. It is important that minimization of error be independent of the coordinate axes; therefore, we use the eigenvector line fitting technique described earlier in this section rather than the conventional MSE technique. We define a line to be the best representation of the major axis if it minimizes the sum of squares of the perpendicular distances from the points in the region to the line.

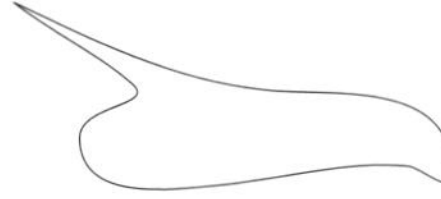

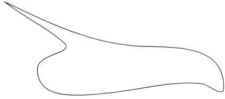

Fig. 9.9. A region with a spur.

Let us assume the region R is described by a set of points $R = \{(x_i, y_i)|i = 1, \ldots, n\}$. Let the point (x_i, y_i) be denoted by the vector $\mathbf{v}_i$, and d_i be the perpendicular distance from $\mathbf{v}_i$ to the major axis. Thus, the major axis of the region is the line which minimizes:

$$d^2 = \sum_{i=1}^{n} d_i^2. \tag{9.36}$$

It is easily shown that the major axis must pass through the center of gravity of the region, thus it is necessary only to find the slope of the axis. Since the line passes through the center of gravity, let us take that point as the origin of our coordinate system. Then the problem becomes: Given n points with zero mean, find the line through the origin which minimizes d^2.

This turns out to be the same eigenvector problem we described earlier. Thus one can find the major axis by:

(1) relocating the origin by subtracting the center of gravity from each point;
(2) finding the principal eigenvector of the scatter matrix of this modified set of points; and
(3) solving for the major axis as the line through the center of gravity parallel to the principal eigenvector.

Having found the major axis, one then treats each point on the boundary as a vector and projects each onto the major axis. The extrema are then those two points on opposite sides of the center of gravity whose projections onto the major axis are maximum in length. (The extrema are not necessarily unique.) This approach yields a solution which is an accurate representation of the "shape" of the region in the least-mean-squared-error sense. It may or may not actually find the two points whose mutual distance is maximum. In many applications, an approximation like this is exactly what is needed. However, occasionally, one encounters regions (Fig. 9.9) with spurs on them, where the spur may be relevant. In this case, it is necessary to use an algorithm which will find the actual extrema (see section 9A.1).

Note that the convex discrepancy is one of those features you are supposed to be figuring out for Assignment 9.1.

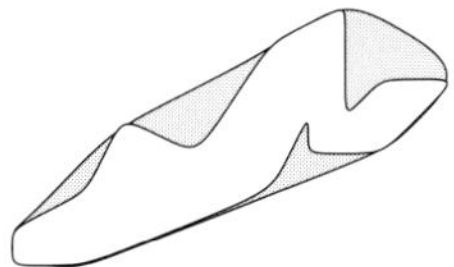

Fig. 9.10. Convex hull of a region. The shaded area is the convex discrepancy.

The convex hull

If one were to stretch a rubber band around a given region, the region that would result would be the *convex hull* (Fig. 9.10).

The difference in area between a region and its convex hull is the convex discrepancy. See Shamos [9.68] for fast algorithms for computing the convex hull, and [9.30] for such algorithms for parallel machines.

We can find the convex hull in $O(n \log n)$ time. Furthermore, finding the convex hull provides us with another simple feature, the *convex discrepancy*, as illustrated in Fig. 9.10.

9.4 Moments

The moments of a shape are easily calculated, and as we shall see, can be robust to similarity transforms.

A moment of order $p + q$ may be defined on a region as

$$m_{pq} = \sum x^p y^q f(x, y). \tag{9.37}$$

If we assume that the region is uniform in gray value and that gray value is arbitrarily set to 1 inside the region and 0 outside, the area of the region is then m_{00}, and we find that the center of gravity is

$$m_x = \frac{m_{10}}{m_{00}} \qquad m_y = \frac{m_{01}}{m_{00}}. \tag{9.38}$$

We may derive a set of moment-like measurements (the central moments) which are invariant to translation by moving the origin:

$$\mu_{pq} = \sum (x - m_x)^p (y - m_y)^q f(x, y). \tag{9.39}$$

By taking into account rotation and zoom, we can continue this sort of derivation and can now define as many features as we wish by choosing higher orders of moments or combinations thereof. From the central moments, we may define the normalized central moments by

$$\eta_{pq} = \frac{\mu_{pq}}{\mu_{00}^{\gamma}}, \qquad \text{where } \gamma = \frac{p+q}{2} + 1.$$

Finally, the *invariant moments* [9.21] have the characteristics that they are invariant to translation, rotation, and scale change, which means that we get the same moment, even though the image may be moved, rotated, or zoomed.[3] They are listed in Table 9.1.

[3] Gonzalez and Wintz [9.21] refer to zooming as "scale change." Since we use the word "scale" in a slightly different way, we refer to this as "zoom."

Table 9.1. *Invariant moments.*

$$\varphi_1 = \eta_{20} - \eta_{02}$$
$$\varphi_2 = (\eta_{20} - \eta_{02})^2 + 4\eta_{11}^2$$
$$\varphi_3 = (\eta_{30} - 3\eta_{12})^2 + (3\eta_{21} - \eta_{03})^2$$
$$\varphi_4 = (\eta_{30} + \eta_{12})^2 + (\eta_{03} + \eta_{21})^2$$
$$\varphi_5 = (\eta_{30} - 3\eta_{12})(\eta_{30} + \eta_{12})[(\eta_{30} + \eta_{12})^2 - 3(\eta_{03} + \eta_{21})^2] + (3\eta_{21} - \eta_{03})(\eta_{03} + \eta_{21})[3(\eta_{30} + \eta_{12})^2 - (\eta_{03} + \eta_{21})^2]$$
$$\varphi_6 = (\eta_{20} - \eta_{02})[(\eta_{30} + \eta_{12})^2 - (\eta_{03} + \eta_{21})^2] + 4\eta_{11}(\eta_{30} + \eta_{12})(\eta_{21} + \eta_{03})$$
$$\varphi_7 = (3\eta_{21} - \eta_{03})(\eta_{30} + \eta_{12})[(\eta_{30} + \eta_{12})^2 - 3(\eta_{03} + \eta_{21})^2] + (3\eta_{12} - \eta_{03})(\eta_{21} + \eta_{03})[3(\eta_{30} + \eta_{12})^2 - (\eta_{03} + \eta_{21})^2]$$

Since their original development by Hu [9.33], the concept has been extended to moments which are invariant to affine transforms by Rothe *et al.* [9.62].

Despite their attractiveness, strategies based on moments do have problems, not the least of which is sensitivity to quantization and sampling [9.45]. (See Assignment 9.9.)

The use of moments is actually a special case of a much more general approach to image matching [9.89] referred to as the *method of normalization*. In this philosophy, we first transform the region into a "canonical frame" by performing a (typically linear) transform on all the points. The simplest such transformation is to subtract the coordinates of the center of gravity (CG) from all the pixels, thus moving the coordinate origin to the CG of the region. In the more general case, such a transformation might be a general affine transform, including translation, rotation, and shear. We then do matching in the transform domain, where all objects of the same class (e.g., triangles) look the same.

Some refinements are also required if moments are to be calculated with gray-scale images, that is, when the f of Eq. (9.37) is not the result of a thresholding operation. All the theory of invariance still holds, but as Gruber and Hsu [9.24] point out, noise corrupts moment features in a data-dependent way.

Once a program has extracted a set of features, some use must be made of this set, either to match two observations or to match an observation to a model. The use of simple features in matching is described in section 13.2.

9.5 Chain codes

A chain code is a feature describing the boundary of a region. In a chain code, we represent a counter-clockwise walk around a region by a sequence of numbers,

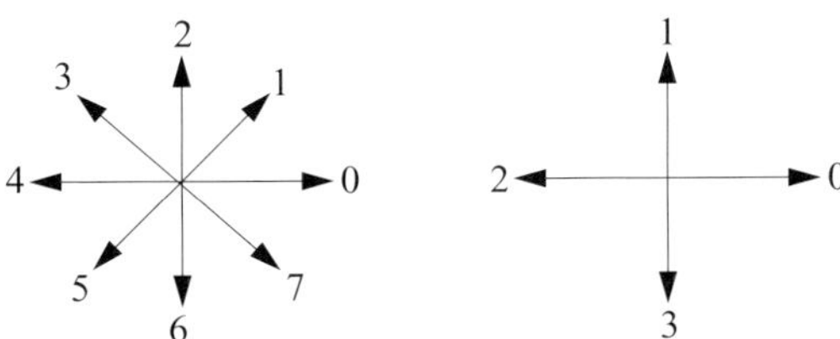

Fig. 9.11. The eight directions (for 8-neighbor) and four directions (for 4-neighbor) in which one might step in going from one pixel to another around a boundary.

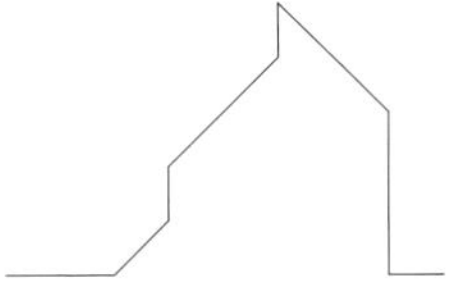

Fig. 9.12. The boundary segment represented by the chain code $0^2121^227^26^30$.

Can you determine what if anything is wrong with the caption on this figure?

all between 0 and 7 (if using eight directions) or between 0 and 3 (if using four directions), designating the direction of each step. The eight and four cardinal directions are defined as illustrated in Fig. 9.11. The boundary of a region may then be represented by a string of single digits. A more compact representation utilizes superscripts whenever a direction is repeated. For example, 0012112776660 could be written $0^2121^227^26^30$, and illustrates the boundary shown in Fig. 9.12. The ability to describe boundaries with a sequence of symbols plays a significant role in the discipline known as "syntactic pattern recognition," and appears frequently in the machine vision literature, including other places in this book.

9.6 Fourier descriptors

A Fourier descriptor is another feature describing the boundary of a region. Given a boundary of a region, we imagine[4] that the region is drawn on the complex plane, and the x coordinate of each point then represents a distance along the real axis, and the y coordinate a distance along the imaginary axis. Each boundary point is therefore viewed as a single complex number. Traversing the boundary then results in a (cyclic) sequence of complex numbers. Taking the Fourier transform of this sequence provides another sequence of complex numbers, one which can be shown to possess a type of invariance as illustrated in Table 9.2.

The following example illustrates these ideas in a somewhat oversimplified way: Suppose we have the Fourier descriptors for two boundaries:

$$f_1 = 0.7, 0.505, 0.304, 0.211, \ldots$$

$$f_2 = 0.87, 0.505, 0.304, 0.211, \ldots$$

We see that these two sequences differ only in the first (DC) term. They therefore represent two encodings of the same boundary which differ only by a translation.

4 Pun alert! (bet you missed that one) – note the kinds of numbers discussed in the paragraph.

Table 9.2. *Equivalence between motions in the image and transform domains.*

In the image	In the transform
A change in size	Multiplication by a constant
A rotation of ϕ about the origin	Phase shift
A translation	A change in the DC term

This example is oversimplified, because in reality the sequence will be a set of complex numbers, not real numbers as illustrated, but the concepts are identical.

Practical considerations in using Fourier descriptors

How we represent the movement from one boundary point to another is critical. Simply using a 4-neighbor chain code produces poor results. Use of an 8-neighbor code reduces the error by 40 to 80%, but is still not as good as using a subpixel interpolation. There are other complications as well, including the observation that the usual parameterization of the boundary (arc length) is not invariant to affine transformations [9.1, 9.96]. Experiments have [9.39] compared affine-invariant Fourier descriptors and autoregressive methods. For more detail on Fourier descriptors, see [9.1].

9.7 The medial axis

In two dimensions, the medial axis of a region is defined as the locus of the centers of "maximal circles." A maximal circle is the largest circle that can be located at a given point in the region. Let's say that a bit more carefully (we all need to learn how to use mathematics to make our wording precise). At point (x, y) inside region Ω, draw a circle of radius R about that point. Make R as large as possible, but such that (1) no points in the circle are outside the region and (2) the circle touches the boundary of the region at no fewer than two points. Any point on the medial axis can be shown (see Assignment 7.12) to be a local maximum of a distance transform (DT). A point with DT values of k is a local maximum if none of its neighbors has a greater value. Fig. 9.13(a) repeats Fig. 7.3. The local maxima of this distance transform are illustrated in Fig. 9.13(b).

Another way to think about the medial axis is as the minimum of an electrostatic potential field. This approach is relatively easy to develop if the boundary happens to be straight lines or, in three dimensions, planes [9.12]. See [9.16] for additional algorithms to efficiently compute the medial axis.

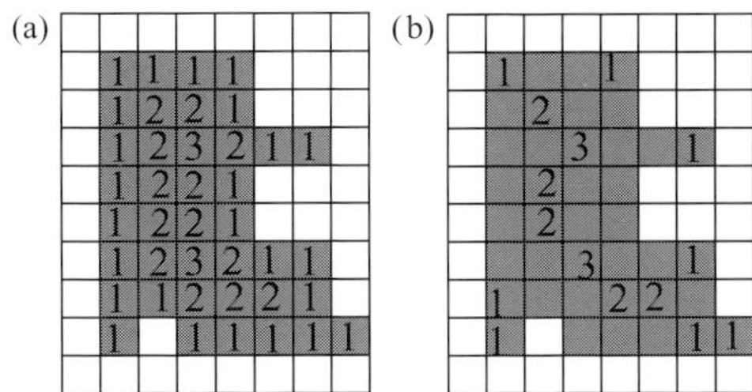

Fig. 9.13. An example region whose DT, computed using 4-neighbors, is shown in (a). The morphological skeleton is illustrated in (b), and simply consists of the local maxima of the DT.

9.7.1 Relating the medial axis and ridges

The medial axis is defined only for binary images, or at least only for images in which each region may be represented by its boundary. The ridge, however, is defined for gray-scale images. We can relate these two characterizations using the following strategy: Use a scale space representation of the image. That is, blur the binary image. This process turns the binary image into a gray-scale image. Now find the ridge. Start with the ridge at high scale (lots of blur); you can call that your initial estimate of the medial axis. Now, reduce the scale slightly and see what new ridges appear. Add them to your estimate. Continue this process over varying scales. See Pizer *et al.* [9.59] for more details on this philosophy.

9.8 Deformable templates

Recall active contours (snakes) from section 8.5? If we think of the region surrounded by a snake, rather than the snake itself, we realize we have a region whose shape can be deformed, and thus have invented "deformable templates." Objects can be tracked utilizing a philosophy which allows for deformation of the template [9.102]. In addition, the concepts of deformable templates may be useful in image data base access. For example, Bimbo and Pala [13.5] do indexing by comparing shapes in the image with a user-drawn sketch, an "iconic index," which is actually a deformable template. The best matching template is written as

$$\phi(s) = \tau(s) + \theta(s) \tag{9.40}$$

where s is (normalized) arc length, $\tau(s)$ is the template as stored in the data base, and $\theta(s)$ is the deformation required to make this particular template match a sequence of boundary points in the image being accessed. We emphasize that $\theta(s)$ is the *difference* between the original and the deformed template. The image in the data base which best matches the template is the image which minimizes "the difference between the original and the deformed template." This can be achieved by minimizing

$$E = \int_0^1 \left(\alpha \left[\left(\frac{d\theta_x}{ds} \right)^2 + \left(\frac{d\theta_y}{ds} \right)^2 \right] + \beta \left[\frac{d^2\theta_x}{ds^2} + \frac{d^2\theta_y}{ds^2} \right] - I_E(\varphi(s)) \right) ds \quad (9.41)$$

which represents (in the first term) how much the template had to be strained to fit the object while the second term represents the energy spent to bend the template. This is thus a deformable templates problem. The optimization problem may be solved numerically [13.5].

A variation on deformable templates is the idea of "geometric flow" to change a given initial curve into a form which is better suited for identification by, say, template matching. The term "geometric" flow means that the flow is completely determined by the geometry of the curve. Pauwels *et al.* [9.58] cast this discipline as the answer to: "Is it possible to use optimization of functionals to crystallize the geometric content of a curve by reducing the noise while at the same time enhancing the salient features?"

9.9 Quadric surfaces

A surface defined by an algebraic equation of degree two is called a *quadric* (see section 8.6.1). The general equation for a quadric is

$$ax^2 + by^2 + cz^2 + 2fyz + 2gzx + 2hxy + px + qy + rz + d = 0. \quad (9.42)$$

This one equation describes all the second-order surfaces, some of which are illustrated in Fig. 9.14.

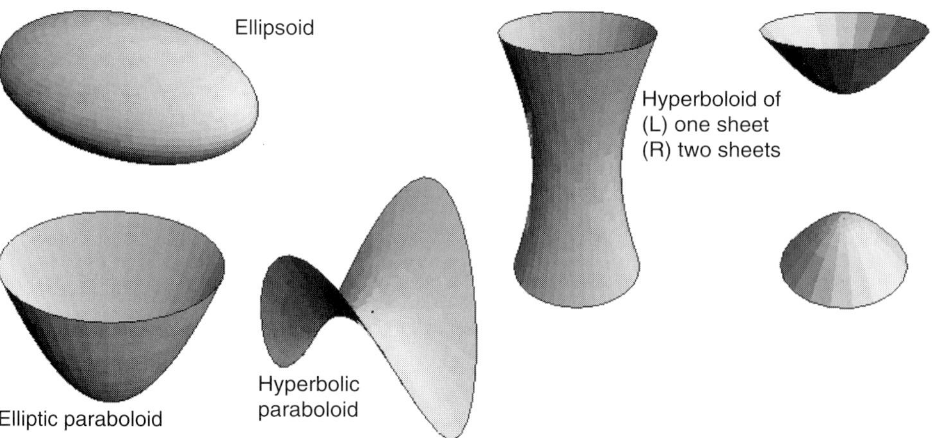

Fig. 9.14. The quadric equation describes a wide variety of surfaces [9.103] (CRC Press. Used with permission).

If the quadric is centered at the origin, and its principal axes happen to be aligned with the coordinate axes, the quadric will take on a particular form. For example, an ellipsoid has the special form

$$\frac{x^2}{a^2} + \frac{y^2}{b^2} + \frac{z^2}{c^2} = 1. \tag{9.43}$$

However, when the axes of the quadric do not align with the coordinate axes, only the general form of Eq. (9.42) occurs.

From range or other surface data, the quadric coefficients may be determined by methods such as those in section 8.6.1. Given the coefficients, the type of quadric may be determined by the following method.[5]

If there is a constant term, d, divide by the constant, redefining the other coefficients (e.g. $a \Leftarrow a/d$), resulting in a form of the quadric equation in which the constant term is unity:

$$ax^2 + by^2 + cz^2 + 2fyz + 2gzx + 2hxy + px + qy + rz + 1 = 0. \tag{9.44}$$

This equation can then be written in the form:

$$[x \quad y \quad z \quad 1]\begin{bmatrix} a & h & g & p \\ h & b & f & q \\ g & f & c & r \\ 0 & 0 & 0 & 1 \end{bmatrix}\begin{bmatrix} x \\ y \\ z \\ 1 \end{bmatrix} = 0. \tag{9.45}$$

Consider the upper left 3×3 submatrix,

$$E = \begin{bmatrix} a & h & g \\ h & b & f \\ g & f & c \end{bmatrix}.$$

Obtain the three eigenvalues, λ_1, λ_2, and λ_3, and find the reciprocal of each which is nonzero, $r_1 = 1/\lambda_1$, $r_2 = 1/\lambda_2$, and $r_3 = 1/\lambda_3$. At least one reciprocal must be positive to have a real surface: If exactly one is positive then the surface is a hyperboloid of two sheets; if exactly two are positive, then it is a hyperboloid of one sheet; if all three are positive, it is an ellipsoid, and the square roots of r_1, r_2, r_3 are the major axes of the ellipsoid. Otherwise the distance between the foci of hyperboloids is determined by the magnitudes of the rs.

[5] The authors are grateful to Dr G. L. Bilbro for his formulation of this method.

9.10 Surface harmonic representations

This approach represents a surface as a linear combination of fixed harmonic basis functions. Harmonic functions are solutions to Laplace's equation

$$\nabla^2 \psi(x, y, z) = 0 \tag{9.46}$$

which is

$$\frac{\partial^2 \psi}{\partial x^2} + \frac{\partial^2 \psi}{\partial y^2} + \frac{\partial^2 \psi}{\partial z^2} = 0 \tag{9.47}$$

in Cartesian coordinates. Most of the work in harmonic representations has not, however, been in Cartesian coordinates, but rather in spherical coordinates. (See Matheny and Goldgof [8.40] for a discussion of other forms.) Any continuous function which can be written as $r = r(\theta, \varphi)$ can be expressed as a linear combination of spherical harmonics. In spherical coordinates, Laplace's equation is

$$\frac{\partial}{\partial r}\left(r^2 \frac{\partial \psi}{\partial r}\right) + \frac{1}{\sin\theta}\frac{\partial}{\partial \theta}\left(\sin\theta \frac{\partial \psi}{\partial \theta}\right) + \frac{1}{\sin^2\theta}\frac{\partial^2 \psi}{\partial \varphi^2} = 0. \tag{9.48}$$

We seek solutions which are separable in the sense that they may be written as the product of functions of a single variable; that is, they have the form

$$\psi(r, \theta, \varphi) = R(r)\Theta(\theta)\Phi(\varphi). \tag{9.49}$$

With this restriction, the partial differential equation may be separated into three ordinary differential equations and the solution can be shown to be

$$P_l^m \cos\theta \sin m\varphi \tag{9.50}$$

where the parameter l is referred to as the "degree," m is an integer less than l, and P is a Legendre polynomial.

Thus, any function may be represented as

$$r(\theta, \varphi) = \sum_{l=0}^{L} \left\{ U_l^0 P_l \cos\theta + \sum_{m=0}^{l} \left[U_l^m P_l^m \cos\theta \cos m\varphi + V_l^m P_l^m \cos\theta \sin m\varphi \right] \right\} \tag{9.51}$$

where the coefficients are found by the process of fitting this form to the data.

9.11 Superquadrics and hyperquadrics

We discussed in section 8.6 the problem of segmenting a range image using the surface function and how to fit functions to data. In this section, we describe how to fit superquadrics and hyperquadrics to range data.

A superquadric is a surface with the equation

$$\left|\frac{x}{a}\right|^{\gamma_1} + \left|\frac{y}{b}\right|^{\gamma_2} + \left|\frac{z}{c}\right|^{\gamma_3} = 1 \tag{9.52}$$

and a hyperquadric has the equation

$$\sum_{i=1}^{N} |A_i x + B_i y + C_i z + D_i|^{\gamma_i} = 1 \tag{9.53}$$

which may be rewritten as $F(x, y, z) = 1$.

Kumar *et al.* [8.32] propose an approach to fit hyperquadrics to range data. It operates as follows: First, recognize that the parameters which minimize

$$\text{EOF} = \sum_{i=1}^{N} (1 - F(x_i, y_i, z_i))^2 \tag{9.54}$$

are presumed to be a good fit. (Since, at every point on the surface F, the value of $F(x, y, z)$ is supposed to be 1.0.) Observe that this minimizes the algebraic distance (see section 8.6.1) from the point to the surface! This distance will be zero if the point lies on the surface, but otherwise, there is not a simple relationship between the Euclidean distance from the point to the surface and the algebraic distance. Kumar *et al.* observe: "This function is biased, especially for oblong objects." Similar complaints exist for essentially all applications of the algebraic distance.

To get a somewhat better fit, in the case of hyperquadrics, the following approach can be used. Suppose we have an initial estimate of the surface, an estimate which is not too bad, but might not be the best estimate. Let that estimate be defined by a set of parameters, A, B, C, and D, defining a function $F(x, y, z)$. For a particular point (x_i, y_i, z_i), substitute these values into F to determine $w_i = F(x_i, y_i, z_i)$. If the point is actually on the surface, then w will be equal to 1. Now, consider the surface defined by $F(x, y, z) = w_i$. The distance normal to this surface can be approximated by the distance d_i in the direction of the gradient,

$$F(x, y, z) = F(x_i, y_i, z_i) + d_i \|\nabla F(x_i, y_i, z_i)\| \tag{9.55}$$

so

$$d_i = \frac{1 - F(x_i, y_i, z_i)}{\|\nabla F(x_i, y_i, z_i)\|}. \tag{9.56}$$

This d_i is really what we want to minimize. It is the distance from a surface through the point (x_i, y_i, z_i) which is in some sense parallel to the surface we wish to determine. These are estimates only; it remains to iteratively refine these estimates until they become the real solutions. To accomplish that, rewrite the squared error in terms of d,

$$E = \sum_{i=1}^{N} \frac{(1 - F(x_i, y_i, z_i))^2}{\|\nabla F(x_i, y_i, z_i)\|} \tag{9.57}$$

and minimize this objective function in the following way.

(1) First, determine an initial estimate, probably by numerically finding the value which minimizes the EOF of Eq. (9.54). This will not be a bad fit, but better fits are possible because of the bias.
(2) Compute, for each data point $w_i = 1/\|\nabla F(x_i, y_i, z_i)\|^2$.
(3) Minimize $\sum_{i=1}^{N}(w_i(1 - F(x_i, y_i, z_i)))^2$.
(4) If the solution is good enough, stop; otherwise go to (2).

Observe that the general idea presented here to fitting an implicit function is applicable to more than just hyperquadrics. The idea of minimizing a function normal to the gradient of a curve through the data points fits many (probably most) problems in fitting of explicit functions.

Dickinson *et al.* [13.11] combine superquadric representations with the concept of aspect graphs.

9.12 Generalized cylinders (GCs)

A cylinder may be described as a circle which translates along a straight line in space, with the plane of the circle perpendicular to the line. Now, suppose that the line is allowed to bend in space, in fact to become an arbitrary space curve, parameterized by arc length, s. Then, the line becomes a vector function $x(s)$, $y(s)$, $z(s)$ of s. Next, allow the radius of the circle to vary with the point along the curve, $R = R(s)$, and you have some idea of what a generalized cylinder is [9.4, 9.21, 9.23, 9.74, 9.75, 9.98]. However, the concepts of GCs are more general even than described above. The object translated does not have to be a circle, it can be any 2D shape.

If we can fit GCs to regions, we can then use the vector function of the line and the radius function as features to describe the shape of the region. However, there are significant challenges to fitting GCs to images. We will not pursue the GC idea any further in this book. The reader can find many interesting papers in the literature, a few of which are listed in the previous paragraph.

9.13 Conclusion

In this chapter, several features were defined that could be used to quantify the shape of a region. Some, like the moments, are easy measurements to make. Others, such as the diameter or the convex discrepancy, require development of fairly sophisticated algorithms to avoid extremely long computation time. There have also been some efforts in automatic learning of shape descriptors [9.11, 9.91].

How DO people do it?

Studies of signal processing in the visual cortex [9.65] have suggested that images are represented in the cortex after being subjected to a log-polar transformation, and that invariance may be provided in this way [9.63, 9.90]. Other transformations [9.87] may provide equivalent or superior representations in computer applications.

In this chapter, again, we have seen the philosophy of consistency invoked by simply saying that a set of pixels is consistent if each point satisfies the same equation, be it quadric or generalized cylinder.

Optimization methods came up quite a few times in this chapter:

Constrained minimization.

- In section 9.2.2 a derivation is presented which finds the straight line which best fits a set of points, in the sense that the sum of perpendicular distances is minimized. To accomplish that, we were required to use constrained minimization with Lagrange multipliers.

Integral squared error.

- In section 9.8 we find a deformation to a template by performing a minimization of an integral squared error.

Pseudo-inverse minimizes the squared error.

- In section 9A.2, we will encounter a problem requiring "inversion" of a nonsquare matrix. Of course, one cannot formally invert such a matrix, and instead we derive the "pseudo-inverse." We also show that the pseudo-inverse is really a minimum-squared-error algorithm.

9.14 Vocabulary

You should know the meanings of the following terms.

Affine transform
Aspect ratio
Basis vector
Center of gravity
Chain code
Compactness
Convex hull
Convex discrepancy
Deformable template
Diameter
Fourier descriptor
Generalized cylinder
Homogeneous transformation matrix
Invariant moment

```
K–L transform
Linear transformation
Medial axis
Metric
Moment
Orthogonal transformation
Principal component
Similarity transform
Thinness
```

Topic 9A Shape description

9A.1 Finding the diameter of nonconvex regions

One fast algorithm for an estimate of the diameter is as follows:

(1) Choose a point on the boundary, call it P_0.
(2) Find the point on the boundary at a maximum distance from P_0. Call it P_1.
(3) Compute the midpoint, M, of P_0, P_1.
(4) Find the point P_2 on the boundary at a maximum distance from M.
(5) Find the boundary point P_3 at a maximum distance from P_2.
(6) If $d(P_2, P_3) \leq d(P_0, P_1)$, stop. P_0 and P_1 are the extrema; else, assign $P_0 \leftarrow P_2$, $P_1 \leftarrow P_3$, go to (3).

This technique is easy to program and converges rapidly. We have had good luck with it. Unfortunately, it is not guaranteed to converge to the global extrema, although there is a high likelihood that it will do so.

An extension of the previous algorithm provides a strategy which is guaranteed to converge. In addition, this algorithm provides a mechanism for rapidly narrowing the search space.

First, define a linear search function $M(P_i, R)$ which returns the point $P_{i+1} = M(P_i, R)$ such that $\forall x \in R, d(P_i, x) \leq d(P_{i+1}, P_i)$.

Choose an arbitrary point $P_1 \in R$, find $P_2 = M(P_1, R - P_1)$. Next, find $P_3 = M(P_2, R - \{P_1, P_2\})$. The relationship between $d(P_1, P_2)$ and $d(P_2, P_3)$ can be one of the following three cases, which are illustrated in Figs. 9.15– 9.17:

Case 1: $d(P_1, P_2) > d(P_2, P_3)$
Case 2: $d(P_1, P_2) = d(P_2, P_3)$
Case 3: $d(P_1, P_2) < d(P_2, P_3)$.

In Case 3, we compute a new P_3 called P_3' so that $d(P_2, P_3') = d(P_2, P_3)$ and P_1, P_2, and P_3' are colinear. We, therefore have a symmetrical lens shape which encloses all the points in R using at most two linear searches.

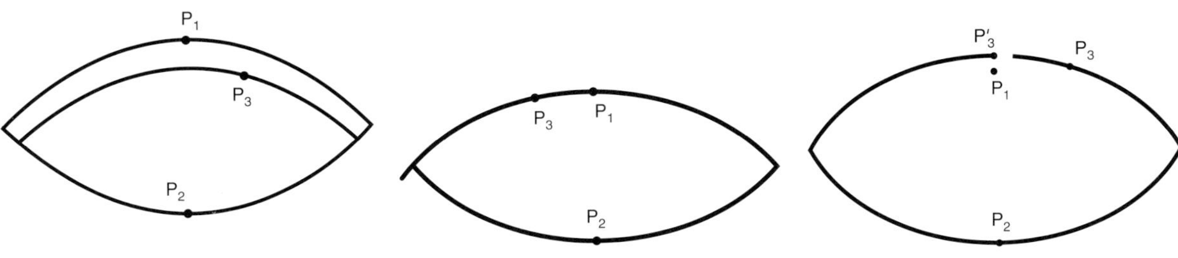

Fig. 9.15. Case 1.

Fig. 9.16. Case 2.

Fig. 9.17. Case 3.

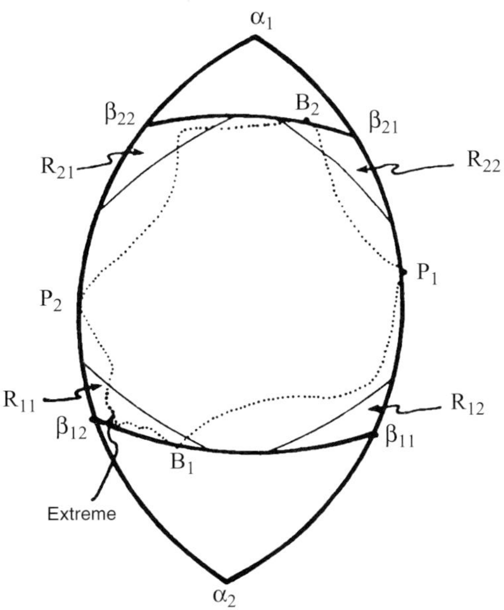

Fig. 9.18. In this figure, B_2 is one of the extrema. The other extreme is an element of R_{11} (redrawn from [9.71]).

Our heuristic states that the "best" direction in which to search is perpendicular to P_1, P_2 (or P_2, P_3' in Case 3).

Compute the apexes α_1 and α_2 as shown in Fig. 9.18. Then find $B_1 = M(\alpha_1, \{R - \{P_1, P_2\}\})$. If $d(\alpha_1, B_1) \leq d(P_1, P_2)$, stop, else partition $R - P_1, P_2$ into two mutually exclusive regions R_1 and R_2, where $R_1 = \{x \in R - \{P_1, P_2\} \| d(x, P_1) > d(P_1, P_2)\}$. Find $B_2 = M(\alpha_2, R - R_1)$, and find $R_2 = \{x \in R - \{P_1, P_2\} - R_1 \| d(x, \alpha_2) > d(P_1, P_2)\}$.

Note that: (1) In general, $R_1 \cup R_2 \subset R$. However, $R - \{R_1 \cup R_2\}$ contains no points of interest. (2) If either $R_1 = \phi$ or $R_2 = \phi$, we can stop and P_1, P_2 is the diameter.

If both R_1 and R_2 are nonempty, we can define them as "antipodal regions" in the following sense: If a diameter exists greater than $d(P_1, P_2)$, one endpoint must lie in R_1 and the other in R_2, which effectively cuts the search space in half.

Two new points, β_{11} and β_{12}, are computed by swinging an arc with center at α_1 and radius $d(\alpha_1, B_1)$ to intersect the lens. Similarly, β_{21} and β_{22} are computed.

Note that $d(\beta_{21}, \beta_{12}) = d(\beta_{11}, \beta_{22})$ and this is an upper bound on the diameter. In a digital picture, having an upper limit on the diameter may allow earlier stopping of the algorithm, since if we know a diameter candidate, and we know the upper bound, if these two values differ by less than 1.414 (the diagonal of a pixel), there is no need to search further.

Compute $r = \text{MAX}(d(B_1, B_2), d(P_1, P_2))$. Use this as a radius with which to draw arcs. Using β_{21} as center and r as radius, partition R_1 into two regions, R_{11} and $R_1 - R_{11}$. Similarly, use β_{22} as center and find $R_{12} \subset R_1$. Note that $R_1 - \{R_{11} \cup R_{12}\}$ contains no points of interest. Similarly, find R_{21} and R_{22} as shown. If $R_{21} = \phi$ and $R_{11} \cap R_{12} = \phi$, then R_{21} and R_{12} is an antipodal pair as is R_{11} and R_{22}. In any case, $R_{21} \cup R_{22}$ is antipodal with $R_{11} \cup R_{12}$. These antipodal regions will be our search space in the next phase.

The strategy to this point has either identified the extrema or it has provided other useful results, specifically:

(1) An upper bound on the diameter has been derived.
(2) All points within a very large area have been eliminated as candidates for extrema.
(3) The remaining points have been partitioned into antipodal regions.

At this point, if fewer than K points remain (which is most often the case), the most appropriate technique is to compute convex hulls. (The optimal choice of K is a function of region topology. It has been our experience that $K = 50$ seems to work well.) This computation is aided by the observations that:

- If either of an antipodal pair of regions is empty, the other may be eliminated.
- The extrema must lie on the opposite sides of the convex hull of the antipodal regions, eliminating a great many more points, since we are able to compute convex hulls of smaller regions.

If, on the other hand, many more points remain, the algorithm can be invoked recursively using the antipodal pairs of regions as subject areas, and choosing new starting points as those points closest to β_{21} and β_{12}, or β_{11} and β_{22}.

For R with N points, the blind exhaustive search takes $O(n^2)$ distance calculations and comparisons. Our search algorithm is exhaustive (hence guarantees convergence), but intelligent by taking the global shape of the region R into consideration. The initial search space R is sequentially divided into smaller mutually exclusive subspaces by eliminating those points that cannot be the end points of the extrema.

Although the number of subspaces is increased by two for each recursive call of the procedure, there are many more points that are eliminated from the search space after each call. Therefore, the search space is rapidly decreasing. How rapid the decreasing rate is depends on the shape of R.

This method is derived from geometric considerations and consequently its rate of convergence is strongly dependent on the geometry of the region. It is, therefore, difficult to accurately access the computational complexity of the method. If used as a pre-processing technique, it operates in $O(4n)$ time, plus the time required to perform the convex hull calculation on the remaining points, or $O(k \log k)$ where k represents the points remaining after pre-processing.

Certainly in the worst case, almost no points are eliminated and convergence operates as convex hull in $O(n \log n)$. It is in fact worse than the convex hull in this case, since the program is more complex.

However, due to the large number of branch points, the algorithm exits early for virtually all regions and it converges very rapidly.

9A.2 Inferring 3D shape from images

Many papers have been written which extract three-dimensional shape from various sources: From silhouettes [9.42, 9.43, 9.44, 9.49, 9.101]; from images of specular reflectors [9.64]; from three orthogonal projections (x-ray projections) [9.81]; making use of the assumption that objects tend to have orthogonality [9.22] or symmetry [9.18]. Ultimately, in all these algorithms one must address the question of visibility [9.83].

Range images, one might think, already contain a complete description of three-dimensional shape, but, of course, you cannot see the entire surface in one image [9.31]. A tough problem is how to integrate several range images to form one description of a three-dimensional object [9.76, 9.93].

They do not intersect because they are in 3D space.

Even though you may have segmented surfaces with some success, those segmentations are almost never perfectly correct. One might think that the equations describing the intersection of edges would be straightforward to find. After all, you have the equations of the surfaces whose intersections determine the edges. Just calculate the intersection! Ah, but, it is never quite that easy. The problems occur when you have vertices, the intersections of edges – the trihedral or multihedral intersection points. Those equations you just derived never intersect at a point. Hoover *et al.* [9.31] address this problem, and extend the solution to nonvisible surfaces.

Another important issue in extracting three-dimensional shape is what representation you choose, including shape from perspective, shape from shading, shape from texture, etc.

9A.2.1 Shape from perspective

We formalize the shape-from-perspective problem as follows: A point is observed with row–column coordinates r_i and c_i, which is actually located in 3-space at $[x_i, y_i, z_i]$. The observation coordinates and true coordinates are related by

$$\begin{bmatrix} r_i \\ c_i \\ 1 \end{bmatrix} = \begin{bmatrix} k_u f & 0 & u_0 & 0 \\ 0 & k_v f & v_0 & 0 \\ 0 & 0 & 1 & 0 \end{bmatrix} \begin{bmatrix} R & T \\ 0 & 1 \end{bmatrix} \begin{bmatrix} x_i \\ y_i \\ z_i \\ 1 \end{bmatrix} \tag{9.58}$$

where k_u, k_v, u_0, v_0 and f are projective properties of the camera, R is a 3×3 rotation matrix and T is a 3×1 translation vector. Assuming we know for each point the actual 3D coordinates and the corresponding 2D observations we should be able to infer the transformations and camera properties. Variations on this problem include [9.95] the Perspective n Point problem (PnP), when n point correspondences are known; the Perspective n Line problem (PnL), when n pairs of corresponding lines are known; and the Perspective n Angle problem (PnA), when

n pairs of corresponding angles are observed. Analytic solutions have been determined for P3P [9.17], P3L [9.14], and PnA [9.95]. A clear explanation of the linear approaches to using uncalibrated cameras (but assuming correct solutions to the correspondence problem) may be found in [9.26].

Some work [9.34] has been done on the harder version of the correspondence problem, the case that the cameras are not only uncalibrated, but are oriented in such a way that the epipolar assumption is not necessarily valid. In this case the stereo matching problem becomes one of search for best matching pairs, using both radiometric and geometric information to narrow the search.

9A.2.2 Shape from shading

Shape from shading was first introduced by Horn, who argued that some knowledge of how light is generated, reflected, and observed could substantially improve the performance of machine vision systems. Consider Fig. 9.19, and assume you know:

- the angle of the light source
- the angle of the observer
- the measured brightness of the pixel
- a law governing how light is scattered
- the albedo of the surface.

Can you find the surface normal? (If you have the normal at every point, how would you determine the surface?)

The solution may be found by solving differential equations. We start by writing the surface normal vector as $\boldsymbol{n} = \boldsymbol{r}/|\boldsymbol{r}|$, where the direction vector

$$\boldsymbol{r} = \left[\frac{\partial z}{\partial x}, \frac{\partial z}{\partial y}, 1\right]^{\mathrm{T}}.$$

In most shape-from-shading literature, the partial derivatives are abbreviated $p \equiv \partial z/\partial x$, $q \equiv \partial z/\partial y$.

Despite the fact that we use the term "brightness" constantly, it actually does not have a rigorous physical definition. Following Horn [9.32], we define *irradiance* as the power per unit area falling on a surface, measured in watts per square meter. Then, we can define *radiance* as the power per unit foreshortened area per unit solid angle. This dependence on

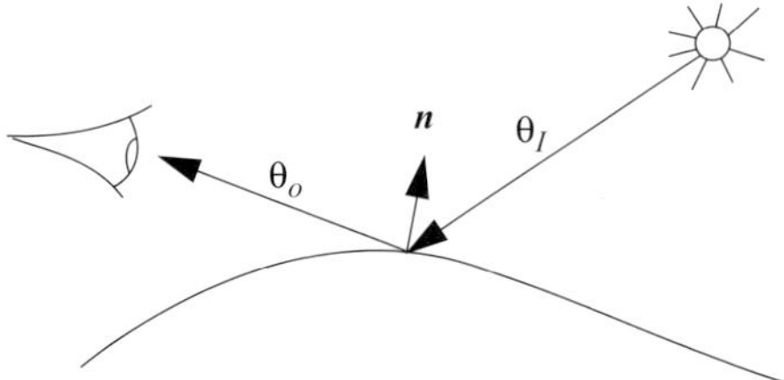

Fig. 9.19. Light strikes a surface at an incident angle, relative to the surface normal, *n*, and is reflected/scattered in another direction.

foreshortened area makes it clear that the angle of observation plays an important role in scene "brightness."

Often, the "reflectivity model" of a surface is known, or may be measured. For example, the brightness observed might be independent of the angle of observation, and depend on the angle of incidence. For example, the reflected brightness might be related to the incident brightness with a relationship such as

$$R(x, y) = aI(x, y)\cos(\theta_I). \tag{9.59}$$

Thus, if we know the incident brightness, I, the albedo (how the surface is painted), a, and the reflected brightness, R, we should be able to solve for θ_I, and from that, infer the surface normal, and from that the surface. The reflectivity function of Eq. (9.59) is known as a Lambertian model. Note that the angle of observation does not enter the Lambertian model. Another familiar reflectivity function is the specular model

$$R(x, y) = aI(x, y)\delta(\theta_I - \theta_o), \tag{9.60}$$

which describes mirrors – you only get a reflection if the angle of illumination equals the angle of incidence. Of course most surfaces, even "shiny" ones, are not perfect specular reflectors, and a more realistic model for a mixed surface might be

$$R(x, y) = aI(x, y)\cos^4(\theta_I - \theta_o). \tag{9.61}$$

Although the use of reflectivity functions requires radiometric calibration of cameras [9.28], that requirement in itself is not the major difficulty. To see the complexity of the problem [9.51], let us expand Eq. (9.59) in terms of the elements of the observation vector and the normal vector.

$$R(x, y) = aI(x, y)\cos(I \bullet N) = aI(x, y)\cos\left(I_x\frac{\partial z}{\partial x} + I_y\frac{\partial z}{\partial y} + I_z N_z\right). \tag{9.62}$$

Assuming we know the angle of observation (which we will actually know only approximately at best), the albedo and the incident brightness, we still have a partial differential equation which we must solve to determine the surface function z.

Many papers and Horn's classic text [9.32] address approaches for solving various special cases of the shape-from-shading problem. A recent paper by Zhang *et al.* [9.100] surveys the field up to 1999. (Remember, Equation (9.62) is itself a special case – it assumes the brightness does not depend on the observation angle.) In the following, we discuss another special case, photometric stereo.

Photometric stereo

In many cases, it is reasonable to model the reflectivity of a surface as proportional to the cosine of the angle between the surface normal vector and the illumination vector:

$$I(x, y) = r_o(N_i \bullet \mathbf{n}) \tag{9.63}$$

where N_i is a unit vector in the direction of light source i. If we are fortunate enough to have an object, a Lambertian reflector, which satisfies this equation, and which possesses the same albedo (r_0), independent of the illumination, we can make use of multiple pictures from multiple angles to determine the surface normal [9.35, 9.94]. Let us illuminate a particular pixel with three different light sources (one at a time) and measure the brightness of that pixel each time. At that pixel, we construct a vector from the three observations

$$\boldsymbol{I} = [I_1, I_2, I_3]^{\mathrm{T}}. \tag{9.64}$$

We know the direction of each light source. Let those directions be denoted by unit vectors from the surface point toward the light source, $\boldsymbol{N}_1$, $\boldsymbol{N}_2$, and $\boldsymbol{N}_3$. Write those three direction vectors in a single matrix by making each vector a row of the matrix.

$$N = \begin{bmatrix} \boldsymbol{N}_1 \\ \boldsymbol{N}_2 \\ \boldsymbol{N}_3 \end{bmatrix} = \begin{bmatrix} n_{11} & n_{12} & n_{13} \\ n_{21} & n_{22} & n_{23} \\ n_{31} & n_{32} & n_{33} \end{bmatrix}, \tag{9.65}$$

and now we have a matrix version of Eq. (9.63):

$$\boldsymbol{I} = r_0 N \boldsymbol{n}. \tag{9.66}$$

Since N is known, and $\boldsymbol{n}$ is a unit vector, we find

$$r_0 = |N^{-1}\boldsymbol{I}|, \tag{9.67}$$

and once we know r_0, we can solve for $\boldsymbol{n}$ using

$$\boldsymbol{n} = \frac{1}{r_0} N^{-1}\boldsymbol{I}. \tag{9.68}$$

Note that this derivation of photometric stereo assumes the albedo (sometimes called the surface reflectance) is the same for every angle of illumination. In the next subsection, we illustrate an application which combines shape from shading with photometric stereo, and does not make this assumption. For example, specular reflectors provide a special condition: the angle of observation is exactly equal to the angle of incidence. This allows special techniques to be used [9.56].

But what if we used more than three light sources? This gives us a wonderful context in which to discuss an important topic, overdetermined systems and the pseudo-inverse.

If we in fact used more than three light sources, we would hope to cancel out some effects of noise and/or measurement error. Suppose we have k light sources. Then Eq. (9.66) is rewritten as $\boldsymbol{I}_{k\times 1} = N_{k\times 3}\boldsymbol{n}_{3\times 1}$ where subscripts are used to emphasize the matrix dimensions, and r_0 is removed for clarity of explanation.

Now, we cannot simply multiply by the inverse of N because N is not square. Instead, as we have done so many times before, let us set up a minimization problem: We will find the surface normal vector $\boldsymbol{n}$ which minimizes the squared sum of differences between measurements $\boldsymbol{I}$ and products of N and $\boldsymbol{n}$. Of course, if Eq. (9.63) is strictly true everywhere, then we do not need to do a minimization. Of course, if Eq. (9.63) were true everywhere, there would be no point to taking more than three measurements. We believe that measurements are not perfect and there is some advantage to taking additional ones. Define an objective function E which incorporates the desire to find an optimal solution:

$$E = \sum_{i=1}^{k} \left(I_i - N_i^{\mathrm{T}}\boldsymbol{n}\right)^2 = (\boldsymbol{I} - N\boldsymbol{n})^{\mathrm{T}}(\boldsymbol{I} - N\boldsymbol{n}). \tag{9.69}$$

Expand the product

$$E = \boldsymbol{I}^{\mathrm{T}}\boldsymbol{I} - 2\boldsymbol{n}^{\mathrm{T}}N^{\mathrm{T}}\boldsymbol{I} + \boldsymbol{n}^{\mathrm{T}}N^{\mathrm{T}}N\boldsymbol{n}. \tag{9.70}$$

We wish to find the surface normal vector $\boldsymbol{n}$ which minimizes this sum-squared difference E, so we differentiate E with respect to $\boldsymbol{n}$,

$$\nabla_{\boldsymbol{n}} E = -2N^{\mathrm{T}}\boldsymbol{I} + 2N^{\mathrm{T}}N\boldsymbol{n} \tag{9.71}$$

and setting the gradient to zero we have

$$N^{\mathrm{T}}N\boldsymbol{n} = N^{\mathrm{T}}\boldsymbol{I} \tag{9.72}$$

or

$$\boldsymbol{n} = (N^{\mathrm{T}}N)^{-1}N^{\mathrm{T}}\boldsymbol{I}. \tag{9.73}$$

(In case you have not recognized it yet, the pseudo-inverse just appeared.) That was a lot of work. Let's see if there is an easier way:

Go back to Eq. (9.66), again omitting the r_0 for clarity, and multiply both sides by N^{T}.

$$N^{\mathrm{T}}\boldsymbol{I} = N^{\mathrm{T}}N\boldsymbol{n}. \tag{9.74}$$

Multiply both sides by $(N^{\mathrm{T}}N)^{-1}$ and we find the same result as Eq. (9.73).

So why did we go to so much trouble – all the work of Eqs. (9.69) through (9.72) seems like a waste. Ah, but there is madness in our method! We have now demonstrated to you that multiplying by the pseudo-inverse $(N^{\mathrm{T}}N)^{-1}N^{\mathrm{T}}$ produces the minimum squared error estimate of an overdetermined linear system. That is a significant result; important in the case of photometric stereo, and in many other applications as well.

Using shape from shading with two light sources

Here we illustrate how to use shape from shading when two light sources are available. We will not go into the physics here, but scanning electron microscopes provide an excellent example application. In such a microscope two images are produced, one from secondary emission (SE) electrons and one from backscattered (BSE) electrons. Reflectivity functions may readily be measured.

Rather than attempting to model the geometry of the microscope exactly, we may simply measure the reflectivity functions R_{SE} and R_{BSE} by placing a sphere in the microscope and imaging it. Assuming the albedo is known (that is an important assumption), the measured brightness is a function of p and q in both the BSE and SE images. For example, a particular brightness value might be represented by the locus of points shown in Fig. 9.20. Although in each image, there are an infinite number of possible p, q values, there are only two possible (p, q) pairs which can explain both the measured brightnesses. We thus solve for $z(x, y)$ by defining an objective function and finding the surface which minimizes that function. To accomplish this, first let $\rho((p_1, q_1), (p_2, q_2))$ define a function which represents the difference in the surface normals defined by the two (p, q) ordered pairs (it could, for example, be something as simple as the cosine of the angle between them $\rho((p_1, q_1), (p_2, q_2)) = p_1 p_2 + q_1 q_2$). Then, we consider the difference between the surface normal defined by the ith (p, q)

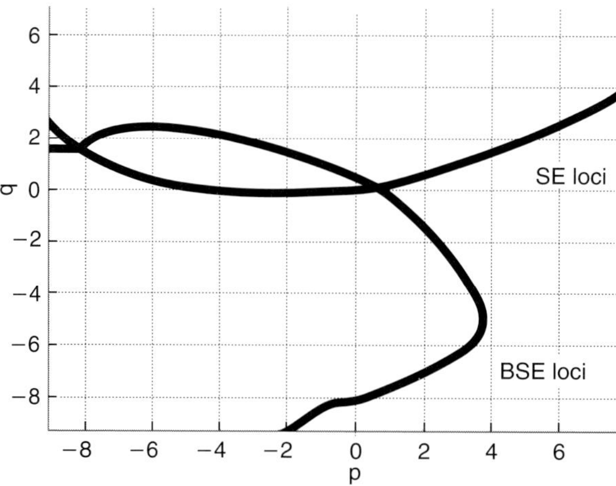

Fig. 9.20. A given measured brightness pair can be created by a locus of p, q values in both SE and BSE images. (The authors are grateful to B. Karacali for this figure.)

pair and the vector which is normal to the surface $z(x, y)$, and write that difference as

$$d_i(x, y) = \rho\left(\left(\frac{\partial z}{\partial x}, \frac{\partial z}{\partial y}\right), (p_i, q_i)\right).$$

Finally, assuming that the two curves of Fig. 9.20 intersect at m points (we make m a function of x, y to remind the reader that all this is being done for a single x, y point), we define an objective function as

$$E = \sum_{x,y}\left(\sum_{i}^{m(x,y)} (d_i(x, y))^{-1}\right)^{-1} + \lambda R, \tag{9.75}$$

allowing R to be a regularization term such as piecewise-linearity.

A variation on the simple model given above occurs when the surface is not totally reflective. For example, in infrared wavelengths (and others, to a lesser extent), the power measured coming from a surface is a combination of energy which is reflected and energy which is emitted (so called "black body radiation") as a result of the temperature of the surface [9.48].

The concepts of generalized cylinders (GCs) were discussed in section 9.12. If you combine GCs with shape from shading, you can improve the functionality of the GC model [9.23].

9A.2.3 Structured illumination

In section 4.2.2, the basic concepts of structured illumination were introduced. The key point is that by controlling the lighting, one or more of the unknowns in the stereopsis problem may be eliminated. Let's look at an example in more detail to see how this might work.

The problem to be solved is an application in robot vision: A robot is to pick up shiny, metallic turbine blades from a rack and place them into a machine for further processing. In order to locate the blades, a horizontal slit of light is projected onto the scene, by passing a laser beam through a cylindrical lens. The geometry of the resulting images is illustrated in

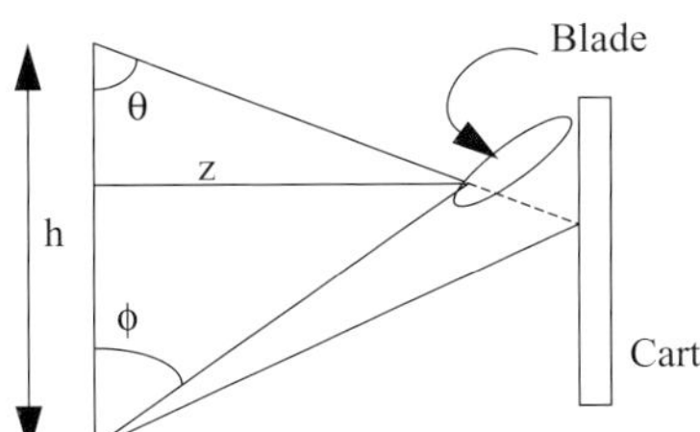

Fig. 9.21. The presence of a blade translates the location of the light stripe vertically in the image.

Fig. 9.21 [9.57]. If there were no blade in the image, the light stripe from the laser would form a horizontal line in the image as it is reflected off the cart. The presence of the blade causes a vertical translation of the light stripe. The number of lines of vertical displacement is directly proportional to an angular difference which produces the angle ϕ. Knowing the two angles and the distance h between the camera and the projector allows a simple calculation of the distance z:

$$z = \frac{h \tan \theta}{\tan \theta + \tan \phi}. \tag{9.76}$$

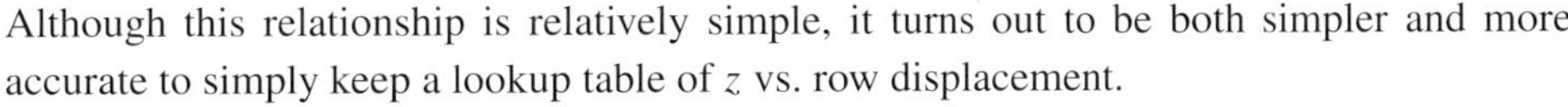

Although this relationship is relatively simple, it turns out to be both simpler and more accurate to simply keep a lookup table of z vs. row displacement.

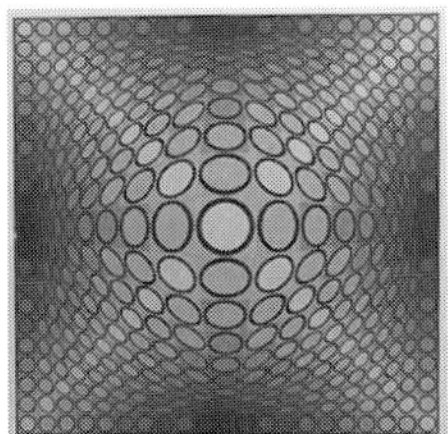

Fig. 9.22. Variations in texture can provide information about shape (from [9.13]).

One practical problem which arises in this problem is the specular nature of the reflections from the turbine blades; the bright spots may be orders of magnitude brighter than the rest of the image. This is dealt with by passing the beam through a polarizing filter. By placing another such filter on the camera lens, the specular spots are considerably reduced in magnitude.

In this example of the use of structured illumination, only one light stripe was projected at a time, so there was no ambiguity about which bright spot resulted from which projector, however, in the more general case, one may have multiple light sources, and then, some method for disambiguation is required [9.7, 9.50].

9A.2.4 Shape from texture

Texture, or rather spatial variations in texture, can be used to characterize 3D shape, as illustrated in Fig. 9.22. For a very nice collection of pictures showing how texture suggests 3D shape, see [9.82]. Of course, in order to perform shape from texture, one must be able to robustly extract texture primitives, as we discussed in section 4A.2.2.

The ideas of shape from texture have also found application in work to recover 3D motion [9.80]; see also [9.77].

9A.2.5 Shape from focus

That one can obtain range from focus is obvious. However, the extension to robust shape from focus is difficult. The principal problem is determining precisely when each pixel is in focus [9.52, 9.79].

9A.3 Motion analysis and tracking

Motion analysis may be viewed as two different problems. The first is the case when the camera is moving and the objects in the world are stationary. In this case, the extraction of camera motion is the challenge. In the second case, the camera is stationary, and objects in the world move. Finally, there is the combination of the two – where both the camera and some objects in the world are moving. Motion analysis has many of the same issues as stereo. As in stereo, correspondence is a major problem, but in this case, the correspondence is between scenes which differ in time rather than in the spatial location of the camera.

One approach to the motion analysis problem is referred to as "optic flow." Consider two images of the same object. Assume the image in frame 2 is the same as the image in frame 1, but displaced

$$f_1(x + \delta) = f_2(x). \tag{9.77}$$

Expand the LHS in a Taylor series

$$f_1(x) + \delta f_1'(x) + \cdots = f_2(x). \tag{9.78}$$

If we truncate this series to two terms and solve for δ, we find

$$\delta = \frac{f_2(x) - f_1(x)}{f_1'(x)}. \tag{9.79}$$

There is a serious problem here. What happens if the gradient is zero? The gradient being zero is really saying that there is no information at this point in the image. Imagine you are looking through a telescope at a semi-tractor–trailer truck which is passing by, and the field of view of your telescope allows you to see only a small area, a few square inches, of the truck. When the bumper passes, you have information, and you know motion has occurred. However, as the top of the trailer is passing, you see no changes for a relatively long time.

Another troublesome issue that arises in computing and applying optic flow is that in 2D, Eq. (9.79) becomes a differential equation (see [9.10]). To deal with these problems, researchers in optic flow have tried various ways to combine information from local measurements to infer global knowledge, such as using clustering to identify sets of points which are moving together [9.41]. As discussed in Chapter 4, one can match either points or boundaries. For example, Quan [9.61] matches conics; Taylor and Kriegman [9.85] match line segments, as does Zhang [9.99]; Smith and Nandha Kumar [4.35] match textures. Motion segmentation is discussed in [4.35, 9.3].

In two dimensions, the δ of Eq. (9.77) becomes a vector. Optic flow algorithms generate a *disparity field*, a vector field which associates a vector with each pixel [9.20]. Efficient implementations for optic flow have been investigated [10.19, 17.13, 17.14], including objects which deform [17.60].

The optic flow at a point in the image plane of a moving camera is

$$u_i = \frac{1}{z_i} A_i T + B_i \Omega \tag{9.80}$$

where the matrices

$$A_i = \begin{bmatrix} -f & 0 & x_i \\ 0 & -f & y_i \end{bmatrix} \quad \text{and} \quad B_i = \begin{bmatrix} x_i y_i & -(1+x_i^2) & y_i \\ (1+y_i^2) & -x_i y_i & -x_i \end{bmatrix} \tag{9.81}$$

are completely determined by the camera. T is the translational velocity of the camera and Ω is the rotational velocity of the camera. z_i is the depth to the image point at camera coordinates (x_i, y_i). The problem of determining T and Ω has been addressed by a number of researchers [9.36, 9.37]. Earnshaw and Blostein [9.15] compare these methods and introduce a new variation.

Motion can be estimated from smear. Chen *et al.* [9.9] make the observation that: "Psychophysical investigation has shown that the human visual system (HVS) integrates retinal images for 120 ms. Due to the integration, motion smear is inevitable. It has been reported that when the HVS is presented with an image blurred due to motion, the amount of smear perceived by human observers increases with observing duration at short durations (up to 20 ms). At longer durations, however, the perceived image becomes sharper. It is our conjecture that the HVS is performing a deblurring, or sharpening, function on the image." This observation leads to a "motion from smear" algorithm [9.9].

Recently, a gradual shift in focus has been observed [9.5] in the motion analysis literature from analyzing the image or camera motion to labeling the action taking place – from "How are things moving?" to "What is happening?"

Shape from motion

The discipline known as shape from motion usually assumes a solution to the correspondence problem: That certain points are identified and corresponded in each frame. Here, we describe the approach to shape from motion presented by Kanade and co-workers [9.60, 9.86]. This description is based on the unrealistic assumption that the points in the image are projected orthographically (perpendicular to) the image plane. We use this derivation because it is simpler to understand. However, if you need to implement this, refer to [9.60] for the modifications required for perspective projection. See also Soatto and Perona [9.72, 9.73] for a recent general study.

In the following derivation, the object is stationary, only the camera is moving. A point at spatial coordinates $\mathbf{s}_p$ is projected onto the imaging at frame f at coordinates (u_{fp}, v_{fp}). The camera moves, and at each frame, the camera is located at $\boldsymbol{t}_f$ and has orientation described by the three vectors $(\boldsymbol{i}_f, \boldsymbol{j}_f, \boldsymbol{k}_f)$. If we knew $\boldsymbol{s}_p$, we would know the camera coordinates of the image of p, from

$$u_{fp} = \boldsymbol{i}_f^{\mathrm{T}}(\boldsymbol{s}_p - \boldsymbol{t}_f) \qquad v_{fp} = \boldsymbol{j}_f^{\mathrm{T}}(\boldsymbol{s}_p - \boldsymbol{t}_f). \tag{9.82}$$

Defining

$$x_f = -\boldsymbol{t}_f^{\mathrm{T}}\boldsymbol{i}_f \qquad y_f = -\boldsymbol{t}_f^{\mathrm{T}}\boldsymbol{j}_f, \tag{9.83}$$

we can rewrite Eqs. (9.82) as

$$u_{fp} = \boldsymbol{i}_f^{\mathrm{T}}\boldsymbol{s}_p + x_f \qquad v_{fp} = \boldsymbol{j}_f^{\mathrm{T}}\boldsymbol{s}_p + y_f. \tag{9.84}$$

Now we have an inverse problem: We know the image coordinates, and from these we must determine the spatial position of both the camera and the object points.

Accumulate all the observations of the image points in a matrix:

$$W = \begin{bmatrix} u_{11} & \dots & u_{1p} \\ \dots & \dots & \dots \\ u_{F1} & \dots & u_{FP} \\ v_{11} & \dots & v_{1P} \\ \dots & \dots & \dots \\ v_{F1} & \dots & v_{FP} \end{bmatrix}, \tag{9.85}$$

given that there are F frames and P points. We observe that each column in this matrix is a listing of the image coordinates of a particular point as it is viewed in different frames, whereas each row is a listing of the locations of all the points in a particular frame. Next, we define a matrix M which is a $2F \times 3$ matrix whose rows are the $\boldsymbol{i}_f$ and $\boldsymbol{j}_f$ vectors, and a matrix S which is a $3 \times P$ "shape matrix" whose columns are the $\boldsymbol{s}_p$ vectors. Finally, define T to be a $2F$-dimensional translation vector with elements x_f and y_f. With these definitions, we can rewrite Eq. (9.84) for all the points as

$$W = MS + T\mathbf{1}_P, \tag{9.86}$$

Is this parenthetical remark correct?

where $\mathbf{1}_P$ is a vector of length P, composed of all ones. (Observe that the product of T with the ROW vector of all ones, an outer product, constructs a matrix of dimension $2F \times P$.)

If we move the origin to the center of gravity of the object (why not? location of the origin is arbitrary), we obtain

$$C = \frac{1}{P}\sum_{p=1}^{P} s_p = 0. \tag{9.87}$$

This location of the origin allows us to solve for T immediately, by observing that since the sum of any row of S is zero, the sum of any row of W is simply PT, and each row of T can be determined as the corresponding row of W divided by P. Now, subtract T from W producing a new matrix $\widehat{W}$ which satisfies

$$\widehat{W} = MS. \tag{9.88}$$

Using singular-valued decomposition, we can find a suitable decomposition of $\widehat{W}$, which we denote by $\widehat{W} = \widehat{M}\,\widehat{S}$. Unfortunately, these are not necessarily the M and S we want, since we could put any product AA^{-1} between them without changing the value of the product. We thus search for a matrix A such that

$$M = \widehat{M}A \qquad S = A^{-1}\widehat{S}. \tag{9.89}$$

To find A, we can make use of the fact that the rows of M are the direction vectors of the camera and are hence orthonormal. With these additional constraints, A is determined, and we know the positions in 3-space of all the P points as well as knowing the camera angles at each frame time.

9A.4 Vocabulary

You should know the meanings of the following terms.

Irradiance
Optic flow
Perspective
Photometric stereo
Pseudo-inverse
Reflectivity
Shape from shading
Structured illumination

Assignment 9.1

For each feature described in section 9.3, determine if that feature is invariant to: (1) rotation in the viewing plane; (2) translation in that plane; (3) rotation out of that plane (an affine transform if the object is planar); and (4) zoom.

Assignment 9.2

Let the Euclidian distance between two points be denoted by the operator $d(P_1, P_2)$ (you may want to use this in the next problem). Design a monotonic metric $R(P_1, P_2)$ which maps all distances to be between 0 and 1. That is, if $d(P_1, P_2) = \infty$, then $R(P_1, P_2) = 1$ and if $d(P_1, P_2) = 0$ then $R(P_1, P_2) = 0$.

For the metric you developed, show how you would prove your measure is a formal metric. Just set up the problem. Extra credit if you actually do the proof.

Assignment 9.3

Following are five points on the boundary of a region. (1,1), (2,1), (2,2), (2,4), (3,2). Use eigenvector methods to fit a straight line to this set of points, thus finding the principal axes of the region. Having found the principal axes, then estimate the aspect ratio of the region.

Assignment 9.4

Write the chain code for the figure below.

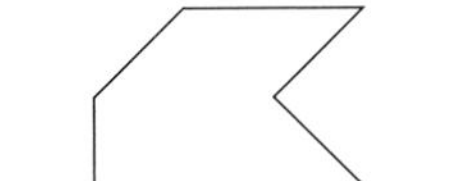

Assignment 9.5

Discuss the following postulate: Let P_1 and P_2 be the two extrema of a region which determine the diameter. Then P_1 and P_2 are on the boundary of the region.

Assignment 9.6

Starting with Eq. (9.19), prove Eq. (9.20).

Assignment 9.7

What is the difference between the intensity axis of symmetry and the medial axis?

Assignment 9.8

In Table 9.1, prove that the invariant moment φ_1 is invariant to zoom.

Assignment 9.9

Your instructor will specify an image containing a single region with unity brightness and zero background.

(1) Compute the seven invariant moments of the foreground region.
(2) Rotate the foreground region about its center of gravity through ten, twenty and forty degrees, and compute the invariant moments of the resulting image. What do you conclude?

Assignment 9.10

Prove that Eq. (9.35) is invariant to: (1) translation; (2) rotation; and (3) zoom.

Assignment 9.11

Is the caption on Fig. 9.12 correct?

Assignment 9.12

Two silhouettes, A and B, are measured and their boundaries encoded. Then, Fourier descriptors are computed. The descriptors are as in Table 9.3.

It is possible that these two objects represent similarity transforms of one another. (A similarity transform is equivalent to a rigid body motion, translation or rotation only.) Could they be affine transformations of one another? (An affine transformation is a linear transformation which includes not only rigid body motion, but also the possibility

Table 9.3. *Complex values of the Fourier descriptor.*

Object A	Object B
$5.00 + i0.00$	$5.83 + i1.80$
$4.2 + i1.87$	$3.69 + i2.57$
$3.86 + i1.00$	$3.48 + i2.00$
$2.95 + i2.05$	$2.30 + i2.77$
$3.19 + i1.47$	$2.70 + i2.24$

for scaling of the coordinate axes. If both axes (in 2D) are scaled by the same amount, you get zoom. If they are scaled by different amounts, you get shear.)

If you decide that these two sets of descriptors represent the same shape, possibly transformed, describe and justify what type of operations convert A into B. If they are not the same shape, explain why.

Assignment 9.13

A cylinder with unit radius and height of ten is oriented vertically about the origin, and is known to have a surface which is a Lambertian reflector. That is, the reflected brightness is independent of the angle of observation, and depends on the angle of incidence following the relationship $f = aI\cos\theta_i$, where a is the albedo and I is the brightness of the source. On this cylinder, the albedo is constant.

The camera is located at $x = 0, y = -2, z = 2$, and the optical axis of the camera is pointed at the origin.

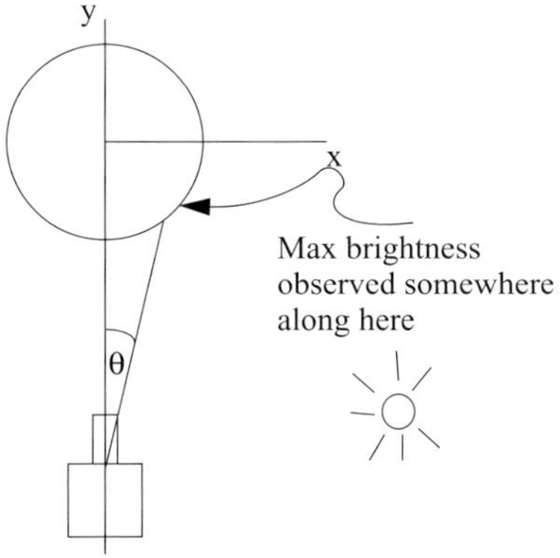

The light source is known to be 4 units from the origin, and is known to be a point source which radiates equally

```
in all directions. The brightest spot in the image of the
cylinder is at an angle of 29 degrees. Where is the light
source? A sketch of the coordinate system is illustrated for
your convenience. (Note: Drawing is NOT to scale, and is not
guaranteed to even be correct.)
```

Bibliography

[9.1] K. Arbter, W. Snyder, H. Burkhardt, and G. Hirzinger, "Application of Affine-invariant Fourier Descriptors to Recognition of 3-D Objects," *IEEE Transactions on Pattern Analysis and Machine Intelligence*, **12**(7), pp. 640–647, 1990.

[9.2] D. Ballard and C. Brown, *Computer Vision*, Englewood Cliffs, NJ, Prentice-Hall, 1982.

[9.3] M. Bichsel, "Segmenting Simply Connected Moving Objects in a Static Scene," *IEEE Transactions on Pattern Analysis and Machine Intelligence*, **16**(11), pp. 1138–1142, 1994.

[9.4] T. Binford, "Visual Perception by Computer," *IEEE Conference on Systems and Control*, Miami, December, 1971.

[9.5] A. Bobick and J. Davis, "The Recognition of Human Movement Using Temporal Templates," *IEEE Transactions on Pattern Analysis and Machine Intelligence*, **23**(3), pp. 257–267, 2001.

[9.6] A. Califano and R. Mohan, "Multidimensional Indexing for Recognizing Visual Shapes," *IEEE Transactions on Pattern Analysis and Machine Intelligence*, **16**(4), pp. 373–392, 1994.

[9.7] D. Caspi, N. Kiryati, and J. Shamir, "Range Imaging with Adaptive Color Structured Light," *IEEE Transactions on Pattern Analysis and Machine Intelligence*, **20**(5), pp. 470–480, 1998.

[9.8] C. Chen, T. Huang, and M. Arrott, "Modeling, Analysis, and Visualization of Left Ventricle Shape and Motion by Hierarchical Decomposition," *IEEE Transactions on Pattern Analysis and Machine Intelligence*, **16**(4), pp. 342–356, 1994.

[9.9] W. Chen, N. Nandhakumar, and W. Martin, "Image Motion Estimation from Motion Smear – A New Computational Model," *IEEE Transactions on Pattern Analysis and Machine Intelligence*, **18**(4), pp. 412–425, 1996.

[9.10] A. Chhabra and T. Grogan, "On Poisson Solvers and Semi-direct Methods for Computing Area Based Optic Flow," *IEEE Transactions on Pattern Analysis and Machine Intelligence*, **16**(11), pp. 1133–1138, 1994.

[9.11] K. Cho and S. Dunn, "Learning Shape Classes," *IEEE Transactions on Pattern Analysis and Machine Intelligence*, **16**(9), pp. 882–888, 1994.

[9.12] J. Chuang, C. Tsai, and M. Ko, "Skeletonization of Three-dimensional Object using Generalized Potential Field," *IEEE Transactions on Pattern Analysis and Machine Intelligence*, **22**(11), pp. 1241–1251, 2000.

[9.13] M. Clerc, "Texture Gradient," http://www.masterworksfineart.com/inventory/vas_originals.htm.

[9.14] M. Dhome, M. Richetin, J. Lapreste, and G. Rives, "Determination of the Attitude of 3-D Objects from a Single Perspective View," *IEEE Transactions on Pattern Analysis and Machine Intelligence*, **11**(12), pp. 1265–1278, 1989.

[9.15] A. Earnshaw and S. Blostein, "The Performance of Camera Translation Direction Estimators from Optical Flow: Analysis, Comparison, and Theoretical Limits," *IEEE Transactions on Pattern Analysis and Machine Intelligence*, **18**(9), pp. 927–932, 1996.

[9.16] A. Ferreira and S. Ubeda, "Computing the Medial Axis Transform in Parallel with Eight Scan Operations," *IEEE Transactions on Pattern Analysis and Machine Intelligence*, **21**(3), pp. 277–282, 1999.

[9.17] M. Fischler and R. Bolles, "Random Sample Consensus: A Paradigm for Model Fitting with Application to Image Analysis and Automated Cartography," *Communications of the Association for Computing Machinery*, **24**(6), 1981.

[9.18] P. Flynn, "3-D Object Recognition with Symmetric Models: Symmetry Extraction and Encoding," *IEEE Transactions on Pattern Analysis and Machine Intelligence*, **16**(8), pp. 814–818, 1994.

[9.19] D. Geiger, A. Gupta, L. Costa, and J. Vlontzos, "Dynamic Programming for Detecting, Tracking, and Matching Deformable Contours," *IEEE Transactions on Pattern Analysis and Machine Intelligence*, **17**(3), pp. 294–302, 1995.

[9.20] S. Ghosal and P. Vanek, "A Fast Scalable Algorithm for Discontinuous Optical Flow Estimation," *IEEE Transactions on Pattern Analysis and Machine Intelligence*, **18**(2), pp. 181–194, 1996.

[9.21] R. Gonzalez and P. Wintz, *Digital Image Processing,* Reading, MA, Addison-Wesley, 1977.

[9.22] A. Gross, "Toward Object-based Heuristics," *IEEE Transactions on Pattern Analysis and Machine Intelligence*, **16**(8), pp. 794–802, 1994.

[9.23] A. Gross and T. Boult, "Recovery of SHGCs from a Single Intensity View," *IEEE Transactions on Pattern Analysis and Machine Intelligence*, **18**(2), pp. 161–180, 1996.

[9.24] M. Gruber and K. Hsu, "Moment-based Image Normalization with High Noise-Tolerance," *IEEE Transactions on Pattern Analysis and Machine Intelligence*, **19**(2), pp. 136–139, 1997.

[9.25] S. Gunn and M. Nixon, "A Robust Snake Implementation: A Dual Active Contour," *IEEE Transactions on Pattern Analysis and Machine Intelligence*, **19**(1), pp. 63–68, 1997.

[9.26] R. Hartley, "Projective Reconstruction and Invariants from Multiple Images," *IEEE Transactions on Pattern Analysis and Machine Intelligence*, **16**(10), pp. 1036–1041, 1994.

[9.27] P. Havaldar and G. Medioni, "Full Volumetric Descriptions from Three Intensity Images," *IEEE Transactions on Pattern Analysis and Machine Intelligence*, **20**(5), pp. 540–545, 1998.

[9.28] G. Healey and R. Kondepudy, "Radiometric CCD Camera Calibration and Noise Estimation," *IEEE Transactions on Pattern Analysis and Machine Intelligence*, **16**(3), pp. 267–276, 1994.

[9.29] D. Heeger and A. Jepson, "Subspace Methods for Recovering Rigid Motion I: Algorithm and Implementation," *International Journal of Computer Vision*, **7**(2), pp. 95–117, 1992.

[9.30] D. Helman and J. JáJá, "Efficient Image Processing Algorithms on the Scan Line Array Processor," *IEEE Transactions on Pattern Analysis and Machine Intelligence*, **17**(1), pp. 47–56, 1995.

[9.31] A. Hoover, D. Goldgof, and K. Bowyer, "Extracting a Valid Boundary Representation from a Segmented Range Image," *IEEE Transactions on Pattern Analysis and Machine Intelligence*, **17**(9), pp. 920–924, 1995.

[9.32] B.K.P. Horn, *Robot Vision*, Cambridge, MA, MIT Press, 1986.

[9.33] M. Hu, "Visual Pattern Recognition by Moment Invariants," *IRE Transactions on Information Theory*, **8**, pp. 179–187, 1962.

[9.34] X. Hu and N. Ahuja, "Matching Point Features with Ordered Geometric, Rigidity, and Disparity Constraints," *IEEE Transactions on Pattern Analysis and Machine Intelligence*, **16**(10), pp. 1041–1049, 1994.

[9.35] Y. Iwahori, R. Woodham, and A. Bagheri, "Principal Components Analysis and Neural Network Implementation of Photometric Stereo," *Proceedings IEEE Conference on Physics-Based Modeling in Computer Vision*, June, 1995, pp. 117–125, 1995.

[9.36] A. Jepson and D. Heeger, "Linear Subspace Methods for Recovering Translational Direction," In *Spatial Vision in Humans and Robots*, ed. L. Harris and M. Jenkin, Cambridge, Cambridge University Press, 1993.

[9.37] K. Kanatani, "Unbiased Estimation and Statistical Analysis of 3-D Rigid Motion from Two Views," *IEEE Transactions on Pattern Analysis and Machine Intelligence*, **15**(1), pp. 37–50, 1993.

[9.38] K. Kanatani, "Comments on 'Symmetry as a Continuous Feature'," *IEEE Transactions on Pattern Analysis and Machine Intelligence*, **19**(3), pp. 246–247, 1997.

[9.39] H. Kauppinen T. Seppänen, and M. Pietikäinen, "An Experimental Comparison of Autoregressive and Fourier-based Descriptors in 2D Shape Classification," *IEEE Transactions on Pattern Analysis and Machine Intelligence*, **17**(2), pp. 201–207, February, 1995.

[9.40] R. Kimmel, A. Amir, and A. Bruckstein, "Finding Shortest Paths on Surfaces Using Level Sets Propagation," *IEEE Transactions on Pattern Analysis and Machine Intelligence*, **17**(6), pp. 635–640, 1995.

[9.41] D. Kottke and Y. Sun, "Motion Estimation via Cluster Matching," *IEEE Transactions on Pattern Analysis and Machine Intelligence*, **16**(11), pp. 1128–1132, 1994.

[9.42] A. Laurentini, "The Visual Hull Concept for Silhouette-based Image Understanding," *IEEE Transactions on Pattern Analysis and Machine Intelligence*, **16**(2), pp. 150–162, 1994.

[9.43] A. Laurentini, "How Far 3D Shapes can be Understood from 2D Silhouettes," *IEEE Transactions on Pattern Analysis and Machine Intelligence*, **17**(2), pp. 188–195, 1995.

[9.44] S. Lavallée and R. Szeliski, "Recovering the Position and Orientation for Free-form Objects from Image Contours Using 3D Distance Maps," *IEEE Transactions on Pattern Analysis and Machine Intelligence*, **17**(4), pp. 378–390, 1995.

[9.45] S. Liao and M. Pawlak, "On Image Analysis by Moments," *IEEE Transactions on Pattern Analysis and Machine Intelligence*, **18**(3), pp. 254–266, 1996.

[9.46] R. Malladi, J. Sethian, and B. Vemuri, "Shape Modeling with Front Propagation: A Level Set Approach," *IEEE Transactions on Pattern Analysis and Machine Intelligence*, **17**(2), pp. 158–175, 1995.

[9.47] C. Manhaeghe, I. Lemahieu, D. Vogelaers, and F. Colardyn, "Automatic Initial Estimation of the Left Ventricular Myocardial Midwall in Emission Tomograms using Kohonen Maps," *IEEE Transactions on Pattern Analysis and Machine Intelligence*, **16**(3), pp. 259–266, 1994.

[9.48] J. Michel, N. Nandhakumar, and V. Velten, "Thermophysical Algebraic Invariants from Infrared Imagery for Object Recognition," *IEEE Transactions on Pattern Analysis and Machine Intelligence*, **19**(1), pp. 41–51, 1997.

[9.49] F. Mokhtarian, "Silhouette-based Isolated Object Recognition through Curvature Scale Space," *IEEE Transactions on Pattern Analysis and Machine Intelligence*, **17**(5), pp. 539–544, 1995.

[9.50] R. Morano, C. Ozturk, R. Conn, S. Dubin, S. Zietz, and J. Nissano, "Structured Light using Pseudorandom Codes," *IEEE Transactions on Pattern Analysis and Machine Intelligence*, **20**(3), pp. 322–327, 1998.

[9.51] S. Nayar and R. Bolle, "Reflectance Based Object Recognition," *International Journal of Computer Vision*, **17**(3), pp. 219–240, 1996.

[9.52] S. Nayar and Y. Nakagawa, "Shape from Focus," *IEEE Transactions on Pattern Analysis and Machine Intelligence*, **16**(8), pp. 824–831, 1994.

[9.53] T. Nguyen and B. Oommen, "Moment-preserving Piecewise Linear Approximations of Signals and Images," *IEEE Transactions on Pattern Analysis and Machine Intelligence*, **19**(1), pp. 84–91, 1997.

[9.54] L. O'Gorman, "Subpixel Precision of Straight-edged Shapes for Registration and Measurement," *IEEE Transactions on Pattern Analysis and Machine Intelligence*, **18**(7), 1996.

[9.55] F. O'Sullivan and M. Qian, "A Regularized Contrast Statistic for Object Boundary Estimation – Implementation and Statistical Evaluation," *IEEE Transactions on Pattern Analysis and Machine Intelligence*, **16**(6), pp. 561–570, 1994.

[9.56] M. Oren and S. Nayar, "A Theory of Specular Surface Geometry," *International Journal of Computer Vision*, **24**(2), pp. 105–124, 1996.

[9.57] N. Page, W. Snyder, and S. Rajala, "Turbine Blade Image Processing System." In *Advanced Software in Robotics*, ed. A. Danthine, Amsterdam, North-Holland, 1984.

[9.58] E. Pauwels, P. Fiddelaers, and L. Van Gool, "Enhancement of Planar Shape Through Optimization of Functionals for Curves," *IEEE Transactions on Pattern Analysis and Machine Intelligence*, **17**(12), 1995.

[9.59] S. Pizer, C. Burbeck, J. Coggins, D. Fritsch, and B. Morse, "Object Shape Before Boundary Shape: Scale Space Medial Axis," *Journal of Mathematical Imaging and Vision*, **4**, pp. 303–313, 1994.

[9.60] C. Poelman and T. Kanade, "A Paraperspective Factorization Method for Shape and Motion Recovery," *IEEE Transactions on Pattern Analysis and Machine Intelligence*, **19**(3), pp. 206–218, 1997.

[9.61] L. Quan, "Conic Reconstruction and Correspondence from Two Views," *IEEE Transactions on Pattern Analysis and Machine Intelligence*, **18**(2), 1996.

[9.62] I. Rothe, H. Süsse, and K. Voss, "The Method of Normalization to Determine Invariants," *IEEE Transactions on Pattern Analysis and Machine Intelligence*, **18**(4), 1996.

[9.63] G. Sandini and V. Tagliasco, "An Anthropomorphic Retina-like Structure for Scene Analysis," *Computer Graphics and Image Processing*, **14**, pp. 365–372, 1980.

[9.64] H. Schultz, "Retrieving Shape Information from Multiple Image of a Specular Surface," *IEEE Transactions on Pattern Analysis and Machine Intelligence*, **16**(2), pp. 195–201, 1994.

[9.65] E. Schwartz, "Computational Anatomy and Functional Architecture of Striate Cortex, Spatial Mapping Approach to Perceptual Coding," *Vision Research*, **20**, pp. 645–669, 1980.

[9.66] J. Sethian, "Curvature and Evolution of Fronts," *Communications in Mathematical Physics*, **101**, pp. 487–499, 1985.

[9.67] J. Sethian, "Numerical Algorithms for Propagating Interfaces: Hamilton–Jacobi Equations and Conservation Laws," *Journal of Differential Geometry*, **31**, pp. 131–161, 1990.

[9.68] M. Shamos, "Geometric Complexity," *7th Annual ACM Symposium on Theory of Computing*, May, 1975, Albuquerque, NM, pp. 224–233, 1975.

[9.69] D. Sinclair and A. Blake, "Isoperimetric Normalization of Planar Curves," *IEEE Transactions on Pattern Analysis and Machine Intelligence*, **16**(8), pp. 769–777, 1994.

[9.70] S. Smith and J. Brady, "ASSET-2: Real-time Motion Segmentation and Shape Tracking," *IEEE Transactions on Pattern Analysis and Machine Intelligence*, **17**(8), 1995.

[9.71] W. Snyder and I. Tang, "Finding the Extrema of a Region," *IEEE Transactions on Pattern Analysis and Machine Intelligence,* **2**, pp. 266–269, 1980.

[9.72] S. Soatto and P. Perona, "Reducing 'Structure from Motion': A General Framework for Dynamic Vision.1. Modeling," *IEEE Transactions on Pattern Analysis and Machine Intelligence*, **20**(9), pp. 933–942, 1998.

[9.73] S. Soatto and P. Perona, "Reducing 'Structure from Motion': A General Framework for Dynamic Vision. 2. Implementation and Experimental Assessment," *IEEE Transactions on Pattern Analysis and Machine Intelligence*, **20**(9), pp. 943–960, 1998.

[9.74] B. Soroka, "Generalized Cylinders from Parallel Slices," *Proceedings of the Conference on Pattern Recognition and Image Processing*, 1979.

[9.75] B. Soroka and R. Bajcsy, "Generalized Cylinders from Serial Sections," *3rd International Joint Conference on Pattern Recognition*, November, Coronado, CA, 1976.

[9.76] M. Soucy and D. Laurendeau, "A General Surface Approach to the Integration of a Set of Range Views," *IEEE Transactions on Pattern Analysis and Machine Intelligence*, **17**(4), pp. 344–358, 1995.

[9.77] J. Stone and S. Isard, "Adaptive Scale Filtering: A General Method for Obtaining Shape from Texture," *IEEE Transactions on Pattern Analysis and Machine Intelligence*, **17**(7), pp. 713–718, 1995.

[9.78] G. Storvik, "A Bayesian Approach to Dynamic Contours through Stochastic Sampling and Simulated Annealing," *IEEE Transactions on Pattern Analysis and Machine Intelligence*, **16**(10), pp. 976–986, 1994.

[9.79] M. Subbarao and T. Choi, "Accurate Recovery of Three-dimensional Shape from Image Focus," *IEEE Transactions on Pattern Analysis and Machine Intelligence*, **17**(3), pp. 266–274, 1995.

[9.80] S. Sull and N. Ahuja, "Integrated 3-D Analysis and Analysis-Guided Synthesis of Flight Image Sequences," *IEEE Transactions on Pattern Analysis and Machine Intelligence*, **16**(4), pp. 357–372, 1994.

[9.81] Y. Sun, I. Liu, and J. Grady, "Reconstruction of 3-D Tree-like Structures from Three Mutually Orthogonal Projections," *IEEE Transactions on Pattern Analysis and Machine Intelligence*, **16**(3), pp. 241–248, 1994.

[9.82] B. Super, and A. Bovik, "Shape from Texture Using Local Spectral Moments," *IEEE Transactions on Pattern Analysis and Machine Intelligence*, **17**(4), pp. 333–343, 1995.

[9.83] K. Tarabanis, R. Tsai, and A. Kaul, "Computing Occlusion-free Viewpoints," *IEEE Transactions on Pattern Analysis and Machine Intelligence*, **18**(3), pp. 279–292, 1996.

[9.84] G. Taubin and R. Ronfard, "Implicit Simplicial Models for Adaptive Curve Reconstruction," *IEEE Transactions on Pattern Analysis and Machine Intelligence*, **18**(3), pp. 321–325, 1996.

[9.85] C. Taylor and D. Kriegman, "Structure and Motion from Line Segments in Multiple Images," *IEEE Transactions on Pattern Analysis and Machine Intelligence*, **17**(11), pp. 1021–1032, 1995.

[9.86] C. Tomasi and T. Kanade, "Shape and Motion from Image Streams under Orthography: A Factorization Method," *International Journal of Computer Vision*, **9**(2), pp. 137–154, 1992.

[9.87] F. Tong and Z. Li, "Reciprocal-wedge Transform for Space-variant Sensing," *IEEE Transactions on Pattern Analysis and Machine Intelligence*, **17**(5), pp. 500–511, 1995.

[9.88] B. Vemuri and Y. Guo, "Snake Pedals: Compact and Versatile Geometric Models with Physics-based Control," *IEEE Transactions on Pattern Analysis and Machine Intelligence*, **22**(5), pp. 445–459, 2000.

[9.89] K. Voss and H. Suesse, "Invariant Fitting of Planar Objects by Primitives," *IEEE Transactions on Pattern Analysis and Machine Intelligence*, **19**(1), pp. 80–84, 1997.

[9.90] C. Weiman and G. Chaikin, "Logarithmic Spiral Grids for Image Processing and Display," *Computer Graphics and Image Processing*, **11**, pp. 197–226, 1979.

[9.91] D. Weinshall and C. Tomasi, "Linear and Incremental Acquisition of Invariant Shape Models from Image Sequences," *IEEE Transactions on Pattern Analysis and Machine Intelligence*, **17**(5), pp. 512–517, 1995.

[9.92] M. Werman and D. Weinshall, "Similarity and Affine Invariant Distances Between 2D Point Sets," *IEEE Transactions on Pattern Analysis and Machine Intelligence*, **17**(8), pp. 810–814, 1995.

[9.93] P. Whaite and F. Ferrie, "Autonomous Exploration: Driven by Uncertainty," *IEEE Transactions on Pattern Analysis and Machine Intelligence*, **19**(3), pp. 193–205, 1997.

[9.94] R. Woodham, "Photometric Method for Determining Surface Orientation from Multiple Images," *Optical Engineering*, **19**, pp. 139–144, 1980.

[9.95] Y. Wu, S. Iyengar, R. Jain, and S. Bose, "A New Generalized Framework for Finding Object Orientation using Perspective Trihedral Angle Constraint," *IEEE Transactions on Pattern Analysis and Machine Intelligence*, **16**(10), pp. 961–975, 1994.

[9.96] R. Yip, P. Tam, and D. Leung, "Application of Elliptic Fourier Descriptors to Symmetry Detection under Parallel Projection," *IEEE Transactions on Pattern Analysis and Machine Intelligence*, **16**(3), pp. 277–286, 1994.

[9.97] H. Zabrodsky, S. Peleg, and D. Avnir, "Symmetry as a Continuous Feature," *IEEE Transactions on Pattern Analysis and Machine Intelligence*, **17**(12), pp. 1154–1166, 1995.

[9.98] M. Zerroug and R. Nevatia, "Three Dimensional Descriptions based on the Analysis of the Invariant and Quasi-invariant Properties of Some Curved-axis Generalized Cylinders," *IEEE Transactions on Pattern Analysis and Machine Intelligence*, **18**(3), pp. 237–253, 1996.

[9.99] Z. Zhang, "Estimating Motion and Structure from Correspondences of Line Segments between Two Perspective Images," *IEEE Transactions on Pattern Analysis and Machine Intelligence*, **17**(12), pp. 1129–1139, 1995.

[9.100] R. Zhang, P. Tsai, J. Cryer, and M. Shah, "Shape-from-shading: A Survey," *IEEE Transactions on Pattern Analysis and Machine Intelligence*, **21**(8), pp. 690–706, 1999.

[9.101] J. Zheng, "Acquiring 3-D Models from Sequences of Contours," *IEEE Transactions on Pattern Analysis and Machine Intelligence*, **16**(2), pp. 163–178, 1994.

[9.102] Y. Zhong, A. Jain, and M. Dubuisson-Jolly, "Object Tracking using Deformable Templates," *IEEE Transactions on Pattern Analysis and Machine Intelligence*, **22**(5), pp. 544–549, 2000.

[9.103] D. Zwillinger, CRC Standard Mathematical Tables and Formulae, 31st edn., CRC Press, 2003.

10 Consistent labeling

On axis, as the planets run,
Yet make at once their circle round the sun;
So two consistent motions act the soul;
And one regards itself and one the whole.

Alexander Pope

The single most challenging problem in all of computer vision is the "local/global inference problem." As in the fable of the blind men and the elephant, the computer must, from a set of local measurements, infer the global properties of what is being observed. In other words, the next level of the machine vision problem is to interpret the global scene (which is composed of individual objects) using local information about each object obtained from segmentation and shape analysis as we have discussed in Chapters 8 and 9. One way to approach the local/global inference problem is to introduce the concept of consistency.

10.1 Consistency

Let's begin with some notation: Define a set of *objects* $\{x_1, x_2, \ldots x_n\}$, and a set of *labels* for those objects $\{\lambda_1, \lambda_2, \ldots \lambda_k\}$, which we assume for now are mutually exclusive (each object may have only one label) and collectively exhaustive (each object has a label). Denote a labeling as the ordered pair (x_i, λ_j). By this notation, we mean that object i has been assigned label j.

As an example of consistent labeling, we will consider the problem of labeling objects in a line drawing. Researchers have been interested in analysis of line drawings since the beginnings of work in machine vision for three reasons: First, humans can obviously look at line drawings and make interpretations with ease. Second, psychological experiments [10.1, 10.6, 10.10] have convincingly demonstrated that it is the points where brightness changes rapidly that convey the most information, and it is relatively easy to convert edges to lines. Third, a line drawing is a dramatic reduction in the amount of data (which is not the same as information) in an image, and perhaps, just perhaps, learning how to process line drawings would make analysis algorithms run faster. Most of the groundwork in line drawing analysis was done in the late 1960s and 1970s [10.5, 10.8], however, progress continues [10.17] to be made.

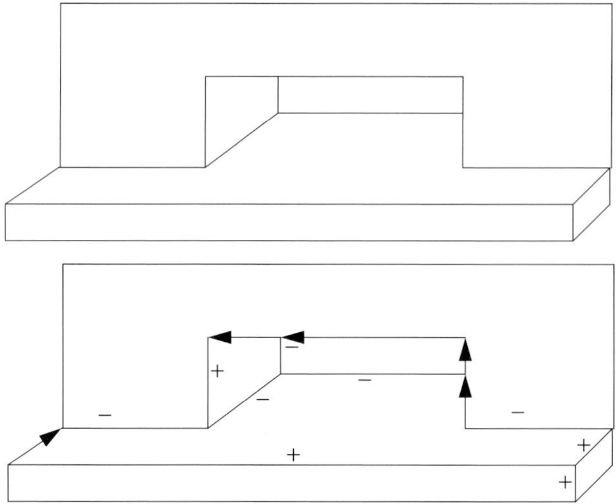

Fig. 10.1. **A line drawing illustrating convex, concave, and occluding lines, and a labeling of that drawing.**

We will allow each line in such a drawing to be labeled as either convex (an edge in three dimensions which points toward the observer, such as the corner of a desk), concave (an edge in three dimensions which points away from the observer, such as the joint between the wall and the floor of a room), or occluding (the edge occurs because one surface is partially hidden by another surface). For example, consider the drawing in Fig. 10.1. In that drawing, lines resulting from convex edges are labeled with a plus sign, concave edges are labeled with a minus sign, and occluding edges are labeled with an arrow. The arrow points in the direction such that the occluded surface is on the left if one moves in the direction of the arrow. You may note that not all the lines in this drawing have been labeled. That is deliberate. Thus we have one type of object, lines, and three types of labels for those lines. Our mission is to learn how to get a computer to do automatically what human beings did so easily when we interpreted the lines in that figure.

The compatibility function $r(.)$ will have differing interpretations through this chapter, depending on the specific labeling algorithm.

Check it out. Did we do the math right? n objects, with each object labeling consistent with the labeling of all other objects.

Before we can do that, we need to address one of those unfortunate ambiguities that natural language introduces into discussions like this. The term *labeling* may have two meanings. It may refer to the label assigned to a single object, to a pair of objects, or to an entire scene. We will try to make sure that the meaning is clear from context in this discussion. In order to accomplish the objective of labeling a scene, we must consider something we call *consistency*. The simultaneous labeling of two objects is represented by some function which we will call *compatibility* and denoted by $r(i, \lambda, j, \lambda')$. This function is defined to have a value of 1 if the two labelings can exist together (mutually compatible) and -1 if they cannot. For example, $r(i, +, i, -) = -1$, since the object i cannot be both concave and convex (consider the drawing in Fig. 10.2). A labeling of an image is said to be *complete and consistent* when $\sum_i \sum_{i \neq j} r(i, \lambda, j, \lambda') = n(n-1)$, where any label values at

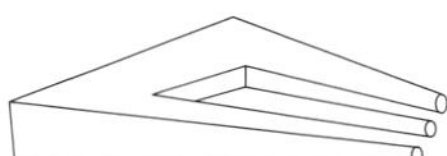

Fig. 10.2. A line which is both concave and convex can exist only in an impossible object.

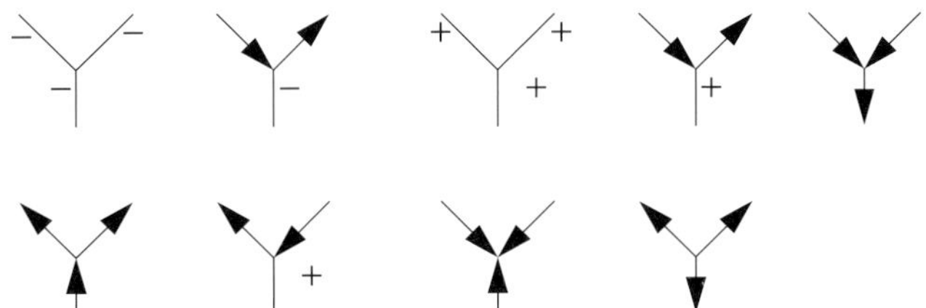

Fig. 10.3. All the physically possible Y vertices.

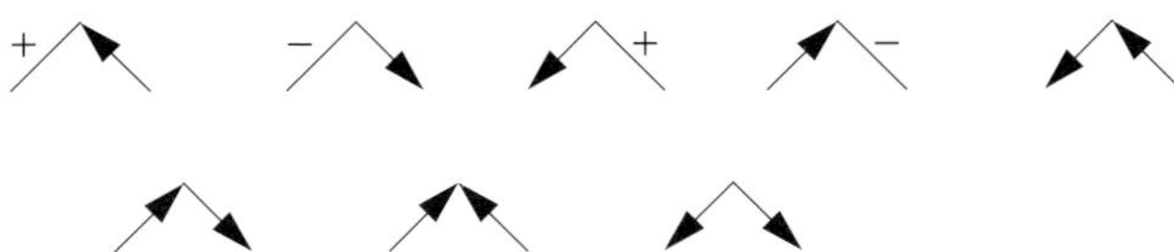

Fig. 10.4. All the physically possible ELL vertices.

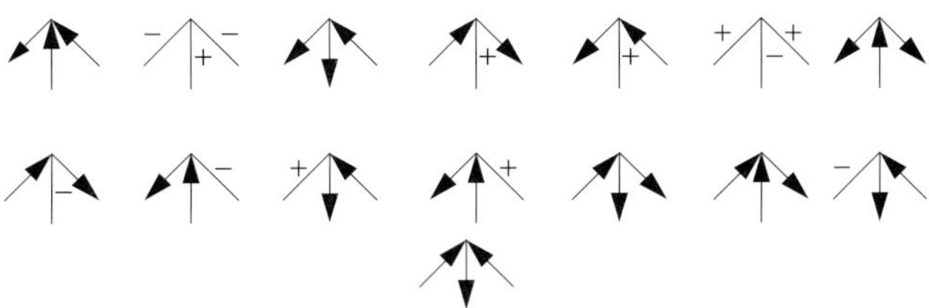

Fig. 10.5. All the physically possible arrow vertices.

all are allowed. In this chapter, we will utilize several different realizations of this compatibility function.

Let's look at the line labeling in more detail to see how this works. Although we are interested in labeling lines, it will turn out to be useful to think about vertices as well. A vertex is where lines meet. If each line can have four labels (concave, convex, arrow in, arrow out), there should be 4^3 ways to label a vertex with three lines meeting. It turns out, however that not all of those combinations are physically possible. In Figs. 10.3 to 10.5, we illustrate all the "Y", "ELL", and "arrow" vertices which are physically possible. There are a variety of ways we could make use of this information. One way is to use depth-first search. The algorithm is as follows:

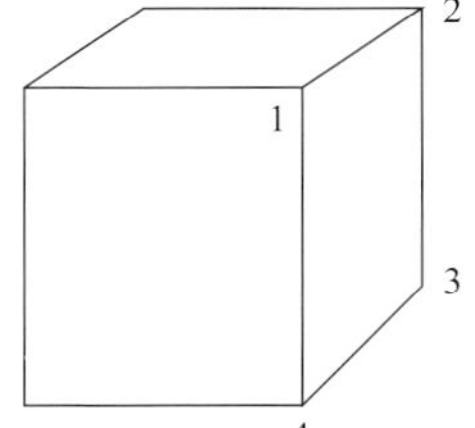

Fig. 10.6. An object which we wish to label consistently.

(1) Choose a starting vertex (call it vertex 1) and label all the lines coming in to it in a physically possible way.
(2) Choose an adjacent vertex (call it vertex 2) and label all lines coming in to it in a physically possible way, such that the labeling is consistent with previous labeling. That is, the line from vertex 1 to 2 can only have one label.
(3) If no consistent labeling is possible, back up.

In Fig. 10.7, we illustrate the labeling process of the 3D object shown in Fig. 10.6, beginning with a choice of one possible labeling for the lines of vertex 1. Given the

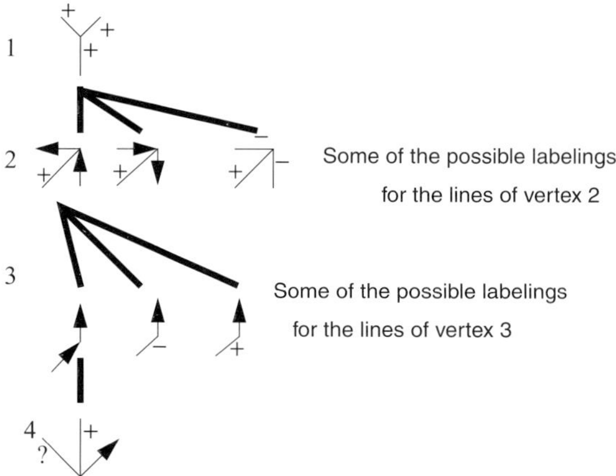

Fig. 10.7. Possible labeling for the object in Fig. 10.6.

labeling of vertex 1, we can choose any of a set of labeling for the lines of vertex 2, but all of them must have a + sign on the line between 1 and 2, as illustrated on the second line of the figure. Now, choose one of those labelings of vertex 2 (let's pick the one on the left), and label vertex 3 in such a way that it is consistent with the labeling of both 1 and 3. Now, we need to assume a "correct" interpretation of vertex 3 in order to label vertex 4. Again, choose the one on the left. In order to label the lines coming into vertex 4, we must choose a labeling which is consistent with the (assumed correct) labeling of vertex 3 and vertex 1. Since two of the lines coming in to vertex 4 are now determined, there is only one way to label vertex 4 consistently, and that is with an arrow coming in on the third line.

Suppose we had reached the labeling of vertex 4, and found there were no physically possible labelings. Then clearly one of the earlier assumptions was incorrect. We now "back up," and choose a new labeling for vertex 3. Suppose we run through all the possible labelings of the lines of vertex 3 and from none of them can we find a consistent labeling of 4, then we back up again and choose a new labeling of vertex 2. We follow this approach until we either find a consistent labeling of the entire object, or we fail to find one. If we fail, then the object cannot be consistently labeled [10.7].

Since line labeling was originally developed, many researchers have added enhancements. For example, Parodi and Piccioli [10.13] demonstrate that if vanishing points can be determined, and the 3D coordinates of a single point are known, the 3D coordinates of all the other labeled points may be found.

10.2 Relaxation labeling

Now that we have seen an application of consistent labeling, let's generalize the idea a bit by allowing a particular object to have more than one label at a time.

To accomplish this, we define the variable $p_i(\lambda_j)$ to represent our confidence that object i has label λ_j. Now some terminology needs to be discussed before we go any further. You may hear $p_i(\lambda_j)$ referred to as the *probability* that object i has label λ_j. While sometimes used, that terminology is incorrect. Yes, p has the range of a probability $0 \leq p_i(\lambda) \leq 1$; and yes, it integrates to one, $\sum_j p_i(\lambda_j) = 1$, but there is no random process here. Rather, $p_i(\lambda_j)$ is our confidence that, knowing what we do right now, object i has label λ_j. As we study the problem more and make more use of consistency, hopefully these ps will change from assuming continuous values between zero and one to becoming only zeros or ones.

10.2.1 Using consistency to update labelings

Linear relaxation

The first way one might approach the idea of using consistency is to set up a linear system that takes into account the initial probabilities and the consistencies. We define the compatibility of object i having label λ with object j having label λ' as $r_L(i, \lambda, j, \lambda')$ (the subscript L denotes that this is used in the linear relaxation algorithm) and require that $0 \leq r_L \leq 1$ and

Be careful! The definition of r will change in the next section.

$$\sum_{\lambda} r_L(i, \lambda, j, \lambda') = 1 \qquad \text{for all } i, j, \lambda'. \tag{10.1}$$

The linear relaxation process iteratively updates the label weights following

$$p_i(\lambda) = \sum_j \sum_{\lambda'} r_L(i, \lambda, j, \lambda') p_j(\lambda'). \tag{10.2}$$

Equation (10.2) is readily seen to be a multiplication of the matrix of consistencies with the vector of weights. Repeated application of this equation can be shown [10.15] to converge to a weight vector that is the eigenvector corresponding to the eigenvalue unity. Under some circumstances, it is possible to learn the consistency matrix [10.14]. Unfortunately, this does not really give us the kind of information we want. How can the result be independent of the initial conditions?

Nonlinear relaxation

In nonlinear relaxation, we again make use of what we know about consistency to update labelings. We will also develop an algorithm which drives the labelings to zero or one, and which does depend on the measurements. First, we define a new update rule. At iteration $k + 1$ of our algorithm, use

$$p_i^{k+1}(\lambda) = \frac{p_i^k(\lambda)\left[1 + q_i^k(\lambda)\right]}{\sum_j p_i^k(\lambda_j)\left[1 + q_i^k(\lambda_j)\right]}. \tag{10.3}$$

Don't worry about the denominator; it is just there to make sure the values of p_i stay within the range of 0–1. The term $q_i(\lambda)$ represents a measure of how compatible $p_i(\lambda_j)$ is with the labeling of all the other objects. While p is strictly positive, q can be either positive or negative. If negative, $q_i(\lambda)$ suggests that the current labeling of p_i by λ_j is incompatible with most other labelings,

$$q_i^k(\lambda) = \sum_j C_{ij} \left[\sum_{\lambda'} r_N(i, \lambda, j, \lambda') p_j^k(\lambda') \right], \tag{10.4}$$

Details of a specific form for r are problem-dependent.

where this is a similar *compatibility* $r(.)$ that we saw earlier in this chapter, but with a subscript N to denote that we will use it in nonlinear relaxation. The only difference is that now we allow r to take on values of not only -1 and 1, but all values in between. If two labelings are completely consistent, their compatibility is said to be 1. If the labelings are totally inconsistent, the compatibility is -1. If the labeling of one simply does not affect the other, the compatibility is 0. Let's examine what Eq. (10.4) means.

The CHANGE in our confidence that object i has label λ is simply the sum of how compatible that label for object i is with all the other labelings currently. Notice that the compatibility is multiplied by the confidence that the other labeling is correct. That is, if we have little confidence in a labeling of object j, we do not really care how much it is compatible with a labeling of object i. Finally, C_{ij} is there for convenience. It simply weights the influence that object j has on object i, without regard to labels. It might be zero, for example, if object i and j have no influence on each other, and we know that ahead of time. C_{ij} is optional; one could just as well incorporate it into the compatibility function.

10.2.2 Example labeling problems

In this section, we illustrate the choice of the compatibility function r for some example problems.

A model matching problem

Assume you have done a segmentation of a range image into planes. Now, you wish to find which of a set of models best matches the object being observed. Assume the segmentation produces patches which are planar, and since the image is a range image, we can compute the orientation of those planes in 3-space. The problem now is to find the set of planar surfaces in the model (or collection of models) which best matches the set of planar surfaces in the image. Patches in the image are the objects, regions in the model(s) are the labels. One way to define compatibility of labeling is as follows. Consider the compatibility of labeling image patch A as model region

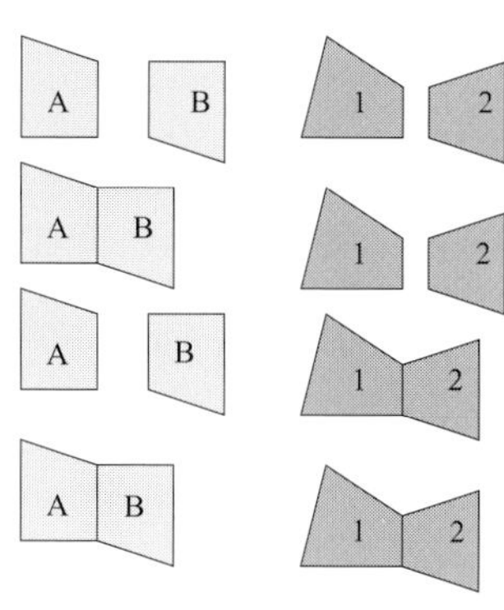

The two patches in the image do not border each other; and neither do the two regions in the model; $r = 0$.

The two patches in the image are adjacent. The two in the model are not; $r = -1$.

The two patches in the image are not adjacent. The two in the model are; $r = -1$.

The two patches in the image are adjacent, and so are the two regions in the model; r is a function of the angle.

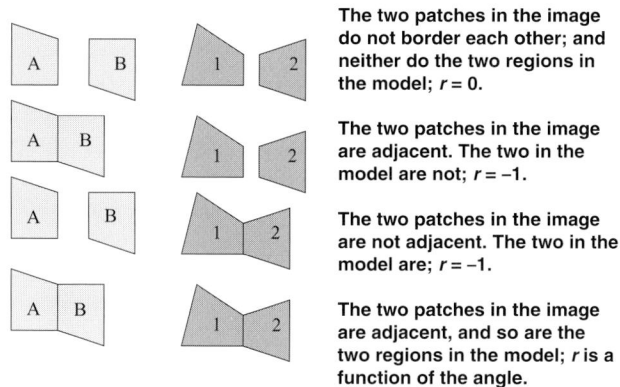

Fig. 10.8. Four possibilities in the definition of compatibility of labeling between two patches and the corresponding two models.

1 with B as 2. Let us now consider: Does patch A border patch B? And does region 1 border region 2? There are four possibilities, illustrated in Fig. 10.8. If the two regions in the image are adjacent, and the two regions in the model are also adjacent, then we *define* the compatibility of two labelings as

$$r_{\mathrm{N}}(A, \lambda_1, B, \lambda_2) = \cos(\theta_{AB} - \theta_{1,2}) \tag{10.5}$$

where θ_{AB} denotes the angle between patch/regions A and B (remember, this is a range image, these angles can be measured).

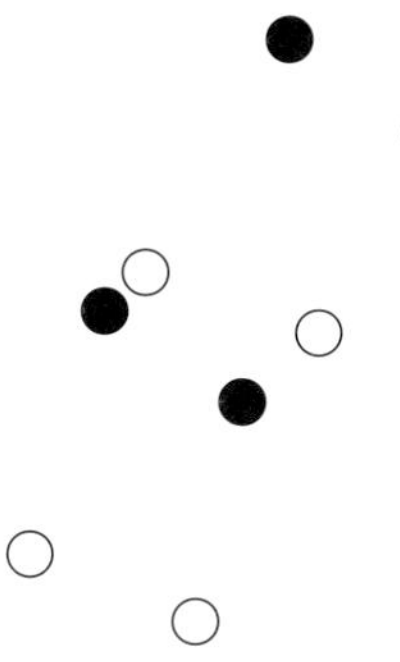

Fig. 10.9. Objects in frame 1 are represented by open circles, objects in frame 2 are filled circles. The problem is to find the most consistent labeling of open circles with filled circles.

The point to emphasize here is that the definition of the compatibility function is totally problem-dependent. The rest of relaxation labeling is a simple iteration defined by Eqs. (10.3) and (10.4).

Another example: tracking of moving objects

Let us assume that we are tracking four-wheeled vehicles as they move. Our camera can only record the position of the tires (it is a pretty weird camera). Our objective then is to determine which tires in one image correspond to which tires in the next image. This is illustrated in Fig. 10.9, in which the location of the tires in frame n are denoted by open circles and the locations in frame $n + 1$ by closed circles. In this application, we have another labeling task. The objects (wheels) in frame n are to be labeled with the labels (also wheels) in frame $n + 1$. So what shall we use as a compatibility function? To figure that out, let's think about some incorrect labelings. For example, Fig. 10.10 illustrates the labeling with an arrow, from object to label. See if you think this makes sense.

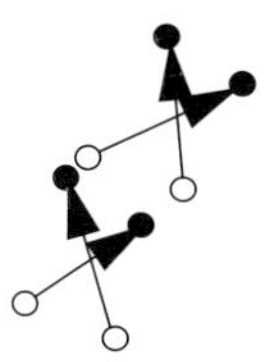

Fig. 10.10. A labeling which probably is not correct.

If Fig. 10.10 is correct, then the front left tire went to the front right, and the rear tires did the same. We can only interpret that as the car flipped over, which, while possible, we certainly hope it did not actually happen. A more reasonable

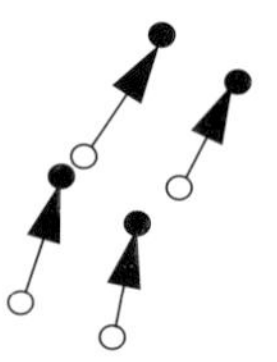

Fig. 10.11. A labeling which makes more sense.

interpretation is in Fig. 10.11, which shows the arrows are almost parallel! Ah, that's it! We can use the cosine of the angle between the arrows,

$$r_{\mathrm{N}}(i, m, j, n) = \cos(\theta(i, m) - \theta(j, p)) \tag{10.6}$$

where i and j are wheels in frame n and m and p are wheels in frame $n + 1$. Equation (10.6) measures the consistency of assuming that wheel i is wheel m and wheel j is wheel p. Although included in a notes version of this book several years earlier, this concept was published in 1995 by Wu [10.19].

10.3 Conclusion

This chapter is all about consistency. We hope to have convinced the student that the best way to fuse information from diverse courses is to seek labelings which are consistent.

Conjugate gradient.

Optimization is formally used only in the next section, where an optimization is set up and solved using a numerical optimization technique called conjugate gradient. The conjugate gradient technique is not explained in this chapter – the reader is referred to standard texts in numerical methods. However, the technique is in many ways similar to gradient descent, but runs much faster.

Researchers continue to work on improving the concepts of consistent labeling, including relaxation labeling [10.4]. Relaxation or similar algorithms have been used in such diverse applications as optical character recognition [10.12] and edge detection [10.15]. See [10.9, 10.16] for underlying theory.

10.4 Vocabulary

You should know the meanings of the following terms.

```
Compatibility
Concave edge
Consistent
Convex edge
Labeling
Linear relaxation
Local/global inference
Nonlinear relaxation
Occluded
Relaxation labeling
```

Assignment 10.1

```
OK. You have seen two examples of compatibility func-
tions. Now you have the opportunity to make up your
own. Here is the problem. You have applied an edge
detector to an image. At every pixel in the image, a
gradient has been taken and you know the magnitude and
direction of that gradient. Recognizing that some of
these measurements may be corrupted by noise and blur,
develop an application of relaxation labeling which
helps determine "real" edge pixels. Hint: A "real" edge
pixel would have its gradient vector point in the same
direction as neighboring edge pixels. Develop a com-
patibility function using this concept. Describe how to
use it. You may use pseudo-code or words or flowcharts,
or all three. Do not write actual software.
```

Topic 10A 3D Interpretation of 2D line drawings

As we have seen, interpretation of line drawings is a difficult problem. A single line drawing shows only one view of a 3D object, and is therefore ambiguous. The ambiguity may be resolved through the use of a set of stored 3D models. This approach requires prior knowledge of what objects are likely to appear in a drawing, and may not give good interpretations of novel drawings which do not correspond to any of the stored models.

Marill [10.11] proposes another approach for interpreting line drawings which requires no models. He uses only one heuristic to generate a 3D wire-frame object from a 2D line drawing, that is, a given 3D interpretation is considered less likely to be correct if some angles between the wires are much larger than others. Specifically, of all the 3D models consistent with a given 2D drawing, the preferred interpretation is the one with the least standard deviation of angles (SDA) as defined in Eq. (10.7), where the angle θ is illustrated in Fig. 10.12.

$$SDA = \sqrt{n \sum_{\theta} \theta^2 - \left(\sum_{\theta} \theta\right)^2}. \tag{10.7}$$

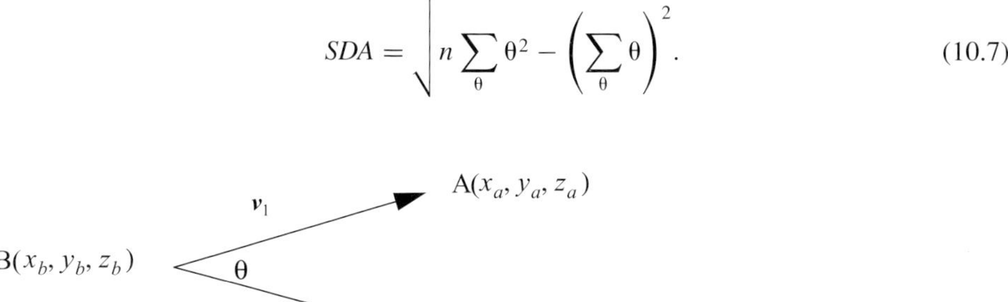

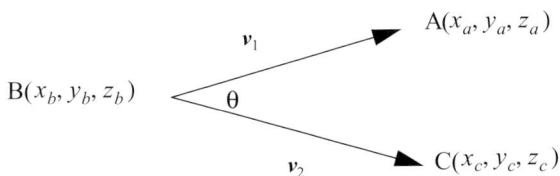

Fig. 10.12. Line and angle.

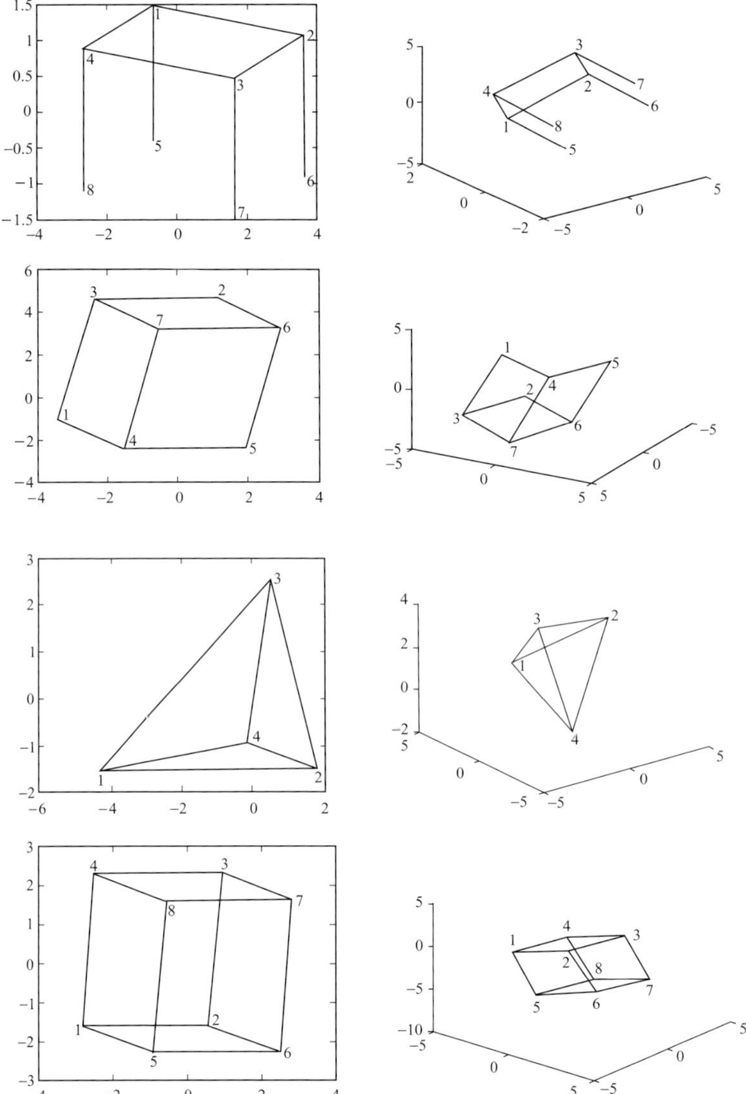

Fig. 10.13. Left column: 2D line drawings. Right: 3D interpretation using the emulation method.

This algorithm can interpret a wide range of line drawings and seems to consistently generate the same interpretations that a human does, even without any explicit models.

To simplify the problem, we square the objective function SDA and call it S . We would like to find the third coordinate (z_i) of the points in the 2D picture which will minimize the objective function S.

$$S = n \sum_{\theta} \theta^2 - \left(\sum_{\theta} \theta \right)^2 . \tag{10.8}$$

The first derivative of S with respect to z_i is derived as Eq. (10.9),

$$\frac{\partial S}{\partial z_i} = 2n \sum_{\theta} \left(\theta \frac{\partial \theta}{\partial z_i} \right) - 2 \sum_{\theta} \theta \sum_{\theta} \frac{\partial \theta}{\partial z_i}, \tag{10.9}$$

where the angle θ, formed by two vectors $\boldsymbol{v}_1$ and $\boldsymbol{v}_2$, can be computed by Eq. (10.10).

$$\theta = \cos^{-1}\left(\frac{\boldsymbol{v}_1 \boldsymbol{v}_2}{\|\boldsymbol{v}_1\|\|\boldsymbol{v}_2\|}\right)$$

$$= \cos^{-1}\left(\frac{(x_a - x_b)(x_c - x_b) + (y_a - y_b)(y_c - y_b) + (z_a - z_b)(z_c - z_b)}{\sqrt{(x_a - x_b)^2 + (y_a - y_b)^2 + (z_a - z_b)^2}\sqrt{(x_c - x_b)^2 + (y_c - y_b)^2 + (z_c - z_b)^2}}\right). \tag{10.10}$$

Some 3D interpretation results of 2D line drawings are shown in Fig. 10.13. The optimization problem is solved using conjugate gradient.

Wang [10.18] has made several improvements on Marill's original algorithm, most of which are less computationally intensive, including the usage of the standard derivation of segment magnitudes (DSM) as the objective function [10.3] and the application of gradient descent to solve the minimization problem [10.2].

References

[10.1] F. Attneave, "Some Informational Aspects of Visual Perception," *Psychology Review*, **61**(3), 1954.

[10.2] L. Baird and P. S. Wang, "3D Object Perception Using Gradient Descent," *International Journal of Mathematical Imaging and Vision*, **5**, pp. 111–117, 1995.

[10.3] E. W. Brown and P. S. Wang, "Why We See Three-Dimensional Objects: Another Approach," *http://www.ccs.neu.edu/home/feneric/msdsm.html.*

[10.4] W. Christmas, J. Kittler, and M. Petrou, "Structural Matching in Computer Vision using Probabilistic Relaxation," *IEEE Transactions on Pattern Analysis and Machine Intelligence*, **17**(8), 1995.

[10.5] M. Clowes, "On Seeing Things," *Artificial Intelligence*, **2**(1), 1971.

[10.6] J. Elder and S. Zucker, "Evidence for Boundary-specific Grouping," *Vision Research*, **38**(1), 1998.

[10.7] D. Hofstadter, *Gödel, Escher, Bach: An Eternal Golden Braid*, New York, Basic Books, Inc., 1979.

[10.8] D. Huffman, "Impossible Objects as Nonsense Sentences," in *Machine Intelligence*, vol. 6, ed. B. Meltzer and D. Michie, Edinburgh University Press, 1971.

[10.9] R. Hummel and S. Zucker, "On the Foundations of Relaxation Labeling Processes," *IEEE Transactions on Pattern Analysis and Machine Intelligence*, **5**(3), 1983.

[10.10] J. Koenderink, "What Does the Occluding Contour Tell Us About Solid Shape?" *Perception*, **13**, pp. 321–330, 1984.

[10.11] T. Marill, "Emulating the Human Interpretation of Line-drawings as Three-dimensional Objects," *International Journal of Computer Vision*, **6**(2), pp. 147–161, 1991.

[10.12] J. Ohya, A. Shio, and S. Akamatsu, "Recognizing Characters in Scene Images," *IEEE Transactions on Pattern Analysis and Machine Intelligence*, **16**(2), 1994.

[10.13] P. Parodi and G. Piccioli, "3D Shape Reconstruction by Using Vanishing Points," *IEEE Transactions on Pattern Analysis and Machine Intelligence*, **18**(2), 1996.

[10.14] M. Pelillo and M. Refice, "Learning Compatibility Coefficients for Relaxation Labeling Processes," *IEEE Transactions on Pattern Analysis and Machine Intelligence*, **16**(9), 1994.

[10.15] A. Rosenfeld, R. Hummel, and S. Zucker, "Scene Labeling Based on the Relaxation Principle," *IEEE Transactions on Systems, Man, and Cybernetics*, **6**, pp. 420–433, 1976.

[10.16] P. Sastry and M. Thathachar, "Analysis of Stochastic Automata Algorithm for Relaxation Labeling," *IEEE Transactions on Pattern Analysis and Machine Intelligence*, **16**(5), 1994.

[10.17] M. Shpitalni and H. Lipson, "Identification of Faces in a 2D Line Drawing Projection of a Wireframe Object," *IEEE Transactions on Pattern Analysis and Machine Intelligence*, **18**(10), 1996.

[10.18] P. S. Wang, "3D Line Drawing Image Analysis and Recognition," *Progress in Image Analysis and Processing*, **111**, pp. 14–21, 1994.

[10.19] Q. Wu, "A Correlation-relaxation-labeling Framework for Computing Optical Flow – Template Matching from a New Perspective," *IEEE Transactions on Pattern Analysis and Machine Intelligence*, **17**(9), 1995.

11 Parametric transforms

Supposing I was on the other side of the glass, wouldn't the orange still be in my right hand?

Lewis Carroll

This chapter discusses another approach to the solution of the local/global inference problem, the use of parametric transformations. In this approach, we assume that the object for which we are searching in the image may be described by a mathematical expression, which in turn is represented by a set of parameters. For example, a straight line may be written in slope–intercept form:

$$y = ax + b, \tag{11.1}$$

where a and b are the parameters describing the line. Our approach is as follows: Given a set of points (or other features), all of which satisfy the same equation, we will find the parameters of that equation. In a sense, this is the same as fitting a curve to a set of points, but as we will discover, the parametric transform approach allows us to find multiple curves, without knowing *a priori* which point belongs to which curve. We begin this process by considering the special case of finding straight lines.

11.1 The Hough transform

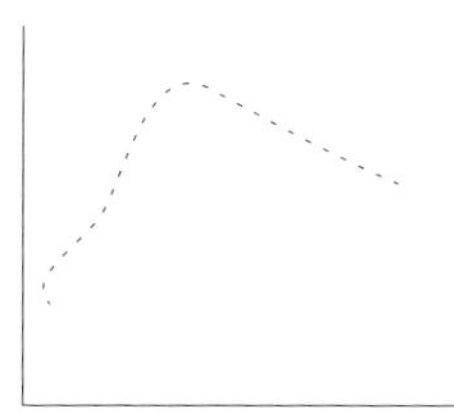

Fig. 11.1. An image which is the output from an edge detector. The human can immediately discern that this boundary consists of two straight segments.

Suppose you are tasked with the problem of finding the straight lines in the image shown in Fig. 11.1. If only one straight line were present in the image, we could use straight line fitting to determine the parameters of the curve. But we have two line segments here. If we could segment this first, then we could fit each segment separately – yes, this is a segmentation problem, but we are segmenting a boundary into boundary segments rather than segmenting an image into regions. In this section, we will learn how to do this.

First, let us prove an illustrative theorem.

Definition

Given a point in a d-space, and a parameterized expression defining a curve in that space, the parametric transform of that point is the curve which results from treating the point as a constant and the parameters as variables. For example, Eq. (11.1) produces the parametric transform

$$b = y - xa \tag{11.2}$$

which is itself a straight line in the 2-space $\langle a, b\rangle$. Given the point $x = 3$, $y = 5$, then the parametric transform is $b = 5 - 3a$.

Theorem

If n points in a 2-space are colinear, all the parametric transforms corresponding to those points, using the form $b = y - xa$ intersect at a common point in the space $\langle a, b\rangle$.

Proof

Suppose n points $\{(x_1, y_1), (x_2, y_2), \ldots (x_n, y_n)\}$ all satisfy the same equation

$$y = a_0 x + b_0. \tag{11.3}$$

Consider two of those points, (x_i, y_i) and (x_j, y_j). The parametric transforms of the points are the curves (which happen to be straight lines)

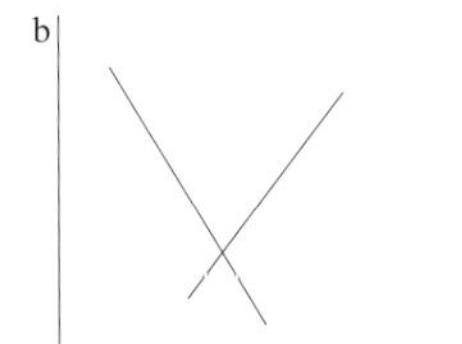

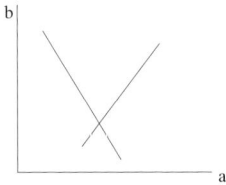

Fig. 11.2. The intersection of two curves in the parameter space.

$$\begin{aligned} y_i &= x_i a + b \\ y_j &= x_j a + b \end{aligned} \tag{11.4}$$

which we rewrite to make clear the fact that a and b are independent variables:

$$\begin{aligned} b &= y_i - x_i a \\ b &= y_j - x_j a. \end{aligned} \tag{11.5}$$

The intersection of those two curves in a, b is illustrated in Fig. 11.2.

Solving the two equations of Eq. (11.5) simultaneously results in

$$y_j - y_i = (x_j - x_i)a \tag{11.6}$$

and therefore $a = \frac{y_j - y_i}{x_j - x_i}$.

We substitute a into Eq. (11.5) to find b,

$$b = y_i - x_i \frac{y_j - y_i}{x_j - x_i} \tag{11.7}$$

and we have the a and b values where the two curves intersect. However, we also know from Eq. (11.3) that all the xs and ys satisfy the same curve. By performing that substitution into Eq. (11.7), we obtain

$$b = (a_0 x_i + b_0) - x_i \frac{(a_0 x_j + b_0) - (a_0 x_i + b_0)}{x_j - x_i} \tag{11.8}$$

which simplifies to

$$b = (a_0 x_i + b_0) - x_i a_0 = b_0. \tag{11.9}$$

Similarly,

$$a = \frac{y_j - y_i}{x_j - x_i} = \frac{(a_0 x_j + b_0) - (a_0 x_i + b_0)}{x_j - x_i} = a_0. \tag{11.10}$$

Thus, for any two points along the straight line parameterized by a_0 and b_0, their parametric transforms intersect at the point $a = a_0$, and $b = b_0$. Since the transforms of any two points intersect at that one point, all such transforms intersect at that common point. QED.

Review of concept: Each POINT in the image produces a CURVE (possibly straight) in the parameter space. If the points all lie on a straight line in the image, the corresponding curves will intersect at a common point in parameter space.

Got that? Now, on to the next problem.

11.1.1 The problem with vertical lines

What about vertical lines? Whoops! The parameter a goes to infinity. This is not good. Perhaps we need a new form for the equation of a straight line. Here is a good one

$$\rho = x \cos\theta + y \sin\theta. \tag{11.11}$$

Pick a value of ρ and θ. Hold those values constant. Then the set of points which satisfy Eq. (11.11) can be shown to be a straight line. There is a geometric interpretation of this equation which is illustrated in Fig. 11.3.

This representation of a straight line has a number of advantages. Unlike the use of the slope, both of these parameters are bounded; ρ can be no larger than the largest diagonal of the image, and θ need be no larger than 2π. A line at any angle may be represented without singularity.

The use of this parameterization of a straight line solves one of the problems which confronts us, the possibility of infinite slopes. The other problem is the calculation of intersections.

There is a *Whoops* here. It is true that the maximum value of ρ is the diagonal. However, when you start calculating ρ while varying θ, you will find points with negative ρ. Rather than testing negative values and reflecting, it is easier to just let ρ be negative; which makes the accumulator twice as big.

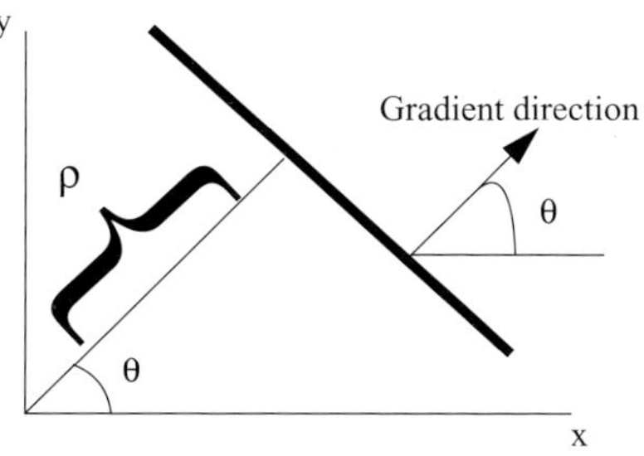

Fig. 11.3. In the ρ, θ representation of a line, ρ is the perpendicular distance of the line from the origin, and θ is the angle made with the x axis.

11.1.2 How to find intersections – accumulator arrays

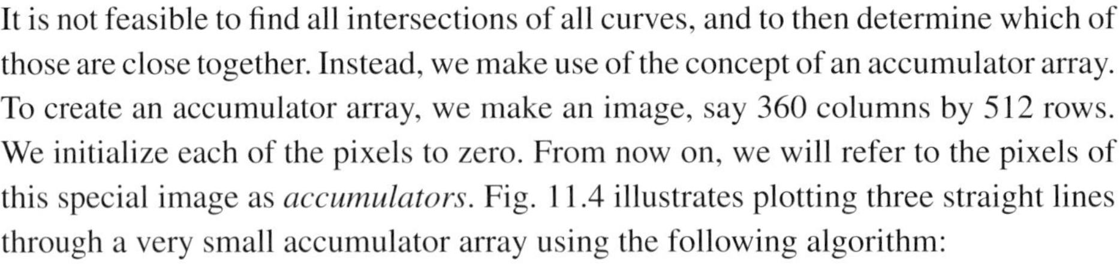

It is not feasible to find all intersections of all curves, and to then determine which of those are close together. Instead, we make use of the concept of an accumulator array. To create an accumulator array, we make an image, say 360 columns by 512 rows. We initialize each of the pixels to zero. From now on, we will refer to the pixels of this special image as *accumulators*. Fig. 11.4 illustrates plotting three straight lines through a very small accumulator array using the following algorithm:

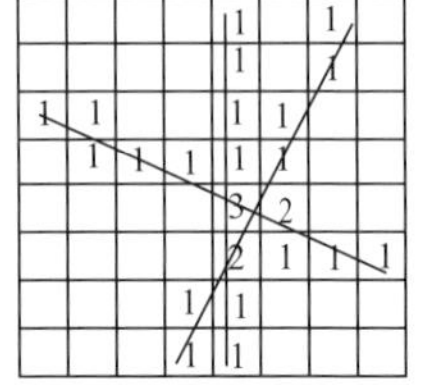

Fig. 11.4. Three lines plotted through an accumulator array. The lines themselves are illustrated. Each accumulator is incremented whenever a line passes through that accumulator.

For each point x_i, y_i in the edge image:

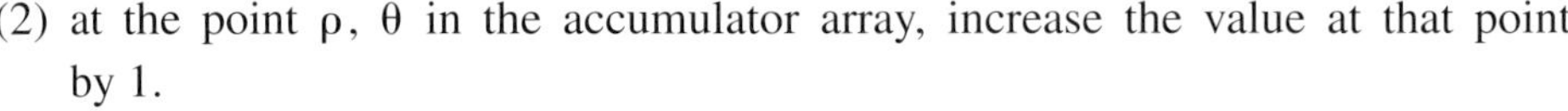

(1) for all values of θ compute ρ
(2) at the point ρ, θ in the accumulator array, increase the value at that point by 1.

This algorithm results in multiple increments of those accumulators corresponding to intersections. Thus the peaks in the accumulator array correspond to multiple intersections, and hence to proper parameter choices.

Figure 11.5 illustrates an image with two straight edges and the corresponding Hough transform where the brightness of each pixel is the value of an accumulator.

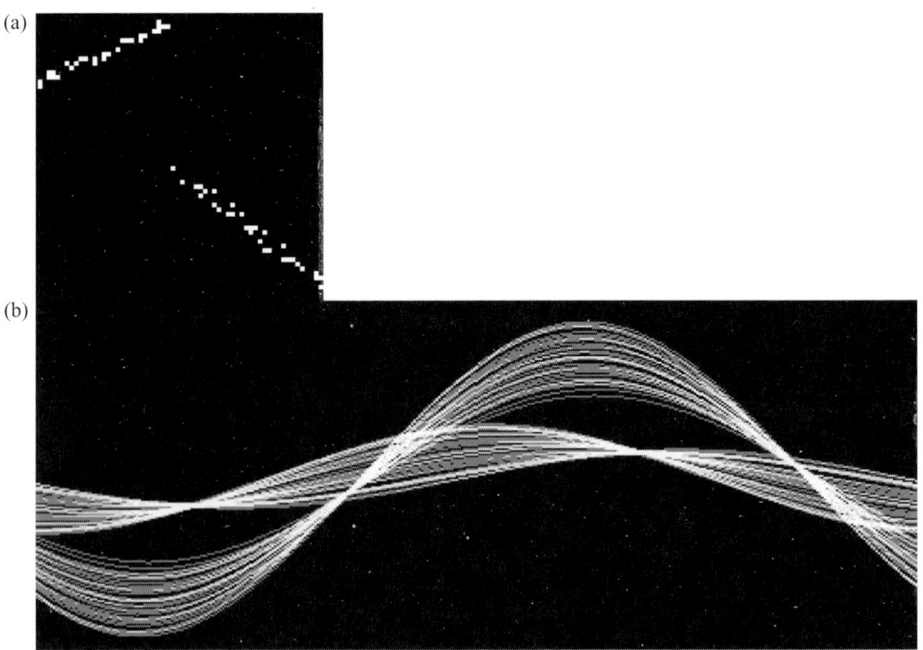

Fig. 11.5. (a) An image with two line segments which have distinctly different slopes and intercepts, but whose actual positions are corrupted significantly by noise. (b) The corresponding Hough transform.

11.2 Reducing computational complexity

Be sure you understand where this accumulator size comes from. Remember, ρ can be negative.

The computational complexity of parametric transforms can be quite high. The Hough is sometimes referred to as the "coffee transform," because you can command your computer to perform a Hough transform, and drink an entire cup of coffee before it finishes. For example, suppose we are performing a Hough transform in which the size of the image is 512×512, and we wish an angular resolution of one degree. Our accumulator array is thus $(512 \times 2\sqrt{2}) \times 360$. Our iteration consists of calculating ρ for each value of θ, so we calculate and increase 360 accumulators for each edge pixel in the image.

11.2.1 Using gradient information

One way to reduce the computation is to observe that the edge points in the image are often the output of some gradient operator, and if we have both the magnitude and direction of that gradient, we can make use of that information to reduce the computational complexity. To see this, refer to Fig. 11.3. If we know the gradient at a point, we know the direction of the edge, we therefore know θ. Thus, we need only compute one value of ρ, instead of 360 of them, and we need increase only one accumulator – a 360 : 1 speedup. Not bad!

Of course, (you should be used to caveats by now) there are some problems with this method. The first is that the gradient direction returned by most gradient operators is not particularly accurate. Therefore, the one cell which you are incrementing may not be precisely located. There is an easy solution to that: The trick is to not increase a point, but to increase a neighborhood. For example, you might increase the point calculated by two, and increment the neighborhood of that point by one. (Yes, that is a Gaussian of sorts.)

You can also use the magnitude of the gradient in the process of incrementing the accumulator array. In the description presented earlier, we suggested that you threshold the gradient image, and increment the accumulator array by 1 at the points corresponding to the edge points in the image. Another approach is to increment by the magnitude of the gradient, or a value proportional to the magnitude. Of course, this requires that you use a floating point representation for the accumulator array, but that should be no problem.

In summary, you can trade off computation in image space (computation of magnitude and direction of the gradient) for computation in parameter space, and this trade off can result in substantial speedups. You will see this again later in this chapter. Other heuristics [11.13] can also result in speedups.

11.3 Finding circles

A generalization of the Hough transform can be used to encompass detection of segments of circles. The practicality of the technique is ensured by use of low-dimensionality accumulator arrays.

11.3.1 Derivation of the location of a circle represented by any three noncolinear pixels

Let the array T contain the brightness values for the candidate edge pixels. The integer Trc represents the magnitude of the edge pixel at row r and column c. Given any three noncolinear points in T, this section will review how the center and radius of a circle passing through those three points can be found. In section 11.3.2 this result will be integrated into a parametric transform which will identify the most likely circular edge segment.

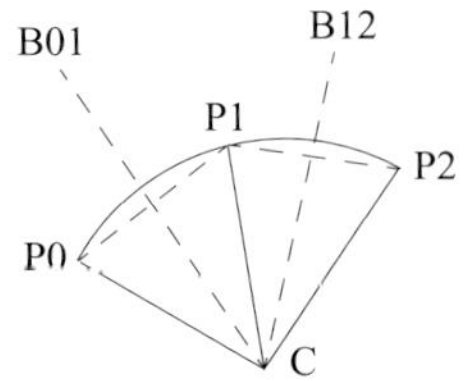

Fig. 11.6. Radii of the circle containing P0, P1, and P2.

Suppose the perpendicular bisectors, B01 and B12, of any two chords intersect at some finite point C (see Fig. 11.6). It is easy to show that C is the center of the circle containing P0, P1, and P2. Let L0C be the line segment from P0 to C, L1C be the line segment from P1 to C, and L2C be the line segment from P2 to C. Since C is on the bisector, B01, P0, and P1 are equidistant from C. Therefore, R, the length of L0C, equals the length of L2C. The three points are equidistant from the same point, C, and hence satisfy the definition of a circle with center C and radius R.

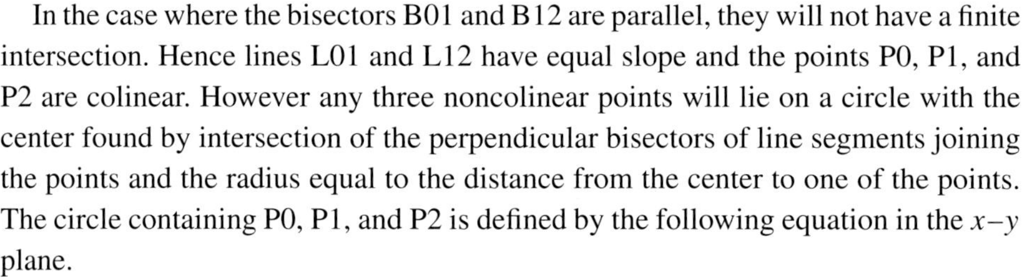

In the case where the bisectors B01 and B12 are parallel, they will not have a finite intersection. Hence lines L01 and L12 have equal slope and the points P0, P1, and P2 are colinear. However any three noncolinear points will lie on a circle with the center found by intersection of the perpendicular bisectors of line segments joining the points and the radius equal to the distance from the center to one of the points. The circle containing P0, P1, and P2 is defined by the following equation in the x–y plane.

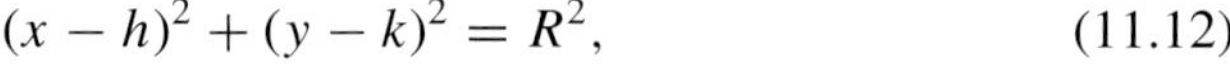

$$(x - h)^2 + (y - k)^2 = R^2, \tag{11.12}$$

where $C = (h, k)$ is the center of this circle with radius R.

But how do we use the techniques of parametric transformations to find the parameters, given that we know something (or nothing, for that matter) about the circle?

11.3.2 Finding circles when the origin is unknown but the radius is known

Equation (11.12) describes an equation in which x and y are assumed to be variables, and h, k, and R are assumed to be constants. As before, let us rewrite this equation interchanging the roles of parameters and variables.

$$(h - x_i)^2 + (k - y_i)^2 = R^2. \tag{11.13}$$

In the space (h, k) what geometric shape does this describe? You guessed it, a circle. Each point in image space (x_i, y_i) produces a curve in parameter space, and if all those points in image space belong to the same circle, where do the curves in parameter space intersect? You should be able to figure that one out by now.

However, that might not be the most efficient way to do it. Think about it . . .

Now, what if R is also unknown? It is the same problem, however, instead of allowing h to range over all possible values and computing k, we must now allow both h and k to range over all values and compute R. We now have a three-dimensional parameter space. Allowing two variables to vary and computing the third defines a surface in this 3-space. What type of surface is this (an ellipse, a hyperboloid, a cone, a paraboloid, a plane)?

11.3.3 Using gradient information to reduce computation in finding circles

Suppose we know we have only one circle, or portion of a circle in the image. How can we find the center with a minimum of computational complexity? Here is the trick: At every edge point, compute the gradient vector. The accumulator is isomorphic to the image. At every edge point, walk out the gradient vector, incrementing the accumulator as you go. Just as before, the maximum of the accumulator will indicate the center of the circle. Fig. 11.7 illustrates this.

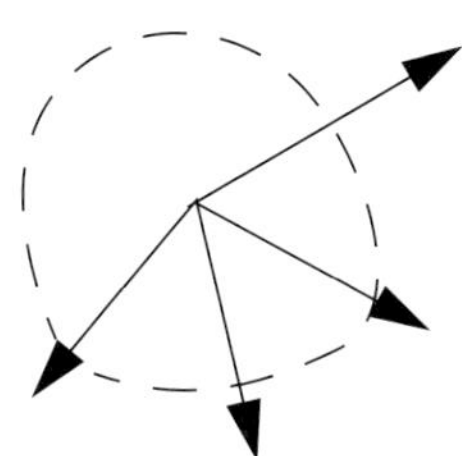

Fig. 11.7. All the extensions of the gradient vectors tend to intersect at a point.

What if you know the radius? If you know the direction (from the gradient), and you know the distance (from knowing the radius), then you know the location of the center of the circle (at least to within an inside/outside ambiguity). Let's say you have a point at (x_i, y_i) which you believe might be on a circle of radius R, and assume the gradient at that point has a magnitude of M and a direction of θ, as illustrated in Fig. 11.8. Then, the location of the center of the circle is at

$$\begin{aligned} x_0 &= x_i - R\cos\theta \\ y_0 &= y_i - R\sin\theta. \end{aligned} \tag{11.14}$$

Again, the accumulator array should be incremented over a neighborhood rather than just a single accumulator, and it will prove efficacious to increment by an amount proportional to M.

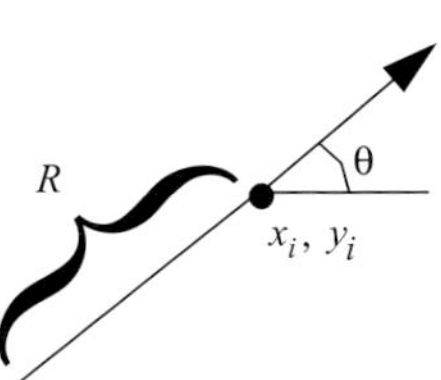

Fig. 11.8. The point x_i, y_i is proposed to lie on a circle of radius R. The gradient points in the direction of the arrow. If the circle is known to be dark relative to the background, the center can be found by taking a step of length R in the direction opposite to the gradient. If the center/surround contrast is opposite, then the step should be taken in the direction of the gradient. If the contrast is not known *a priori*, steps can be taken (and accumulators incremented) in both directions.

11.4 The generalized Hough transform

So far, we have assumed that the shape we are seeking can be represented by an analytic function, representable by a set of parameters. The concepts we have been using, of allowing data components which agree to "vote," can be extended to generalized shapes. Initially we suppose we have an arbitrarily shaped region, and suppose we know the orientation, shape, and zoom. Our first problem is to figure out how to represent this object in a manner which is suitable for use by Hough-like methods. The following is one such approach [11.2]:

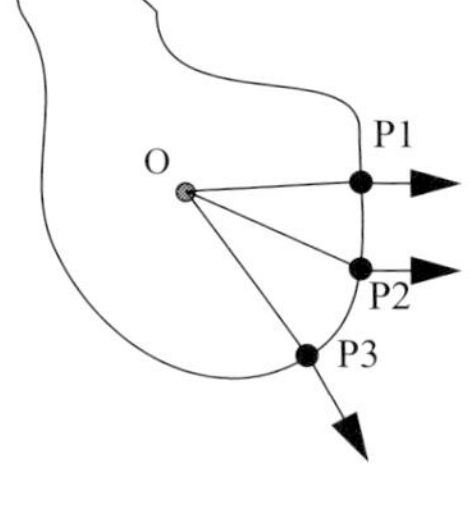

Fig. 11.9. Boundary points P1 and P2 have gradient vectors which have identical directions, and therefore correspond to entries on the same row of the *R*-table. P3 is an entry on the second row of the *R*-table.

First, define some reference point. The choice of a reference point is arbitrary, but the center of gravity is convenient. Call that point O. For each point P_i on the boundary, calculate both the gradient vector at that point and the vector $\overrightarrow{OP_i}$. from the reference to the boundary point. Quantize the gradient direction into, say, n values, and create a table with n rows. Each time a point P_j on the boundary has a gradient direction with value $G_i (i = 1, \ldots n)$, a new column is filled in on row i, containing $\overrightarrow{OP_j}$. Thus, the fact that multiple points on the boundary may have identical gradient directions is accommodated by placing a separate column in the table for each entry. In Fig. 11.9, shape is shown and three entries in the R-table are illustrated in Table 11.1.

To utilize such a shape representation to perform shape matching and location, we use the following algorithm.

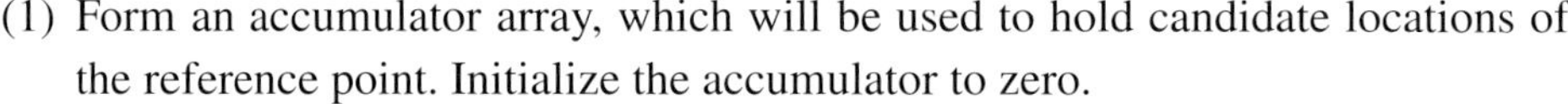

(1) Form an accumulator array, which will be used to hold candidate locations of the reference point. Initialize the accumulator to zero.
(2) For each edge point, P_i, do the following.
  (2.1) Compute gradient direction, and determine which row of the R-table corresponds to that direction.
  (2.2) For each entry, j, on that row:
    (a) compute the location of the candidate center by adding the stored vector to the boundary point location: $A = T[i, j] + P_i$
    (b) increment accumulator determined by A.

Table 11.1. *The R-table corresponding to the three boundary points of Figure 11.9.*

Gradient direction (in degrees)	Vector from boundary point to reference point	Vector from boundary point to reference point
0	$-1, -0.1$	$-1, 0.5$
300	$-0.6, 1.1$	empty

11.5 Conclusion

Accumulator arrays enforce consistency.

In this chapter, another approach to analysis of consistency was introduced in the form of the Hough transform and its descendants, all of which use accumulator arrays. An accumulator array allows for easy detection of consistency, since "things" which are consistent all add to the same accumulator, or at least to nearby accumulators. By constructing the accumulator array by adding hypotheses, we also gain a measure of noise immunity, since inconsistent solutions tend not to contribute to the globally consistent solution.

11.6 Vocabulary

You should know the meanings of the following terms.

```
Accumulator array
Generalized Hough transform
Hough transform
Parametric transform
```

Topic 11A Parametric transforms

11A.1 Finding parabolae

Wechsler and Sklansky [11.12] have developed one approach to the problem of finding parabolic curves in images, as described below.

A parabola is the locus of points each of whose distance from a fixed point, the focus, is equal to its distance from a fixed straight line, the *directrix*, as illustrated in Fig. 11.10.

$$x^2 = (x - 2a)^2 + y^2 \tag{11.15}$$

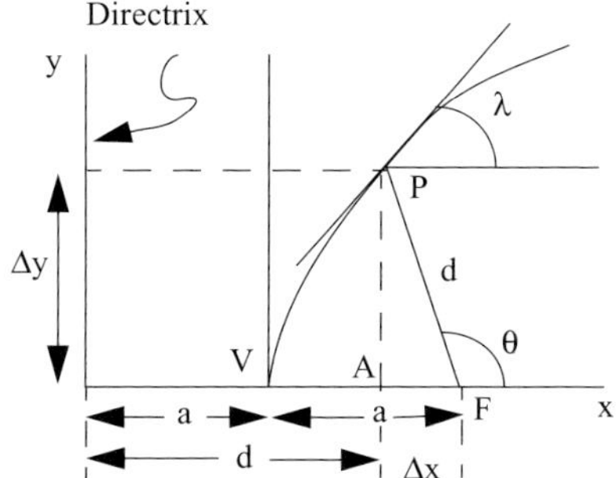

Fig. 11.10. A parabola.

or

$$y^2 = 4a(x - a). \tag{11.16}$$

Differentiate Eq. (11.16) with respect to x:

$$\beta = \tan\lambda = \frac{dy}{dx} = \sqrt{\frac{a}{x-a}} \tag{11.17}$$

and solve for a:

$$a = \frac{x\beta^2}{1+\beta^2}. \tag{11.18}$$

In Figure 11.10, use the right-angled triangle PAF to determine

$$\tan\theta = -\tan(\pi - \theta) \tag{11.19}$$

and

$$\tan\theta = \frac{y}{x-2a} = \frac{\sqrt{4a(x-a)}}{x-2a}. \tag{11.20}$$

Substituting for a,

$$\tan\theta = \frac{2\tan\lambda}{1-\tan^2\lambda} = \tan 2\lambda. \tag{11.21}$$

The solution to Eq. (11.21) is

$$\theta = 2\lambda. \tag{11.22}$$

Referring again to Figure 11.10, obtain

$$\Delta x = d\cos(\pi - \theta) = d\cos\theta \tag{11.23}$$

and

$$\Delta y = -d\sin(\pi - \theta) = -d\sin\theta \tag{11.24}$$

$$x_F = x_p + \Delta x \tag{11.25}$$

$$y_F = y_p + \Delta y. \tag{11.26}$$

x_F and y_F can be used as parameters for an accumulator array.

The derivation given in Eqs. (11.17)–(11.26) provides a mechanism for using accumulation arrays to detect parabolae. It is, however, very restrictive. The assumptions are as follows:

- there is only one parabola in the field of view
- the parabola is symmetric about a horizontal line.

Equation (11.16) assumes the focus lies at $x = a$. This technique will work if only one parabola lies in the field of view, for then the position of the origin is arbitrary. In the more general case, however, we must make the location of the origin explicit.

The derivation of Wechsler and Sklansky [11.12] can be generalized as follows to overcome these difficulties.

Initially, we continue to assume the parabola is symmetric about a horizontal line, but having an origin at an arbitrary point $\langle x_0, y_0 \rangle$. The parabolic equation becomes

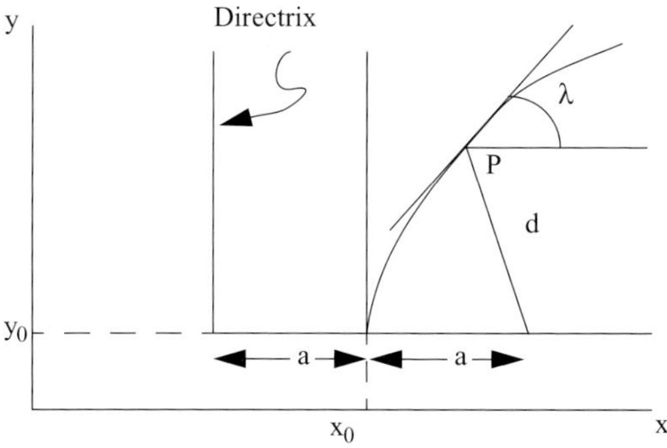

Fig. 11.11. Parabola with arbitrary reference.

(see Fig. 11.11)

$$(y - y_0)^2 = 4a(x - x_0). \tag{11.27}$$

Differentiating, as in Eq. (11.17), yields

$$\beta = \tan \lambda = \frac{dy}{dx} = \sqrt{\frac{a}{x - x_0}} \tag{11.28}$$

and

$$a = \beta^2(x - x_0). \tag{11.29}$$

Substituting Eq. (11.29) into Eq. (11.27),

$$(y - y_0)^2 = 4\beta^2(x - x_0)^2. \tag{11.30}$$

Taking the square root of both sides,

$$(y - y_0) = 2\beta(x - x_0). \tag{11.31}$$

Equation (11.31) describes a straight line in the x_0, y_0 parameter space. Thus, by making use of the local derivative, the 3-parameter problem of Eq. (11.27) has been reduced to a much more tractable two-dimensional problem.

It should be obvious that shapes other than circles and parabolae can be found using this approach [11.3].

11A.2 Finding the peak

The peak of the accumulator array is, hopefully, at or near the true parameter value, but in general may be displaced by noise in the original data. There are several ways to approach this problem, including some sophisticated methods [11.4, 11.11] and some simple ones. In general, however you look at it, this is a clustering problem [11.4], so any algorithm which does clustering well is also good at peak finding. Simple techniques such as k-means, described in section 15.2.2, perform well.

Clustering is the process of finding natural groupings in data. In this chapter the obvious application is to find the best estimate of the peak(s) of the accumulator; however, these ideas are much broader in applicability than just finding the mode of a distribution. For example, McLean and Kotturi [11.9] use clustering [11.5] to locate the vanishing points in an image. Clustering is discussed in detail in Chapter 15.

11A.3 The Gauss map

The so called "Gauss map" provides an effective way to represent operations in range images. It is philosophically a type of parametric transform. The concept is quite simple: First, tessellate the surface of a sphere into whatever size patches you wish. This tesselation will determine the resolution of the map. Associate a counter (accumulator) with each cell of the tesselation. Now consider a range image. In the range image, calculate the surface normal at each pixel, and increment an accumulator of the Gauss map which has the same surface normal. This provides essentially a histogram of normal vectors, which in turn can be used to recognize the orientation of the object in the range image.

Since curvature represents rates of change of normal direction, the Gauss map can be related back to curvature. The invertability of the Gauss map and invariance under rotation and translation are discussed in more detail in [11.7]. The map also finds application in identifying the vanishing points in an image [11.8].

11A.4 Parametric consistency in stereopsis

As we discussed in section 4.2.2, the correspondence problem prohibits simple use of stereopsis to provide three-dimensional information about the world. In this section, we take a closer look at this problem and provide a partial solution, based on accumulator arrays.

Consider the two-camera stereo problem. Recall from Fig. 4.9 that the disparity is the distance in pixels between two corresponding pixels, z is the distance to the point corresponding to those two pixels, and B is the baseline (the distance between the two cameras). Then

$$d = \frac{BF}{z} \tag{11.32}$$

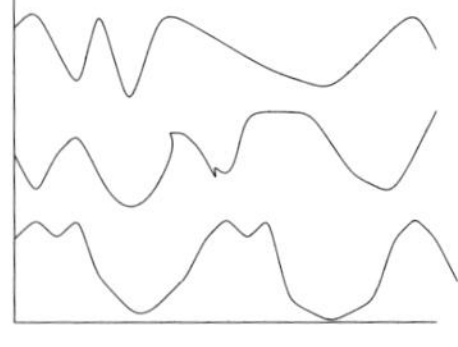

Fig. 11.12. The result of matching a template extracted from one image in a stereo pair with a second image, as a function of the location of the inverse distance to the matching point.

where F is the focal distance of either of the cameras (which are assumed to have the same focal length). The hard problem, of course, is to determine which pixels correspond to the same point in 3-space. Suppose we extract a small window from the leftmost image and use the sum of squared differences (SSD) to template match that small window along a horizontal line in the other image. We could graph that objective function vs. the disparity, or the inverse distance ($1/z$). We find it convenient to use the inverse distance and we will typically find that the match function has multiple minima, as illustrated in Fig. 11.12. If we take an image from a third or fourth camera, with a different baseline from camera 1, we find similar nonconvex curves. However, all the curves will have minima at the same point, the correct disparity. Immediately, we have a consistency! We form a new function, the sum of such curves taken from multiple baseline pairs, and find that this new function (called the SSD-in-inverse-distance) has a sharp minimum at the correct answer. Okutomi and

Kanade [11.10] have proven that this function always exhibits a clear minimum at the correct matching position, and that the uncertainty of the measurement decreases as the number of baseline pairs increases.

11A.5 Conclusion

The general concept of the parametric transform seeks consistency! That is the key here. Many points which are in some sense consistent contribute to the same cell in the accumulator – they vote, in a sense. Hopefully, noise averages out in this voting process, and we can arrive at consistent solutions.

In computed tomography (CT), the signal measured is a line integral along the ray from the x-ray source to the x-ray detector. The line along which the integral is performed may be represented using the $\rho - \theta$ parameterization of a straight line:

$$R(\rho, \theta) = \int (\rho - (x(s)\cos\theta + y(s)\sin\theta))\, ds. \qquad (11.33)$$

Examination of Eq. (11.33) leads one to the immediate conclusion that the Hough transform may be formally represented by the Radon transform, and that, except for applications, they are the same transform.

In addition to the use of these transforms for identifying specific shapes, Leavers [11.6] has shown that if one considers the shape of the parameter space, rather than simply the location of the peaks, one may determine the convex hull and several shape parameters of regions.

A fascinating alternative to the Hough transform was proposed by Aghajan and Kailath [11.1], using wavefront propagation. The idea is to think of each pixel as a radio transmitter emitting signals which are detected by receivers at the end of each row. Using the mathematics of *direction of arrival* signal processing, they show it is possible to detect straight lines with a computational complexity significantly less than the conventional Hough transform. This idea of wavefront propagation makes sense as a paradigm for how the brain detects straight lines.

11A.6 Vocabulary

You should know the meanings of the following terms.

Gauss map
Parabola
Radon transform
SSD

Assignment 11.1

In the directory named "leadhole" are a set of images of wires coming through circuit board holes. The holes are roughly circular and black. Use parametric transform methods to find the centers of the holes. This is a project and

Table 11.2. *The R-table.*

?	$p1(x, y)$	$p2(x, y)$

will require a formal write up. Process as many images as possible. If your method fails in some cases, discuss why.

Assignment 11.2

You are to use the generalized Hough transform approach to both represent an object and to search for that object in an image. It turns out that the object is a perfect square, centered at the origin, with sides two units long, but you do not know that ahead of time. You only have five points, those at (0,1), (1,0), (1, 0.5), (-1,0), and (0, -1). Fill out the "*R*-table" which will be used in the generalized Hough transform of this object. (Table 11.2 contains four rows; that is just a coincidence. You are not required to fill them all in, and if you need more rows, you can add them.)

Assignment 11.3

Let P1 $= [x_1, y_1] = [3, 0]$ and P2 $= [x_2, y_2] = [2.39, 1.42]$ be two points, both of which lie approximately on the same disk. We do not know *a priori* whether the disk is dark inside or bright. The image gradients at P1 and P2 are $5\angle 0$ and $4.5\angle\pi/4$ (using polar notation).

Use Hough methods to estimate the location of the center of the disk, and radius of the disk, and determine whether the disk is darker or brighter than the background.

References

[11.1] H. Aghajan and T. Kailath, "SLIDE: Subspace-based Line Detection," *IEEE Transactions on Pattern Analysis and Machine Intelligence*, **16**(11), 1994.

[11.2] D. Ballard, "Generalizing the Hough Transform to Detect Arbitrary Shapes," *Pattern Recognition*, **13**(2), 1981.

[11.3] N. Bennett, R. Burridge, and N. Saito, "A Method to Detect and Characterize Ellipses Using the Hough Transform," *IEEE Transactions on Pattern Analysis and Machine Intelligence*, **21**(7), 1999.

[11.4] Y. Cheng, "Mean Shift, Mode Seeking, and Clustering," *IEEE Transactions on Pattern Analysis and Machine Intelligence*, **17**(8), 1995.

[11.5] T. Hofmann and J. Buhmann, "Pairwise Data Clustering by Deterministic Annealing," *IEEE Transactions on Pattern Analysis and Machine Intelligence*, **19**(1), 1997.

[11.6] V. Leavers, "Use of the Two-dimensional Radon Transform to Generate a Taxonomy of Shape for the Characterization of Abrasive Powder Particles," *IEEE Transactions on Pattern Analysis and Machine Intelligence*, **22**(12), 2000.

[11.7] P. Liang and C. Taubes, "Orientation-based Differential Geometric Representations for Computer Vision Applications," *IEEE Transactions on Pattern Analysis and Machine Intelligence*, **16**(3), 1994.

[11.8] E. Lutton, H. Maître, and J. Lopez-Krahe, "Contribution to the Determination of Vanishing points using the Hough Transform," *IEEE Transactions on Pattern Analysis and Machine Intelligence*, **16**(4), 1994.

[11.9] G. McLean and D. Kotturi, "Vanishing Point Detection by Line Clustering," *IEEE Transactions on Pattern Analysis and Machine Intelligence*, **17**(11), 1995.

[11.10] M. Okutomi and T. Kanade, "A Multiple-Baseline Stereo", *IEEE Transactions on Pattern Analysis and Machine Intelligence*, **15**(4), 1993.

[11.11] J. Princen, J. Illingworth, and J. Kittler, "Hypothesis Testing: A Framework for Analyzing and Optimizing Hough Transform Performance," *IEEE Transactions on Pattern Analysis and Machine Intelligence*, **16**(4), 1994.

[11.12] H. Wechsler and J. Sklansky, "Finding the Rib Cage in Chest Radiographs," *Pattern Recognition*, **9**, pp. 21–30, 1977.

[11.13] Ylä-Jääski and N. Kiryati, "Adaptive Termination of Voting in the Probabilistic Circular Hough Transform," *IEEE Transactions on Pattern Analysis and Machine Intelligence*, **16**(9), 1994.

12 Graphs and graph-theoretic concepts

> Functions are born of functions, and in turn, give birth or death to others. Forms emerge from forms and others arise or descend from these
>
> L. Sullivan

You have already seen the use of graph-theoretic terminology in connected component labeling in Chapter 8. The way we used the term "connected components" in the past was to consider each pixel as a vertex in a graph, and think of each vertex as having four, six, or eight edges to other vertices (that is, four-connected neighbors, six neighbors if hexagonal pixel is used, and eight-connected neighbors). However, we did not build elaborate set-theoretic or other data structures there. We will do so in this chapter. The graph-matching techniques discussed in this chapter will be used a great deal in Chapter 13.

12.1 Graphs

A graph is a relational data structure. It consists of data elements, referred to as *vertices* or *nodes*, and relationships between vertices, referred to as *edges*.

Graphs may be completely described by sets. The set of vertices is a simple set, and the edges form a set of ordered pairs. For example, let $G = \langle V, E \rangle$ represent a graph, where the vertex set $V = \{a, b, c, d, e, f\}$ and the edge set $E = \{(a, b), (b, c), (a, c), (b, e), (d, f)\}$. Graphs may also be represented pictorially. The pictorial representation of this example is drawn in Fig. 12.1. There are two connected components in this graph.

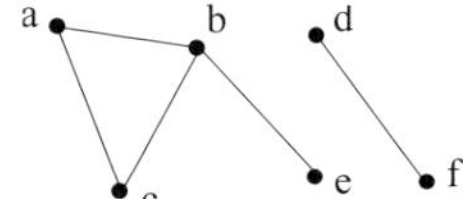

Fig. 12.1. A graph with six nodes, five edges, and two connected components.

This example allows us to elaborate on the concept of an edge just a bit more. In general, graphs may be directed. That is, the relations which are represented by edges may have a direction. Consider for example the relations "above" and "adjacent to." Clearly ABOVE$(a, b) \neq$ ABOVE(b, a), and so ABOVE would be represented by a directed graph, whereas the relation ADJACENT_TO has no preferred direction. The pictorial version of a directed graph may be thought of as having arrowheads on the ends of the edges. Let's redo the graph example given above, and include ordered pairs with both directions in them: $G = \langle V, E \rangle$ where $V = \{a, b, c, d, e, f\}$ and $E = \{(a, b), (b, a), (b, c), (c, b), (a, c), (c, a), (b, e), (e, b), (d, f), (f, d)\}$. This is a properly done, undirected graph. We define a graph as undirected if

$\forall(a, b \in V)[(a, b) \in E \Leftrightarrow (b, a) \in E]$. Otherwise the graph is directed (or, in some special cases, partially directed – a seldom-used term).

12.2 Properties of graphs

In this section, we define the vocabulary we will need to further our discussion of graphs.

- The *degree* of a node is the number of edges coming into that node.
- A *path* between nodes v_0 and v_l is a sequence of nodes $v_0, v_1, \ldots v_l$ such that there exists an edge between v_i and v_{i+1} for all is.
- A graph is *connected* if there exists a path between any two nodes.
- A *clique* (remember this word, you will see it again) is a subgraph in which there exists an edge between any two nodes.
- A *tree* is a graph which contains no loops. Tree representations have found application in speeding up Markov random field applications [12.8].

12.3 Implementing graph structures

The first data structure used to implement graph structures in computers was the linked list as shown in Fig. 12.2. Such a data structure contains two types of data:

- *nodes*, which consist of two pointers (addresses), and
- *atoms*, which are data.

Atoms are distinguished from nodes by a single bit, usually the most significant bit of the computer word. Certain nodes contain a zero in their right half, indicating the end of the list. Such nodes are indicated in Fig. 12.2 by the cross in the right-hand half. The linked list can also be used to store computer instructions, thus providing a powerful mechanism for automatic programs. This was the foundation of the programming language LISP.

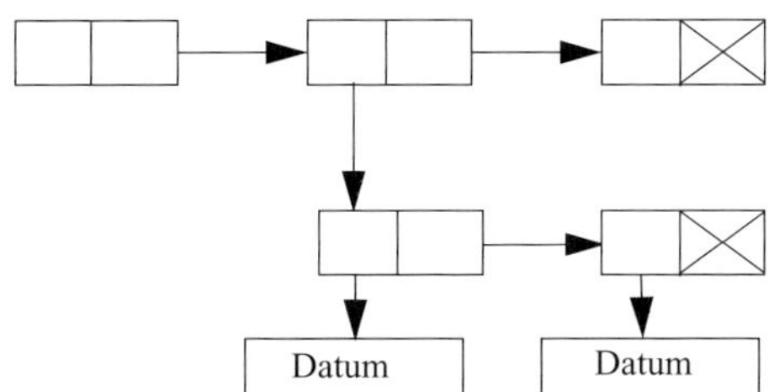

Fig. 12.2. In a linked list, each node contains two pointers. A pointer with the value of zero indicates the end of a list.

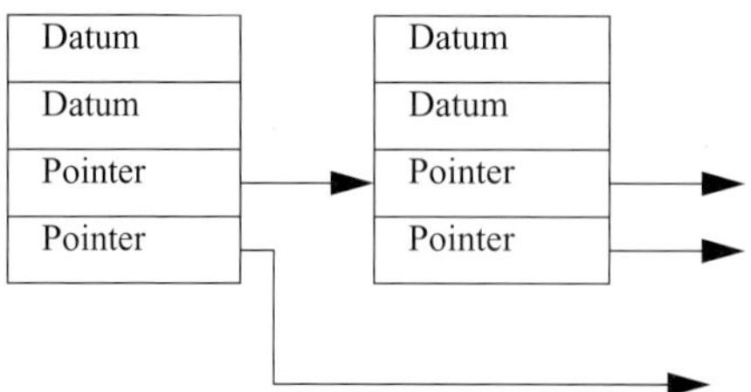

Fig. 12.3. A more general data structure may contain both data and pointers.

The concept of the linked list was incorporated into more modern programming languages and extended to allow more generality. For example, a structure such as that illustrated in Fig. 12.3 can contain data and pointers to other data structures of the same or different types. For example the following C definition describes the data structure of Fig. 12.3.

```
struct patch
{
int area;
int perimeter;
struct *patch;
struct *patch;
}
```

The C programmer will recognize the *patch as denoting a pointer to another structure of type *patch*.

12.4 The region adjacency graph

In model matching, we will make use of the region adjacency graph (RAG) as a means for identifying how the regions in a segmented image match (or do not match) faces in a three-dimensional model. A model of a polyhedron with six faces is illustrated in Fig. 12.4.

The RAG for the object of Fig. 12.4 is illustrated in Fig. 12.5; Fig. 12.6 shows another example.

A polyhedron has all FLAT faces.

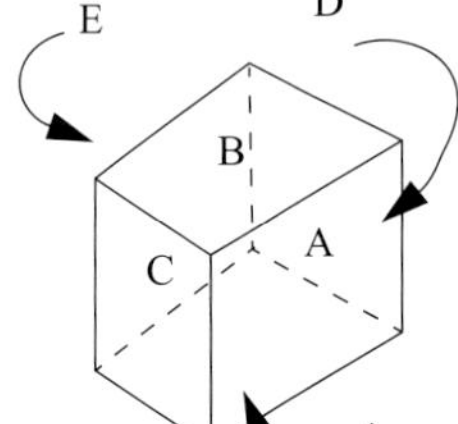

Fig. 12.4. A polyhedron with six faces.

What a mess! Do you think there is a planar way to draw this graph? That is, can you draw it without any lines crossing?

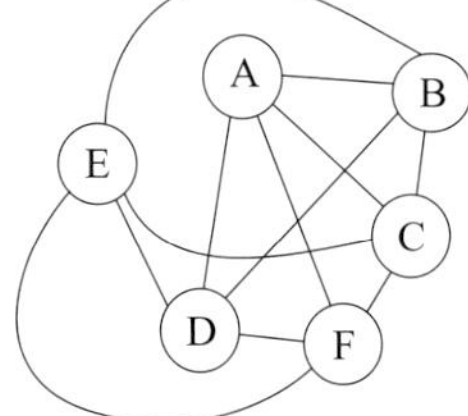

Fig. 12.5. The RAG of the three-dimensional object in Fig. 12.4.

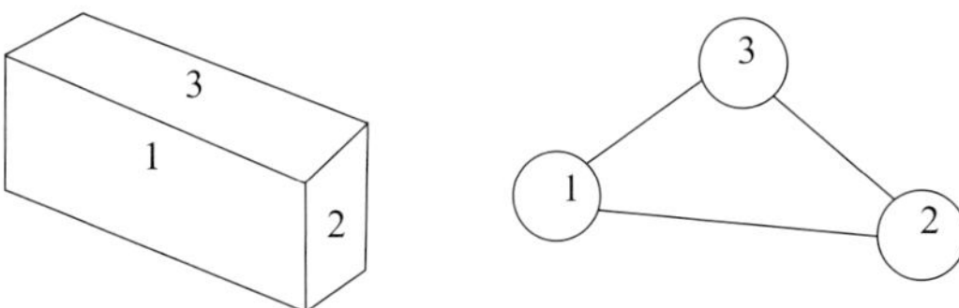

Fig. 12.6. The image and its RAG.

Now is the problem: Given an observation, and the RAG derived from the observation, and given a collection of models and their corresponding graphs, which model best matches the observation? We will address this matching problem later.

Other graph representations are possible and often useful, for example, the constructive solid geometry (CSG) community uses a collection of primitives subjected to transformations to represent input to automatic parts manufacturing systems. The primitives are objects like spheres and cylinders. Methods have been developed [13.8] to match scenes to models constructed from CSG representations as well as to RAG representations.

12.4.1 The scene graph

First, let's introduce a term to help with vocabulary. When talking about models, we use the term "region" to represent an identifiable surface/face, etc. When we are talking about the output of the segmenter, that is, what one might consider a "region" in the observation, we generally use the term "patch."

The RAG does not contain a lot of information, only that there is a region and it is adjacent to certain other regions. We can produce a graph containing much more information by constructing an augmented RAG, which we call the *scene graph*. In a scene graph, nodes have properties, such as the area of the corresponding patch, the color, the albedo, etc. An example is shown in Fig. 12.7.

Furthermore, the scene graph may be multiply traversed. In a simple RAG, the only edges between nodes are edges that correspond to the predicate ADJACENT_TO, however, we can also have edges corresponding to the property JUST_LARGER_THAN, and thus have a mechanism for traversing the graph in order of patch size, or whatever other property seems convenient for a particular application. Generally,

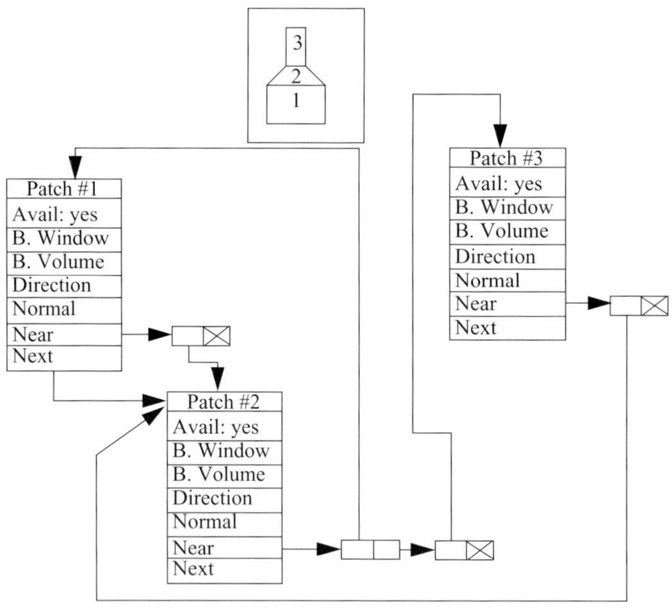

Fig. 12.7. Scene graph for an image with one object segmented into three patches.

Notice the use of "cardinality" rather than area. Why do you suppose this is? Think about a range image viewed obliquely.

- the patch list is sorted by cardinality (number of pixels in a patch), and
- each node has a pointer to a list of adjacent patches.

12.5 Using graph-matching: The subgraph isomorphism problem

There are several kinds of problems in computer science, defined by how long they take to run as a function of the size of the data set, represented by the variable n.

- *Polynomial time problems.* The time required may be written as a polynomial in n, e.g. $t \propto n^3$.
- *Exponential time problems.* The time required may be written as an exponential, e.g. $t \propto e^n$. Observe that as n becomes large, n^k is tiny compared to k^n.
- *NP-hard problems.* Here, NP says nonpolynomial, but it really means exponential. NP-hard problems are provably exponential in complexity. That is, one can prove that there does not exist any way to compute it in polynomial time.
- *NP-complete problems.* These are problems for which there does not exist any known algorithm to compute the answer in polynomial time. However, there is also no proof that no such algorithm exists. Any algorithm (or machine) which could convert an NP-complete problem to polynomial time would be a dramatic breakthrough, since all NP-complete problems can be shown to be isomorphic. That is, any algorithm which could solve one NP-complete problem could be applied to any other such problems.

Subgraph isomorphism is one approach to matching a scene graph to a RAG. The idea is simple. We know the scene graph, assuming the segmenter worked correctly (a terrible assumption in realistic images). Furthermore, we have a collection of RAGs, one for each model in our data base. To accomplish matching, we simply look for where the scene graph is isomorphic to some subgraph of the model graph. Unfortunately, constructing all subgraphs of a model graph is an NP-complete problem, with exponential complexity. Still, the use of heuristics can provide dramatic improvements in performance [12.4].

We will discuss matching in more detail in Chapter 13.

12.6 Aspect graphs

Consider a polyhedron with its center of gravity located at the origin, and think about the image formed by a camera pointed at the origin and located at the 3D spatial location parameterized by spherical coordinates $[\rho, \theta, \varphi]$ where θ denotes rotation about the x axis and φ denotes rotation about the y axis, as illustrated in Fig. 12.8.

If ρ is a constant, then the locus of possible camera positions is a sphere about the origin. For now, we will only consider this case. Thus, the camera may be thought of as moving on a sphere. As can be seen from Fig. 12.9, two different viewpoints may produce very different views. But consider only a very small camera motion. Except in rare conditions, small camera motions cause only slight changes in the image.

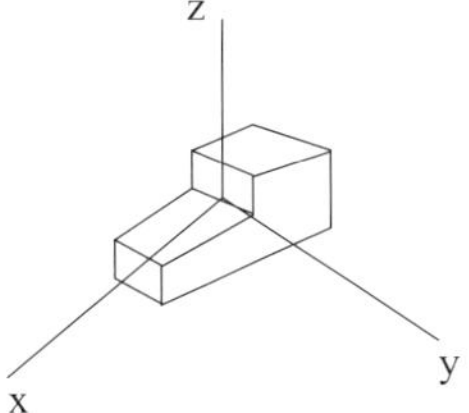

Fig. 12.8. The coordinate system is defined so that the object to be characterized has its center of gravity located at the origin.

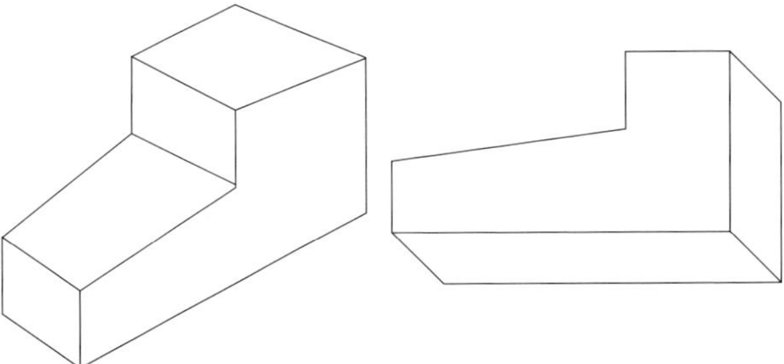

Fig. 12.9. Two different aspects of the same object produce very different two-dimensional images.

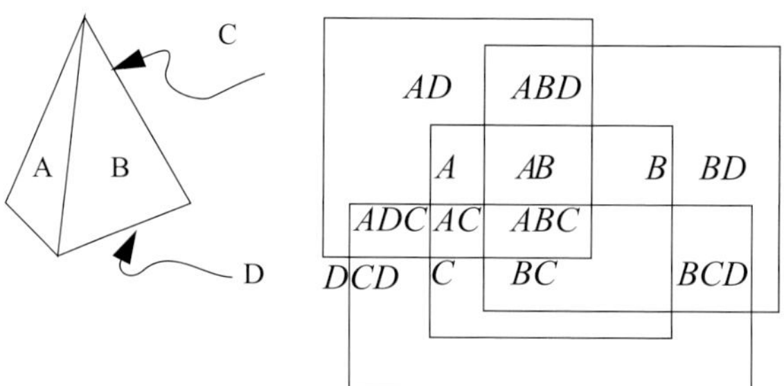

Fig. 12.10. Each partition of the VSP is identified by a list of the surfaces visible from that set of viewpoints (redrawn from [12.5]).

In those rare camera movements, however, radical things happen to the image – surfaces disappear or appear. That is, the topological structure of the image changes.

Two viewpoints V_1 and V_2 are defined to be *aspect equivalent*, denoted $V_1 \sim V_2$, if and only if there exists a sequence of infinitesimal camera motions, a path, from V_1 to V_2 such that the topology of the viewed image does not change. The aspect equivalent property is obviously symmetric, reflexive, and trivially transitive. It is therefore an equivalence relation and thus imposes a partition, denoted the *viewpoint space partition* (*VSP*), on the set of points on the sphere. Each element of this partition is referred to as a *viewing region*. Fig. 12.10, from [12.5], illustrates the VSP for a tetrahedron with faces A, B, C, and D. The aspect graph (referred to originally as the visual potential) is the dual of the VSP.

Computing the aspect graph is accomplished [12.2] by first constructing the labeled image structure graph (LISG), which is an augmented graph in which each node in the graph corresponds to a vertex in the line drawing, and each arc corresponds to the line segment between those nodes. The arcs are augmented with the labels corresponding to the properties $+$, $-$, and $\rightarrow$, as we used them in Chapter 10 to mean convex, concave, or occluding. The algorithm in [12.2] partitions the viewing sphere such that all points in a partition have isomorphic LISGs.

For arbitrary (potentially nonconvex) polyhedra, using orthographic projection, the algorithm is of high (but still polynomial) computational complexity. For a polyhedron with n faces, the total worst case time complexity is $O(n^8)$.

Aspect graphs were originally developed by Koenderink and Van Doorn [12.3], and extended by a variety of authors; Bowyer and Dyer [12.1] provide a good survey of work prior to 1990. More recent work has focussed on such nasty problems as the fact that we are dealing with sampled data [12.7].

12.7 Conclusion

Consistent labeling searches a tree of interpretations.

The concepts of graphs occur throughout the machine vision literature. See [12.6] for more on scene structure graphs and Bayesian networks which use them. As we now understand, the search algorithm used in section 10.1 to label line drawings actually searched a tree of interpretations.

12.8 Vocabulary

You should know the meanings of the following terms.

```
Aspect graph
Clique
Connected
Degree
Edge
Isomorphic
Node
NP—complete
Path
RAG
Scene graph
Tree
Vertex
```

References

[12.1] K. Bowyer and C. Dyer, "Aspect Graphs: An Introduction and Survey of Recent Results," *International Journal of Imaging Systems and Technology*, **2**, pp. 315–328, 1990.

[12.2] Z. Gigus and J. Malik, "Computing the Aspect Graph for Line Drawings of Polyhedral Objects," *IEEE Transactions on Pattern Analysis and Machine Intelligence*, **12**(2), 1990.

[12.3] J. Koenderink and A. Van Doorn, "The Internal Representation of Solid Shape with Respect to Vision," *Biological Cybernetics*, **32**, pp. 211–216, 1979.

[12.4] B. Messmer and H. Bunke, "A New Algorithm for Error-tolerant Subgraph Isomorphism Detection," *IEEE Transactions on Pattern Analysis and Machine Intelligence*, **20**(5), 1998.

[12.5] H. Plantinga and C. Dyer, "Visibility, Occlusion, and the Aspect Graph," *International Journal of Computer Vision*, **5**(2), pp. 137–160, 1990.

[12.6] S. Sarkar and P. Soundararajan, "Supervised Learning of Large Perceptual Organization: Graph of Spectral Partitioning and Learning Automata," *IEEE Transactions on Pattern Analysis and Machine Intelligence*, **22**(5), 2000.

[12.7] I. Shimshoni and J. Ponce, "Finite-resolution Aspect Graphs of Polyhedral Objects," *IEEE Transactions on Pattern Analysis and Machine Intelligence*, **19**(4), 1997.

[12.8] C. Wu and P. Doerschuk, "Tree Approximations to Markov Random Fields," *IEEE Transactions on Pattern Analysis and Machine Intelligence*, **17**(4), 1995.

13 Image matching

One of these things is not like the other

Sesame Street

In this chapter we will consider issues associated with matching – matching observed images with models as well as matching images with each other. We will consider matching iconic representations as well as matching graph-theoretic representations.

Matching establishes an interpretation. That is, it puts two representations into correspondence.

- Both representations may be of the same form. For example, correlation matches an observed image with a template. Similarly, subgraph isomorphism matches a region adjacency graph to a subgraph of a model graph.
- Both representations might be of different forms. For example, one image matches one paragraph describing something. In most such applications, we find ourselves matching an equation to some data, and in this case, "fitting" might be a better word.

In the remainder of this chapter we address all of these matching problems except fitting, which was discussed earlier in this book.

13.1 Matching iconic representations

13.1.1 Template matching

A template is a representation for an image (or subimage) which is itself a picture. A template is typically moved around the target image until a location is found which maximizes some match function. The most obvious function is the squared error

$$SE(x, y) = \sum_{\alpha=1}^{N} \sum_{\beta=1}^{N} (f(x - \alpha, y - \beta) - T(\alpha, \beta))^2, \qquad (13.1)$$

(assuming the template is $N \times N$) which provides a measure of how well the template (T) matches the image (f) at point x, y. If we expand the square and carry the summation through, we find

The relationship between correlation and squared error.

$$SE(x, y)=\sum_{\alpha=1}^{N}\sum_{\beta=1}^{N} f^2(x-\alpha, y-\beta) - 2\sum_{\alpha=1}^{N}\sum_{\beta=1}^{N} f(x-\alpha, y-\beta)T(\alpha, \beta) + \sum_{\alpha=1}^{N}\sum_{\beta=1}^{N} T^2(\alpha, \beta). \quad (13.2)$$

Let's look at these terms: The first term is the squared sum of the image brightness values at the point of application. It says nothing about how well the image matches the template (although it IS dependent on the image). The third term is simply the sum of the squared elements of the template, and is a constant, no matter where the template is applied. The second term obviously is the key to matching, and that term is the correlation.

In matching using an optimization criterion, the assumption is made that the quality of match can be described by a set of parameters $\mathbf{a} = \{a_1, a_2, \ldots, a_n\}$, which could be the pixels themselves. We define a merit function $M(\boldsymbol{a}, f(\boldsymbol{x}))$ which quantifies the quality of the match between the template and the local image. Matching consists of determining $\boldsymbol{a}$ so that M is maximized. Typically, $\boldsymbol{a}$ is the x–y coordinates specifying where the template is placed.

If M is monotonic in $\boldsymbol{a}$, we can maximize M by solving

$$M_{a_j} = \frac{\partial M}{\partial a_j} = 0 \quad \text{for } j = 1, \ldots, n. \quad (13.3)$$

If M is not monotonic, the process of finding points where the partial derivatives are zero can terminate in a local maximum. Furthermore, as we have discussed earlier, it is probably not possible to find an analytic solution to Eq. (13.3). In that case, we could use hill-climbing:

The equation for hill-climbing should be compared with that for gradient descent.

$$a_j^k = a_j^{k-1} + cM_{a_j}. \quad (13.4)$$

This strategy is generally a difficult approach to template matching, since $\partial M/\partial \mathbf{a}$ is difficult to use as it often contains multiple minima and broad peaks. However, if we happen to be close to the matching point, it is possible that hill-climbing will lead to an optimal match. The width of the peaks can be reduced somewhat by using heuristics such as phase-only matching [13.7].

Assuming a correct segmentation, one could match 2D binary shapes which have been transformed by first finding the affine (or similarity) transform that best matches the two shapes [17.61], and then simply doing template matching on the transformed regions. Generally, the complexity of template matching can be reduced somewhat by masking out impossible matches [13.37]. As we have seen in section 11A.4, iconic matching is a critically important part of any stereopsis system [13.31].

One of the principal problems in any template matching system is choosing the optimal size of the template. If the template chosen is too small, it will not cover enough variation in the image to make an accurate estimate. If it is too large, the window covers too large an area to match at all. Some adaptive algorithms have been reported [13.27]. In addition, one must consider what objective is being minimized when matching. The squared error and the signal-to-noise ratio are not always the best [13.40].

13.1.2 Point matching

Consider an image as just a set of points with known distances between each other (one could consider this an instance of the springs and templates problem discussed in the next section, with trivial templates). One example where this type of data occurs is in the recognition of targets in Synthetic Aperture Radar (SAR) images. In this situation one may approach the problem by assuming a 3D model of the object, and finding the transformation from 3D to 2D which best describes the observation [13.53]; see also [13.3].

Matching in the stereo environment can make use of the epipolar constraint as well as characteristics of edges and vertices, and can be treated probabilistically [13.47].

13.1.3 Segment matching

The problem of finding a match between short arcs and pieces of a long arc has been investigated in the literature [13.3, 13.39, 13.42]. The corresponding 3D problem has not received much attention largely because it is difficult to extract 3D curves; [13.22] provides one approach.

13.1.4 Eigenimages

The eigenimage approach has been an effective solution to problems like object identification and recognition [13.49, 13.50], where the image of an unknown object is compared to images of known objects in a data base (or a training set) and the unknown object can be identified or recognized when a close match is found. We can surely do a pixel-by-pixel comparison. However, this is very time-consuming, especially when the size of the image is large and the number of images included in the data base is large as well.

The eigenimage approach has its origin in principal component analysis (PCA), which is a popular technique for dimensionality reduction. In section 9.2.1, one type of PCA, the K–L transform, is described in detail. PCA constructs a representation of the data with a set of orthogonal basis vectors that are the eigenvectors of the covariance matrix generated from the data. By projecting the data onto the dominant eigenvectors (corresponding to the larger eigenvalues), the dimension of the original data set can be reduced with minimal loss of information. Similarly, in the eigenimage approach, each image is represented as a linear combination of a set of dominant principal components (the eigenimages). Matching is then conducted based on the coefficients of the linear combination (or the weight of projections onto the eigenimages) which greatly speeds up the process. The projection preserves most of the energy, and thus captures the highest amount of variation in the data base. Here, we discuss the calculation of the eigenimages in detail.

Let $f_1, f_2, \ldots, f_p$ represent a set of images of known objects in the data base. Without loss of generality, assume these images are of the same dimension $m \times n$. The following steps lead us to the eigenimages.

Step 1. Construct a set of vectors $\{I_1 \ldots I_p\}$ where each $I_i (i = 1, \ldots, p)$ is the lexicographical representation of the corresponding image f_i minus the average image, $I_i = f_i - A$, and $A = \frac{1}{p}\sum_{i=1}^{p} f_i$. Note that each vector is $mn \times 1$.

Step 2. Calculate the covariance matrix of this set by

$$C = \frac{1}{p-1}\sum_{i=1}^{p} I_i I_i^{\mathrm{T}}, \qquad (13.5)$$

where C is an $mn \times mn$ matrix.

Step 3. Use eigenvalue decomposition techniques to obtain the eigenvectors and eigenvalues of C:

$$C = E\Lambda E^{\mathrm{T}}$$

where E is an $mn \times mn$ matrix with each column vector an eigenvector of C (or an eigenimage of C), Λ is a diagonal matrix

$$\Lambda = \begin{bmatrix} \lambda_1 & \ldots & 0 \\ 0 & \ldots & 0 \\ 0 & \ldots & \lambda_{mn} \end{bmatrix}$$

with the eigenvalues of C on the diagonal of Λ, and $\lambda_1 > \lambda_2 > \cdots > \lambda_{mn}$.

Step 4. Suppose among the mn eigenvalues, the first k values are much larger than the other values, that is, $\sum_{j=1}^{k} \lambda_j / \sum_{j=1}^{mn} \lambda_j = \eta$ is very close to 1. Then we can use the first k eigenvalues to reconstruct the original image without losing too much information. Hopefully, $k \ll mn$.

Step 5. Calculate the projection coefficients of image f_i onto the selected eigenimages for comparison purpose by

$$W_i = I_i^{\mathrm{T}} \times [E_1 \ldots E_k] \tag{13.6}$$

where E_j is the jth eigenimage, and W_i is a $p \times k$ matrix, containing the projection coefficient of the original image onto each eigenimage. The comparison process can be easily carried out by the following: Given an image of unknown object in the test set f_{test}, calculate its projection coefficients by

$$W_{\text{test}} = I_{\text{test}}^{\mathrm{T}} \times [E_1 \ldots E_k].$$

Compare the distance between W_{test} and all the W_is in the data base (a Euclidean distance might be the simplest approach); the one with the closest distance is selected as a match.

We now show an interesting example of applying the eigenimage approach to face recognition [13.51]. Assume we have three images in the data base (Lena, Einstein, and the Clock), the unknown image is Monalisa. Following Steps 1 through 3 described above, we can derive 64×64 eigenimages. We use only two of the dominant ones since the ratio between the summation of the first two eigenvalues and the summation of all the eigenvalues is close to 1, as stated in Step 4. Fig. 13.1 shows all the original images and the two eigenimages derived. Following Step 5, we compute the projection coefficients of all four original images on the two eigenimages; these are listed in Fig. 13.1 as well. Based on a simple Euclidean distance calculation, it turns out that the closest match to Monalisa is Einstein. Is that a surprise? Not really. In the "eyes" of the computer, these two images are indeed more alike than Monalisa and Lena.

Reducing computational complexity

Even though the eigenimage approach has great potential for image matching, from the procedure described above, we see that the most time-consuming step is the derivation of the eigensystem. When the size of images is large, the calculation of the covariance matrix (which is $mn \times mn$) can take up a lot of computation resources or be completely infeasible. For more efficient calculations, readers are referred to [13.34, 13.35].

We illustrate one approach to reducing computation through an example. Assume that each image has only three pixels and that there are only two such images in the set. Let them be

$$f_1 = [1 \quad 2 \quad 3]^{\mathrm{T}}$$
$$f_2 = [5 \quad 8 \quad 9]^{\mathrm{T}}.$$

Training images Test image

Original images

I_1 I_2 I_3 I_{test}

The 1st and 2nd eigenimages

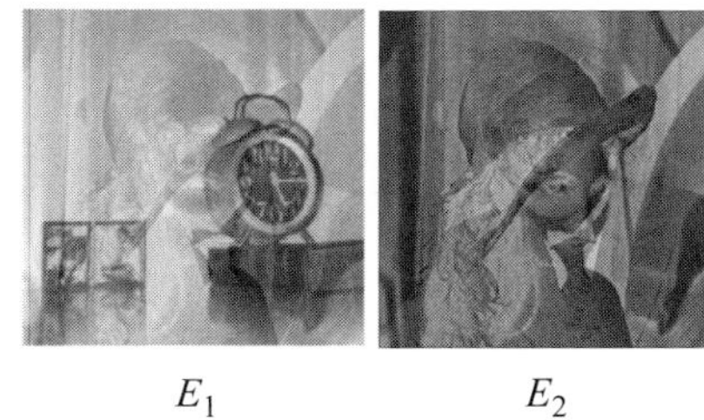

E_1 E_2

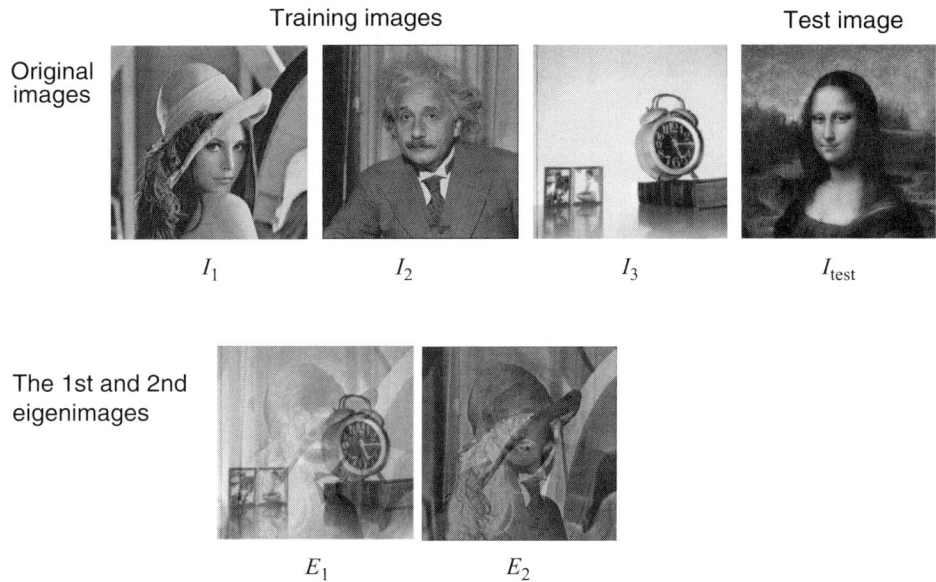

The projection coefficients

$$W_1 = [-1.422\ -0.8161],\ W_2 = [-1.7015\ 0.7688],\ W_3 = [3.1235\ 0.0473]$$

$$W_{test} = [-1.4399\ 0.4710]$$

The distance

$$d_{1,\,test} = 1.2872,\ d_{2,\,test} = 0.3963,\ d_{3,\,test} = 3.5831$$

Fig. 13.1. Demonstration of the eigenimage approach on a data base of three images (Lena, Einstein, Clock) and a test image (Monalisa). The images have been rescaled for display purpose.

Then the mean $A = [3\ \ 5\ \ 6]^{\mathrm{T}}$, and

$$\begin{aligned} I_1 &= [-2 \quad -3 \quad -3]^{\mathrm{T}} \\ I_2 &= [2 \quad 3 \quad 3]^{\mathrm{T}} \end{aligned}.$$

Construct the matrix $I = \left[I_1 \ldots I_p\right]$, in which the ith column of I is one of the images I_i, and consider the product $S = II^{\mathrm{T}}$. In this example,

$$I = \begin{bmatrix} -2 & 2 \\ -3 & 3 \\ -3 & 3 \end{bmatrix} \quad \text{and} \quad S = \begin{bmatrix} 8 & 12 & 12 \\ 12 & 18 & 18 \\ 12 & 18 & 18 \end{bmatrix}.$$

Observe that if $p < mn$, then S is the scatter matrix, identical to the covariance except for the multiplicative scale factor. S is huge – if the image is 256×256, then S is $256^2 \times 256^2$. However, if there are only, say, five images in the set, I is $256^2 \times 5$, and $I^T I$ is 5×5. Suppose μ_i is one of the eigenvectors of $I^T I$. That is,

$$I^T I \mu_i = \lambda_i \mu_i. \tag{13.7}$$

Here's the mathemagic: Multiply on both sides of Eq. (13.7) by I to obtain $I I^T I \mu_i = I \lambda_i \mu_i = \lambda_i I \mu_i$, and we notice that $I \mu_i$ is an eigenvector of $I I^T$. Since the size of $I^T I$ is dramatically smaller than the size of $I I^T$, we obtain a similarly dramatic decrease in complexity in the process of determining the eigenvectors. Thus, if e_is are the eigenvectors of S, we can obtain them using

$$e_i = I \mu_i = \sum_{k=1}^{p} \mu_{ik} I_k. \tag{13.8}$$

13.2 Matching simple features

The most straightforward way to use the simple features we described in Chapter 9 is to use them in a pattern classifier. To do this, we will extract a statistical representation for the model and the object and match those representations. The strategy is as follows.

- Decide which measurements you wish to use to describe the shape. For example, one might build a system which measures seven invariant moments and the aspect ratio, for a total of eight "features." The best collection of features is application-dependent, and methods for optimally choosing feature sets is beyond the scope of this book (see [14.4, 14.11, 18.30] which are just a few of many texts in statistical methods). Organize these eight features into a vector, $\boldsymbol{x} = [x_1, x_2, \ldots, x_8]^T$.
- Describe a "model" object using a collection of example images (called a "training set"), from which feature vectors have been extracted. Continuing the example with eight features, we could collect a set of n images of axes, measure the feature vector of each axe, and characterize the model axe by its average over this set $\mu_{\text{axe}} = \frac{1}{n} \sum_{x \in \text{axe}} \boldsymbol{x}$. Hatchets might be similarly characterized by an average over a set of sample hatchets.
- Now, given an unknown region, characterized by its feature vector $\boldsymbol{x}$, shape matching consists of finding the model which is "closest" in some sense to the observed region. Probably the simplest definition of "close" uses the Euclidian distance

The unknown is distinguished from the examples by lack of a subscript.

$$\begin{aligned} d(\text{modelaxe, observation}) &= \sqrt{(\boldsymbol{x} - \mu_{\text{axe}})^T (\boldsymbol{x} - \mu_{\text{axe}})} \\ d(\text{modelhatchet, observation}) &= \sqrt{(\boldsymbol{x} - \mu_{\text{hatchet}})^T (\boldsymbol{x} - \mu_{\text{hatchet}})} \end{aligned}.$$

We make the decision based on which of these distances is smaller.

This example is rather naive for several reasons. There are better models for a set of examples than just the mean, and there are better matching metrics than the

Euclidian distance. In Chapter 14, the concepts in this discipline of statistical pattern classification are covered in more detail.

13.3 Graph matching

Recall the formal definition of a metric. This paragraph refers to subgraph isomorphism as a metric. Is that correct?

In this section, we consider the problem of matching image representations which are fundamentally graph-based. However, we allow the data stored at a node in the graph to include images or templates.

Recall that a *clique* of size N is a totally connected subgraph of size N. We also will want to use matching metrics, such as the mean squared error or correlations mentioned in section 13.1.1. A *matching metric* measures the "goodness" of a match.

In a totally graph-based representation, one matching metric might be subgraph isomorphism. But, subgraph isomorphism does not really allow for close but not perfect matches. Most machine vision specialists would say that it is too inflexible.

The graph-matching problem can be approached using annealed neural networks [13.8].

As we saw earlier, relaxation labeling can also provide a mechanism for a type of graph matching. In the example we saw in section 10.2.2, we can match a subgraph of the scene graph (i.e. the two surfaces in the example given) with a subset of the model graph. Other variations are possible, for example, Gold and Rangarajan [13.14] describe a variation on graph matching which utilizes a nonlinear optimization method which they say runs much faster and more accurately than relaxation labeling.

Two other approaches are described here – *association graphs* and *spring-loaded templates*.

These are methods which will produce matchings on hybrid representations, that is, those which are fundamentally graph-based, but which include image information.

13.3.1 Association graphs

Association graphs embody a methodology which is less restrictive than isomorphism, and which may converge more rapidly. It will converge to a solution which is consistent, but not necessarily optimal (depending, of course, on the criteria for optimality used in any particular application).

The method matches a set of nodes from the model to a set of nodes (extracted) from the image.

Definition

Here, a graph is denoted $G = \langle V, P, R \rangle$, where V represents a set of nodes, P represents a set of unary predicates on nodes, and R represents binary relations between nodes.

A predicate is a statement which takes on only the values TRUE or FALSE. For example let x denote a region in a range image. Then CYLINDRICAL(x) is a predicate which is true or false depending on whether all the pixels in x lie on a cylindrical surface.

A binary relation describes a property possessed by a pair of nodes. It may be considered as a set of ordered pairs $R = \{(a_1, b_1), (a_2, b_2), \ldots (a_n, b_n)\}$. In most applications, order is important. It is possible to think of a relation as a predicate, since for any given pair, say (a_k, b_k), either it is an element of the set R or it is not. However, it seems more descriptive to use the word *relation* in this context.

Given two graphs, $G_1 = \langle V_1, P, R \rangle$ and $G_2 = \langle V2, P, R \rangle$, we construct the association graph G by:

- for each $v_1 \in V_1$ and $v_2 \in V_2$, if v_1 and v_2 have the same properties, construct a node of G labeled (v_1, v_2);
- if $r \in R$ and $r(v_1, v_1') \Leftrightarrow r(v_2, v_2')$, connect (v_1, v_2) to (v_1', v_2').

The best match of G_1 to G_2 is the largest clique of G.

Like every other technique in machine vision, we need to ask, "How good is this method?" Some problems arise when attempting to answer that question:

Problem 1. Is the largest clique the best match? The largest clique is the largest set of consistent matches. Is this really the best match?

Problem 2. Computational complexity. Like the subgraph isomorphism problem, the problem of finding the largest clique is NP-complete. That is, no algorithm is known which can solve this problem in less than exponential time.

An example of using association graphs to match a scene to a model

In Fig. 13.2 we illustrate an observation in which a segmentation error, oversegmentation, has occurred. That is, regions B and C are actually part of the same region, but due to some measurement or algorithmic error, have been labeled as two separate regions. In this example, the unary predicates are labels *spherical, cylindrical*, and *planar*. Regions A and 1 are spherical, while B, C, D, 2, and 3 are cylindrical. The only candidates for matches are those with the same predicate. So only A can match 1. We now construct a graph in which all candidate matches are the nodes. We then have the nodes of the association graph, as illustrated in Fig. 13.3.

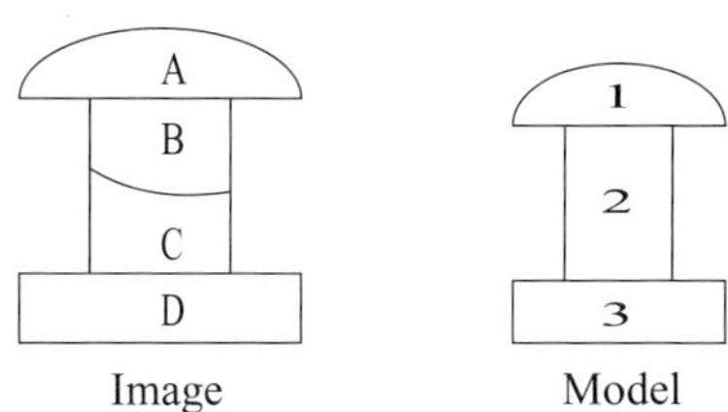

Fig. 13.2. A range camera has observed a scene and segmented it into segments which satisfy the same equation, however, an error has occurred.

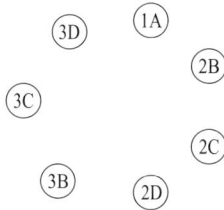

3D 1A 2B 3C 2C 3B 2D

Fig. 13.3. Candidate matches.

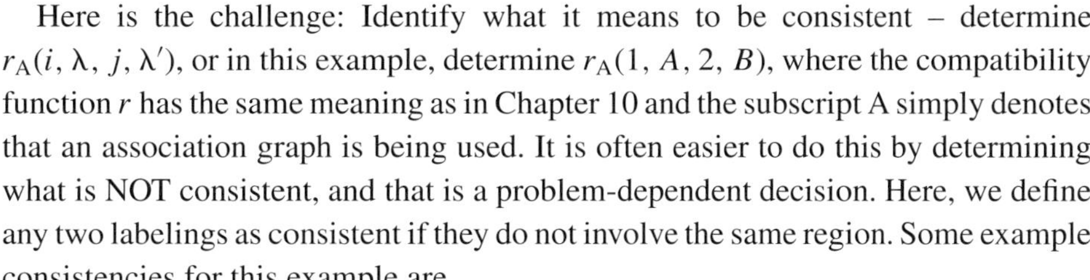

Here is the challenge: Identify what it means to be consistent – determine $r_A(i, \lambda, j, \lambda')$, or in this example, determine $r_A(1, A, 2, B)$, where the compatibility function r has the same meaning as in Chapter 10 and the subscript A simply denotes that an association graph is being used. It is often easier to do this by determining what is NOT consistent, and that is a problem-dependent decision. Here, we define any two labelings as consistent if they do not involve the same region. Some example consistencies for this example are

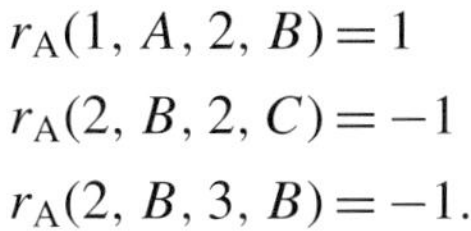

$$r_A(1, A, 2, B) = 1$$
$$r_A(2, B, 2, C) = -1$$
$$r_A(2, B, 3, B) = -1.$$

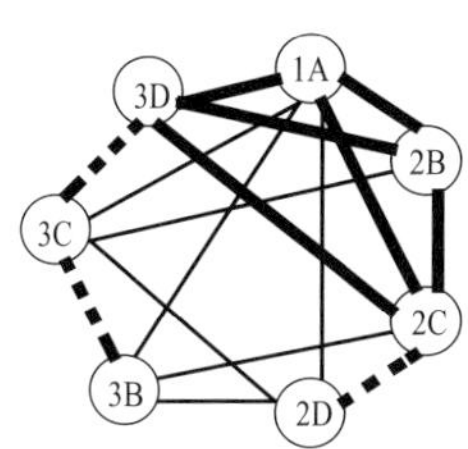

Fig. 13.4. Solid lines denote the edges which are present if we assume the segmenter does not fail. The dotted lines are added if we believe the segmenter can fail by oversegmentation. The heavy lines indicate a maximal clique.

The second line says that patch B in the image could not be region 2 in the model while simultaneously, patch C in the image is the same region. In both examples, inconsistencies are really based on the assumption that the segmenter is working correctly. However, one could allow the segmenter to fail. In that case, new edges are added, because new relationships are now consistent. For example, $r_A(3, C, 3, D) = 1$ since we believe that two patches could be part of the same regions (the segmenter can fail by oversegmentation), however, $r_A(2, D, 3, D) = -1$ still holds because we still believe the segmenter will not merge patches (fail by undersegmentation). Allowing for oversegmentation produces the association graph of Fig. 13.4.

Note one other type of inconsistency which prevents an edge from being constructed: 3D and 3B are not connected since B and D do not border. That is, we believe that if the segmentation fails by oversegmentation, the segmenter will not introduce an entire new patch between the two. We must emphasize that how you develop these rules is totally problem-dependent!

The maximal clique is not unique.

Once you have the allowable consistencies, the matching is straightforward. Simply find all maximal cliques. The maximal clique is not unique, since there may be several cliques of the same size.

In this case, there are at least two maximal cliques, two of which are: {(1, A)(2, B)(2, C)(3, D)} and {(1, A)(3,B)(2, C)(2, D)}.

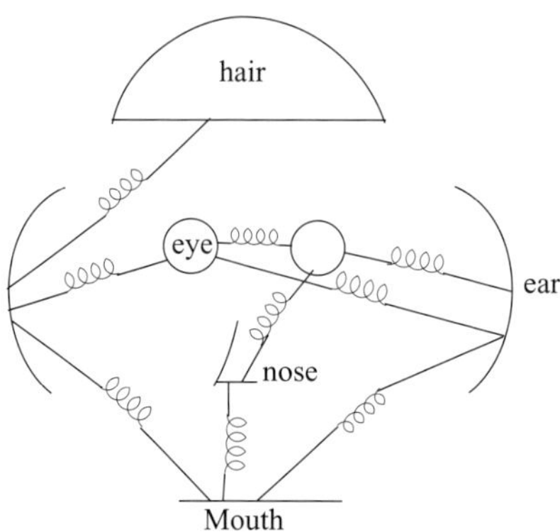

Fig. 13.5. A springs and templates model of a face.

13.3.2 Springs and templates

Another approach to the local–global problem is provided by the "springs and templates philosophy" [13.13]. This is a hybrid model-matching paradigm which involves both graph-structured matching and template matching. The model is a set of rigid "templates" connected by "springs" which characterize how much deformation must be applied to the model to make it match the image. Fig. 13.5 illustrates this concept using a simple model of a human face.

Specific features such as eyes are matched by iconic methods such as template matching. However, in order to match an entire face, distances between the best matching locations of templates are also recorded. A match is then based on minimizing a total cost, as follows.

$$\begin{aligned} \text{Cost} = & \sum_{d \in \text{templates}} \text{TemplateCost}(d, F(d)) \\ & + \sum_{d,e \in \text{ref} \times \text{ref}} \text{SpringCost}(F(d), F(e)) \\ & + \sum_{c \in (R_{\text{missing}} \times R_{\text{missing}})} \text{MissingCost}(c). \end{aligned} \tag{13.9}$$

In Eq. (13.9), d is a template and $F(d)$ is the point in the image where that template is applied. TemplateCost is therefore a function indicating how well a particular template matches the image when applied at its best matching point. SpringCost is a measure of how much the model must be distorted (the springs stretched) to apply those particular templates at those particular locations. Finally, it may be that not every template can be located – in some images, the left eye may not be visible – and a cost may be imposed for things missing. All these costs are empirically determined. However, once they are determined, it becomes relatively easy to determine how well any given image matches any given model.

There is one significant (among several others) problem with spring matching: The number of elements matched affects the magnitude of the cost. The costs are summed, so a poor match of only a few things may be less than a good match of many (and therefore better, in a minimal cost algorithm).

This is a problem which is not unique to springs and templates. The usual solution is to normalize the calculations, using a technique such as

$$\text{Cost} = \frac{\text{SpringCost(unary and binary)}}{\text{Total number of springs}} + \frac{\text{constant}}{\text{Total number of references matched}}.$$

13.4 Conclusion

Association graphs are a kind of consistent labeling.

Association graphs use the concepts and formalisms of consistent labeling directly. The advantage of using a graph structure is that the search for largest clique is aided by a body of available software for performing such searches as quickly as computational complexity allows. Similarly, the springs and templates ideas measure both consistency and deviation from consistency. The springs and templates concepts also illustrate both how one might construct an appropriate objective function, and a problem that can easily arise if one does not pay attention to interpretation of the objective function – if we are summing match quality, a good match of many things (adding up lots of small numbers) may be more than (and therefore worse than) a poor match of only a few things (adding up just a few rather large numbers).

Objective functions need suitable normalization. SSD is a common objective function.

We began this chapter by pointing out that formal optimization methods, either descent or "hill-climbing," are hard to apply to image matching because the search space is littered with local minima. However, if we initialize the algorithm sufficiently close to the solution, such techniques will work. We used the sum of squared differences (SSD), also sometimes called the sum-squared error, as the objective function.

Eigenimages are lower dimensionality representations of the original images. The projections are chosen by minimizing the error between the original data and the projected data.

13.5 Vocabulary

You should know the meanings of the following terms.

Association graph
Clique

Correspondence
Deformable template
Eigenimage
Hill-climbing
Matching metric
PCA
Template

Assignment 13.1

In this chapter, we stated that the problem of finding the largest clique is NP-complete. What does that really mean? Suppose you have an association graph with ten nodes, interconnected with 20 edges. How many tests must you perform to find all cliques (which you must do in order to identify which of these are maximal)? You ARE permitted (encouraged!) to look up clique-finding in a graph theory text.

Assignment 13.2

In section 13.3.1, an example problem is presented which involves an association graph which allows for segmentation errors. The result of that graph is two maximal cliques, which (presumably) mean two different interpretations of the scene. Describe in words these two interpretations.

Assignment 13.3

In the bibliography for this chapter, there is an incomplete citation to Olson [13.36]. First, locate a copy of that paper. You may use a search engine, the Web, the library, or any other resource you wish. In that paper, the author does template matching in a different way: Using a binary (edge) image and a similar template, he does not ask "Does the template match the image at this point?" Instead he asks, "At this point, how far is it to the nearest edge point?"

How does he perform this operation, apparently a search, efficiently?

Once he knows the distance to the nearest edge point, how does he make use of that information to compute a quality of match measure?

Assignment 13.4

In an image-matching problem, we have two types of objects, lions and antelope (which occupy only one pixel each).

- A scene may contain only lions and antelope.
- Lions hunt in packs, so if you see one lion, you will see at least one other lion, usually about 5 pixels away.
- Antelope stay as close to one another as possible.
- Except for certain rare, and (for the antelope) unpleasant events, lions and antelope are VERY far apart.

We wish to use relaxation labeling to solve this assignment problem. All the formulae are in the book except the formula for the consistency $r(a, \lambda_1, b, \lambda_2)$, where a and b are points of interest in the image and the lambdas are labels for either "antelope" or "lion." Invent an r function for this problem. That is, tell how to compute values for:

(1) $r(a$, lion, b, antelope)
(2) $r(a$, lion, b, lion)
(3) $r(a$, antelope, b, lion)
(4) $r(a$, antelope, b, antelope).

Assignment 13.5

Do you think the concepts of springs and templates would be applicable to Assignment 13.4? Discuss.

Assignment 13.6

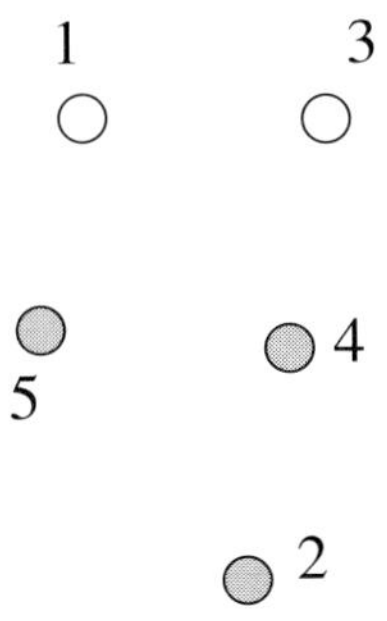

Still thinking about lions and antelope, you observe the scene opposite: The sketch is not to scale. However, for your convenience, we have tabulated the distance between each pair of animals (Table 13.1).

Lions are yellow (which is denoted by a dotted interior) or brown (denoted by a black interior — that's right; there aren't any). Antelope are white (denoted by a white interior) or yellow. You wish to use association graph methods to solve this problem; and since this technique is not as powerful as nonlinear

Table 13.1. *Distances between pairs of animals.*

Pair	Distance (arbitrary units)
1, 2	5.5
1, 3	2
1, 4	3
1, 5	2
2, 3	2
2, 4	3
2, 5	4
3, 4	2
3, 5	3.8
4, 5	3.4

relaxation, you talked to a botanist[1] who gave you some improved information: Lions NEVER get closer to each other than 3 pixels, and antelope never get farther from another antelope than 3 pixels.

Draw an association graph for this problem. (Denote the nodes in the association graph by pairs, e.g., 1L means "interpret node 1 as a lion".) Indicate the maximal clique by circling the nodes in that clique.

Topic 13A Matching

13A.1 Springs and templates revisited

Recall the correspondence problem, described in section 4.2.2, which may be restated as "given a set of features in two images, identify which feature in image 1 corresponds to which feature in image 2." There is little difference between the model-matching problem and the correspondence problem, except for the fact that in the correspondence problem, both of the images may be corrupted by noise.

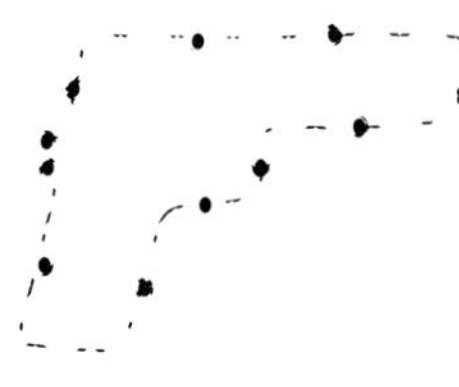

Fig. 13.6. Boundary of an object. Points where the sign of the curvature changes are marked.

To solve the correspondence problem, we use a consistency-based philosophy. In this section, we describe this philosophy through an example, which we will use again in section 13A.2. The first step is to identify feature points which are relatively distinctive. Then the algorithm will make use of relationships between these points. Points on the boundary of a region where the curvature of the boundary changes sign meet this requirement, as illustrated in Fig. 13.6.

1 Yes, a botanist! He was out of his field.

This example derivation was first described by Shapiro and Brady [13.46] who use eigenvector methods as follows.

As in the original springs and templates formulation, we are finding which of a collection of feature point sets best matches one particular set. Let d_{ij} be the Euclidian distance between feature points x_i and x_j, and construct a matrix of weights

$$H = [H_{ij}], \text{ where } H_{ij} = \exp\left(-\frac{d_{ij}^2}{2\sigma^2}\right). \tag{13.10}$$

The matrix H is diagonalized using standard methods into the product of three matrices

$$H = E \Lambda E^{\mathrm{T}} \tag{13.11}$$

where E is a matrix with the eigenvectors of H as its columns, and Λ is a diagonal matrix with the eigenvalues on the diagonal. Let us assume the rows and columns of E and Λ are sorted so that the eigenvalues are sorted along the diagonal in decreasing size. We think of each row of E as a feature vector, denoted F_i. Thus

$$E = \begin{bmatrix} F_1 \\ \dots \\ F_m \end{bmatrix}.$$

Suppose we have two images, f_1 and f_2, and suppose f_1 has m feature points while f_2 has n feature points, and suppose $m < n$. Then by treating each set of feature points independently, we have $H_1 = E_1 \Lambda_1 E_1^{\mathrm{T}}$ for image f_1, and $H_2 = E_2 \Lambda_2 E_2^{\mathrm{T}}$ for image f_2. Since the images have different numbers of points, the matrices H_1 and H_2 have different numbers of eigenvalues. We therefore choose to use only the most significant k features for comparison purposes.

It is important that the directions of the eigenvectors to be matched be consistent, but changing the sign does not affect the orthonormality. We choose E_1 as a reference and then orient the axes of E_2 by choosing the direction that best aligns the two sets of feature vectors; see [13.46] for details. After aligning the axes, a matrix Z characterizing the match between image 1 and image 2 is defined by

$$Z_{ij} = (F_{i1} - F_{j2})^{\mathrm{T}}(F_{i1} - F_{j2}). \tag{13.12}$$

The best matches are indicated by the elements of Z which are the smallest in their row and column. We will revisit this example in the next section.

Sclaroff and Pentland [13.44] present a further alternative to the springs and templates formulation: First, compute a description of the entire shape which is robust to sampling and parameterization error. Then, using this description of the entire shape, find a coordinate system which effectively describes the shape. Doing this on the image and the model makes it straightforward to determine cardinal directions.

Wu [10.19] takes the problem of computing optic flow and uses relaxation labeling to find consistent template matches.

The concept of deformable templates can be combined with graph representations to produce an approach [13.1] to matching of objects which are similar, but not identical in shape (e.g. x-rays of hands). The idea of deformable templates can be viewed as an extension of MAP methods. See [13.26] for a well-written concise description. Methods like this also find applications in target tracking and automatic target recognition (ATR) [13.12].

13A.2 Neural networks for object recognition

We have already discussed the fact that pattern recognition techniques provide for us a means of making "what is it?" decisions, when we have been presented with a set of measurements (features) which describe the item being observed. There are many ways to develop classifiers, and methods which follow the neural networks paradigm have been among the most successful. Neural networks accept features as inputs and produce decisions as outputs. They are based on mathematical abstractions of what we know about how individual neurons compute.

There are two types of neural networks which can perform matching, feedforward and recurrent.

13A.2.1 Feedforward neural networks

In a feedforward neural network, each computational element (which we will henceforth refer to as a "neuron") has a large number of inputs, and a single output. Although many variations have been explored in the literature, the most common version of the computation performed by a single neuron is $y = S(\sum_i w_i x_i)$, where the function S is a sigmoid, the x_i are the inputs and the w_i are weights which modify the significance of the various inputs to the neuron. Fig. 13.7 illustrates the architecture and naming conventions for a single neuron.

Almost arbitrary functions can be computed by using layers of neurons, as illustrated in Fig. 13.8. The principal problem in neural network design is determining the weights in order to make the input produce an appropriate output. In the feedforward case, information about the model(s) is embedded in the weights, and when presented with image information, perhaps represented by shape features, the network produces a "yes" answer if the feature vector matches the model. The principal method for determining weights involves finding the weights which solve a gradient descent problem, perhaps minimizing the mean squared difference between what the output is and what it should be. Such an algorithm implements

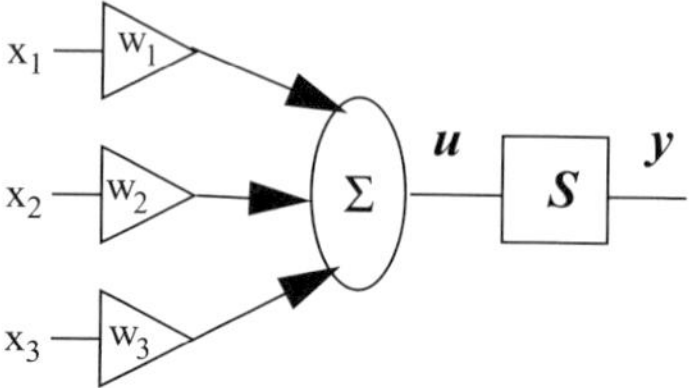

Fig. 13.7. Computation performed by a single neuron. Each input (x_i) is multiplied by a weight (w_i) and the results are added, producing a signal u, which is passed through a sigmoid-like nonlinearity function (S) producing the neuron output y.

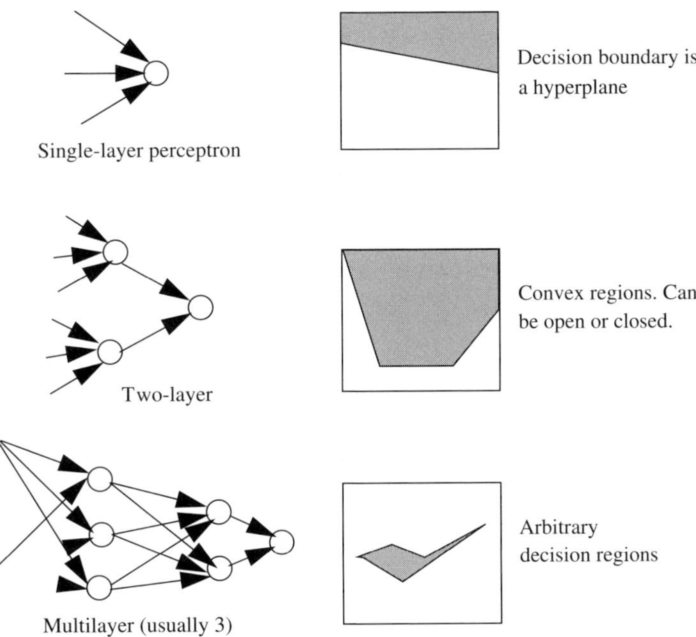

Fig. 13.8. Types of feedforward neural networks, and the decision regions which they can implement.

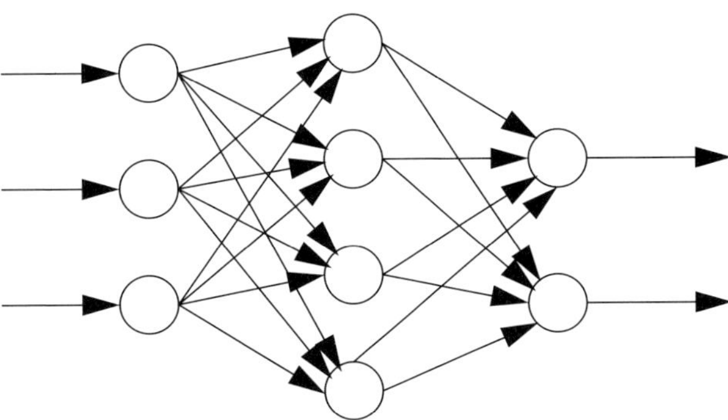

Fig. 13.9. A feedforward neural network with three inputs and two outputs. Each circle denotes a neuron. Weights are not explicitly shown, but exist on the connections.

the familiar gradient descent rule

$$w_{ij}(t + \Delta t) = w_{ij}(t) - c_k \frac{\partial}{\partial w_{ij}} MSE.$$

Using the three-level neural network model illustrated in Fig. 13.9, the gradient descent rule may be readily implemented by making use of the chain rule for derivatives. Hussain and Kabuka [13.24] demonstrate use of a neural network for character recognition.

13A.2.2 Recurrent neural networks

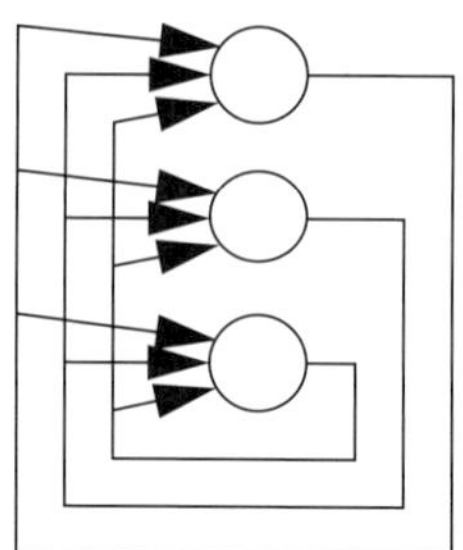

Fig. 13.10. A recurrent neural network with 3 neurons. The weights are not shown, but each input to each neuron has an associated weight.

A recurrent neural network (NN) is one which feeds the output back to the input at run time, as illustrated in Fig. 13.10. Following the same notation used earlier, in the steady state, the output of neuron i satisfies

$$v_i = S(y_i) = S\left(\sum_{j=1}^{d} w_{ij} v_j - I_i\right). \tag{13.13}$$

This model of the behavior of a neuron is true only in the steady state. That is, since the output is dependent on the input, which is the output, which is dependent . . . (*to iterate is human, to recurse, divine*[2]). But such a description is woefully inadequate when things are changing. In that case, we need some model of the dynamics of the system. Many different models can be used, and the reader is referred to the literature [13.15, 13.20, 13.23] for a closer examination. Here, we consider a single, rather simple model, one in which the rate of change of output from the summer is dependent on the input, and can be represented by a first-order differential equation:

$$\frac{d}{dt} y_i(t) = -\beta y_i(t) + \sum_{j=1}^{n} w_{ij} S(u_j) - I_i \tag{13.14}$$

where the y_is are the neuron outputs, the w_is are the weights, as before, and the I_is are inputs to each neuron from the external world (not shown in the figure). Thus the change is proportional to the current state, the inputs from all the other neurons, and the external input.

In operation, a recurrent NN is presented with a particular input, and then allowed to run. They should converge to a particular state.

This model was described by Hopfield [13.23], among others. In Hopfield's model, the rate constant, β, resulted from a lumped-constant model of capacitance and resistance in an operational-amplifier implementation of such a recurrent network.

Now, forget about Eq. (13.14) for a moment, and consider the objective function below, which we wish to minimize:

$$E = -\frac{1}{2}\sum_i \sum_j w_{ij} v_i v_j + \sum_i \beta \int_0^{v_i} S^{-1}(\upsilon)\, d\upsilon + \sum_i I_i v_i. \tag{13.15}$$

If we are to find the vs which minimize this, we need to differentiate E with respect to those vs. Doing that, we find

$$\frac{\partial E}{\partial v_i} = -\sum_j w_{ij} v_j + \beta S^{-1}(v_i) + I_i. \tag{13.16}$$

Now, we observe that the derivative of E with respect to the variables v has the same form as the dynamics of a Hopfield neural network, or

$$\frac{\partial E}{\partial v_i} = -\frac{du_i}{dt}. \tag{13.17}$$

Think about the steady state of the system described by Eq. (13.17). When the network has finished changing (all the derivatives with respect to time are zero), all the partials of the

[2] L. P. Deutsch.

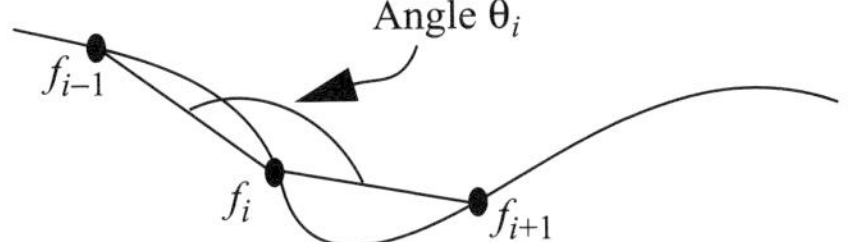

Fig. 13.11. The angle between neighboring feature points is a measurement local to feature point *i*.

energy function are also zero. So we are at an extreme. It is relatively easy to show that one may ignore that annoying integral in Eq. (13.15), and therefore *a Hopfield neural network finds the set of variables v_i which minimize an objective function of the form described by Eq. (13.15)* (without the integral). We illustrate the use of such a network for matching through an example.

Using the same set of features as the previous section, the zero crossings of the boundary curvature, we assign a local measure to each feature point, in this case, the angle between the vectors to neighboring points, as illustrated in Fig. 13.11. We will use this and a more global feature, the distance between feature points, to solve the correspondence problem.

Assume image 1 (which you can think of as a model, if you wish) has n feature points, and image 2 has m feature points. We define a matrix of neurons which has n columns and m rows. The neuron at row i, column j should have a value between zero and one depending on the degree to which feature point i in the first image matches feature point j in the second image.

The matching process is posed as minimizing the expression

$$E = -\frac{A}{2}\sum_i\sum_j\sum_k\sum_l C_{ijkl}V_{ik}V_{jl} + \frac{q}{2}\left(\sum_i\sum_k\sum_{k\neq l} V_{ik}V_{il} + \sum_k\sum_l\sum_{i\neq j} V_{ik}V_{jk}\right). \tag{13.18}$$

The first term quantifies the compatibility of matches ik and jl. The last two terms are included to encourage uniqueness of matches. This form is chosen to allow for occlusions. The compatibility coefficient is the sum of three terms

$$C_{ijkl} = \omega_1\Gamma(\theta_i, \theta_k) + \omega_2\Gamma(\theta_j, \theta_l) + \omega_3\Gamma(r_{ij}, r_{kl}) \tag{13.19}$$

where

$$\Gamma(a, b) = \begin{cases} 1 & \text{if } (|a-b| < T) \\ -1 & \text{otherwise} \end{cases},$$

for threshold T; θ_i is a measurement local to feature point i as illustrated in Fig. 13.11; and r_{ij} is a measure of similarity of relational measures between feature points. For example, if the distance between points i and j is the same as the distance between points k and l, then labelings ik and jl are consistent.

Proper manipulation of Eq. (13.18) allows it to be put in the form of Eq. (13.15), enabling minimization by a neural network. More details are available in [13.28, 13.54]. In Fig. 13.12 an outline is shown of a pistol partially occluded by a hammer. A neural network using these principles is able to identify both objects from this image.

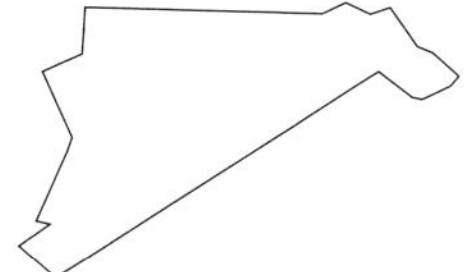

Fig. 13.12. Silhouette of a gun partially occluded by a hammer (redrawn from [13.28]).

13A.3 Image indexing

Up to this point, we have considered the process of image matching as searching a data base of models for the model which best matches the observation. We have not addressed the process of "search" itself. One could, of course, simply try all models, but that could be prohibitively time-consuming, particularly in instances which involve large data bases of models. In applications like automatic target recognition, where matching requires both high speed and large data bases [13.45], better methods are required. The alternate paradigm, *indexing* (sometimes called *image hashing*) is analyzed in [9.6]. In an indexing scheme, a set of parameters are extracted from the image. Obviously, such parameters need to be invariant to as many image transformations as possible and also need to be robust [13.1]. This resulting parameter vector is then used as indices into a lookup table containing references to models. The lookup process returns a list of candidate models consistent with this particular parameter vector. To see how this works, consider the following algorithm.

Begin by looking at local areas around the boundary and attempting to match each local area with a data base of feature descriptors such as lines, circular arcs, and minima and maxima of curvature. Assuming a successful segmentation of an unoccluded object, we start with an edge image, where the edges are not required to be connected. About some point $[x_0, y_0]$ on the edge, we sample the edge in that neighborhood using a sampling scheme[3] which is invariant to zoom. Form all possible combinations of that point with two other nearby points and generate an invariant parameter vector similar to that described in [9.37]. That parameter vector is then used to index a data base of local shapes. For each entry selected, a feature instance is extracted and after all the triples have been considered, the feature instance with the highest number of votes is selected.

Now, the boundary is represented by a sequence of feature instances, and the indexing method may be repeated, using a look-up table of object models which are indexed by geometry and occurrence of feature instances.

Numerous other approaches to indexing exist [13.5, 13.32]; an excellent review is included in [13.48]. The space requirements for some indexing schemes are analyzed in [13.25].

As data bases get larger, one must consider the image indexing problem in the context of the entire digital library. The reader is directed to an entire special issue of *IEEE Transactions on Pattern Analysis and Machine Intelligence* (August, 1996) which addresses this.

13A.4 Matching geometric invariants

We start with simply finding a set of numbers which are invariant. The approach will be to find five points in the 3D model and calculate from them some properties which uniquely characterize them in an invariant way. Then, we will find five points in the image and determine which model they best match.

Choose a set of five feature points $\{X_1, X_2, X_3, X_4, X_5\}$ from the 3D model, at least four of which are noncoplanar. Since five points cannot be linearly independent, we can write one of them as a linear combination of the others. We choose to represent point X_5 in this way,

[3] To avoid cluttering this description of the indexing paradigm with lots of details, we ask the reader to tolerate the omission of some details. They are in the cited paper.

using homogeneous coordinates (section 9.1)

$$X_5 = aX_1 + bX_2 + cX_3 + dX_4. \tag{13.20}$$

We make use of the observation that the determinant of a matrix of points is invariant to rigid body motions,[4] and write the determinant which is constructed from any four of the five points, using as a subscript, the index of the point we omitted. For example,

$$M_1 = |X_2\ X_3\ X_4\ X_5\ |. \tag{13.21}$$

From the linear dependence of X_5 in Eq. (13.20), we substitute for X_5 in each case, deriving

$$\begin{aligned} M_1 &= a|X_2\ X_3\ X_4\ X_1| + b|X_2\ X_3\ X_4\ X_2| \\ &\quad + c|X_2\ X_3\ X_4\ X_3| + d|X_2\ X_3\ X_4\ X_4|. \end{aligned} \tag{13.22}$$

This can be simplified by observing that the determinant of a matrix which has two identical columns is zero:

$$M_1 = a|X_2\ X_3\ X_4\ X_1|. \tag{13.23}$$

But this can be simplified even more by observing that if you interchange two columns, you flip the sign of the determinant.

$$M_1 = (-a)|X_1\ X_3\ X_4\ X_2| = a|X_1\ X_3\ X_2\ X_4| = (-a)|X_1\ X_2\ X_3\ X_4|. \tag{13.24}$$

So

$$M_1 = -aM_5. \tag{13.25}$$

Similarly

$$\begin{aligned} M_2 &= bM_5 \\ M_3 &= -cM_5 \\ M_4 &= dM_5. \end{aligned} \tag{13.26}$$

From this, we can write an expression for the coefficients:

$$a = -\frac{M_1}{M_5} \qquad b = \frac{M_2}{M_5} \qquad c = -\frac{M_3}{M_5} \qquad d = \frac{M_4}{M_5}. \tag{13.27}$$

In 2D, the same five points project to a set of 3-vectors (again, using homogeneous coordinates), and

$$x_5 = ax_1 + bx_2 + cx_3 + dx_4. \tag{13.28}$$

We construct 3×3 matrices by leaving out two indices, and denoting by subscript the indices left out:

$$m_{12} = |x_3\ x_4\ x_5|. \tag{13.29}$$

4 In fact, absolute invariants of linear forms are always ratios of powers of determinants [13.19].

At this point, we simplify the notation, get rid of the xs and just keep track of the subscripts, rewriting the definition of m_{12}:

$$m_{12} = |3\ 4\ 5|. \tag{13.30}$$

As above, we can do algebra to relate the determinants and the coefficients, for example,

$$\begin{aligned} m_{12} &= a|3\ 4\ 1| + b|\ 3\ 4\ 2| \\ &= a|1\ 3\ 4| + b|2\ 3\ 4| \\ &= am_{25} + bm_{15} \end{aligned} \tag{13.31}$$

and

$$\begin{aligned} m_{13} &= am_{35} - cm_{15} \\ m_{14} &= am_{45} + dm_{15}. \end{aligned} \tag{13.32}$$

We have determined forms for the coefficients in terms of the M_is, and adding those relations into the equations we just derived produces

$$\begin{aligned} M_5 m_{12} + M_1 m_{25} - M_2 m_{15} &= 0 \\ M_5 m_{13} + M_1 m_{35} - M_3 m_{15} &= 0 \\ M_5 m_{14} + M_1 m_{45} - M_4 m_{15} &= 0. \end{aligned} \tag{13.33}$$

These relations are invariant to both 3D and 2D motions except for a multiplicative scale which affects all the M_is the same. We can eliminate this dependence by using ratios and define 3D invariants

$$I_1 = \frac{M_1}{M_5} \qquad I_2 = \frac{M_2}{M_5} \qquad I_3 = \frac{M_3}{M_5} \tag{13.34}$$

and 2D invariants

$$i_{12} = \frac{m_{12}}{m_{15}} \qquad i_{13} = \frac{m_{13}}{m_{15}} \qquad i_{25} = \frac{m_{25}}{m_{15}} \qquad i_{35} = \frac{m_{35}}{m_{15}}. \tag{13.35}$$

The denominators are not zero, since they are the determinants of matrices which we know are nonsingular. Look at Eq. (13.33) and divide the top line by M_5:

$$\frac{M_5}{M_5} m_{12} + \frac{M_1}{M_5} m_{25} - \frac{M_2}{M_5} m_{15} = 0, \tag{13.36}$$

which simplifies to

$$m_{12} + I_1 m_{25} - I_2 m_{15} = 0. \tag{13.37}$$

Similarly divide by m_{15} to produce two independent equations:

$$\begin{aligned} i_{12} + I_1 i_{25} - I_2 &= 0 \\ i_{13} + I_1 i_{35} - I_3 &= 0. \end{aligned} \tag{13.38}$$

So if we have 2D invariants we have two equations for the 3D invariants. The two equations of Eq. (13.38) do not, unfortunately, determine the three 3D invariants. Still those two equations determine a space line in the 3-space of the Is.

How do we use an idea like this? Given a 3D model of an object, and any five points, four of which are not coplanar, we can find I_1, I_2, and I_3, a point in a 3D space. To perform recognition, we first extract from the 2D image (generally several) 5-tuples of feature points and from them construct the 2D invariants. Each 5-tuple gives rise to two equations in I_1, I_2, I_3 space, that is, a straight line in the 3D invariant space. If a 5-tuple in the 2D image is a projection of some 5-tuple in 3D, then the line so obtained will pass through the single point representing the model. If we have a different projection of those five points, we get a different straight line, but it still passes through the model point.

Implementing this for realistic scenes is slightly more complicated than this description because one must actually make use of projective geometry rather than assuming orthogonal projections. Other complications arise in determining a suitable way to choose 5-tuples, and a means for dealing with the fact that the line may "almost" pass through the point. Weiss and Ray [13.52] address these issues.

13A.5 Conclusion

13A.5.1 Which model to use?

So far, we have described quite a collection of representations for objects, but certainly not all that are in the literature. Other methods include variations on deformable models [13.10, 13.11], especially for range images [13.21].

Fig. 13.13. A set of points produced by an edge detector which might come from a circle or a polygon.

Consider Fig. 13.13: Should you match it to a circle or a six-sided polygon? Clearly, there is no simple answer to this question. If you have prior, problem-specific knowledge that you are always dealing with circular objects, you might choose to use the circular model, which is certainly less complex than that of a polygon. The idea of minimum description length (MDL) provides some help along these lines. The MDL paradigm states that the optimal representation for a given image may be determined by minimizing the combined length of the encoding of the representation and the residual error. Interestingly, a MAP representation can be shown [13.9, 13.30] to be equivalent to the MDL representation where the prior truly represents the signal.

Schweitzer [13.43] uses the MDL philosophy to develop algorithms for computing the optic flow, and Lanterman [13.29] uses it to characterize infrared scenes in ATR applications – "if there are several descriptions compatible with the observed data, we select the most parsimonious" [13.29].

Rissanen [13.41] suggests that the quality of an object/model match could be represented by

$$L(x, \theta) = -\log_2 P(x|\theta) + L(\theta) \qquad (13.39)$$

where x is the observed object, θ is the model, represented as a vector of parameters, $P(x|\theta)$ is the conditional probability of making this particular measurement given the model, and $L(\theta)$ denotes the number of bits required to represent the model. The logarithm of the conditional probability is then a measure of how well the data fits the model. We thus may trade off a more precise fit of a more complex model with a less accurate fit of a simpler model [13.6].

Ultimately, machine vision is not going to be solved by one program, one algorithm, or one set of mathematical concepts. Ultimately, its solution will depend on the ability to build systems which integrate a collection of specialists. The jury is still out on how to accomplish this. Regrettably, only a few papers have undertaken this formidable task. For example, Grosso and Tistarelli [13.18] combine stereopsis and motion. Bilbro and Snyder [13.4] fuse luminance and range to improve the quality of the range imagery, and Pankanti and Jain [13.38] fuse stereo, shading, and relaxation labeling. Zhu and Yuille [8.80] incorporate the MDL approach, including active contours and region growing, into a unified look at segmentation. Gong and Kulikowski [13.16] use a planning strategy, primarily in the medical application area.

13A.5.2 Consistency and optimization in matching

In sections 13A.1 and 13A.2, the first step is to identify feature points which are relatively distinctive. Then the algorithm makes use of relationships between these points, relying on consistency to find the best match.

A recurrent neural network is an optimization engine!

In the discussion of recurrent neural networks, we showed that such a network achieves a stable state which is in fact the minimum of the objective function of Eq. (13.15).

13A.6 Vocabulary

You should know the meanings of the following terms.

```
Eigenvector
Feedforward neural net
Geometric invariant
Image indexing
Recurrent neural net
```

Bibliography

[13.1] Y. Amit and A. Kong, "Graphical Templates for Model Registration," *IEEE Transactions on Pattern Analysis and Machine Intelligence*, **18**(3), 1996.

[13.2] K. Astrom, "Fundamental Limitations on Projective Invariants of Planar Curves," *IEEE Transactions on Pattern Analysis and Machine Intelligence*, **17**(1), 1995.

[13.3] B. Bhanu and O. Faugeras, "Shape Matching of Two Dimensional Objects," *IEEE Transactions on Pattern Analysis and Machine Intelligence*, **6**(2), 1984.

[13.4] G. Bilbro and W. Snyder, "Fusion of Range and Luminance Data," *IEEE Symposium on Intelligent Control*, Arlington, August, 1988.

[13.5] A. Bimbo and P. Pala, "Visual Image Retrieval by Elastic Matching of User Sketches," *IEEE Transactions on Pattern Analysis and Machine Intelligence*, **19**(2), 1997.

[13.6] J. Canning, "A Minimum Description Length Model for Recognizing Objects with Variable Appearances (The VAPOR Model)," *IEEE Transactions on Pattern Analysis and Machine Intelligence*, **16**(10), 1994.

[13.7] Q. Chen, M. Defrise, and F. Deconinck, "Symmetric Phase-only Matched Filtering of Fourier–Mellin Transforms for Image Reconstruction and Recognition," *IEEE Transactions on Pattern Analysis and Machine Intelligence*, **16**(12), 1994.

[13.8] T. Chen and W. Lin, "A Neural Network Approach to CSG-based 3-D Object Recognition," *IEEE Transactions on Pattern Analysis and Machine Intelligence*, **16**(7), 1994.

[13.9] T. Darrell and A. Pentland, "Cooperative Robust Estimation Using Layers of Support," *IEEE Transactions on Pattern Analysis and Machine Intelligence*, **17**(5), 1995.

[13.10] D. DeCarlo and D. Metaxas, "Blended Deformable Models," *IEEE Transactions on Pattern Analysis and Machine Intelligence*, **18**(4), 1996.

[13.11] S. Dickinson, D. Metaxas, and A. Pentland, "The Role of Model-based Segmentation in the Recovery of Volumetric Parts From Range Data," *IEEE Transactions on Pattern Analysis and Machine Intelligence*, **19**(3), 1997.

[13.12] M. Dubuisson Jolly, S. Lakshmanan, and A. Jain, "Vehicle Segmentation and Classification using Deformable Templates," *IEEE Transactions on Pattern Analysis and Machine Intelligence*, **18**(3), 1996.

[13.13] M. Fischler and R. Elschlager, "The Representation and Matching of Pictorial Structures," *IEEE Transactions on Computers*, **22**(1), 1973.

[13.14] S. Gold and A. Rangarajan, "A Graduated Assignment Algorithm for Graph Matching," *IEEE Transactions on Pattern Analysis and Machine Intelligence*, **18**(4), 1996.

[13.15] R. Golden, *Mathematical Methods for Neural Network Analysis and Design*, Cambridge, MA, MIT Press, 1996.

[13.16] L. Gong and C. Kulikowski, "Composition of Image Analysis Processes Through Object-centered Hierarchical Planning," *IEEE Transactions on Pattern Analysis and Machine Intelligence*, **17**(10), 1995.

[13.17] F. Goudail, E. Lange, T. Iwamoto, K. Kyuma, and N. Otsu, "Face Recognition System Using Local Autocorrelation and Multiscale Integration," *IEEE Transactions on Pattern Analysis and Machine Intelligence*, **18**(10), 1996.

[13.18] E. Grosso and M. Tistarelli, "Active/Dynamic Stereo Vision," *IEEE Transactions on Pattern Analysis and Machine Intelligence*, **17**(9), 1995.

[13.19] G. Gurevich, *Foundations of the Theory of Algebraic Invariants*, Transl. Raddock and Spencer, Groningen, The Netherlands, Nordcliff Ltd, 1964.

[13.20] S. Haykin, *Neural Networks, A Comprehensive Foundation*, Englewood Cliff, NJ, Prentice-Hall, 1999.

[13.21] M. Hebert, K. Ikeuchi, and H. Delingette, "Spherical Representation for Recognition of Free-form Surfaces," *IEEE Transactions on Pattern Analysis and Machine Intelligence*, **17**(7), 1995.

[13.22] D. Heisterkamp and P. Bhattachaya, "Matching of 3D Polygonal Arcs," *IEEE Transactions on Pattern Analysis and Machine Intelligence*, **19**(1), 1997.

[13.23] J. Hopfield, "Neural Networks and Physical System with Emergent Collective Computational Abilities," *Proceedings of the National Academy of Science*, 79, pp. 2554–2558, 1982.

[13.24] B. Hussain and M. Kabuka, "A Novel Feature Recognition Neural Network and its Application to Character Recognition," *IEEE Transactions on Pattern Analysis and Machine Intelligence*, **16**(1), 1994.

[13.25] D. Jacobs, "The Space Requirements of Indexing Under Perspective Projections," *IEEE Transactions on Pattern Analysis and Machine Intelligence*, **18**(3), 1996.

[13.26] A. Jain, Y. Zhong, and S. Lakshmanan, "Object Matching Using Deformable Templates," *IEEE Transactions on Pattern Analysis and Machine Intelligence*, **18**(3), 1996.

[13.27] T. Kanade and M. Okutomi, "A Stereo Matching Algorithm with an Adaptive Window: Theory and Experiment," *IEEE Transactions on Pattern Analysis and Machine Intelligence*, **16**(9), 1994.

[13.28] J. Kim, S. Yoon, and K. Sohn, "A Robust Boundary-based Object Recognition in Occlusion Environment by Hybrid Hopfield Neural Networks," *Pattern Recognition*, **29**(12), 1996.

[13.29] A. Lanterman, "Minimum Description Length Understanding of Infrared Scenes," *SPIE Automatic Target Recognition VIII*, **3371**, April 1998.

[13.30] Y. Leclerc, "Constructing Simple Stable Descriptions for Image Partitioning," *International Journal of Computer Vision*, **3**, pp. 73–102, 1989.

[13.31] M. Lew, T. Huang, and K. Wong, "Learning and Feature Selection in Stereo Matching," *IEEE Transactions on Pattern Analysis and Machine Intelligence*, **16**(9), 1994.

[13.32] S. Li, K. Chan, and C. Wang, "Performance Evaluation of the Nearest Feature Line Method in Image Classification and Retrieval," *IEEE Transactions on Pattern Analysis and Machine Intelligence*, **22**(11), 2000.

[13.33] T. Mitchell, *Machine Learning*, New York, McGraw-Hill, 1997.

[13.34] H. Murakami and B. Kumar, "Efficient Calculation of Primary Images from a Set of Images," *IEEE Transactions on Pattern Analysis and Machine Intelligence*, **4**(5), 1982.

[13.35] H. Murase and M. Lindenbaum, "Partial Eigenvalue Decomposition of Large Images Using the Spatial Temporal Adaptive Method," *IEEE Transactions on Image Processing*, **4**(5), 1995.

[13.36] C. Olson, "Maximum Likelihood Template Matching," Completing this citation is a homework assignment (Assignment 13.3).

[13.37] D. Paglieroni, G. Ford, and E. Tsujimoto, "The Position-orientation Masking Approach to Parametric Search for Template Matching," *IEEE Transactions on Pattern Analysis and Machine Intelligence*, **16**(7), 1994.

[13.38] S. Pankanti and A. Jain, "Integrating Vision Modules: Stereo, Shading, Grouping, and Line Labeling," *IEEE Transactions on Pattern Analysis and Machine Intelligence*, **17**(9), 1995.

[13.39] B. Parsi, A. Margalit, and A. Rosenfeld, "Matching General Polygonal Arcs," *Computer Vision, Graphics, and Image Processing. Image Understanding*, **53**(2), pp. 227–234, March, 1991.

[13.40] K. Rao and J. Ben-Arie, "Optimal Edge Detection Using Expansion Matching and Restoration," *IEEE Transactions on Pattern Analysis and Machine Intelligence*, **16**(12), 1994.

[13.41] J. Rissanen, "A Universal Prior for Integers and Estimation by Minimum Description Length," *Annals of Statistics*, **11**(2), pp. 416–431, 1983.

[13.42] J. Schwartz and M. Sharir, "Identification of Partially Obscured Objects in Two and Three Dimensions by Matching Noisy Characteristic Curves," *International Journal of Robotics Research*, **6**(2), 1987.

[13.43] H. Schweitzer, "Occam Algorithms for Computing Visual Motion," *IEEE Transactions on Pattern Analysis and Machine Intelligence*, **17**(11), 1995.

[13.44] S. Sclaroff and A. Pentland, "Model Matching for Correspondence and Recognition," *IEEE Transactions on Pattern Analysis and Machine Intelligence*, **17**(6), 1995.

[13.45] K. Sengupta and K. Boyer, "Organizing Large Structural Modelbases," *IEEE Transactions on Pattern Analysis and Machine Intelligence*, **17**(4), 1995.

[13.46] L. Shapiro and J. M. Brady, "Feature-based Correspondence: an Eigenvector Approach," *Image and Vision Computing*, **10**(5), 1992.

[13.47] X. Shen and P. Palmer, "Uncertainty Propagation and Matching of Junctions as Feature Groupings," *IEEE Transactions on Pattern Analysis and Machine Intelligence*, **22**(12), 2000.

[13.48] A. Smeulders, M. Worring, S. Santini, G. Gupta, and R. Jain, "Content-based Image Retrieval at the End of the Early Years," *IEEE Transactions on Pattern Analysis and Machine Intelligence*, **22**(12), 2000.

[13.49] D. Swets and J. Weng, "Using Discriminant Eigenfeatures for Image Retrieval," *IEEE Transactions on Pattern Analysis and Machine Intelligence*, **18**(8), 1996.

[13.50] M. Turk, and A. Pentland, "Eigenfaces for Recognition," *Journal of Cognitive Neuroscience*, **3**(1), pp. 71–86, 1991.

[13.51] X. Wang and H. Qi, "Face Recognition Using Optimal Non-orthogonal Wavelet Basis Evaluated by Information Complexity," *International Conference on Pattern Recognition*, vol. 1, pp. 164–167, Quebec, Canada, August, 2002.

[13.52] I. Weiss and M. Ray, "Model-based Recognition of 3D Objects from Single Images," *IEEE Transactions on Pattern Analysis and Machine Intelligence*, **23**(2), 2001.

[13.53] M. Yang and J. Lee, "Object Identification from Multiple Images Based on Point Matching under a General Transformation," *IEEE Transactions on Pattern Analysis and Machine Intelligence*, **16**(7), 1994.

[13.54] S. Yoon, *A New Multiresolution Approximation Approach to Object Recognition*, Ph.D. Thesis, North Carolina State University, 1995.

14 Statistical pattern recognition

Statistics are used much like a drunk uses a lamppost: for support, not illumination

Vin Scully

The discipline of statistical pattern recognition by itself can fill textbooks (and in fact, it does). For that reason, no effort is made to cover the topic in detail in this single chapter. However, the student in machine vision needs to know at least something about statistical pattern recognition in order to read the literature and to properly put the other machine vision topics in context. For that reason, a brief overview of the field of statistical methods is included here. To do serious research in machine vision, however, this chapter is not sufficient, and the student must take a full course in statistical pattern recognition. For texts, we recommend several: The original version of the text by Duda and Hart [14.3] included both statistical pattern classification and machine vision, however, the new version [14.4] is pretty much limited to classification, and we recommend it for completeness. The much older text by Fukanaga [14.6] still retains a lot of useful information, and we recommend [14.11] for readability.

14.1 Design of a classifier

Recall the example described in section 13.2. In that example, we are given models for axes and hatchets which were derived statistically by computing averages of samples known to be either axes or hatchets. We called these collections "training sets." In section 13.2, we compared an unknown, represented by a feature vector, with both models and decided to assign the unknown to the class of the model it most closely resembled, where "closely resembled" was simple Euclidian distance between the unknown feature vector and the two models. In this chapter, we demonstrate that this "closest mean" decision rule is actually a simplification of maximum likelihood assuming a Gaussian probability density for the classes. Further, we show that decision rules other than closest mean may be used, and may perform better and/or be more effectively computed.

We begin by discussing some of the options in decision rules.

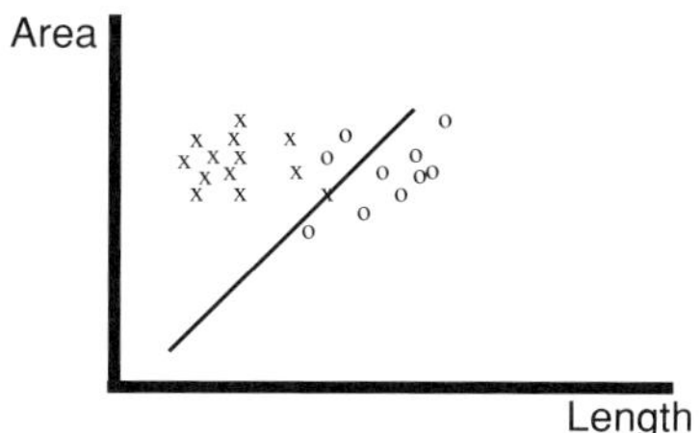

Fig. 14.1. A linear decision boundary.

14.1.1 Choice of a decision rule

Remember from Figure 1.2 that the image is measured and features are the output of the measurement process. The features are a set of numbers that characterize some property of the image. We use those features in a classifier.

Designing a classifier essentially consists of developing a method to implement a decision rule. Several options exist for determining both the *form* of the decision rule and the *parameters* which characterize that rule. We describe those options briefly here.

Linear decision rules

Figure 14.1 illustrates the result of a large number of measurements of two different industrial parts. The features, area and length have both been measured and each measurement indicated in the figure by a single mark. (A chart like this is called a "scatter graph.") The x points represent one class, flanges and the o points represent a second class, gaskets. A linear *decision boundary* has been drawn on the figure. A linear decision rule would be as follows: "Decide the unknown object is a flange if the result of the measurements lies on the left of the decision boundary otherwise decide it is a gasket." Linear decision rules are particularly attractive because they can be implemented by *linear machines* which have a great deal of potential parallelism and therefore high speed. As can be seen from the figure these two classes are not *linearly separable*. That is, there does not exist any one straight line which completely partitions the two classes. The choice of the best such straight line is the result of the linear classifier design process. A variation on linear machines is given in section 14A.2, where we introduce support vector machines.

14.1.2 Maximum likelihood classifier

We have already seen the term *maximum likelihood* in section 6.2, in the context of image restoration. Here, we describe the use of this particular set of mathematics in classifier design.

Conditional probabilities play a critical role in maximum likelihood functions.

In the maximum likelihood approach to classifier design, statistical representations are used to describe the probability that an object having a certain set of measurements belongs to a particular class; that is, we estimate $P(w_i|\mathbf{x})$, the conditional probability that the object being measured belongs to class w_i given a particular measurement (vector) $\mathbf{x}$. We compute such a probability for each w_i and the decision rule is then "assign the object to the class i, if $P(w_i|\mathbf{x}) > P(w_j|\mathbf{x})$ for all j."

It should be noted that the term "maximum likelihood" does not necessarily mean best performance. Maximum likelihood algorithms are typically based on assumptions concerning the form of probability density functions: Assumptions which may be invalid for many of the data points within a class. Designing a maximum likelihood classifier entails both the process of choosing the form for the probability density functions for each class (e.g., Gaussian density functions) and the process of choosing the parameters which describe those density functions (e.g., mean and variance).

14.1.3 Determining descriptions for classifiers

A training set is a set of samples which are drawn from each class and which may be used to statistically characterize that class. The method which we will use to determine the parameters of our pattern classifier will depend upon whether we have training sets available.

Supervised learning

If we are given one training set for each class and from those training sets we can develop the statistical representations of the classes, then this process is known as "supervised learning." The word *supervised* refers to the fact that each data point is independently labeled according to the class to which it belongs. Each class may then be characterized either statistically by its mean, variance or other statistical measures, or by some other parametric representation. Fig. 14.1 illustrates the data distribution resulting from supervised sampling, i.e., the x points are identified as belonging to one class and the o points to another. The example we have used previously in section 13.2, of distinguishing axes from hatchets, is a supervised learning problem, since it was assumed that we had training sets of both classes.

Unsupervised learning

Fig. 14.2 represents a possible result of an unsupervised learning process. In that figure the samples are not identified by their class. However, two distinct *clusters* may readily be seen. An example might be a multispectral satellite image of a field of wheat. If the field contains two distinct types of wheat, two clusters could occur

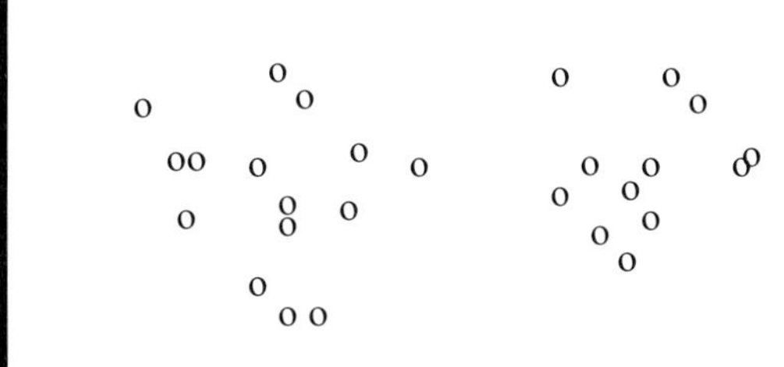

Fig. 14.2. In unsupervised learning, only one set of measurements is taken; however, such data may fall into natural clusters.

in the measurements. Clustering seems, from this example, to be a very simple problem – the two clusters are obvious, just looking at them. However, remember, not all measurement vectors are two dimensional, and humans have a much harder time dealing with, for example, seven-dimensional data. An unsupervised learning or *clustering* algorithm will automatically identify those two (or more) clusters and associate samples with the nearest (in some sense) cluster.

As we proceed through this material, you will learn how to determine the measurements needed in order to implement these classification algorithms. Clustering turns out to be useful in other machine vision algorithms as well, such as locating the peaks in the Hough transform accumulator array. It is covered in more detail in Chapter 15.

14.2 Bayes' rule and the maximum likelihood classifier

In this section we will design a classifier based upon assumptions concerning the statistical nature of the training sets. If these assumptions should be valid for the training sets in question, then the classifier which results will give the best performance. We will also investigate the performance of such a classifier, including error rates.

The attentive student will note similarities between the descriptions of statistical concepts presented here and those presented in Chapter 6. The similarities are correct and deliberate. In that chapter, we sought an image which minimized certain properties. In this chapter we seek a decision: To which class an object belongs.

14.2.1 Bayes' rule

We define $P(w_i)$ to represent the *a priori* probability that class w_i occurs, that is, the probability of class w_i occurring before any measurements are made. For example, suppose we have a factory which manufactures flanges and gaskets, but makes nine

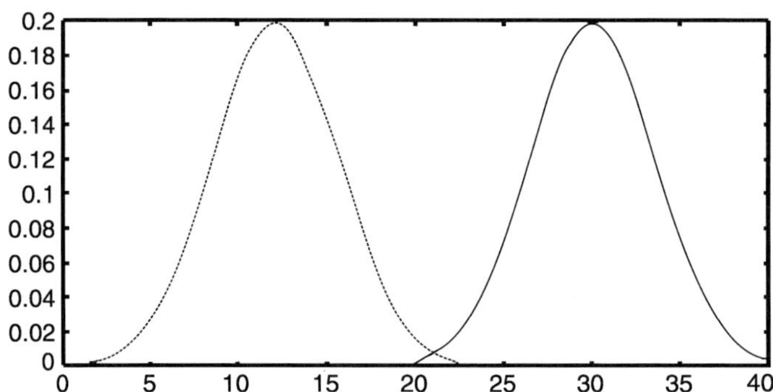

Fig. 14.3. The same measurement on two objects may have different average values, but due to noise in the measurement, or actual variation, these values may overlap.

times as many flanges as gaskets. Flanges and gaskets may come down the conveyor at random times. But because of our *a priori* knowledge that the plant manufactures nine times as many flanges as gaskets we know that we are much more likely to see a flange than a gasket if we chose to look at the conveyor at some random time. Thus the *a priori* possibility of flanges is 0.9 and the *a priori* probability of gaskets is 0.1.

We define $p(\mathbf{x}|w_i)$ to represent the *conditional probability density* of a measurement $\mathbf{x}$ occurring given that the sample is known to come from class w_i. For a particular w_i, $p(\mathbf{x}|w_i)$ should be thought of as a function of $\mathbf{x}$. Suppose we have a factory which manufactures axes and hatchets. Then we might find the probability densities for the lengths of axes and hatchets to be represented by Fig. 14.3. In that figure we see that an axe is most likely to be 30 inches long and a hatchet most likely to be 12 inches, but that some variation in length can occur.

The probability density function may be characterized in several possible ways. One is by simply tabulating the number of times a particular value occurs for each possible value of the variable, in this case, length. Such a tabulation is referred to as a *histogram* of the variables. Properly normalized, a histogram can be a useful representation of a probability density function, but of course requires that only a finite number of possible values may exist. One may also describe a density function in a parametric way using some analytic function (e.g., the Gaussian) to represent the density.

Finally we define $P(w_i|\mathbf{x})$, the *posterior conditional probability*, to represent the conditional probability that the object being observed belongs to a class w_i given a measurement $\mathbf{x}$. $P(w_i|\mathbf{x})$ is what we are looking for. We will use it as our decision rule or, more correctly, as our *discriminant function*. Our decision rule will then be as follows: For a measurement $\mathbf{x}$ made on an unknown object, compute $P(w_i|\mathbf{x})$ for each class, that is, for each possible value of i. Then decide that the unknown belongs to the class i for which $P(w_i|\mathbf{x})$ is greater than $P(w_j|\mathbf{x})$ for all $i \neq j$.

The term *likelihood* is used rather than *probability* because we will actually maximize some function of the probability.

When we make a classification decision based on $P(w_i|\mathbf{x})$, we are using a **maximum likelihood classifier**.

We can relate the three functions just defined by using *Bayes' rule*:

$$P(w_i|\mathbf{x}) = \frac{p(\mathbf{x}|w_i)P(w_i)}{Something} \tag{14.1}$$

$$Something = p(\mathbf{x}) = \sum_{j=1}^{c} p(\mathbf{x}|w_j)P(w_j). \tag{14.2}$$

In Eq. (14.1) we used "something" to represent the denominator for the conditional probability density. We used the word "something" to call attention to the fact that this number represents the probability density of that value of **x** occurring, independent of the class of the observation. Since this number is independent of the class, and is the same for all classes, it therefore does not provide us any help in distinguishing which class is most likely. Instead, it is a normalization constant which we use to ensure that the number $P(w_i|\mathbf{x})$ has the desirable properties of a probability; that is, it lies between 0 and 1 and when summed over all the classes, it sums to 1 (the observed object belongs to at least one of the classes which we are considering).

In a sense, Equation (14.1) solves the pattern recognition problem. It tells us how to make a decision, assuming we know each of the components of the RHS. In the next section we consider how one goes about determining those components.

14.2.2 Parametric pattern classifiers

In the previous section we demonstrated that a maximum likelihood classifier could make a decision based upon knowledge of two things: The *a priori* probability of the class being considered and the *conditional probability densities* for each class. We have said nothing up to this point about the form which the probability density function may take. In Fig. 14.3 the density functions take the familiar "bell curve" shape. This shape could be described by a table of values or by an analytic function. For scalar-valued **x** the analytic function of the "bell curve" has the form

$$p(x) = \frac{1}{\sqrt{2\pi}\sigma} \exp\left[-\frac{1}{2}\left(\frac{x-\mu}{\sigma}\right)^2\right]. \tag{14.3}$$

Here $p(x)$ is known as the **univariate Gaussian (normal) density**. The word "univariate" refers to the fact that x is a scalar, a single variable. Here, the mean, μ and standard deviation, σ comprise the elements of a "parameter vector:"

$$\theta_i = \begin{bmatrix} \mu_i \\ \sigma_i \end{bmatrix}. \tag{14.4}$$

These two numbers serve to completely describe the conditional probability density of the variable x for class i assuming that the density has a Gaussian form.

If, instead of a single measurement, we have taken a set of measurements, then $\mathbf{x}$ becomes a vector quantity and $p(\mathbf{x}|w_i)$ becomes the "multivariate Gaussian density."

$$p(\mathbf{x}|w_i) = \frac{1}{(2\pi)^{d/2}|K_i|^{1/2}} \exp\left[-(1/2)(\mathbf{x}-\boldsymbol{\mu}_i)^{\mathrm{T}} K_i^{-1}(\mathbf{x}-\boldsymbol{\mu}_i)\right]. \tag{14.5}$$

In Eq. (14.5), d is the dimensionality of the vector $\mathbf{x}$, $\boldsymbol{\mu}_i$ is the mean vector which represents the average (vector-valued) value of the random vector of measurements and K_i is a $d \times d$ covariance matrix.

Thus, in the univariate Gaussian case, we can represent the class conditional density of the measurements by two numbers, and in the multivariate case by a d-dimensional vector and a d by d matrix. Given these parameters, we can easily substitute them into Eq. (14.5) and then into Eq. (14.1) and compute the most likely class for an unknown object. Unfortunately, in most applications we are not given the mean and covariance but rather must estimate them from training sets.

Take the log of the RHS of Eq. (14.5). That will eliminate the exponential, and, since the logarithm is monotonic, will result in an expression involving the measurement vector $\mathbf{x}$, and the statistics $\boldsymbol{\mu}_i$ and K_i which characterize class i. This expression is maximized in the same way as the original probability. Classification is now straightforward: Just substitute $\mathbf{x}$ into each equation you can generate like this using all the different means and covariances. This gives you c (assuming you have c classes) different functions, called discriminant functions, all of which have the same form but which have different parameters. Assign $\mathbf{x}$ to the class for which the corresponding discriminant function is the largest.

14.2.3 Density estimation

Since the Gaussian (normal) density occurs so often in actual distributions of random variables, and is so convenient to work with, we will use it often, and treat it as a special case.

To design a pattern classifier using supervised methods, we need to estimate the parameters of the density from a training set of samples. We will denote the parameter set by the vector θ.

The univariate Gaussian case

Let X_i represent a set of n_i elements where each element represents a measurement made on an object known to belong to class i. Because we are considering here only the univariate case, the elements of X_i will be scalar-valued. We will consider vector-valued elements and multivariate densities in the next subsection.

We assume that the samples in X_i give us no information about the parameters in any class other than class i. This permits us to work with each class separately.

Assuming that the samples are drawn independently, the probability of drawing the entire set X_i is determined by

$$p(X_i) = \prod_{k=1}^{n_i} p(x_{ik}). \tag{14.6}$$

Since this probability is intrinsically dependent upon the parameters θ_i, we rewrite Eq. (14.6) to make this dependence clear:

$$p(X_i|\theta_i) = \prod_{k=1}^{n_i} p(x_{ik}|\theta_i). \tag{14.7}$$

A second use of the term "maximum likelihood."

The maximum likelihood estimate of θ_i is then defined as that value θ_i which maximizes $p(X_i|\theta_i)$. Eq. (14.7) describes the likelihood of any particular training set occurring, given that the probability distribution is described by the parameter vector θ_i. Since we are dealing with the Gaussian density, we rewrite Eq. (14.7) as

$$p(X_i|\theta_i) = \prod_{k=1}^{n_i} \frac{1}{\sqrt{2\pi}\sigma_i} \exp\left(-\frac{(x_{ik}-\mu_i)^2}{2\sigma_i^2}\right). \tag{14.8}$$

Now an important observation: The value of θ_i which maximizes $p(X_i|\theta_i)$ also maximizes $\ln\left[p(X_i|\theta_i)\right]$. This is true because the natural logarithm is a monotonically increasing function. Thus, we have our choice of finding the parameter vector θ_i which maximizes either the density or its logarithm. The logarithm will be much easier to use. Taking the log of the RHS we find

$$\ln(p(X_i|\theta_i)) = \sum_{k=1}^{n_i} \ln\left(\frac{1}{\sqrt{2\pi}\sigma_i}\right) - \sum_{k=1}^{n_i} \frac{1}{2}\left(\frac{x_{ik}-\mu_i}{\sigma_i}\right)^2. \tag{14.9}$$

14.2.4 Estimating the mean

We denote our estimate of μ by $\hat{\mu}$. To find the value of μ which maximizes Eq. (14.9) we differentiate with respect to μ and set the result to 0.

$$\sum_{k=1}^{n_i} \frac{x_{ik}-\hat{\mu}_i}{\sigma_i} = 0 \tag{14.10}$$

which simplifies to

$$\sum_{k=1}^{n_i} x_{ik} - \sum_{k=1}^{n_i} \hat{\mu}_i = 0 \tag{14.11}$$

$$\hat{\mu}_i = \frac{1}{n_i}\sum_{k=1}^{n_i} x_{ik}. \tag{14.12}$$

$\hat{\mu}_i$ is known as the *sample mean* and the fact that it is equal to the average value is certainly intuitively satisfying.

14.2.5 Estimating the variance

We perform the derivation as before, but consider both mean and variance as unknowns. In the following, we drop the subscript i denoting the class, for notational simplicity. Initially, consider the univariate case, then, the parameter vector θ is a 2-vector:

$$\theta = \begin{bmatrix} \mu \\ \sigma^2 \end{bmatrix}. \tag{14.13}$$

Rewrite Eq. (14.9), to make it a bit simpler and the log of the probability becomes

$$L = -\sum \ln\left(\sqrt{2\pi}\right) - \sum \ln \sigma - \frac{1}{2}\sum\left(\frac{x_k - \mu}{\sigma}\right)^2 \tag{14.14}$$

and we find (from $\partial L/\partial \mu = 0$),

$$\sum_{k=1}^{n} \frac{1}{\sigma^2}(x_k - \hat{\mu}) = 0 \tag{14.15}$$

and from $\partial L/\partial \sigma$,

$$-\sum_{k=1}^{n} \frac{1}{\hat{\sigma}} + \sum_{k=1}^{n} \frac{(x_k - \hat{\mu})^2}{\hat{\sigma}^3} = 0. \tag{14.16}$$

We simplify Eq. (14.15) by multiplying by $\hat{\sigma}^2$ and find

$$\sum_{k=1}^{n} x_k = \hat{\mu} \sum_{k=1}^{n} 1 \tag{14.17}$$

and

$$\hat{\mu} = \frac{1}{n}\sum_{k=1}^{n} x_k \tag{14.18}$$

as before.

Equation (14.16) simplifies similarly to yield

$$\frac{n}{\hat{\sigma}} = \frac{1}{\hat{\sigma}^3}\sum_{k=1}^{n}(x_k - \hat{\mu})^2, \tag{14.19}$$

and therefore

$$\hat{\sigma}^2 = \frac{1}{n}\sum_{k=1}^{n}(x_k - \hat{\mu})^2. \tag{14.20}$$

Thus we see that the best estimate for the parameters of a normal density are the familiar sample mean and sample variance.

The multivariate Gaussian density

We may treat the (vector-valued) parameters of the multivariate normal density in a manner similar to the previous subsection, and find the maximum likelihood estimates of those parameters:

$$\hat{\mu}_i = \frac{1}{n_i} \sum_{k=1}^{n_i} \boldsymbol{x}_{ik} \tag{14.21}$$

Look at how K is defined. Is it a matrix? A scalar? A vector?

$$K_i = \frac{1}{n_i} \sum_{k=1}^{n_i} (\boldsymbol{x}_{ik} - \hat{\mu}_i)(\boldsymbol{x}_{ik} - \hat{\mu}_i)^{\mathrm{T}}. \tag{14.22}$$

Thus we have essentially the same results for the multivariate case as for the univariate case: That the best estimate (in the maximum likelihood sense) of the mean and variance of a Gaussian are the sample mean and sample (co)variance.

Now what is there to remember from this chapter? To perform the maximum likelihood estimate of a set of parameters, given a training set, assume independence (if you can) and write the probability of the entire training set occurring as a product. Take logs, differentiate, and set to zero to produce a set of simultaneous equations which, when solved, will be the best estimates of the parameters. This approach works for any distribution, not just the Gaussian. However, for some cases, the process of solving the system of simultaneous equations may be intractable.

Finally, there are other ways to find parameters, other than maximum likelihood; techniques which space and time do not permit us to cover here.

14.2.6 The likelihood ratio

We wish to make a decision which maximizes the likelihood that we are correct. To accomplish this, we choose the class which maximizes the *a posteriori* probability – the probability of our decision being correct, given the measurement.

That is, choose i to maximize $P(w_i|\mathbf{x})$. To accomplish this, recall Bayes' rule, for class 1:

$$P(w_1|\mathbf{x}) = \frac{p(\mathbf{x}|w_1)P(w_1)}{p(\mathbf{x})}. \tag{14.23}$$

As mentioned above, remember that $p(\mathbf{x})$ is the same regardless of whether $\mathbf{x}$ belongs to class 1 or 2. Since this denominator is unaffected by the classification decision, we can ignore it in making that decision.

In the two-class case, we choose class 1 if $P(w_1|\mathbf{x}) > P(w_2|\mathbf{x})$, or, substituting Bayes' rule, we choose class 1 if

$$p(\mathbf{x}|w_1)P(w_1) > p(\mathbf{x}|w_2)P(w_2), \tag{14.24}$$

that is,

$$\frac{p(\mathbf{x}|w_1)}{p(\mathbf{x}|w_2)} > \frac{P(w_2)}{P(w_1)}. \tag{14.25}$$

The expression on the left is known as the *likelihood ratio*. The relationship in Eq. (14.25) provides a true–false relationship between the likelihood ratio and the prior information. If it is false, we choose class two. Observe that this form was derived by making the decision which maximizes the probability of making a correct decision, using knowledge of the measurement and the prior probability of the classes. We could use other criteria as well. For example, instead of maximizing the probability, we could choose to minimize the conditional risk.

14.3 Decision regions and the probability of error

The effect of any decision rule is to partition the feature space into c decision regions $\Omega_1, \ldots \Omega_c$. Suppose we define a set of discriminant functions $g_i(\mathbf{x})$. Then, if $g_i(\mathbf{x}) > g_j(\mathbf{x})$ for all $j \neq i$, then $x \in \Omega_i$ and we decide w_i. The equation of a decision boundary is $g_i(\mathbf{x}) = g_j(\mathbf{x})$ when Ω_i borders Ω_j.

In the two-class case, we may compute the probability of error in terms of these decision regions by

$$P(error) = P(\mathbf{x} \in \Omega_2, w_1) + P(\mathbf{x} \in \Omega_1, w_2). \tag{14.26}$$

That is, an error occurs if $\mathbf{x}$ falls in Ω_1, but it is really w_2, or if $\mathbf{x}$ falls in Ω_2, but it is really w_1. Since these events cannot both be true,

$$\begin{aligned} P(error) &= P(\mathbf{x} \in \Omega_2|w_1)P(w_1) + P(\mathbf{x} \in \Omega_1|w_2)P(w_2) \\ &= \int_{\Omega_2} p(\mathbf{x}|w_1)P(w_1)d\mathbf{x} + \int_{\Omega_1} p(\mathbf{x}|w_2)P(w_2)d\mathbf{x}. \end{aligned} \tag{14.27}$$

We also use the notation $P(error|w_2)$ to mean the probability that we make an incorrect decision when w_2 is the true state. Fig. 14.4 illustrates the *a posteriori* probability density of two classes, the process of deriving the decision boundary, and the probability of error.

In general, if $p(\mathbf{x}|w_1)P(w_1) > p(\mathbf{x}|w_2)P(w_2)$, we should decide that $\mathbf{x}$ is in Ω_1, so that the smaller term contributes to the error integral. That is exactly what Bayes' decision rule does.

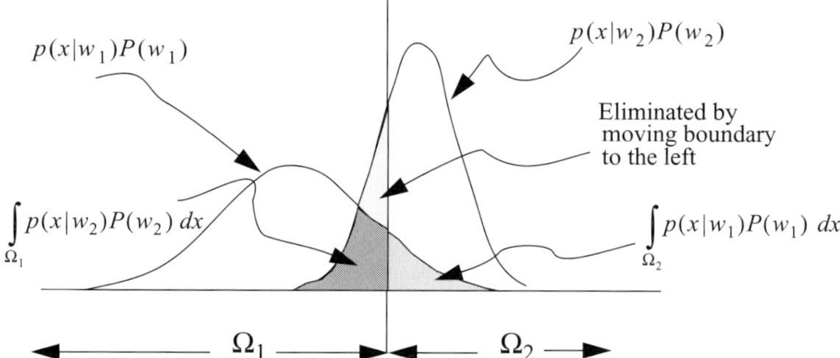

Fig. 14.4. The *a posteriori* probability density function of two classes as Gaussian, the decision boundary, and the probability of error.

In the multiclass case, since there are more ways to be wrong than right, it is simpler to compute the probability of being correct:

$$
\begin{aligned}
P(correct) &= \sum_{i=1}^{c} P(\mathbf{x} \in \Omega_i, w_i) \\
&= \sum_{i=1}^{c} p(\mathbf{x} \in \Omega_i | w_i) P(w_i) = \sum_{i=1}^{c} \int_{\Omega_i} p(\mathbf{x}|w_i) P(w_i)\, d\mathbf{x}.
\end{aligned} \tag{14.28}
$$

This result will be valid no matter how the feature space is partitioned. The Bayes' classifier maximizes this probability by choosing regions which maximize the integrals.

14.4 Conditional risk

Let $W = \{w_1, w_2, \ldots, w_c\}$ be a finite set of possible classes, and let $A = \{\alpha_1, \alpha_2, \ldots, \alpha_a\}$ be the set of possible activities – action α_i will be taken if the object lies in class w_i. We will, in general, incur some loss if we take an incorrect action. Let us assume that we have some way to quantify these losses. That is, let C_{ij} be the loss incurred for taking action α_i when the object really is in class w_j.

In order to continue with this section, you need to recall the definition of the expected value of a random variable. You should know what this is, but just to refresh your memory, let's suppose a random variable, x, can take on four different values, 0, 1, 2, and 3. Now, suppose that we know the probabilities of each of those values, and suppose they are

$$p(x = 0) = 0,\ p(x = 1) = 0,\ p(x = 2) = 1.0,\ p(x = 3) = 0.$$

What value of x would you *expect* to observe? Pretty obvious, isn't it? Of course you will always observe $x = 2$. Now let's repeat this for a slightly more complicated case,

$$p(x = 0) = 0,\ p(x = 1) = 0,\ p(x = 2) = 0.5,\ p(x = 3) = 0.5.$$

Would you agree that the expected value of x is 2.5? That is, half the time one would expect to see x equal to 2, and half the time, x will be 3.

Now, generalize that concept to lots of possible values with a probability associated with each value. If the number of possible values is finite, the expected value is a sum

$$\langle x \rangle = \sum x P(x). \tag{14.29}$$

In the more general case, when x is continuous, we will replace the probability with a density, and replace the summation with an integral.

Watch out for this replacement of summation by integral to occur. We are not going to warn you . . .

Suppose we observe $\mathbf{x}$ and contemplate taking action α_i. If the true state of nature is w_j, we incur loss C_{ij}. Since $P(w_j|\mathbf{x})$ is the probability that w_j is true, the expected loss associated with a_i is

$$r_i = \sum_{j=1}^{c} C_{ij} P(w_j|\mathbf{x}). \tag{14.30}$$

The expected loss is called the *risk*. We write r_i as $r(\alpha_i|\mathbf{x})$ to make it clear that this is the *conditional risk*.

We wish to minimize the total risk, which we denote as r by choosing the best α_i. A *decision rule* is a function $\alpha(\mathbf{x})$ that tells us what action to take to minimize r.

For every $\mathbf{x}$ the decision function $\alpha(\mathbf{x})$ assumes a value $\alpha_1, \ldots, \alpha_a$. The overall risk is associated, then, with the decision rule.

Since r_i or $r(\alpha_i|\mathbf{x})$ is the conditional risk, the overall risk is

$$r = \int r(\alpha(\mathbf{x})|\mathbf{x}) p(\mathbf{x})\, d\mathbf{x}. \tag{14.31}$$

If we choose $\alpha(\mathbf{x})$ so that $r(\alpha(\mathbf{x})|\mathbf{x})$ is as small as possible for every value of $\mathbf{x}$, we minimize the overall risk. Thus, to minimize r, compute r_i and select the action α_i for which r_i is minimal. The resulting overall risk is called the *Bayes' risk*.

14.4.1 Derivation of the likelihood ratio when risk is a consideration

The two-class derivation is presented here in detail. The generalization of multiple classes, if one understands this, is straightforward. Of course, students should remind themselves that the likelihood RATIO is only going to work for the two-class case.

Let's look at the previous derivation of the minimum risk criterion in more detail, for the two-class case.

As you read these terms, consider the concepts of "false negative," "false positive," "true positive," and "true negative." Which of these terms represents the total amount of each of these?

Reminder: C_{ij} is the cost of deciding i when reality is j.

If there are only two classes, we have four possibilities, two that we made the correct decision, and two that we were wrong. The total risk therefore is:

$$\begin{aligned} r = & \int_{\Omega 1} C_{11} P(w_1|x)\, dx \\ & + \int_{\Omega_1} C_{12} P(w_2|x)\, dx \\ & + \int_{\Omega_2} C_{21} P(w_1|x)\, dx \\ & + \int_{\Omega_2} C_{22} P(w_2|x)\, dx. \end{aligned} \tag{14.32}$$

That is, the probability that we guessed it was in class 1, integrated over the region of x in which our decision rule says we SHOULD guess class 1, plus . . .

Rewriting all that, we get

$$\begin{aligned} r = & \int_{\Omega_1} [C_{11} P(w_1|x) + C_{12} P(w_2|x)]\, dx \\ & + \int_{\Omega_2} [C_{21} P(w_1|x) + C_{22} P(w_2|x)]\, dx. \end{aligned} \tag{14.33}$$

Now, we observe that the integral over Ω_1 of any density function, say q, is simply one minus the integral over Ω_2 of the same function. Thus we can write both integrals in terms of region Ω_1, and obtain

$$\begin{aligned} r = & C_{11} \int_{\Omega_1} P(w_1|x)\, dx + C_{12} \int_{\Omega_1} P(w_2|x)\, dx \\ & + 1 - C_{21} \int_{\Omega_1} P(w_1|x)\, dx + 1 - C_{22} \int_{\Omega_1} P(w_2|x)\, dx \end{aligned} \tag{14.34}$$

which reorganizes to

$$r = 2 + \int_{\Omega_1} ((C_{11} - C_{21}) P(w_1|x) + (C_{12} - C_{22}) P(w_2|x))\, dx. \tag{14.35}$$

Our objective is to minimize this quantity (remember, it is the risk incurred in making all four possible decisions).That is, the decision rule is really the determination of the decision region(s). In this case, since there are only two decision regions, and everywhere that we do not decide class one, we decide class two, all we need to do is to determine the region Ω_1. To accomplish that, first, we need to make an assumption. We assume that the cost of making a correct decision is always less than the cost of an error. So $(C_{11} - C_{21}) < 0$, etc.

For some reason, students get upset when we say "make the integrand negative everywhere." Remember, we are optimizing by finding the limits on the integral!

How do we choose the limits of an integral (and remember, the limits of the integral are in fact the boundaries of the region where we decide class 1) such that the integral is maximally small? Simply choose the decision region such that the integrand is negative everywhere. Doing so produces the condition required for region 1 to be chosen: Choose Ω_1 such that

$$(C_{11} - C_{21})P(w_1|x) + (C_{12} - C_{22})P(w_2|x) < 0. \tag{14.36}$$

Replace the posterior probabilities with the product of the conditional densities and prior probabilities to get

$$(C_{11} - C_{21})p(x|w_1)P(w_1) < (C_{22} - C_{12})p(x|w_2)P(w_2), \tag{14.37}$$

which after appropriate algebraic manipulation becomes the decision rule: Choose class 1 if

$$\frac{p(x|w_1)}{p(x|w_2)} > \frac{(C_{12} - C_{22})P(w_2)}{(C_{21} - C_{11})P(w_1)}, \tag{14.38}$$

else choose class 2.

Try this: Substitute the symmetrical cost function into Eq. (14.38) and see how the likelihood ratio test simplifies.

This expression is called the "likelihood ratio test."

Consider the symmetrical loss function:

$$C_{ij} = \begin{cases} 0 & i = j \\ 1 & i \neq j \end{cases} \tag{14.39}$$

so that all errors are equally costly and there is no cost for a correct decision. We may now rewrite the conditional risk, the cost of making decision i

$$r_i = \sum_{j=1}^{c} C_{ij} P(w_j|\mathbf{x}) = \sum_{j \neq i} P(w_j|\mathbf{x}) = 1 - P(w_i|\mathbf{x}). \tag{14.40}$$

Thus, to minimize the average probability of error, we select i as the i which maximizes the *a posteriori* probability $P(w_i|\mathbf{x})$. That is, for minimum cost, we decide w_i, if $P(w_i|\mathbf{x}) > P(w_j|\mathbf{x})$ for all $i \neq j$, which we have already seen is the simple maximum likelihood classifier. Thus, we see that the maximum likelihood classifier minimizes the Bayes' risk associated with a symmetric cost function.

14.5 The quadratic classifier

Consider the general multivariate Gaussian classifier, with two classes. As in Assignment 14.1, if we take logs, we can work out a decision rule based on a likelihood ratio: Decide class 1 if

$$\ln \frac{|K_1|}{|K_2|} + (\mathbf{x} - \boldsymbol{\mu}_1)^{\mathrm{T}} K_1^{-1} (\mathbf{x} - \boldsymbol{\mu}_1) - (\mathbf{x} - \boldsymbol{\mu}_2)^{\mathrm{T}} K_2^{-1} (\mathbf{x} - \boldsymbol{\mu}_2) < Threshold; \tag{14.41}$$

else decide class 2; where $Threshold = -2\ln \left\{ \frac{P(w_2)(C_{12} - C_{22})}{P(w_1)(C_{21} - C_{11})} \right\}$.

If we define

$$A = K_1^{-1} - K_2^{-1}, \mathbf{b} = 2(K_2^{-1}\boldsymbol{\mu}_2 - K_1^{-1}\boldsymbol{\mu}_1), \quad \text{and}$$
$$c = \boldsymbol{\mu}_1^{\mathrm{T}} K_1^{-1} \boldsymbol{\mu}_1 - \boldsymbol{\mu}_2^{\mathrm{T}} K_2^{-1} \boldsymbol{\mu}_2 + \ln \frac{|K_1|}{|K_2|}, \tag{14.42}$$

we can rewrite Eq. (14.41) using

$$g(\mathbf{x}) \equiv \mathbf{x}^{\mathrm{T}} A \mathbf{x} + \mathbf{b}^{\mathrm{T}} \mathbf{x} + c. \tag{14.43}$$

And the decision rule becomes: Decide class 1 if $g(\mathbf{x}) < T$. In this formulation, we see clearly why the Gaussian parametric classifier is known as a *quadratic classifier*.

The Mahalanobis distance has the properties of a metric. Can you prove that? (Do you recall the defnition of a metric?)

Let's examine the implications of this rule. Consider the quantity $(\mathbf{x} - \boldsymbol{\mu}_1)^{\mathrm{T}} K_1^{-1} (\mathbf{x} - \boldsymbol{\mu}_1)$. This is some sort of measure involving a measurement, $\mathbf{x}$, and a class parameterized by mean vector and a covariance matrix. This quantity is known as the *Mahalanobis distance*.

First, let's look at the case that the covariance is the identity. Then, the Mahalanobis distance simplifies to $(\mathbf{x} - \boldsymbol{\mu}_1)^{\mathrm{T}} (\mathbf{x} - \boldsymbol{\mu}_1)$. That is, take the difference between the measurement and the mean. That is a vector. Then take the inner product of that vector with itself, which is, of course, the squared magnitude of that vector. What is this quantity? Of course! It is just the (squared) Euclidean distance between the measurement and the mean of the class. If the prior probabilities are the same and we use symmetric costs, *Threshold* works out to be zero, and the decision rule simplifies to: Decide class 1 if

$$(\mathbf{x} - \boldsymbol{\mu}_1)^{\mathrm{T}} (\mathbf{x} - \boldsymbol{\mu}_1) - (\mathbf{x} - \boldsymbol{\mu}_2)^{\mathrm{T}} (\mathbf{x} - \boldsymbol{\mu}_2) < 0 \tag{14.44}$$

else decide class 2. If the measurement is closer to the mean of class 1 than class 2, this quantity is less than zero. Therefore, we refer to this (very simplified) classifier as a *nearest mean* classifier, or *nearest mean* decision rule.

Now, let's complicate the rule a bit. We no longer assume the covariances are equal to the identity, but do assume they are equal to each other ($K_1 = K_2 \equiv K$). In this case, look at Eq. (14.42) and notice that the A matrix becomes zero. Now, the operations are not quadratic any more. We have a *linear classifier*.

We could choose to ignore the ratio of the determinates of the covariance matrices, or, more appropriately, to include that number in the threshold T. Then we have a minimum distance decision rule, but now the distance used is not the Euclidean distance. We refer to this as a *minimum Mahalanobis distance classifier*.

Here is another special case: What if the covariance is not only equal, but diagonal? Now, the Mahalanobis distance takes on a special form. We illustrate this by using a three-dimensional measurement vector, and letting the mean be zero:

$$[x_1\ x_2\ x_3]\begin{bmatrix}\frac{1}{\sigma_{11}} & 0 & 0\\ 0 & \frac{1}{\sigma_{22}} & 0\\ 0 & 0 & \frac{1}{\sigma_{33}}\end{bmatrix}\begin{bmatrix}x_1\\ x_2\\ x_3\end{bmatrix}$$

which we expand to

$$\frac{x_1^2}{\sigma_{11}} + \frac{x_2^2}{\sigma_{33}} + \frac{x_3^2}{\sigma_{33}}.$$

Do you recall seeing this ellipse discussion somewhere else in this book?

This is the equation of an ellipsoid, centered at the origin (or, in the case that the mean is not zero, centered at the mean) with axes located along the coordinate axes. In the more general case, with covariance which is not diagonal, the only thing that happens is that this ellipsoid may rotate. So, the equation which represents the Mahalanobis distance from a point to a class produces an ellipsoid.

Here is one more interesting case: Suppose the covariances are the same, diagonal, and proportional to the identity $K_i = \sigma^2 I$. Now, the discriminant function for class i takes on the form

$$g_i(\mathbf{x}) = \frac{2\boldsymbol{\mu}_i^{\mathrm{T}}\mathbf{x}}{\sigma^2} - \frac{|\boldsymbol{\mu}_i|^2}{\sigma^2} + 2\ln P(w_i). \tag{14.45}$$

Remember! An inner product computes a projection.

Assume further that the magnitudes of all the means are the same. That is, all the means are located on a hypersphere centered at the origin. Then, we do not need to consider the second term in Eq. (14.45), and the discriminant function simplifies to

$$g_i(\mathbf{x}) = \boldsymbol{\mu}_i^{\mathrm{T}}\mathbf{x} = \sum_{k=1}^{d} \mu_{ik}x_k, \tag{14.46}$$

which we refer to as the *inner product classifier*.

14.6 The minimax rule

Sometimes the *a priori* probabilities are unknown. In this case, a fixed decision rule will not yield the minimum risk, so we use the *minimax rule* and attempt to minimize the maximum possible risk. Suppose we have c classes; then the overall expected risk is

$$r = \sum_i P(w_i) \sum_j \int_{\Omega_i} C_{ij}\, p(\boldsymbol{x}|w_j)\, d\boldsymbol{x}, \tag{14.47}$$

that is, the probability of a particular state of nature times all the decisions we could make if that were the state of nature, and the cost associated with those decisions.

To see what this means clearly, think about the two-class case and let $i = 1$. We observe that r is linear in $P(w_i)$. Thus, r is maximized at one extreme of $P(w_1)$ or the other, e.g., $P(w_1) = 0$ or $P(w_1) = 1$. If we let $C_{11} = C_{22} = 0$ then the maximum of r becomes either

$$\int_{\Omega_1} C_{12} p(x|w_2)\, dx \tag{14.48}$$

or

$$\int_{\Omega_2} C_{21} p(x|w_1)\, dx. \tag{14.49}$$

Since $\Omega_1 \cup \Omega_2$ is the complete space, then

$$\max\left(\int_{\Omega_1} C_{12} p(x|w_2)\, dx, \int_{\Omega_2} C_{21} p(x|w_1)\, dx\right) \tag{14.50}$$

takes its minimum value when

$$\int_{\Omega_1} C_{12} p(x|w_2)\, dx = \int_{\Omega_2} C_{21} p(x|w_1)\, dx. \tag{14.51}$$

So, if $C_{21} = C_{12}$, the minimax rule says to choose Ω_1 and Ω_2 so that the probabilities of the two types of errors are the same. That is, we have chosen a condition which, in the absence of prior information about the classes, will minimize our maximum risk.

14.7 Nearest neighbor methods

In previous sections, we have assumed we have a model for the density, usually a Gaussian. However, if we simply have a training set, that data may or may not fit a Gaussian (or any other parametric model for that matter). A simple heuristic which we might use is called the "nearest neighbor rule" – assign the unknown to the same class as the class of the nearest neighbor in the training set.

In this section, we extend the nearest neighbor rule, and show that it is equivalent to a maximum likelihood classifier with the density estimated by the extended nearest neighbor rule.

This method utilizes a volume[1] V around the unknown. We simply count the number of points from the various classes which occur. Then the class-conditional density is estimated by

$$p(\boldsymbol{x}|\omega_m) = \frac{k_m}{n_m V}, \tag{14.52}$$

where k_m is the number of samples in class m inside the volume V centered at $\boldsymbol{x}$, and n_m is the total number of samples in class m in the training set.

Use of a constant volume is a problem, because in regions which are densely populated (many training set points nearby) the volume will contain many points, resulting in too much smoothing, whereas in more sparsely populated areas, the same volume results in estimates which are not sufficiently representative. The simple solution is to let the volume depend on the data. For example, to estimate $p(\mathbf{x})$ from n samples, one can center a cell about $\mathbf{x}$ and let it grow until it contains k_n samples, where k_n is some (yet to be specified) function of n. If the density of samples near $\mathbf{x}$ is high, then the volume will be small, resulting in good resolution. If the density is small, then the region will grow, providing smoothing. Duda *et al.* [14.4] point out that $k_n = \sqrt{n}$ provides one form for k_n which behaves in a reasonable way.

The k-nearest-neighbor (k-NN) rule can be extended slightly to allow us to use the strategy directly for classification. Given c training sets, we combine all the sample points from all the training sets into one data set of n points, where now

$$n = \sum_{i=1}^{c} n_i \tag{14.53}$$

where n_i is the number of samples in training set i.

Now, given a point $\boldsymbol{x}$ at which we wish to determine the statistics, we find the hypersphere of volume V which just encloses k points from the combined set. If, within that volume, k_m of those points belong to class ω_m, then we estimate the density as before for class ω_m by

$$p(\boldsymbol{x}|\omega_m) = \frac{k_m}{n_m V} \tag{14.54}$$

and

$$P(\omega_m) = \frac{n_m}{n} \tag{14.55}$$

$$p(\boldsymbol{x}) = \frac{k}{nV}. \tag{14.56}$$

If we apply Bayes' rule to Eqs. (14.54)–(14.56), we find

$$P(\omega_m|\boldsymbol{x}) = \frac{k_m}{k}. \tag{14.57}$$

[1] Of course, in more than three dimensions, this is a hypervolume. For simplicity, we will continue to use the word "volume" with the understanding that no limit on dimensionality exists.

This rule tells us to look in a neighborhood of the unknown feature vector for k samples. If, within that neighborhood, more samples lie in class i than any other class, we assign the unknown as belonging to class i. We thus have the *k-nearest-neighbor classification rule.*

The student should note that in the k-NN strategy, we have never defined precisely how *nearest* should be computed. The Euclidean metric is generally assumed to be the most reasonable measure for distance, but others may certainly be used.

In the authors' own experience in classifying large data sets of industrial data, we have found nearest neighbor algorithms to work surprisingly well.

A major practical disadvantage of the k-NN strategy for classification is the fact that all the data must be stored. This can be a massive storage burden, especially when compared with parametric methods which require only a few points. The computational burden associated with the k-NN techniques can likewise be significant, since, in order to find the k ***nearest*** neighbors, the distance from the unknown to *all* the neighbors must be determined. Heuristics have been published which speed up this process significantly, and the student is referred to the literature for suggestions. See, for example, the condensed-nearest-neighbor rule described in Hand [15.7].

14.8 Conclusion

Maximum likelihood and minimum squared error.

In this brief introduction to statistical pattern recognition, you have seen how statistical methods can assist in the process of making decisions. You have also noticed how pervasive the optimization approach to problem solving is. Probability densities are estimated using maximum likelihood methods, where the likelihood is a product of probabilities. For Gaussian forms, maximum likelihood simplifies to sum-squared error.

Minimize the risk.

You learned how to find decision regions which minimize the total risk, even when different decisions have different costs, by finding limits of integration which make the argument of the integral negative. Even in the case that the risk cannot be computed, we can develop a scheme which minimizes the maximum risk.

Minimize the maximum risk.

Minimum distance.

Classification is often considered a "minimum distance" process. That is, we make the decision which minimizes some sort of distance, and you have seen several examples in this chapter.

14.9 Vocabulary

You should know the meaning of the following terms.

```
Bayes' rule
Cluster
Conditional density
```

Decision boundary
Decision rule
Discriminant function
Feature vector
Likelihood ratio
Linear machine
Linearly separable
Maximum likelihood
Minimax
Multivariate
Prior probability
Quadratic classifier
Risk
Supervised learning
Training set
Univariate
Unsupervised learning

Assignment 14.1

Assume class 1 and 2 are well represented by Gaussian densities with the following parameters: Class 1 mean = 0, variance = 1. Class 2 mean = 3, variance = 4. Substitute the forms for the Gaussian into Eq. (14.25) and derive an equation which gives the range of x in which class 1 is chosen. You will need to make a reasonable assumption about prior probabilities (equal probabilities are often chosen).

Hint: After doing the substitution, take natural logarithms of both sides.

Assignment 14.2

In a one-dimensional problem, the conditional density for class 1 is Gaussian with mean 0 and variance 2; for class 2, the conditional density is also Gaussian with mean 3 and variance 1. That is:

$$p(x|w_1) = \frac{1}{\sqrt{2}\sqrt{2\pi}} \exp\left(-\frac{1}{2}\left(\frac{x}{\sqrt{2}}\right)^2\right)$$

$$p(x|w_2) = \frac{1}{\sqrt{2\pi}} \exp\left(-\frac{1}{2}(x-3)^2\right)$$

(1) Sketch the two densities on the same axis.
(2) What is the likelihood ratio?

(3) Assuming that $P(w_1) = P(w_2) = 0.5$, $C_{11} = C_{22} = 0, C_{12} = 1$ and $C_{21} = \sqrt{3}$, use the integral form for the probability of error assuming a Bayes' decision rule.

Assignment 14.3

In a one-dimensional problem the class-conditional densities for a feature x are

$$p(x|w_1) = \begin{cases} \exp(-(x-r)) & x \geq r \\ 0 & \text{otherwise} \end{cases} \quad \text{and}$$

$$p(x|w_2) = \begin{cases} \exp(x-3) & x < 3 \\ 0 & \text{otherwise} \end{cases},$$

where $P(w_1) = P(w_2) = 0.5$.

(1) Assume that $r < 3$, and sketch the densities. Determine the decision rule that minimizes the probability of error, and indicate what that decision rule means by marking a point on the x axis.
(2) Find the value of r which minimizes $P(\text{error}|w_2)$.

Topic 14A Statistical pattern recognition

14A.1 Matching feature vectors using statistical methods

Statistical pattern recognition, as mentioned above, is a process worthy of an entire book, and in fact many books have been written on the topic. Here, through a simple example [13.17], we will present just a glimpse of what the discipline entails. Our problem is to recognize faces. Let's first collect images which contain only faces (and thereby avoid the segmentation problem) by requiring the subjects all wear black clothing and stand against a black wall. We acquire relatively low-resolution images, 180 × 120 pixels. We then scan over the image with a collection of feature extractors, shown in Fig. 14.5. Each feature extractor operates on the neighborhood of each pixel, in much the same way that a kernel operator does, but instead of a sum of products, this operator returns the product of the image pixels corresponding to the black pixels in the kernel. First, we observe that each kernel, used in this way, is returning a very local autocorrelation of the image, in a particular direction. Denote the result of applying kernel i to the neighborhood of pixel j by $\mathbf{\Phi}_{ij}$. Then, the sum

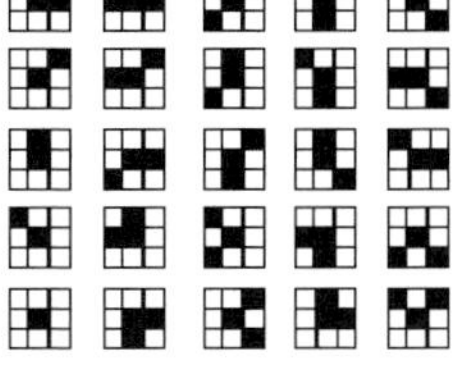

Fig. 14.5. The collection of 25 kernels used to extract a 25-element feature vector from an image.

$$x_i = \sum_{j=1}^{n} \mathbf{\Phi}_{ij} \tag{14.58}$$

is computed, producing a 25-element vector which in some sense describes the image.

So for every image, we have a vector consisting of 25 numbers. Using that 25-vector, the challenge is to properly make a decision. The first step is to reduce the dimensionality to something more manageable than 25.

We look for a method for reducing the dimensionality from, in general, d dimensions, to $c-1$ dimensions, where we are hoping to classify the data into c classes. (Somehow, we must know c, which in this example is the number of individual faces.) The following strategy is an extension of a method known in the literature as "Fisher's linear discriminant."

Assume we have c different classes, and a training set, X_i of examples from each class. Thus, this is a supervised learning problem. Define the *within-class scatter matrix* to be

$$S_W = \sum_{i=1}^{c} S_i \tag{14.59}$$

where

$$S_i = \sum_{\mathbf{x}\in X_i} (\mathbf{x}-\boldsymbol{\mu}_i)(\mathbf{x}-\boldsymbol{\mu}_i)^{\mathrm{T}}, \tag{14.60}$$

and $\boldsymbol{\mu}_i$ is the mean of class i. Thus, S_i is a measure of how much each class varies from its average.

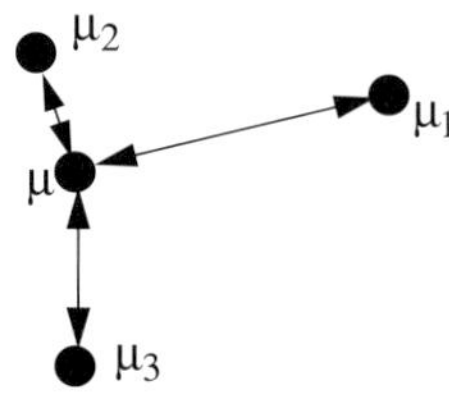

Fig. 14.6. The between-class scatter is a measure of the total distance between the class means and the overall mean.

$$\boldsymbol{\mu}_i = \frac{1}{n_i}\sum_{\mathbf{x}\in X_i} \mathbf{x}. \tag{14.61}$$

We define the *between-class scatter matrix* as

$$S_B = \sum_{i=1}^{c} n_i(\boldsymbol{\mu}_i - \boldsymbol{\mu})(\boldsymbol{\mu}_i - \boldsymbol{\mu})^{\mathrm{T}}, \tag{14.62}$$

where $\boldsymbol{\mu}$ is the mean of all the points in all the training sets and n_i is the number of samples in class i. To see what this means, consider Fig. 14.6. The *between-class scatter* is a measure of the sum of the distances between each of the class means and the overall sample mean. Maximization of some measure of S_{B} will push the class means apart, away from the overall mean.

The idea is to find some projection of each data vector $\mathbf{x}$ onto a vector $\mathbf{y}$,

$$\mathbf{y} = W\mathbf{x} \tag{14.63}$$

such that first, $\mathbf{y}$ is of lower dimension than $\mathbf{x}$, and second, the classes are better separated after they are projected.

The projection from d-dimensional space to $c-1$ dimensional space is accomplished by $c-1$ linear discriminant functions

$$y_i = \mathbf{w}_i^{\mathrm{T}}\mathbf{x}. \tag{14.64}$$

If we view the y_i as components of a vector, and the vectors $\mathbf{w}_i$ as columns of a matrix W, we can describe all the discriminant functions by a single matrix equation

$$\mathbf{y} = W^{\mathrm{T}}\mathbf{x}. \tag{14.65}$$

We now define a criterion function which is a function of W and measures the ratio of between-class scatter to within-class scatter. That is, we want to maximize S_{B} relative to S_{W}, or rather, to maximize some measure of $S_W^{-1}S_B$. The trace of $S_W^{-1}S_B$ is the sum of the spreads of $S_W^{-1}S_B$ in the direction of the principal components of $S_W^{-1}S_B$. We can see clearly what this means in the two-class case.

$$J = \mathrm{tr}S_W^{-1}S_B = \frac{n_1 n_2}{n_1+n_2}\mathrm{tr}S_W^{-1}(\mu_1-\mu_2)(\mu_1-\mu_2)^T = \frac{n_1 n_2}{n_1+n_2}D^2 \tag{14.66}$$

where D^2 is the squared Mahalanobis distance between the two classes.

Rather than maximize the trace of the matrix, we might redefine J using the determinant:

$$J(W) = \frac{|W^T S_B W|}{|W^T S_W W|}. \tag{14.67}$$

Since the determinant is the product of the eigenvalues, it is therefore the product of the "spread" in the principal directions. As in the case of Fisher's linear discriminant, the solution to this equation can be found by eigenvector analysis. The columns of the optimal W are the eigenvectors which correspond to the largest eigenvalues in

$$S_B\mathbf{w}_i = \lambda_i S_W \mathbf{w}_i. \tag{14.68}$$

One can find the eigenvalues as the roots of the characteristic equation

$$|S_B - \lambda_i S_W| = 0 \tag{14.69}$$

and then solve

$$(S_B - \lambda_i S_W)\mathbf{w}_i = 0 \tag{14.70}$$

for the eigenvectors $\mathbf{w}_i$.

This generalization of Fisher's linear discriminant is the method most widely implemented in computer packages. It is often used for selection of a subset of the original variables rather than for dimensionality reduction via transformation. This is done by examining the resulting eigenvalues and noting the relative sizes (positive or negative) corresponding to the variables. A large eigenvalue means that the corresponding variable makes an important contribution to the between-group separability, as we learned when we studied the K–L transform.

We have reduced the dimensionality of the 25-element vector to something smaller (yes, this assumes the number of classes is less than 25). In this lower dimensionality space, we will design a classifier. How? It is easy at this point. Use Eq. (14.5), or rather the logarithm of the RHS of Eq. (14.5) (no, YOU do the math!) to derive a discriminant function which is maximal when the example is assigned to the correct class.

14A.2 Support vector machines (SVMs)

Pattern classifiers based on the concepts of support vectors are relatively new. They were first introduced by Vapnik [14.12] based on the concept of minimizing structural risk. We present them here because they appear to provide performance superior to most other pattern classification methods, and to illustrate the approach – to set up an optimization problem which maximizes the margin – a derivation of the simplest support vector approach is provided.

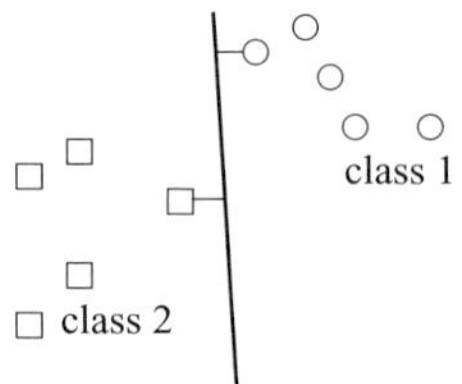

Fig. 14.7. A poor choice of the dividing hyperplane (a line in this 2D example) produces a margin which is small.

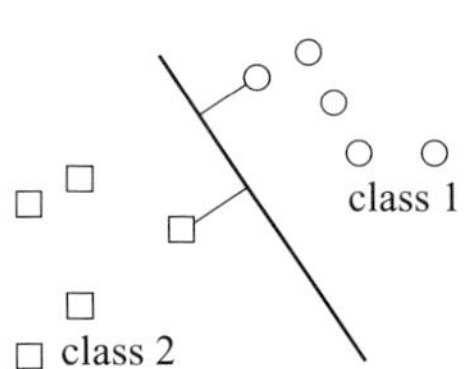

Fig. 14.8. A good choice of the dividing hyperplane results in a large margin.

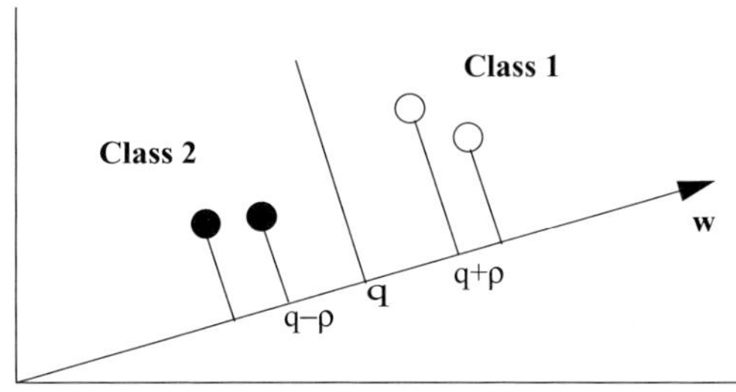

Fig. 14.9. Two classes denoted by 1 and 2 are both projected onto the same line, w. The line dividing the sets is orthogonal to w, and we seek the points in the training sets which are closest to that line.

14A.2.1 Derivation of SVMs assuming separability

In this section, we assume the training sets are separable by a linear surface. In the next subsection, we discuss this assumption and demonstrate how to ensure it.

As before, we seek to divide the feature space by a hyperplane into two segments, in which the examples of the training sets are linearly separable. Define the distance from the closest point in class 1 to the hyperplane as d_1. Similarly let d_2 be the distance from the closest point in class 2 to the hyperplane. Define the *margin* as $d_1 + d_2$. We will seek the hyperplane which maximizes the margin (see Figs. 14.7 and 14.8).

We don't know φ yet. Finding it is the challenge.

Given an example $\mathbf{x}$, we project this sample onto some unit vector φ, and make a decision using the rule *decide class 1 if* $\varphi^{\mathrm{T}}\mathbf{x} - q > 0$ where q is a constant.

I_1 is the set of points in the training set representing class 1. Ω is the decision region for class 1.

Let $\mathbf{x}_1$ be a point in class 1, and similarly, let $\mathbf{x}_2$ be a point in class 2. Since we are seeking the points closest to the decision line, we wish to choose $\mathbf{x}_1$ and $\mathbf{x}_2$ so that their projections onto φ are as close together as possible. As illustrated in Fig. 14.9, we seek points $\mathbf{x}_1$ and $\mathbf{x}_2$ such that with minimal positive ρ

$$\varphi^{\mathrm{T}}\mathbf{x}_i = q + \rho \qquad \varphi^{\mathrm{T}}\mathbf{x}_2 = q - \rho. \tag{14.71}$$

Define I_i to be the set of training examples from class i. For any point in I_1, $\varphi^{\mathrm{T}}\mathbf{x} - q > \rho$, and for any point in I_2, $\varphi^{\mathrm{T}}\mathbf{x} - q < \rho$. We need to find two things. (1) A pair of points, one[2] in each class, which are as close together as possible. We will call these points *support vectors.* (2) A vector onto which to project the support vectors so that their projections are maximally far apart. We solve this problem as follows.

Any other vector in the same direction, actually.

Recall that φ was a unit vector. It is thus equal to *some* other vector in the same direction divided by its magnitude, $\varphi = \mathbf{w}/\|\mathbf{w}\|$. We will look for one such vector, with certain properties which will be introduced in a moment: For now, let $\mathbf{x}_1$ denote any point in I_1, not necessarily a support vector, and similarly for $\mathbf{x}_2$. Then

$$\left(\frac{\mathbf{w}}{\|\mathbf{w}\|}\right)^{\mathrm{T}}\mathbf{x}_1 - q \geq \rho \qquad \left(\frac{\mathbf{w}}{\|\mathbf{w}\|}\right)^{\mathrm{T}}\mathbf{x}_2 - q \leq \rho \tag{14.72}$$

2 It is possible to have more than one support vector in each class, since two points might both be precisely the same distance from the hyperplane.

which leads to

$$\mathbf{w}^{\mathrm{T}}\mathbf{x}_1 - q\|\mathbf{w}\| \geq \rho\|\mathbf{w}\| \qquad \mathbf{w}^{\mathrm{T}}\mathbf{x}_2 - q\|\mathbf{w}\| \leq \rho\|\mathbf{w}\|. \tag{14.73}$$

Define $b = -q\|\mathbf{w}\|$, and we then add a constraint to $\mathbf{w}$ to require that its magnitude have a particular property:

$$\|\mathbf{w}\| = 1/\rho. \tag{14.74}$$

Now we have two equations which describe behavior for any points in class 1 or class 2:

$$\mathbf{w}^{\mathrm{T}}\mathbf{x}_1 + b \geq 1 \qquad \mathbf{w}^{\mathrm{T}}\mathbf{x}_2 + b \leq 1. \tag{14.75}$$

From this point on in this derivation, the subscript on the $\mathbf{x}$ no longer denotes the class to which $\mathbf{x}$ belongs, but rather just its index, as an element of the training set.

Since we wish to find the line which maximizes the margin, ρ, from Eq. (14.74), we see that this is the same as finding the projection vector $\mathbf{w}$ whose magnitude is minimal, thus we seek a minimizer $\mathbf{w} = \arg\min(\frac{1}{2}\mathbf{w}^{\mathrm{T}}\mathbf{w})$. Unfortunately, the null vector would minimize this, so we need to add some constraints to avoid this trivial solution.

Let y_i be the label for point $\mathbf{x}_i$, and define the labels as

$$y_i = \begin{cases} 1 & \text{if} \quad x_i \in I_1 \\ -1 & \text{if} \quad x_i \in I_2 \end{cases} \tag{14.76}$$

and consider the expression $y_i = (\mathbf{w}^{\mathrm{T}}\mathbf{x}_i + b)$. This will always be greater than or equal to 1, regardless of the class of $\mathbf{x}_i$. We thus have a constraint, and our minimization problem becomes: Find the $\mathbf{w}$ which minimizes $\mathbf{w}^{\mathrm{T}}\mathbf{w}$ such that $y_i(\mathbf{w}^{\mathrm{T}}\mathbf{x}_i + b) \geq 1$.

This can be accomplished by setting up the following constrained optimization problem:

$$L(\mathbf{w}, b, \lambda) = \frac{1}{2}\mathbf{w}^{\mathrm{T}}\mathbf{w} - \sum_{i=1}^{l} \lambda_i (y_i(\mathbf{w}^{\mathrm{T}}\mathbf{x}_i + b) - 1) \tag{14.77}$$

where l is the number of the samples in the training set.

The Lagrange multipliers will all be positive. Take the partial derivative with respect to $\mathbf{w}$,

$$\frac{\partial L}{\partial \mathbf{w}} = \mathbf{w} - \sum \lambda_i \mathbf{x}_i y_i \tag{14.78}$$

and setting this to zero we find

$$\mathbf{w} = \sum \lambda_i \mathbf{x}_i y_i. \tag{14.79}$$

Similarly, take the derivative with respect to b,

$$\frac{\partial L}{\partial b} = -\sum \lambda_i y_i = 0. \tag{14.80}$$

With these two observations, the equation for L is simplified to

$$L = \frac{1}{2}\sum_i \sum_j \lambda_i y_i \lambda_j y_j \mathbf{x}_i^{\mathrm{T}}\mathbf{x}_j - \sum_i \sum_j \lambda_i y_i \lambda_j y_j \mathbf{x}_i^{\mathrm{T}}\mathbf{x}_j - b\sum_i \lambda_i y_i + \sum_i \lambda_i \tag{14.81}$$

where the first and second term are the same except for the 1/2. The third term is zero.

Defining the matrix A by

$$A = \left[y_i y_j \mathbf{x}_i^{\mathrm{T}} \mathbf{x}_j\right] \tag{14.82}$$

allows us to write L in a matrix form

$$L = -\frac{1}{2}\Lambda^{\mathrm{T}} A \Lambda + \boldsymbol{1}^{\mathrm{T}} \Lambda, \tag{14.83}$$

where $\boldsymbol{1}$ denotes a vector of all ones.

Finding the vector of Lagrange multipliers (Λ) which minimizes L is a quadratic optimization problem. There are several numerical packages available which perform such operations. Once we have the set of Lagrange multipliers, the optimal projection vector is found by Eq. (14.79), which we observe requires a summation over all the elements of the training set. To solve for b, we need to make use of the Kuhn-Tucker [14.5] conditions:

$$\lambda_i (y_i (\mathbf{w}^{\mathrm{T}} \mathbf{x}_i + b) - 1) = 0 \quad \forall i. \tag{14.84}$$

In principle, Eq. (14.84) may be solved for b using any i, but it is numerically better to use an average.

Similarly, we note the dimension of A is the same as the number of samples in the training set. Thus, unless some "filtering" is done on the training set prior to building the SVM, the computational complexity can be substantial.

14A.2.2 Nonlinear support vector machines

Instead of dealing with the actual samples, we apply a nonlinear transformation which produces a vector of higher dimension, $\mathbf{y}_i = \vartheta(\mathbf{x}_i)$. For example, if $\mathbf{x} = [x_1, x_2]^{\mathrm{T}}$ is of dimension 2, $\mathbf{y}$ might be

$$\mathbf{y}_i = \left[x_1^2 x_2^2 x_1 x_2 x_1 x_2 \ 1\right]^{\mathrm{T}}, \tag{14.85}$$

which is of dimension 6. Surprisingly, this increase in dimension does not seem to destroy the capability of the classifier. How can an expansion to more "degrees of freedom" improve both accuracy on the training set and generalizability? Concerning this question, Burges [14.2] says:

> Usually, mapping your data to a "feature space" with an enormous number of dimensions would bode ill for the generalization performance of the resulting machine. After all, the set of all hyperplanes $\{\boldsymbol{w}, \mathrm{b}\}$ are parameterized by $\dim(H) + 1$ numbers. Most pattern recognition systems with billions, or even an infinite number, of parameters would not make it past the start gate. How come SVMs do so well? One might argue that, given the form of solution, there are at most $l + 1$ adjustable parameters (where l is the number of training samples), but this seems to be begging the question. It must have something to do with our requirement of maximum margin hyperplanes that is saving the day. . . . a strong case can be made for this claim.

An expansion of the form described in Eq. (14.85) increases the dimensionality of the space, and increases the likelihood that the classes will be linearly separable in the higher dimensional space (for reasons beyond the scope of this brief explanation). It also provides a simple mechanism for incorporating nonlinear mixtures of the information from the measurements. The polynomial form of Eq. (14.85) is but one way of expanding the dimensionality of the measurement vector. A more interesting collection of ways comes to mind when one looks at Eq. (14.83), and observes that to compute the optimal separating hyperplane, one does not need to know the vectors themselves, but only the scalars which result from computing all possible inner products. Thus, we do not need to map each vector to a high-dimensionality space and then take the inner product of those vectors, not if we can figure out ahead of time what those inner products should be.

Kernels and inner products

We seek the best separating hyperplane in the higher dimensional space defined by $\mathbf{y}_i = \vartheta(\mathbf{x}_i)$, $\vartheta : (\Re^d \to \Re^m)$ where $m > d$. Then, the equations for the elements of A become functions of the inner products of these new vectors $A = [y_i\ y_j\ \vartheta^{\mathrm{T}}(\mathbf{x}_i)\ \vartheta(\mathbf{x}_j)]$. For notational convenience (and to lead to a really clever result), define a kernel operator, $K(\mathbf{x}_i, \mathbf{x}_j)$, which takes into account both the nonlinear transformation ϑ and the inner product. Instead of asking "What nonlinear operator should I use?" let's ask a different question: "Given a particular kernel, is there any chance it represents the combination of a nonlinear operator and an inner product?" The amazing answer is yes, under certain conditions that is true. These conditions are known as *Mercer's conditions*: Given a kernel function $K(\boldsymbol{a}, \boldsymbol{b})$ of two vector-valued arguments, if for any $g(\boldsymbol{x})$ which has finite energy (that is, $\int (g(\boldsymbol{x}))^2\, d\boldsymbol{x}$ is finite), then $\int K(\boldsymbol{a}, \boldsymbol{b}) g(\boldsymbol{a}) g(\boldsymbol{b})\, d\boldsymbol{a}\, d\boldsymbol{b} \geq 0$, then there exists a mapping ϑ and a decomposition of K of the form

$$K(\boldsymbol{a}, \boldsymbol{b}) = \sum_i \vartheta_i(\boldsymbol{a}) \vartheta_i(\boldsymbol{b}). \tag{14.86}$$

In Eq. (14.86), the subscript i denotes the ith element of the vector-valued function ϑ. Thus, that expression represents an inner product. Notice that Mercer's conditions simply state that if K satisfies these conditions, then K may be decomposed into an inner product of two instances of a function ϑ. It does not say what ϑ is, nor does it say what the dimensionality of ϑ is. But that's OK. We do not have to know. In fact, the vector ϑ may have infinite dimensionality. That's still OK.

One kernel which is known to satisfy Mercer's condition, and which is very popular in the SVM literature, is the radial basis function

$$K(\boldsymbol{a}, \boldsymbol{b}) = \exp\left(-\frac{(\boldsymbol{a} - \boldsymbol{b})^{\mathrm{T}}(\boldsymbol{a} - \boldsymbol{b})}{2\sigma^2}\right). \tag{14.87}$$

In the literature, SVMs have been applied to various problems such as face recognition [14.10] and breast cancer detection [14.1]. In previous studies [14.7] and in comparative analysis in the literature, they have empirically been shown to outperform classical classification tools such as neural networks and nearest neighbor rules [14.9, 14.13, 14.14]. Interestingly, in a comparison with a classifier based on hyperspectral data, a SVM-based classifier using multispectral data (derived from the original hyperspectral data by filtering) performed better than classifiers based on the original data [14.8].

14A.3 Conclusion

Statistical methods provide tools for making decisions, based on measurements. If the measurements are sufficiently discriminating that simple thresholds may be used, sophisticated statistical methods may not be required. On the other hand, most collections of measurements are not sufficient to make such decisions based on trivial feature comparisons.

We have seen in section 14A.1 an example of what is in the discipline of statistical pattern recognition, and just one application to machine vision. There is NOT enough information in this book to teach you all you need to know about statistical methods. You really need to take a full course. We hope this chapter has given you enough motivation to do that.

In section 14A.1 we derived an objective function which, if maximized, would result in projected data with classes maximally separated. It turned out that this maximization problem turns into an eigenvalue problem.

Maximize the margin.

A SVM finds the decision boundary which maximizes the margin, where the margin is the distance between the closest points and the decision boundary, and the derivation of the machine requires use of constrained optimization with Lagrange multipliers.

14A.4 Vocabulary

You should know the meaning of the following terms.

```
Between-class scatter
Fisher's linear discriminant
Margin
Mercer's conditions
Support vector
Within-class scatter
```

References

[14.1] P. S. Bradley, U. M. Fayyad, and O. L. Mangasarian, "Mathematical Programming for Data Mining: Formulations and Challenges," *INFORMS Journal on Computing*, **11**(3), pp. 217–238, 1999.

[14.2] C. J. C. Burges, "A Tutorial on Support Vector Machines for Pattern Recognition," *Data Mining and Knowledge Discovery*, Vol. 2, No. 2, Dordrecht, Kluwer, 1998.

[14.3] R. Duda and P. Hart, *Pattern Recognition and Scene Analysis*, New York, Wiley, 1973.

[14.4] R. Duda, P. Hart, and D. Stork, *Pattern Classification, Second Edition*, New York, Wiley, 2001.

[14.5] R. Fletcher, *Practical Methods of Optimization*, New York, Wiley, 1987.

[14.6] K. Fukanaga, *Introduction to Statistical Pattern Recognition*, New York, Academic Press, 1972.

[14.7] B. Karacali and H. Krim, "Fast Minimization of Structural Risk Using the Nearest Neighbor Rule," *IEEE Transactions on Neural Networks*, **14**(1), Jan. 2003.

[14.8] B. Karacali and W. Snyder, "On-the-fly Multispectral Automatic Target Recognition," *Combat Identification Systems Conference*, Colorado Springs, June, 2002.

[14.9] D. Li, S. M. R. Azimi, and D. J. Dobeck, "Comparison of Different Neural Network Classification Paradigms for Underwater Target Discrimination," *Proceedings of SPIE, Detection and Remediation Technologies for Mines and Minelike Targets* V, **4038**, pp. 334–345, 2000.

[14.10] E. Osuna, R. Freund, and F. Girosi, "Training Support Vector Machines: An Application to Face Detection," *Proceedings of Computer Vision and Pattern Recognition*, Puerto Rico, June, 1997.

[14.11] C. Therrien, *Decision, Estimation, and Classification*, New York, Wiley, 1989.

[14.12] V. Vapnik, *The Nature of Statistical Learning Theory*, Berlin, Springer, 1995.

[14.13] M. H. Yang and B. Moghaddam, "Gender Classification Using Support Vector Machines," *Proceedings of IEEE International Conference on Image Processing*, Vancouver, BC, Canada, September, 2000, vol. 2, pp. 471–474, 2000.

[14.14] Y. Yang and X. Liu, "Re-examination of Text Categorization Methods," *Proceedings of the 1999 22nd International Conference on Research and Development in Information Retrieval (SIGIR'99)*, Berkeley, CA, pp. 42–49, 1999.

15 Clustering

Woes cluster; rare are solitary woes

Edward Young

In this chapter, we approach the problem alluded to in Chapter 14 where the training set simply contains points, and those points are not marked in any way to indicate from which class they may have come. As in the previous chapter, we present only a brief overview of the field, and refer the reader to other texts [14.4, 15.7] for more thorough coverage. One very important area which we omit here is the use of biologically inspired models for clustering [15.4, 15.5, 15.6], and the reader is strongly encouraged to look into these.

We will discuss the issues of clustering in a rather general sense, but note one particular application, which is identification of peaks in the Hough transform array.

Consider this example from satellite pattern classification: We imagine a downward-looking satellite orbiting the earth, which, at each observed point, makes a number of measurements of the light emitted/reflected from that point on the earth's surface. Typically, as many as seven different measurements might be taken from a given point, each measurement in a different spectral band. Each "pixel" in the resulting image would then be a 7-vector where the elements of this vector might represent the intensity in the far-infrared, the near-infrared, blue, green, etc. Now suppose we have labeled training sets indicating examples of pixels containing wheat, corn, grass, and trees. With these training sets, it would seem that we should be able to build a pattern classifier; and indeed we can. Furthermore, the problem as stated so far is a supervised learning problem. Let's consider for a moment however, the class which we call "trees." This class consists of evergreen and deciduous trees and, depending upon the time of year, these two subclasses will give radically different spectral signatures. We thus have a pattern classification problem which is not easily approached with parametric classifiers. While we could use a non-parametric approach, parametric methods are very attractive. An alternative to nonparametric classifiers is to consider methods for determining the existence of the subclasses, assigning points in the training set to the correct subclass and then representing that subclass parametrically. Fig. 15.1 illustrates a two-dimensional problem in which the existence of two classes within the same training set is readily apparent. We will

Fig. 15.1. A training set with two clusters.

refer to these subclasses as "clusters" for the duration of this discussion. Each cluster could be represented fairly accurately by a 2D Gaussian. The entire measurement space, however, is obviously bimodal.

Such clustering is easy (for us humans) to visualize in a 2-space, and essentially impossible to visualize in problems with more than three dimensions.

15.1 Distances between clusters

It is relatively easy to define, and to visualize, measures of distance between points, including the Euclidian distance $d(a, b) = |a - b|$, and the city block distance $\sum_i |a_i - b_i|$. These ideas can be readily extended to measures for distance between a point and a cluster. One such measure is the Euclidean (or city block) distance between the point x and the mean of the cluster μ. Another such measure, which takes into account the distribution of the cluster, is the familiar Mahalanobis distance (see section 14.5) between a point x and a cluster A

$$d_{\mathrm{mah}}(x, A) = (x - \mu_A)^{\mathrm{T}} K_A^{-1} (x - \mu_A), \tag{15.1}$$

where K_A^{-1} is the inverse of the covariance which best represents cluster A.

We have not yet however considered measures of distance between two clusters and for our discussion of clustering algorithms we will need such a measure. It is not quite obvious (as we will see from the following discussion) how one should define the distance between two clusters since each cluster contains potentially many points. The simplest such definition would be just the distance between the sample means of the two clusters (also called the centroid distance)

$$d_{\mathrm{mean}}(A, B) = |\mu_A - \mu_B|. \tag{15.2}$$

One might also consider some generalization of the Mahalanobis distance from (point, cluster) to (cluster, cluster) defining

$$d_{\mathrm{fisher}}(A, B) = \frac{|\mu_A - \mu_B|}{\sigma_A^2 + \sigma_B^2}, \tag{15.3}$$

where the sigmas are computed by first projecting the two clusters onto the line between the two means. The sample mean and sample variances of the projected data are the parameters of Eq. (15.3).

We can provide a more formal statement about how to define a distance between two clusters by first stating the desirable properties of such a distance. We require

$$\begin{aligned} d(A, B) &\geq 0 \\ d(A, B) &= 0 \quad \text{if } A = B \\ d(A, B) &= d(B, A). \end{aligned} \tag{15.4}$$

There are many measures which satisfy these conditions; for example, we could integrate the densities over all the sample space to obtain the *divergence*,

$$d_{\text{div}}(A, B) = \int [p(\mathbf{x}|A) - p(\mathbf{x}|B)] \ln \frac{p(\mathbf{x}|A)}{p(\mathbf{x}|B)} \, d\mathbf{x}. \tag{15.5}$$

In the case of multivariate Gaussians, this becomes

$$\left((\mu_1 - \mu_2)^{\mathrm{T}} \left(K_1^{-1} + K_2^{-1}\right)[\mu_1 - \mu_2] + \operatorname{tr}\left(K_1^{-1} K_2 + K_2^{-1} K_1 - 2I\right)\right)/2, \tag{15.6}$$

which simplifies to Eq. (15.7) if the two covariance matrices are equal to each other.

$$\Delta^2 \equiv (\mu_1 - \mu_2)^{\mathrm{T}} K^{-1} (\mu_1 - \mu_2). \tag{15.7}$$

The Chernoff distance is

$$d_{\text{ch}}(A, B) = -\ln \int (p(\mathbf{x}|A))^{1-s} (p(\mathbf{x}|B))^s \, d\mathbf{x} \tag{15.8}$$

which, for multivariate Gaussians, becomes

$$\begin{aligned} &\frac{1}{2} s(1 - s)(\mu_A - \mu_B)^{\mathrm{T}} [(1 - s)K_A + sK_B]^{-1} (\mu_A - \mu_B) \\ &+ \frac{1}{2} \left(\ln \frac{|(1 - s)K_A + sK_B|}{|K_A|^{(1-s)} |K_B|^s} \right). \end{aligned} \tag{15.9}$$

In the case that s = 1/2, the Chernoff distance turns into the Bhattacharyya distance:

$$\frac{1}{8} (\mu_A - \mu_B)^{\mathrm{T}} \left(\frac{K_A + K_B}{2} \right)^{-1} (\mu_A - \mu_B) + \frac{1}{2} \ln \frac{\left|\frac{1}{2}(K_A + K_B)\right|}{|K_A|^{1/2} |K_B|^{1/2}}. \tag{15.10}$$

The most commonly occurring cluster distance metric in the literature is the nearest neighbor measure

$$d_{\min}(A, B) = \min_{a,b} d(a, b) \quad \text{for } a \in A, b \in B. \tag{15.11}$$

That is, over all pairs of points, one from cluster A and one from cluster B, choose those two points which are closest together and define that to be the distance between the two clusters.

Similarly, one may define the furthest neighbor distance

$$d_{\max}(A, B) = \max_{a \in A, b \in B} d(a, b). \tag{15.12}$$

Each of the definitions given above simply gives us a scalar measure for representing, in some sense, how far apart two clusters are.

Any time you use a distance measure on vector-valued quantities, you should pay attention to the possibility that scaling of the coordinate axes might change the results. For example, consider the set of points shown in Fig. 15.2. Another example that fairly well shows the impact of clustering is a classification problem involving the vector $[a, b]^{\mathrm{T}}$, where a represents population and b represents the number of

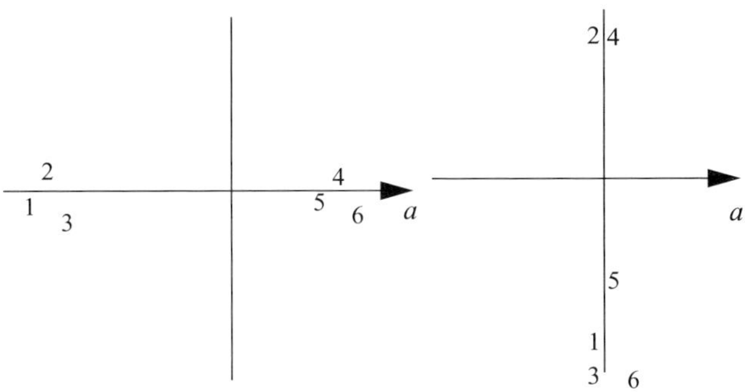

Fig. 15.2. Simple scaling of the coordinate axes can change the apparent clusters.

universities. Now, consider the distance between a vector x and a vector y,

$$d(x, y) = \sqrt{(x_a - y_a)^2 + (x_b - y_b)^2} \approx x_a - y_a \tag{15.13}$$

and in this case, the second feature is essentially not considered because it is so small in magnitude compared with the first feature. A common normalization which is often used to deal with such problems is to divide each feature by its standard deviation.

15.2 Clustering algorithms

We will consider only two clustering algorithms here, agglomerative clustering and k-means, and will also formulate the clustering problem as an optimization problem. There are a number of other algorithms in the literature, but these are less often used in applied systems.

15.2.1 Agglomerative clustering

In agglomerative clustering, we begin by assigning each data point in the training set to a separate cluster. If there are N data points, then at the beginning, we have N clusters. Next, we perform iterations of: Merge the two "closest" clusters.

By "merge" we mean: (1) find the two closest clusters (each cluster may be viewed as a set); (2) create a new set, consisting of the union of the two; and (3) remove the original two. Fig. 15.3 shows an example.

This process continues until there are only c clusters. Presumably c is known beforehand.

When we begin, every cluster consists of a single point, and the distance between clusters is the same as the measure used for the distance between points. After we have begun the process, however, we will be forced to make use of measures of distances between clusters as described earlier.

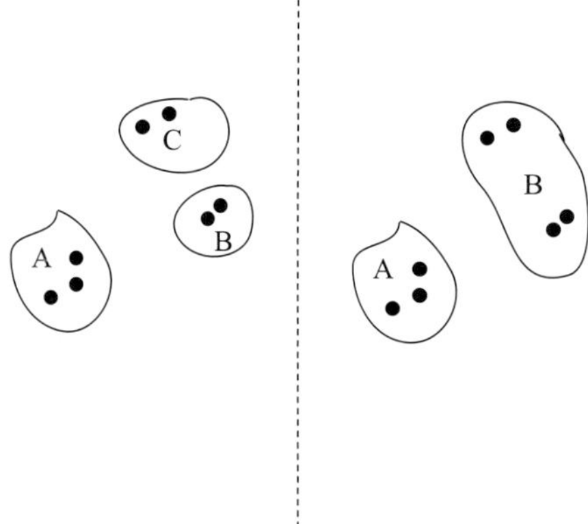

Fig. 15.3. Before the clustering iteration (left), the data points have already been assigned to three clusters. Clusters *B* and *C* are determined (somehow) to be closer than *A*, *B* or *A*, *C*, so *B* and *C* are merged and renamed to become a new cluster *B*.

Properties of agglomerative clustering

The clusters which result from this algorithm are highly dependent on the measure of cluster distance which is used. If, for example, one uses d_{min}, and illustrates the distances used by drawing a line in Fig. 15.4, one gets the minimum spanning tree (MST) of the graph which represents the data points by simply continuing the algorithm until there is only one cluster. If we want three clusters, then we need only cut the longest two edges in this graph. One then realizes that if formal graph theoretic operations result from clustering, then the converse is likewise true: Whatever we know about graphs may help us in designing clustering algorithms. In particular, the following algorithm constructs the MST of a graph very quickly:

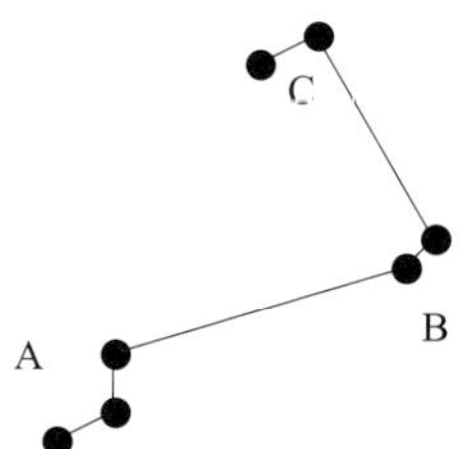

Fig. 15.4. Illustrating the distances chosen at each iteration by a line, we arrive at the minimum spanning tree.

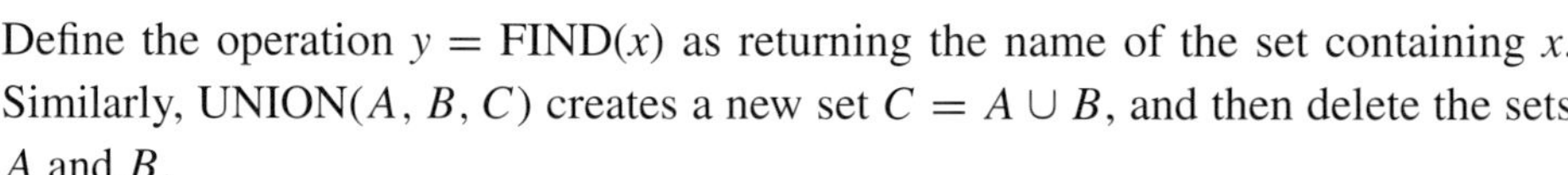

Define the operation $y = \text{FIND}(x)$ as returning the name of the set containing x. Similarly, UNION(A, B, C) creates a new set $C = A \cup B$, and then delete the sets A and B.

Algorithm: Finding the Minimum Spanning Tree

(1) Compute all edges in the graph (in clustering, this means finding all distances between pairs of points – but not all possible clusters; just the points).
(2) Sort the edges by length (if there are *N* points, we have N^2 edges).
(3) Beginning with the shortest edge, for each edge, between nodes u and v, perform the following operations:
 (3.1) $A = \text{FIND}(u)$
 (3.2) $B = \text{FIND}(v)$
 (3.3) IF $(A \neq B)$ THEN $C = \text{UNION}(A, B, C)$, and erase sets *A* and *B*.

Generally, in step 3.3, we index each set by an integer index, and rather than discarding indices when A and B are erased, we use the index for A or for B (whichever is smaller) as the index for the new set C.

As has been discussed in the literature [15.1, 15.8] there exist parallel algorithms for performing the union–find operations in constant time. The parallel algorithm assumes the existence of a lookup table which performs the FIND operation. Then a UNION in parallel hardware is implemented as follows: to do a UNION–FIND operation on points u and v:

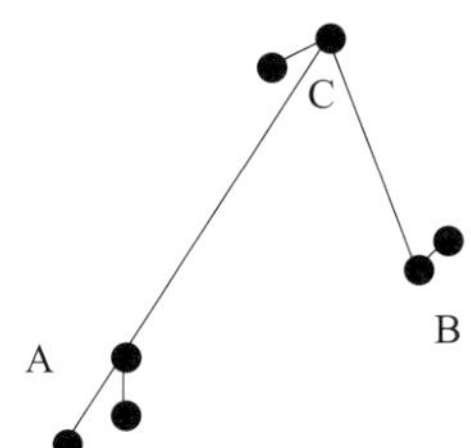

Fig. 15.5. Using d_{max} and illustrating the distances chosen at each iteration by a line, we derive a clustering very similar to that using d_{min}.

(1) Register 1 <= lookup (u)
(2) Register 2 <= lookup (v)
(3) If XOR(register1, register2), *rewrite* (minimum (u, v), maximum (u, v)), where *rewrite* is a parallel operation.

rewrite (x, y) will set all locations of the lookup table which contain y to x.

Now you know how agglomerative clustering works. You know also that agglomerative clustering with the d_{min} distance function will result in the minimum spanning tree of the set of points, and furthermore that this tree may be computed using graph-theoretic and/or parallel algorithms.

It is interesting to note how d_{max} would cluster the same data. This result is shown in Fig. 15.5. However, as illustrated by Figs. 15.6 to 15.8, this is not always the

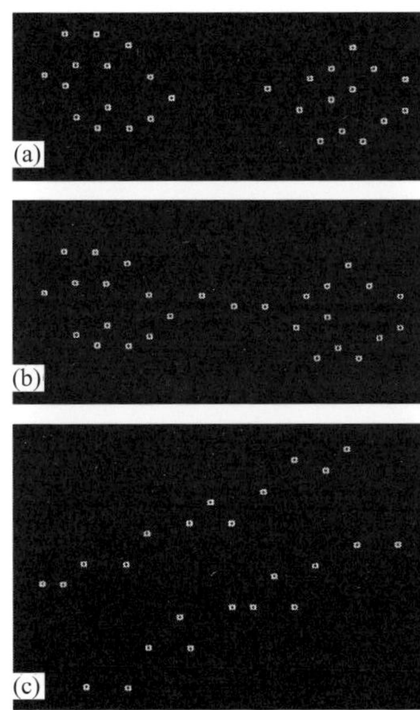

Fig. 15.6. Three examples of two-dimensional clustering problems (from [14.3]). Used with permission.

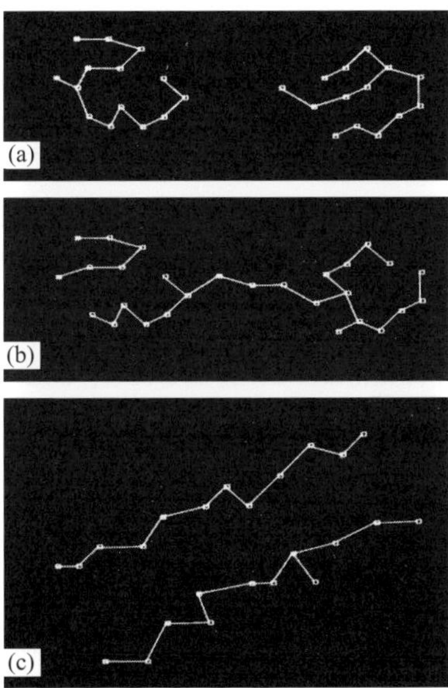

Fig. 15.7. The result of using minimum distance metric on the example of Fig. 15.6 (from [14.3]). Used with permission.

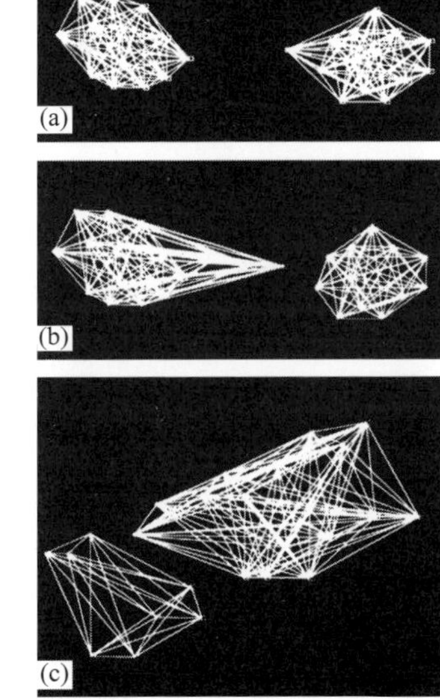

Fig. 15.8. The result of using the maximum distance metric d_{max}, to cluster the data of Fig. 15.6 (from [14.3]). Used with permission.

case. In particular, $d_{\min}$ will tend to choose clusters which are long and thin, and $d_{\max}$ will choose clusters which are basically round. Students often get confused right here, concerning the maximum and minimum criteria, so let us spend just a few extra words to reiterate what we are doing: In this algorithm (the agglomerative clustering algorithm) we are ALWAYS merging the two CLOSEST clusters. We get into the maximum distance thing when we define the distance between the clusters. $d_{\max}$ says to use as a measure of distance between the clusters, the distance between those points, in those clusters, whose mutual distance is maximum.

15.2.2 *k*-means clustering

There is, of course, another way to do clustering. In fact there are several ways. The *k*-means algorithm is probably the most popular, and is described as follows:

Algorithm: *k*-means clustering

Step 1. In an arbitrary way, assign samples to clusters. Or, if you don't like being that arbitrary, choose an arbitrary set of cluster centers and then assign all the samples to the nearest cluster. How you pick the cluster centers is problem-dependent. For example, if you were clustering points in a color space, where the dimensions are red, green, and blue, you might scatter all your cluster centers uniformly over this 3-space, or you might put all of them along the line from 0, 0, 0 to maxred, maxgreen, maxblue.

Step 2. Compute the mean of each cluster.

Step 3. Reassign each sample as belonging to the cluster with the nearest mean.

Step 4. If nothing changed this iteration, exit, otherwise go to step 2.

In Fig. 15.9 we illustrate the use of *k*-means to identify the peaks in a Hough accumulator array. We use the same accumulator array illustrated in Fig. 11.5. Each point in the accumulator array is treated as if it contains a number of points equal to the value of the number stored at that point. The initial cluster centers were chosen well separated and far from the actual centers. The simplest implementation of *k*-means does not work in this application because there are very many points in which the accumulator array contained only one point. All those ones add up to place the mean at a point not near the peak we seek. The solution is to simply ignore points with low values. In this example, any point not containing at least three points was ignored. Other heuristics are possible (see Assignment 15.4).

The ISODATA algorithm [15.2] extends *k*-means. ISODATA allows the algorithm to pick its own version of the number of clusters, and provides for more flexibility in specifying maximum and minimum cluster sizes.

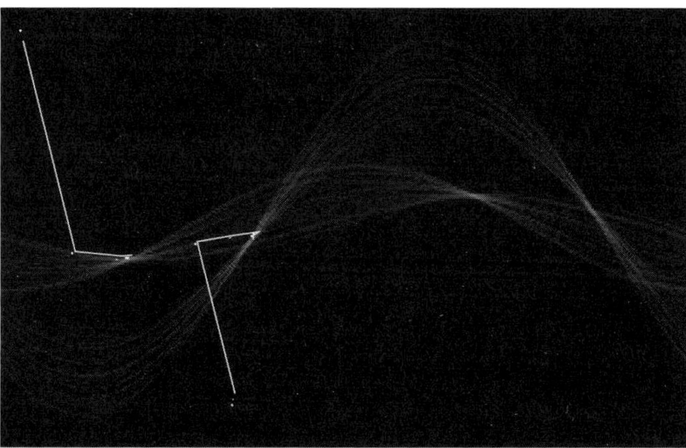

Fig. 15.9. Path followed by two cluster centers initialized far from their final positions in a Hough accumulator array. The length of the lines indicates the move from one estimate of the center to the estimate calculated in the next iteration.

15.3 Optimization methods in clustering

Let's see if we can figure out how to do clustering in a rigorous, formal way, by posing and solving an optimization problem.

We want to find the best clusterING. That is, that assignment of points to clusters which is in some sense the best assignment. In order to begin to approach this problem, we must first invent a scalar measure of the quality of an assignment. As a reasonable start on defining such a measure, consider the within-class scatter

$$S_{\mathrm{w}} = \sum_{i=1}^{c} \sum_{x \in X_i} (x - \mu_i)(x - \mu_i)^{\mathrm{T}} \tag{15.14}$$

where μ_i is the mean (centroid) of cluster X_i. S_{w} provides a good measure of average deviation of training set points from their means. However, in order to do minimization, we need a scalar measure of S_{w}, such as the trace or determinant.

The trace of S_{w}

$$\mathrm{Tr}(S_{\mathrm{w}}) = \sum_{i=1}^{c} \sum_{x \in X_i} \mathrm{Tr}((x - \mu_i)(x - \mu_i)^{\mathrm{T}}) = \sum_{i=1}^{c} \sum_{x \in X_i} (x - \mu_i)^{\mathrm{T}}(x - \mu_i) \tag{15.15}$$

is simply the sum of the squared deviations of the points from their means. The principle disadvantage of the trace criterion is that when used in a clustering algorithm, it can yield different results when the variables are scaled.

The determinant of S_{w} is invariant to axis scaling, but use of the determinant imposes the assumption that all clusters have roughly the same shape.

Once we have chosen an optimization criterion, whether it be $\mathrm{Tr}(S_w)$, $\det(S_w)$, or other criterion, we need to search the space of all possible clusterings of the points, in order to find the best clustering. Thus, we need to consider optimization methods. In this section, we consider only the method known as "branch and bound," but another, possibly more important method is given in section 2.3.3, simulated annealing.

15.3.1 Branch and bound

Suppose we have chosen a *scalar-valued* criterion function to minimize, let that criterion be called J. (Why are these things always called either J or H? You don't know? Go look up Hamiltonians and Jacobians.) We will present this optimization method by an example. Let's suppose we have four things, I, J, K, and L to assign into two clusters, cluster 1 and cluster 2. Define some notation.

1112 means we have assigned I, J, and K to cluster 1 and L to cluster 2; and
112X means we have assigned I and J to cluster 1, K to cluster 2, and have not yet decided to which cluster to assign L.

Assumption
(required to make this technique work). Adding a point to a clustering always *increases* the cost, independent of what that decision is. That is, suppose we have evaluated the cost of 112X; that is, assume J(112X) $= a$. When we make the decision about L, we know that the cost increases. J(1121) $> a$ and J(1122) $> a$.

With that assumption, we can define the branch and bound search algorithm. Since we have no better way to search, let's just start evaluating possibilities in order – let's try 1XXX, 11XX, 111X, 1111, 1112, 112X, etc. Suppose, along the way, we have determined that J(12XX) is greater than J(1112). Then, there is no point in evaluating any of the children of 12XX, since, if adding a decision to a cluster can only INCREASE the criterion function, we can only get a higher answer than we already have. That is the essence of the branch and bound algorithm. It's a search of all possibilities, enumerated in sequential order, but remembering the lowest value encountered and culling the search tree based on the assumption given above.

15.3.2 Vector quantization

In the research area known as "vector quantization," the computer is presented with a set of n vectors, and is to find natural groupings among the vectors. Said another way, the computer is to find a set of c "reference vectors" which represent the set of n in some optimal way. If you think this sounds like clustering you are correct. It is in fact precisely clustering; so let's call it that.

15.3.3 Winner-take-all approaches

The winner-take-all (also called *competitive learning*) approach originally resulted from researchers who were interested in modeling the cognitive process we know as "generalization," and therefore each reference vector/cluster center was represented by a mathematical construct which modeled a neuron. It will not be necessary for us to go into the physiological model of neurons here. Instead, we use the term "cluster center" for what some other material might refer to as a "neuron."

Each cluster center ω_i has an associated weight vector which we will refer to by the same name $\omega_i = [\omega_{ij}]^{\mathrm{T}}$, $j = 1, \ldots, d$. Note that the vector ω describes a location in a d-space. To do clustering, we present an input vector $[v_1, v_2, \ldots v_d]^{\mathrm{T}}$. Define the *winner* as the cluster center closest (in whatever sense is appropriate) to the input vector. That is,

$$d(\omega_i, v_j) \leq d(\omega_k, v_j) \quad \forall (k \neq i). \tag{15.16}$$

Compare this with the step size in gradient descent. Can you speculate on a relationship between the two algorithms?

Suppose cluster center α wins. Then, adjust the weights of α using

$$\omega_{\alpha j} \Leftarrow \omega_{\alpha j} + \varepsilon(v_j - \omega_{\alpha j}) \tag{15.17}$$

where the scalar ε is known as a "learning parameter." Typical values of this parameter are small, on the order of 0.01. The input data is presented to the algorithm repeatedly, in randomized order. Each presentation of the entire data set is called an *epoch*. After several epochs, the cluster centers will move to accurately represent the clusters in the data.

Often, a single cluster is always chosen. Since this is clearly not desirable, to allow other cluster centers to sometimes be chosen, we include a parameter known as *loneliness*. On each epoch, any cluster center that did not win any points has its loneliness increased slightly. The choice of cluster centers is then made with the decision strongly biased by the distance between the cluster center and slightly by the loneliness.

Kohonen feature maps

A Kohonen map is a clustering algorithm which is an extension of the winner-take-all algorithm. In this extension, a problem-dependent *topological distance* is assumed to exist between each pair of the cluster centers. Then, when the winning cluster center is updated, so are its neighbors in the sense of this topological distance. The primary cluster, ω_i, (the winner) is updated as in Equation (15.17). Other clusters are updated by

$$\omega_j^{k+1} = \omega_j^k + F(\eta, d_{ij})\big(v - \omega_j^k\big), \tag{15.18}$$

where v is the data presented to the algorithm at this iteration; F is some nonincreasing scalar function of d_{ij}, a measure of the distance between clusters i and j; and η is a maximum on that distance. This algorithm is easily programmed and converges to excellent clusterings.

15.4 Conclusion

As we have noted, the form of the clustering algorithm significantly affects the results of the clustering. Some attempts to reduce the dependency on algorithm form have been made [15.3], but this remains a fertile area for new ideas.

We view clustering as a collection of methods for determining consistency (as in determining the peaks of the Hough transform) but not for using consistency to solve other problems.

Minimize the trace of the scatter matrix.

Clustering algorithms are totally dependent on optimization methods. In section 15.3, we minimized the trace of the scatter matrix to find a good clustering measure.

We used branch and bound to speed up the combinatorial problem which results when points must simply be switched between clusters.

Gradient descent.

Although couched in the terminology of neural networks, the winner-take-all methods of section 15.3.3 use Eq. (15.17) (which is quite reminiscent of gradient descent) to find the best cluster center.

15.5 Vocabulary

You should know the meaning of the following terms.

Agglomerative clustering
Bhattacharyya distance
Branch and bound
Chernoff distance
Cluster
Competitive learning
Distance
Euclidean distance
Furthest neighbor distance
k-means
Kohonen map
Mahalanobis distance
Minimum spanning tree
Nearest neighbor distance
Union–find

Assignment 15.1

Prove that for the equal-covariance case, the Bhattacharyya distance becomes the same as the measure given in Eq. (15.7).

Assignment 15.2

The following points are to be partitioned into two clusters:

[0,0],[0,1],[0,2],[0,3],[0,4],[0,5],[0,7],[0,8].

Sketch the points and indicate the minimum spanning tree.

Using d_{min}, find and identify the two clusters. You may do this graphically if you wish.

Using d_{max} also find and identify the two clusters.

Can you suggest another distance measure to use in this case? Discuss.

Assignment 15.3

In your images directory are three images, called

facered.ifs
faceblue.ifs

and (can you guess?)

facegreen.ifs

These are the red, blue, and green components of a full color image. Each pixel can be represented by 8 bits of red, 8 of green and 8 of blue. Therefore, there are potentially 2^{24} colors in this image. Unfortunately, your workstation (probably) only has 8 bits of color, for a total of 256 possible colors. Your mission (should you choose to accept it) is to figure out a way to display this picture in full color on your workstation.

Approach: Use some kind of clustering algorithm. Find 128 clusters which best represent the color space, and assign all points to one of those colors. Then, make a file with the following data in it:

```
brightness_value red green blue
Example
1 214 9 3
```

This means that if a pixel has brightness 1, it should be displayed on the screen as 214 of red, 9 of green and 3 of blue. Such a pixel will show up as almost pure red. Thus, each cluster center is represented by a color. In the example above, cluster center 1 is the almost pure red point. Now, make an image in which every pixel has the brightness equal to the cluster number of the nearest cluster.

Assignment 15.4

In Fig. 15.9 a Hough accumulator is illustrated. That same accumulator array is available on the CDROM as hough.ifs. Invent a new way to find the peaks using a clustering algorithm. (Do NOT simply find the brightest point.) Suggestions might include weighting each point by the square of the accumulator value, doing something with the exponential of the square of the value, etc.

References

[15.1] R. Anderson and H. Woll, "Wait-free Parallel Algorithms for the Union-Find Problem," *Proc. 22nd ACM Symposium on Theory of Computing*, pp. 370–380, New York, ACM Press, 1991.

[15.2] G. Ball, "Data Analysis in the Social Sciences: What about the Details?" *Proc. AFIPS Fall Joint Computer Conference*, Washington, DC, Spartan Books, 1965.

[15.3] G. Beni and X. Liu, "A Least Biased Fuzzy Clustering Method," *IEEE Transactions on Pattern Analysis and Machine Intelligence*, **16**(9), 1994.

[15.4] G. Carpenter and S. Grossberg, "A Massively Parallel Architecture for a Self-organizing Neural Pattern Recognition Machine," *Computer Vision, Graphics, and Image Processing*, **37**, pp. 54–115, 1987.

[15.5] G. Carpenter and S. Grossberg, "ART-2: Stable Self-organization of Stable Category Recognition Codes for Analog Input Patterns," *Applied Optics*, **26**(23), p. 4919, 1987.

[15.6] G. Carpenter and S. Grossberg, "ART-3: Hierarchical Search using Chemical Transmitters in Self-Organizing Pattern Recognition Architectures," *Neural Networks*, **3**, pp. 129–152, 1990.

[15.7] D. Hand, *Discrimination and Classification*, New York, Wiley, 1989.

[15.8] W. Snyder and C. Savage, "Content-Addressable Read–Write Memories for Image Analysis," *IEEE Transactions on Computers*, **31**(10), pp. 963–967, 1982.

16 Syntactic pattern recognition

Ours is the age of substitution:
Instead of language, we have jargon;
Instead of principles, slogans;
and instead of genuine ideas, bright suggestions.

Eric Bentley

In this chapter, we discuss a completely different approach to pattern recognition, a methodology based on an analogy to language understanding. These methods have not often been used recently, primarily because they are very sensitive to noise and distortion. However, for certain applications, they may be appropriate, and the student is advised to learn enough about this topic to recognize potential applications.

Consider a boundary segment represented by a chain code. Each step in that chain code is a symbol, an integer between 0 and 7, so that the boundary segment is represented by a string of symbols: What makes syntactic methods work is the analogy between this string of symbols and the string of symbols which show up in the description of a formal language.

16.1 Terminology

To make more progress in this area, we need to define some terminology. The definitions are in reference to analysis of strings of symbols, such as occur in language analysis.

Terminal symbol.

A *terminal symbol*: A word, like "horse," "aardvark," "professor," "runs," "grades." Terminal symbols may also be line segments, parts of a picture, or other features. Generally, we denote terminal symbols using lower case. Most often, terminal symbols are denoted by a single symbol, like "a" or "0" but in the example of words from English, the terminal symbols are words, not letters.

Nonterminal symbol.

A *nonterminal symbol*: A symbol that describes a grammatical construct, like "noun," "verb," "verb phrase," "adverb phrase," etc. Generally, we denote terminal symbols with an upper case character such as "A" or a string of upper case characters like "VP" abbreviating "Verb Phrase."

Table 16.1. *Productions in a simple grammar. The start symbol is S.*

S	>	NP VP
VP	>	VP ADV
VP	>	V
NP	>	ADJ N
NP	>	ADJ NP
NP	>	ART N
N	>	horse
N	>	professor
V	>	runs
V	>	sleeps
ADV	>	quickly
ADJ	>	green
ART	>	the

Grammar.

A *grammar*: A set of terminal symbols, a set of nonterminal symbols, and a set of rules that generates a set of strings. The critical technical component of syntactic pattern recognition is the fact that for every grammar there exists a machine (e.g. finite state machine, pushdown automaton, Turing machine) that recognizes the language generated by that grammar.

Production.

A rewriting rule, or a *production*, is an allowable substitution, including an arrow. The string of symbols to the left of the arrow may be replaced by the string on the right.

For a more thorough understanding of syntactic pattern recognition, see [16.5].

We will discuss "traditional" formal languages (as defined by Chomsky [16.2]) in this section. However, current research continues, including interest in stochastic grammars [16.3], in which the rewriting rules have probabilities associated with them. For a more thorough description of the theory of formal languages, see [16.7].

An example of a grammar is given below (see Table 16.1) for a limited set of English sentences.

Terminals: {horse, boy, professor, runs, green, the, quickly, sleeps}.

Start symbol: S.

Nonterminals: {S, VP, NP, N, V, ADV, ART, ADJ}.

Derivation.

In the application of a grammar, any production may be applied any number of times, in any order. A *derivation* is one instance of an application of the productions to the start symbol. For example,

S > NP VP > ART N VP > ART N V ADV > The professor sleeps quickly.

is a valid derivation in this grammar.

Language.

Several observations are in order about this grammar. The first is that it generates an infinite number of strings. This is demonstrated most clearly by the productions with the same nonterminal on both the left and right, such as NP > ADJ NP. The set of all strings which could possibly be generated by a grammar is called the *language* generated by the grammar.

If you know the grammar, you know how to build a recognizer.

For every grammar, there is a machine which will recognize the language generated by that grammar. Furthermore, the rules for building those machines are very straightforward. By recognize, we mean: *If the machine is provided an input which is a string in the language, then the machine will stop and indicate "yes, this string could have been generated by this particular grammar.*" Here is the key point. If one can devise a grammar which generates a particular set of strings, then one can easily build a recognizer, and it is usually easier for us humans to devise generators than recognizers.

16.2 Types of grammars

The set of possible grammars can be divided into four categories, depending on restrictions on the types of allowable productions.

16.2.1 Type 0 grammars

In a type 0 grammar, any rewrite rule is allowable. The left-hand side of the production may contain any mix of terminal and nonterminal symbols. For example

abAaBc > abAaCCc

is allowable. Let's reiterate what this means: If, in the course of a derivation, the string abAaBc occurs, it may be replaced by abAaCCc. e.g. aardvabAaBcark could be replaced by aardvabAaCCcark.

No, we are not going to explain what a Turing machine is. Just assume that it is a machine which could be simulated by a computer program.

For any type 0 grammar, there is a Turing machine which recognizes the language generated by that grammar.

16.2.2 Type 1 grammars[1]

Any rewrite rule is allowable, except that strings are not permitted to get shorter. The left-hand side of the production may contain any mix of terminal and nonterminal

1 Type 1 grammars are sometimes called "context-sensitive," but we will not use that term because some students find it confusing.

symbols. For example,

abAaBc > aabCC

is not allowable, because the resulting string is shorter, however

bAaBc > aabbCC

is allowable.

No, we are not going to explain what an LR-automaton is. Just assume that it is a machine which could be simulated by a computer program.

For any type 1 grammar, there is an LR automaton which recognizes the language generated by that grammar.

16.2.3 Type 2 grammars[2]

The left-hand side of the production may contain only a single nonterminal symbol. For example

Bc > aabCC

is not allowable, because the LHS contains more than one symbol, however

B > aabbCC

is allowable.

Actually, we ARE going to explain what a pushdown automaton is. However, it is a machine which could be simulated by a computer program.

For any type 2 grammar, there is a pushdown automaton which recognizes the language generated by that grammar. We present one example of a type 2 grammar which you might find interesting.

EXAMPLE: TYPE 2 GRAMMAR.

Terminals {0,1}
Nonterminals: {S}
Start symbol: S

Table 16.2. *Productions in an example type 2 grammar.*

S	>	0S1
S	>	01

Is it ALL such strings? In particular, is the string 01 generated by this grammar?

What is the language generated by this grammar? Is it obvious to you? It is the set of strings of zeros followed by exactly the same number of ones, denoted $0^n 1^n$.

In this chapter, only type 2 and type 3 grammars are discussed in detail.

2 Also sometimes called "context-free" grammars.

16.2.4 Type 3 grammars

The left-hand side of the production may contain only a single nonterminal symbol. The right-hand side many contain only strings of the form "a" or "aA," that is, a single terminal symbol, or a single terminal followed by a single nonterminal.

For example,

$$\mathrm{B} > \mathrm{aCCb}$$

is not allowable, because the RHS is not of an allowable form, however

$$\mathrm{B} > \mathrm{aC}$$

is allowable.

16.3 Shape recognition using grammatical structure

In this section, we illustrate a few examples of the use of the syntactic approach to image shape recognition.

16.3.1 Type 3 grammars

The reader familiar with FSMs will recognize this as defining a Mealy machine (or is it a Moore machine? You look it up).

For any type 3 grammar, there is a finite state machine (FSM) which recognizes the language generated by that grammar. A FSM is a system which may exist in a finite number of states[3], denoted by upper case letters, and has a finite number of input symbols, denoted here by lower case letters or numbers. Its operation is governed by a set of transition rules of the form $\delta(A, a) = B$; that is, when the machine is in state A and receives input a, it will change to state B. The output of the machine depends only on its state. The machine produces an output when it is in an "accepting" state.

Nondeterministic machines.

To derive the machine from the grammar requires two steps, first, we construct a strange artifice called a *nondeterministic finite state machine*. Then, we convert that machine into one which we can actually build. Let's do that for a grammar which describes regular ECGs.[4] First, a bit about how your heart beats.

A normal heart beat is sketched in Fig. 16.1. It illustrates several waves. The P wave is generated by the electrical signal, "depolarization," which initiates the contraction of the two atria, the smaller chambers of the heart. After the atrial beat, the ECG returns to the isoelectric line, denoted i, for a short period of time, to allow the larger chambers, the ventricles, to fill. Then the ventricles depolarize, creating the QRS (denoted by just R) signal. In a healthy heart, the signal again returns to the isoelectric line until the ventricles repolarize, creating the T wave. Another isoelectric period occurs until the next P wave. So a healthy heart produces

[3] "A finite state machine is a system which may exist in a finite number of states." We can't believe we actually said that!

[4] An ECG is an electrocardiogram. The familiar expression "EKG" comes from the German spelling of the word.

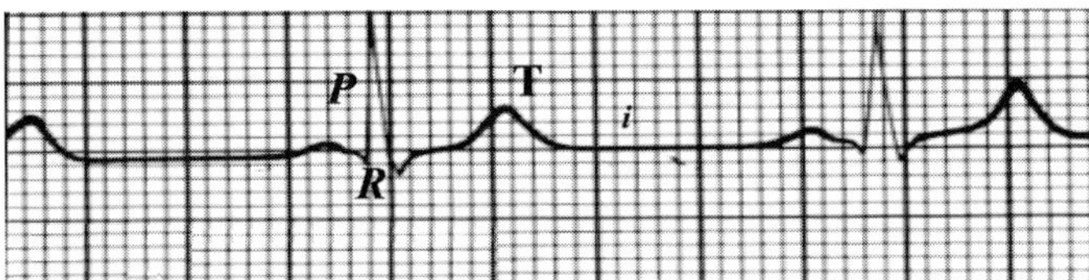

Fig. 16.1. A normal heart beat.

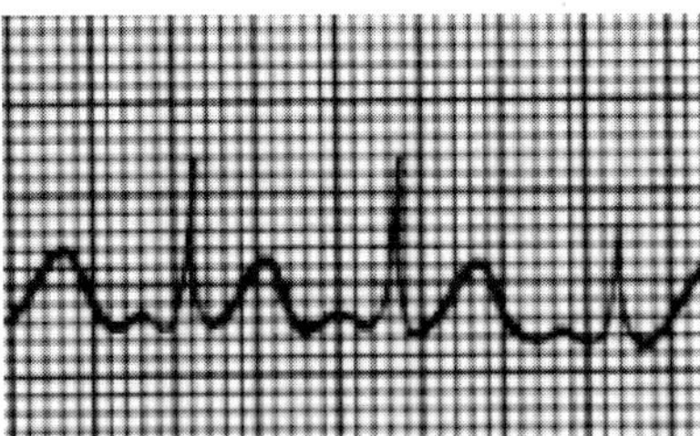

Fig. 16.2. Sinus Tachycardia: Insufficient time between *T* and *P* ***piritip***.

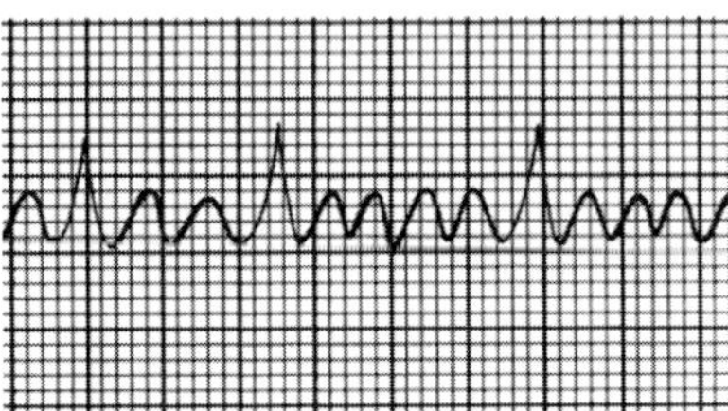

Fig. 16.3. Atrial flutter: Uncontrolled *P* waves, ***pipipirpipipir***.

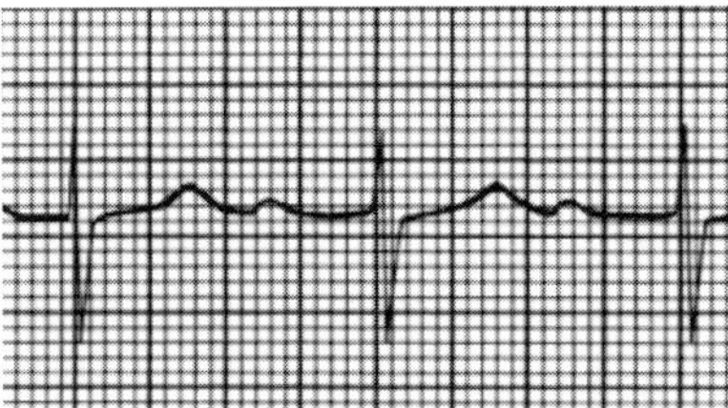

Fig. 16.4. Atrioventricular block: Delay between the *P* and the *R*, ***piiritipiiiritip***.

a sequence of symbols like **piritiipiritiipiritii** (repeat forever, or for 90 or so years anyway). A few of the things that could go wrong include the conditions illustrated in Figs. 16.2 to 16.5.

Of course, this is oversimplified, but it gives us a useful example to work on. Table 16.3 illustrates a grammar that generates normal heart beats.

To build the machine, which we will name M, we follow the following steps: Each nonterminal symbol in the grammar becomes a state in the machine. In addition one extra state is needed, denoted Q. Then, for each production of the form $A > aB$, construct a state change of the form $\delta(A, a) = B$. For each production of the form

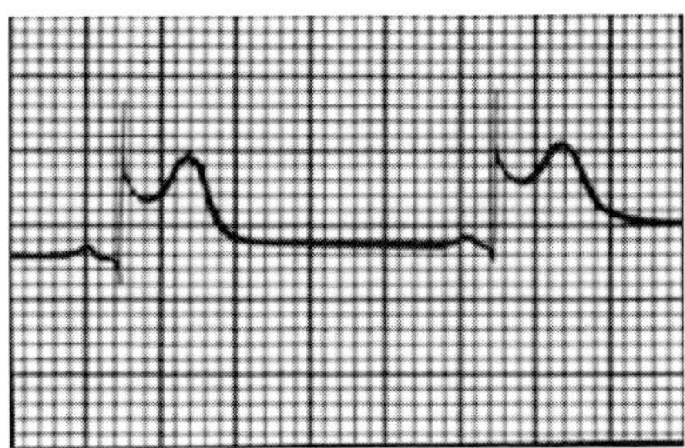

Fig. 16.5. Myocardial infarction: Signal does not return to isoelectric between *R* and *T*, ***pirtii***.

Table 16.3. *A type 3 grammar which generates normal ECGs.*

S	>	pA
A	>	iC
C	>	rD
D	>	iE
E	>	tF
F	>	iG
G	>	i
G	>	iH
H	>	i
H	>	iS

Table 16.4. *A nondeterministic FSM which recognizes ECGs.*

$\delta(S, p) = A$	$\delta(A, i) = C$	$\delta(C, r) = D$
$\delta(D, i) = E$	$\delta(E, t) = F$	$\delta(F, i) = G$
$\delta(G, i) = \{H, Q\}$	$\delta(H, i) = \{S, Q\}$	$\delta(Q, i) = \varphi$

$A > a$, construct a state change of the form $\delta(A, a) = Q$. Finally, if a is any input symbol, $\delta(Q, a) = \varphi$, where φ denotes the "empty" symbol.

The state change description of the machine which recognizes the language generated by this grammar is shown in Table 16.4.

Do you see why this is called "nondeterministic?" This machine goes from *H* to *Q* under input *i* or *H* to S under the same input *i*. We do not mean sometimes it goes to *Q* and sometimes to *S*. We mean it really does both, which of course is impossible in a physically realizable machine.

To convert this into something that could be built, we construct a machine M′ as follows.

The states of the new machine are all possible subsets of states of the original machine, including φ (but not all will necessarily be used). In this example, there

Table 16.5. *A deterministic FSM.*

$\delta([S], p) = [A]$	$\delta([A], i) = [C]$	$\delta([C], r) = [D]$
$\delta([D], i) = [E]$	$\delta([E], t) = [F]$	$\delta([F], i) = [G]$
$\delta([G], i) = [H, Q]$	$\delta([H], i) = [S, Q]$	$\delta([Q], i) = \varphi$
$\delta([H, Q], i) = [S]$	$\delta([S, Q], i) = A$	

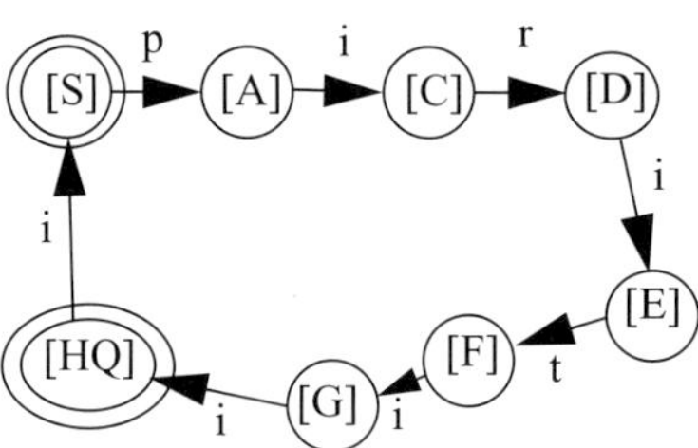

Fig. 16.6. Simple deterministic FSM which recognizes normal ECGs. Accepting states are circled.

are 2^9 such states. These states will be denoted by square brackets and a list of the original state names. If a transition involves such a set-valued state on the left, the new state will be the union of the states to which the original machine went. (Whew! was that awkward enough?) The accepting states of the new machine are any states involving Q or any states which were accepting states in the original machine. The accepting states of the new machine are any states containing accepting states of the original machine. This process produces the physically implementable machine illustrated in Table 16.5 and Fig. 16.6. In this example, although there are 2^9 states in the new machine, only a few of them are used.

Thus, we have a machine which recognizes heart rhythms. We hope you have observed that we have only listed what to do if the "normal" or expected input occurs. Now, just for the fun of it, we could modify the state diagram (Fig. 16.6) to include pathological conditions. For example, we could add $\delta(D, t) = Y$ which would cause a transition to an alarm state indicating the patient might be having a myocardial infarction (or aortic dissection or other bad stuff).

Next, we will give one more example of a type 3 grammar, this one using a chain code. But first, another way to describe regular languages: Regular expressions.

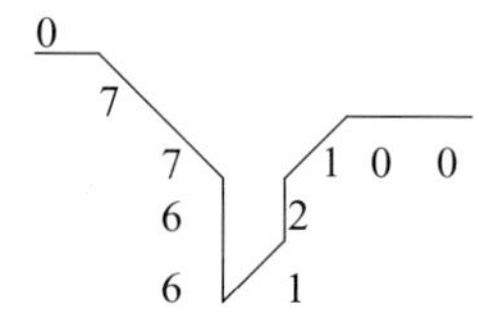

Fig. 16.7. Boundary segment corresponding to the chain code 0776612100.

Given a set of terminal symbols, T, a regular expression is a string constructed by concatenating elements of T and the symbol * (denoting repetition), with parentheses to delineate order of operation, and comma to denote the logical OR operation.

In this section, we will use the terminals $\{0, 1, 2, 3, 4, 5, 6, 7\}$ (the elements of a chain code).

One element of the language generated by (0, 7)(0, 7)*(7, 6)(7, 6)(61, 72)(1, 2)(1, 2)0(0, 1)* is illustrated in Fig. 16.7. The FSM which recognizes the set of all strings generated by this regular expression is given in Fig. 16.8.

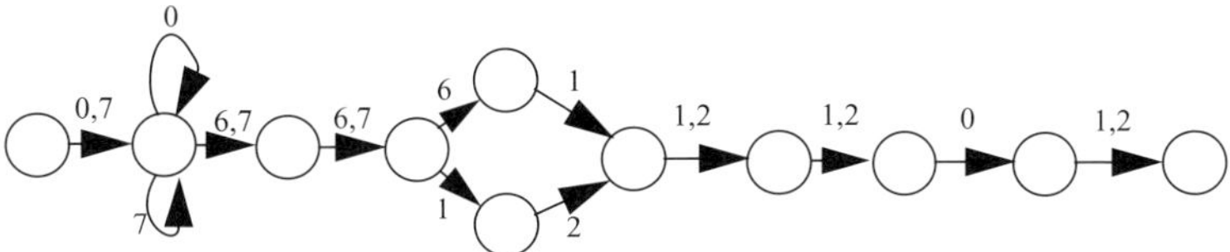

Fig. 16.8. State diagram for the nondeterministic FSM that recognizes strings generated by the regular expression above. Two numbers separated by a comma denote that either input will cause this transition. Any other input will cause a transition to an error state which is not shown.

In this example, we illustrate how a boundary segment may be represented by a chain code, and how a possibly infinite number of similar boundary segments may be recognized by the same machine.

There are many ways to represent images as strings other than chain codes, including for example the curve code [16.4, 16.8].

16.3.2 Type 2 grammars

Although type 3 grammars provide simple implementations as simple finite state machines, which can be built using only flip-flops and combinational logic, they do not have sufficient generality to solve many problems. In some applications, other types of grammars may be more appropriate. In this subsection we present two examples of shape recognition using type 2 grammars.

Recognition of chromosomes

The following example is abstracted from the text by Gonzalez and Thomason [16.6], based originally upon the work of Ledley *et al.* [16.9], and illustrates the use of a context-free grammar to recognize types of chromosomes.

The terminals in this grammar are boundary segments, denoted by a, b, c, d, and e, and illustrated in Fig. 16.9. In the recognition setting, these might be called boundary *primitives*. A chromosome will be described by a sequence of symbols a–e. Note that

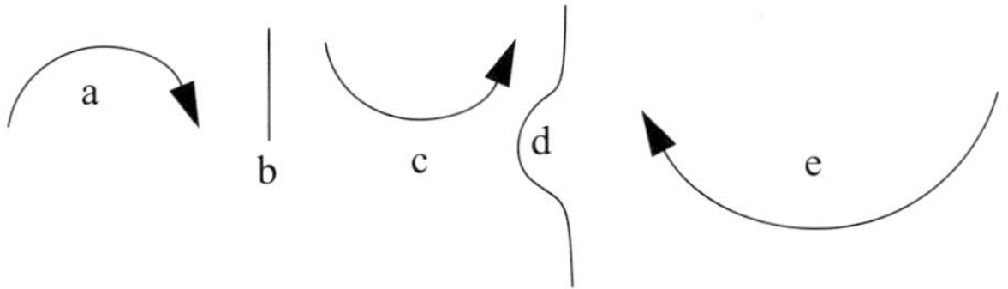

Fig. 16.9. Primitive boundary segments used for syntactic pattern recognition. Note that segment size and direction are important.

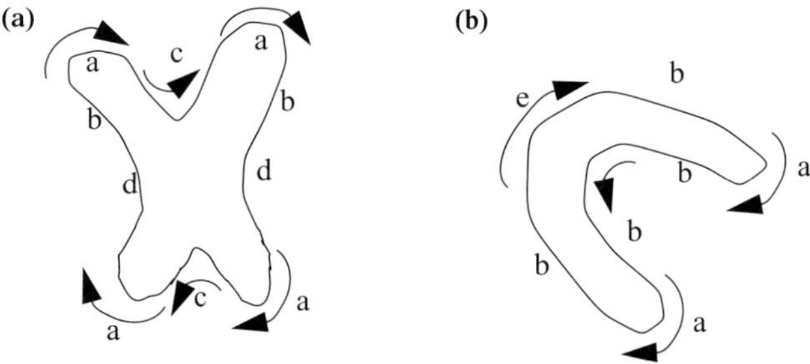

Fig. 16.10. (a) A submedian chromosome. (b) A telocentric chromosome. (Redrawn from [16.6].)

Table 16.6. *Productions to generate images of chromosomes.*

Submedian			Telocentric		
S	>	S1	S>	>	S2
S1	>	AA	S2	>	BA
A	>	CA	A	>	AC
A	>	DE	A	>	FD
B	>	bB	B	>	Bb
C	>	Cb	C	>	bC
D	>	Db	D	>	bD
E	>	cD	F	>	Dc
B	>	e	C	>	b
C	>	d	D	>	a

except for the symbol d, which can appear either way, the symbols have associated directions.

There are two types of chromosomes recognized by this grammar, telocentric and submedian, as illustrated in Fig. 16.10. Each may be described by a sequence of boundary segments. The following grammar (Table 16.6) will generate either type of chromosome.

These productions were not invented without some thought. The first two productions, those involving the start symbol, S, control which type of chromosome image is being generated: S1 denotes a submedian chromosome, and S2 designates a telocentric chromosome. In addition, the other symbols connote components of the chromosome boundary. That is, A will result in generation of *armpair*, B will result in generation of *bottom*, C will result in generation of *side*, D will result in generation of *arm*, E will result in generation of *rightside*, F will result in generation of *leftside*.

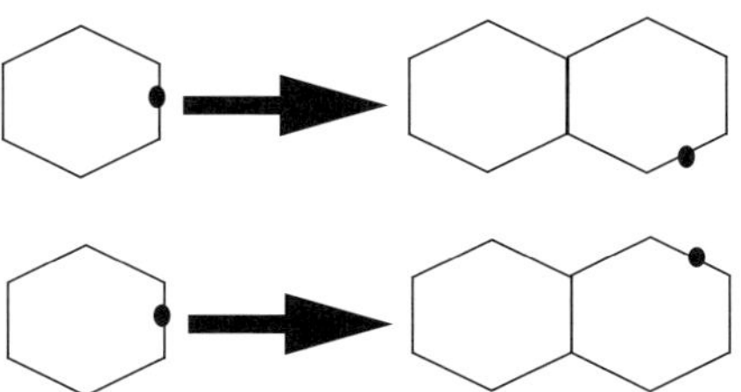

Fig. 16.11. Two of the productions used to generate a hexagonal texture. (Redrawn from [9.2].)

Shape grammars

Finally one last example from [9.2] makes use of shape grammars [17.57] to generate and recognize textures. In a shape grammar, both the set of terminal symbols, V_T, and the set of nonterminal symbols, V_N, are sets of shapes, with the restriction that $V_T \cap V_N = \emptyset$.

In this example, the set of terminals contains only one element, a hexagon:

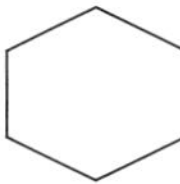

The set of nonterminals likewise contains a single element, a dot: •

The productions describe how to expand nonterminals, which are in the context of terminals, into additional shapes. For example, a single dot in the context of a hexagon might be against one specific side of the hexagon, and the production could allow replication of the hexagon at the point where the dot occurs. Many such rules make up the grammar, as illustrated by only two rules in Fig. 16.11.

Olstad and Torp [16.10] make use of syntactic methods to extend the capabilities of active contour methods – one of very few recent papers which make use of these methods in imaging – to distinguish occlusions.

Recognizers for type 2 languages

Pushdown automata.

Type 2 languages are recognized by *pushdown automata*. A pushdown automaton is a finite state machine which has been augmented with an unbounded memory in the form of a pushdown stack. Such a memory is a last-in-first-out memory. The operation of storing information in such a memory is called PUSH, and the operation of reading whatever is on the top of the stack is called POP. Note that POP does two things: It reads whatever is stored there, and it changes the memory, so the next POP operation will return the next-from-top value.

To implement a pushdown automaton, we add the symbol on the top of the stack to the criteria for the state change. That is: $\delta(A, i, j) = (C, q)$ now means "if the state is A, the input is i, and the symbol on the top of the stack is j, change to state

C and push the symbol q onto the stack." Using such a machine, it is now possible to recognize languages such as the example in section 16.2.3, where the number of ones must equal the number of zeros. The idea is simple: Every time we see a zero, push a zero onto the stack and stay in the same state. When we first see a one, change state and pop the stack. Subsequently, every time we see a one, pop the stack. If we ever see another zero, go to an error state. Else, when the stack goes empty, the number of ones is equal to the number of zeros, and go to an accepting state.

One of the principal concerns of practitioners of syntactic pattern recognition is illustrated well by both examples of chromosome recognition and ECG recognition. Both systems assume that a recognizer exists which can identify the primitives such as T waves. The underlying assumption is that such a primitive preprocessor would be simple, perhaps a template matcher, and robust to noise. In practice, this may be difficult to accomplish, and may require that the grammar itself be designed for some degree of noise tolerance. The reader is referred to textbooks [16.5, 16.6] on syntactic methods for details.

16.4 Conclusion

Syntactic methods do not rely on optimization methods or consistency except in how the terminal symbols are classified. A lower level algorithm is required to extract these features from images. This is in fact the principal weakness of syntactic methods. Because they must rely on other algorithms to provide input, they can fail in two different ways. The symbol recognizers may fail due to noise, blur, occlusion or other unexpected variation. Second, the object may simply not "look like" what the grammar was designed for. It may look similar, but there is no simple way to incorporate "similarity" into a grammar.

16.5 Vocabulary

You should know the meaning of the following terms.

Derivation
Finite state machine
Grammar
Nonterminal symbol
Primitive
Production
Pushdown automaton
Regular expression

Regular grammar
Shape grammar
Terminal symbol

Assignment 16.1

Show that the string representing the submedian chromosome in Fig. 16.10(a) can be generated by the grammar of Table 16.6.

Assignment 16.2

The statement was made earlier that the grammar of Table 16.6 is a type 2 grammar. Prove or disprove this statement.

References

[16.1] M. Chen, A. Kundu, and J. Zhou, "Off-line Handwritten Word Recognition using a Hidden Markov Model Type Stochastic Network," *IEEE Transactions on Pattern Analysis and Machine Intelligence*, **16**(5), 1994.

[16.2] N. Chomsky, "Three Models for the Description of Language," *IRE Transactions on Information Theory*, **2**(3), 1956.

[16.3] A. Corazza, R. De Mori, R. Gretter, and G. Satta, "Optimal Probabilistic Evaluation Functions for Search Controlled by Stochastic Context-free Grammars," *IEEE Transactions on Pattern Analysis and Machine Intelligence*, **16**(10), 1994.

[16.4] C. Fermuller and W. Kropatsch, "A Syntactic Approach to Scale-space-based Corner Description," *IEEE Transactions on Pattern Analysis and Machine Intelligence*, **16**(7), 1994.

[16.5] K.S. Fu, *Syntactic Pattern Recognition and Applications*, Englewood Cliffs, NJ, Prentice-Hall, 1982.

[16.6] R. Gonzalez and M. Thomason, *Syntactic Pattern Recognition*, Reading, MA, Addison-Wesley, 1978.

[16.7] J. Hopcroft and J. Ullman, *Introduction to Automata Theory, Languages, and Computation*, Reading, MA, Addison-Wesley, 1979.

[16.8] W. Kropatsch, "Curve Representation in Multiple Resolution," *Pattern Recognition Letters*, **6**(3), 1987.

[16.9] R. Ledley, L. Rotolo, R. Kirsch, M. Ginsberg, and J. Wilson, "FIDAC: Film Input to Digital Automatic Computer and Associated Syntax-directed Pattern-recognition Programming System," In *Optical and Electro-optical Information Processing*, ed. J. Tippet, D. Beckowitz, L. Clapp, C. Koester, and A. Vanderburgh, Cambridge, MA, MIT Press, 1965.

[16.10] B. Olstad and A. Torp, "Encoding of a priori Information in Active Contour Models," *IEEE Transactions on Pattern Analysis and Machine Intelligence*, **18**(9), 1996.

17 Applications

Example isn't another way to teach, it is the only way to teach

Albert Einstein

Machine vision has found a wide set of applications from astronomy [17.44] to industrial inspection, to automatic target recognition. It would be impossible to cover them all in the detail which they deserve. In this chapter, we choose to provide the reader with more of an annotated bibliography than a pedagogical text. We mention a few applications very briefly, and provide a few references. In the next chapter, we choose one application discipline, automatic target recognition, to cover in a bit more detail.

17.1 Multispectral image analysis

The strategy of multispectral image analysis combines spatial and spectral representations in a representation in which each pixel is a vector, an ordered set of measurements. Color, where the elements of the vector are $[r, g, b]$, is the obvious example, and there is a great deal of work in the literature in color processing. Most of the reported work has been intended for image quality enhancement. Only some recent papers elaborate on the use of color for recognition [17.14, 17.18, 17.53, 17.58].

The methods we have studied for univariate images, for example using Markov random field methods to remove noise, are applicable to multispectral images [17.3]. Often, all that is necessary is to use a vector description instead of scalar pixels.

17.2 Optical character recognition (OCR)

Despite our love for this topic and the huge number of papers devoted (e.g., in this paragraph, we cite only a few references [16.1, 17.32, 17.64]), we cannot take the space to cover it in the kind of detail it deserves. One of the first problems in OCR is automatic zoning [17.28]: that is, locating the text on the page [17.37]. Many OCR papers are special-purpose applications of techniques we have already covered, for example, thinning or skeletonizing.

17.3 Automated/assisted diagnosis

There is a growing application of (and need for) machine vision techniques in medicine. Tagari *et al.* [7.40] use Voronoi diagrams to develop a graph-based representation for understanding the topology of the components of the heart, relating 2D and 3D views. Measuring ejection fraction, the ratio of fluid ejected during a single beat to total ventricular capacity (particularly for the left ventricle) has attracted considerable attention with both methods which use nuclear medicine images [17.11] and those which use multiple conventional x-ray images [17.52].

Some representations are convenient for representing specific body parts and motions. For example, harmonic representations have found considerable application in representing the heart [8.40]. Gong and Kulikowski [13.16] use a planning strategy to identify features in MRI images.

Thermal imaging was really hot[1] in the early 1970s as it seemed to provide potential for identifying and possibly diagnosing conditions such as breast cancer. However, interest soon cooled as it became clear that the accuracy was not sufficiently high and the false positive rate was too high for the method to be useful. Therefore, despite its deployment in many areas of industry and military, the usage of thermal imaging in medicine declined [17.27]. Recently, several papers and studies have been published to reappraise the use of thermal infrared (IR) imaging in medicine [17.31] for the following three reasons. (1) We have much improved infrared technology. Largely due to developments in night vision systems for the Army, new generations of IR cameras have much enhanced accuracy. (2) We have much better capabilities in image processing. Advanced techniques including image enhancement, restoration and segmentation have been effectively used in processing IR images. (3) We have a deeper understanding of how IR images represent underlying pathophysiology of human systems [17.12, 17.42].

17.4 Inspection/quality control

One would suspect that the area of manufacturing inspection would be intensely researched, as industry strives to utilize the latest technology to gain competitive advantage. However, this seems not to be the case. For example, a conference was sponsored by the US National Science Foundation in the spring of 2000 to discuss

[1] We couldn't resist the pun!

how industry and universities might collaborate more effectively. Over two hundred CEOs of machine vision companies (most of which are involved in manufacturing inspection) were invited, and fewer than 30 showed up. Why was there such a poor turnout when the topic is seemingly so important?

One possible answer is that most machine vision companies are small. But one might ask, "Why are so many machine vision companies so small?" We speculate that the uniqueness of this field comes in the answer to that question. The capital investment required to set up a machine vision company is actually quite small. One can get into the business with a computer, some inexpensive hardware, and some good ideas. Unless you go into custom hardware or sophisticated specializations, you might be able to get your company running without venture capital. You do not need to be big to be in the machine vision business. Because the companies are so small, they are intensely market-driven, and do not see that basic academic research can help them in the short term. Sometimes they are right.

Still, some basic research in industrial inspection is getting done, such as registration using fiducial marks [17.61], automatic extraction of features [17.62], recognition of overlapping parts [17.21], as well as applications in assembly [17.43].

If a company has been in the business of manufacturing a particular line of products for many years, it is likely that many of those product designs were not entered into CAD data bases. Reverse engineering is the process of going from legacy designs into modern data bases. It may require reading of blueprints [17.13], and it may also require generation of CAD models and data bases [17.59] from actual objects, which in turn may require extraction of geometric primitives such as spheres, cylinders, cones, etc. from range data [17.40], or other coordinate measuring machines.

Microscopy is another application area in which machine vision plays an important role. For example, Pap smears are often screened using automated systems, and white blood cell counts may be done by computers as well. Tracking tubular molecules in epi-fluorescence microscopy [17.46] has been recently reported in the research literature.

Many industrial parts are specular reflectors, and their shape and roughness can be extracted using multiple light sources [17.17, 17.56].

However it is done, and whatever the application, machine vision requires system building and work on sensor modeling, together with hypothesis generation [17.73].

17.5 Security and intruder identification

Intruder detection may involve identifying an individual using a combination of cues [17.6] as well as facial appearance. There has been a lot of research on face recognition which we do not have space to discuss [17.26].

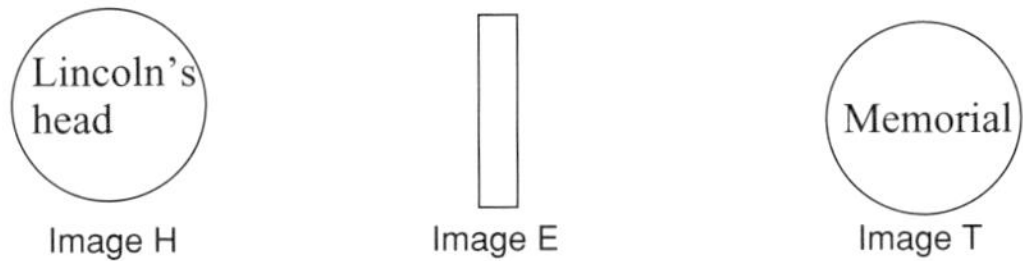

Fig. 17.1. Three possible views of a penny. Although possible, the view of a penny standing on edge is so unlikely as to be discarded.

17.6 Robot vision

Robot vision combines everything we have studied. Camera calibration [4.37] is an important first component, but the robot must not only identify, it must navigate, and it must therefore build maps [17.4, 17.60]. Sogo *et al.* recently [17.54] focused on qualitative maps, which imbed such information as INFRONTOF, but without calibrated position data.

In industrial automation, particularly robotics parts handling, a great deal of use is made of the concept of "stable states," which are best illustrated by an example. Three images of a penny which we might observe are shown in Fig. 17.1. Clearly, if we see image H or T, we are likely to recognize it as a penny, but image E, although possible, is so unlikely a view of a penny that any "reasonable" machine vision system (including a human) would reject it as an interpretation (unless other knowledge were provided). Thus, a penny has two "stable states." Rather than treating all possible views of a penny, a machine vision system could have two models in its database, PENNYH and PENNYT, and treat them as separate objects. This concept of stable states works, of course, only for objects, like coins, which may exist in only a few possible configurations. The term "aspect" is often used for such configurations, (see section 12.6, where the aspect graph, a data structure representing all possible aspects of an object, is discussed in more detail).

This idea of stability may be generalized [17.71] to take into account the likelihood and stability of viewpoints, including what position is optimal in order to take the next look [9.93], which the authors refer to as "autonomous exploration," a type of active vision [17.39]. Another approach to tracking using active vision makes use of the assumption that all motion is in the plane [17.5].

17.6.1 Robot surgery

Robot-assisted surgery is becoming more important in applications where precise placement requirements are in excess of what a human can achieve. A common application is in brain surgery, where the head may be rigidly and precisely held [17.23, 17.34, 17.38]. In robot-assisted surgery it is necessary to match 3D medical images (MRI or CT) with 2D x-ray projections. This can be formulated as the estimation of the spatial pose of a 3D object from 2D images [17.36]. Recent robotic surgical successes include coronary artery bypass graft on the beating heart [17.9], stomach surgery [17.7], and gall bladder surgery [17.24.]

17.6.2 Robot driving

Robot navigation includes identifying and avoiding obstacles [17.74] as well as navigating on roads and off. Finding road edges from a moving vehicle has been approached using a variety of imaging modes [17.20, 17.65, 17.72] as well as ground level millimeter-wave radar [17.35].

We reiterate, robotic vision is really a system science. It draws components from all of the techniques described elsewhere in this book. For example, optic flow may be used to analyze camera motion and keep the camera trained on the target [17.49], Grosso and Tistarelli [13.18] combine stereopsis and motion. Zhang *et al.* [17.74] make use of the assumption that the robot is moving in the ground plane.

In fact, if one were to design a single project which would teach students the most about engineering, robotics would probably be the optimal topic.

The reader interested in robot vision should peruse the proceedings of the *IEEE International Conference on Robotics and Automation*.

Bibliography

[17.1] M. Anbar, *Quantitative Dynamic Telethermometry in Medical Diagnosis and Management*, Boca Raton, FL, CRC Press, 1994.

[17.2] M. Barzohar and D. Cooper, "Automatic Finding of Main Roads in Aerial Images by Using Geometric-stochastic Models and Estimation," *IEEE Transactions on Pattern Analysis and Machine Intelligence*, **18**(7), 1996.

[17.3] M. Berman, "Automated Smoothing of Image and Other Regularly Spaced Data," *IEEE Transactions on Pattern Analysis and Machine Intelligence*, **16**(5), 1994.

[17.4] Ö. Bozma and R. Kuc, "A Physical Model-based Analysis of Heterogeneous Environments using Sonar – ENDURA method," *IEEE Transactions on Pattern Analysis and Machine Intelligence*, **16**(5), 1994.

[17.5] K. Bradshaw, I. Reid, and D. Murray, "The Active Recovery of 3D Motion Trajectories and Their Use in Prediction," *IEEE Transactions on Pattern Analysis and Machine Intelligence*, **19**(3), 1997.

[17.6] R. Brunelli and D. Falavigna, "Person Identification Using Multiple Cues," *IEEE Transactions on Pattern Analysis and Machine Intelligence*, **17**(10), 1995.

[17.7] G. Cadiere, J. Himpens, M. Vertruyen, J. Bruyns, O. Germay, G. Leman, and R. Izizaw, "Evaluation of Telesurgical (Robotic) NISSEN Fundoplication." *Surg. Endosc.*, **15**(9), pp. 918–923, 2001.

[17.8] V. Caglioti, "Uncertainty Minimization in the Localization of Polyhedral Objects," *IEEE Transactions on Pattern Analysis and Machine Intelligence*, **16**(5), 1995.

[17.9] I. W. Chitwood and L. Nifong, "Minimally Invasive Videoscopic Mitral Valve Surgery: The Current Role of Surgical Robotics," *J. Card. Surg.*, **15**(1), pp. 61–75, 2000.

[17.10] I. Cox and S. Hingorani, "An Efficient Implementation of Reid's Multiple Hypothesis Tracking Algorithm and Its Evaluation," *IEEE Transactions on Pattern Analysis and Machine Intelligence*, **18**(2), 1996.

[17.11] X. Dai, W. Snyder, G. Bilbro, R. Williams, and R. Cowan, " Left-Ventricle Boundary Detection from Nuclear Medicine Images," *Journal of Digital Imaging*, February, 1998.

[17.12] N. Diakides, ed., Special Issue on Infrared Imaging, *IEEE EMBS Magazine*, June, 2000.

[17.13] D. Dori, "Vector-based Arc Segmentation in the Machine Drawing Understanding System Environment," *IEEE Transactions on Pattern Analysis and Machine Intelligence*, **17**(11), 1995.

[17.14] F. Ennesser and G. Medioni, "Finding Waldo, or Focus of Attention Using Local Color Information," *IEEE Transactions on Pattern Analysis and Machine Intelligence*, **17**(8), 1995.

[17.15] C. Fan, N. Namazi, and P. Penafiel, "A New Image Motion Estimation Algorithm Based on the EM Technique," *IEEE Transactions on Pattern Analysis and Machine Intelligence*, **18**(3), 1996.

[17.16] D. Fleet and K. Langley, "Recursive Filters for Optical Flow," *IEEE Transactions on Pattern Analysis and Machine Intelligence*, **17**(1), 1995.

[17.17] J. Franke and W. Snyder, "Determination of Part Pose with Unconstrained Moving Lighting," *International Conference on Robotics*, Atlanta, 1984.

[17.18] B. Funt and G. Finlayson, "Color Constant Color Indexing," *IEEE Transactions on Pattern Analysis and Machine Intelligence*, **17**(5), 1995.

[17.19] D. Geman and B. Jedynak, "An Active Testing Model for Tracking Roads in Satellite Images," *IEEE Transactions on Pattern Analysis and Machine Intelligence*, **18**(1), 1996.

[17.20] F. Gibbs and B. Thomas, "The Fusion of Multiple Image Analysis Algorithms for Robot Road Following," in *Image Processing and its Applications*, IEE, Edinburgh, 1995.

[17.21] W. Grimson and T. Lozano-Perez, "Localizing Overlapping Parts by Searching the Interpretive Tree," *IEEE Transactions on Pattern Analysis and Machine Intelligence*, **9**(4), 1987.

[17.22] Z. Haddad and S. Simanca, "Filtering Image Records Using Wavelets and the Zakai Equation," *IEEE Transactions on Pattern Analysis and Machine Intelligence*, **17**(11), 1995.

[17.23] S. Hayati and M. Mirmirani, "Improving the Absolute Positioning Accuracy of Robot Manipulators," *Journal of Robotic Systems*, 2, 1985.

[17.24] J. Himpens, G. Leman, and G. Cardiere. "Telesurgical Laparoscopic Cholecystectomy," *Surg. Endosc.*, **12**, p. 1091, 1998.

[17.25] Y. Huang and C. Suen, "A Method of Combining Multiple Experts for the Recognition of Unconstrained Handwritten Numerals," *IEEE Transactions on Pattern Analysis and Machine Intelligence*, **17**(1), 1995.

[17.26] X. Jia and M. Nixon, "Extending the Feature Vector for Automatic Face Recognition," *IEEE Transactions on Pattern Analysis and Machine Intelligence*, **17**(12), 1995.

[17.27] B. F. Jones, "A Reappraisal of the Use of Infrared Thermal Image Analysis in Medicine," *IEEE Transactions on Medical Imaging*, **17**(6), pp. 1019–1027, 1998.

[17.28] J. Kanai, S. Rice, T. Nartker, and G. Nagy, "Automated Evaluation of OCR Zoning," *IEEE Transactions on Pattern Analysis and Machine Intelligence*, **17**(1), 1995.

[17.29] T. Kanungo, R. Haralick, H. Baird, W. Stuezle, and D. Madigan, "A Statistical, Nonparametric Methodology for Document Degradation Model Validation," *IEEE Transactions on Pattern Analysis and Machine Intelligence*, **22**(11), 2000.

[17.30] A. Katz and P. Thrift, "Generating Image Filters for Target Recognition by Genetic Learning," *IEEE Transactions on Pattern Analysis and Machine Intelligence*, **16**(9), 1994.

[17.31] J. R. Keyserlingk, P. D. Ahlgren, E. Yu, N. Belliveau, and M. Yassa, "Functional Infrared Imaging of the Breast," *IEEE Engineering in Medicine and Biology*, May/June, pp. 30–41, 2000.

[17.32] G. Kopec and P. Chou, "Document Image Decoding using Markov Source Models," *IEEE Transactions on Pattern Analysis and Machine Intelligence*, **16**(6), 1994.

[17.33] A. Kumar, Y. Bar-Shalom, and E. Oron, "Precision Tracking Based on Segmentation with Optimal Layering for Imaging Sensors," *IEEE Transactions on Pattern Analysis and Machine Intelligence*, **17**(2), 1995.

[17.34] Y. Kwoh, J. Hou, E. Jonckheere, and S. Hayati, "A Robot with Improved Absolute Positioning Accuracy for CT Guided Stereotactic Brain Surgery," *IEEE Transactions on Biomedical Engineering*, **35**(2), 1988.

[17.35] S. Lakshmanan and D. Grimmer, "A Deformable Template Approach to Detecting Straight Edges in Radar Images," *IEEE Transactions on Pattern Analysis and Machine Intelligence*, **18**(4), 1996.

[17.36] S. Lavallée and R. Szeliski, "Recovering the Position and Orientation of Free-form Objects from Image Contours Using 3D Distance Maps," *IEEE Transactions on Pattern Analysis and Machine Intelligence*, **17**(4), 1995.

[17.37] K. Lee, Y. Choy, and S. Cho, "Geometric Structure Analysis of Document Images: A Knowledge-based Approach," *IEEE Transactions on Pattern Analysis and Machine Intelligence*, **22**(11), 2000.

[17.38] P. Le Roux, H. Das, S. Esquenzai, and P. Kelly, "Robot-assisted Microsurgery; Feasibility in a Rat Microsurgical Model," *Neurosurgery*, **48**, 2001.

[17.39] E. Marchand and F. Chaurmette, "Active Vision for Complete Scene Reconstruction and Exploration," *IEEE Transactions on Pattern Analysis and Machine Intelligence*, **21**(1), 1999.

[17.40] D. Marshall, G. Lukacs, and R. Martin, "Robust Segmentation of Primitives from Range Data in the Presence of Geometric Degeneracy," *IEEE Transactions on Pattern Analysis and Machine Intelligence*, **23**(3), 2001.

[17.41] N. Merlet and J. Zerubia, "New Prospects in Line Detection by Dynamic Programming," *IEEE Transactions on Pattern Analysis and Machine Intelligence*, **18**(4), 1996.

[17.42] J. Michel, N. Nandhakumar, and V. Velten, "Thermophysical Algebraic Invariants from Infrared Imagery for Object Recognition," *IEEE Transactions on Pattern Analysis and Machine Intelligence*, **19**(1), 1997.

[17.43] J. Miura and K. Ikeuchi, "Task-oriented Generation of Visual Sensing Strategies in Assembly Tasks," *IEEE Transactions on Pattern Analysis and Machine Intelligence*, **20**(2), 98.

[17.44] R. Molina, "On the Hierarchical Bayesian Approach to Image Restoration: Applications to Astronomical Images," *IEEE Transactions on Pattern Analysis and Machine Intelligence*, **16**(11), 1994.

[17.45] H. Nogawa, Y. Nakajima, Y. Sato, and S. Tamura, "Acquisition of Symbolic Description from Flow Fields: A New Approach Based on Fluid Model," *IEEE Transactions on Pattern Analysis and Machine Intelligence*, **19**(1), 1997.

[17.46] B. Parvin, C. Peng, W. Johnson, and F. Maestre, "Tracking of Tubular Molecules for Scientific Applications," *IEEE Transactions on Pattern Analysis and Machine Intelligence*, **17**(8), 1995.

[17.47] H. Qi, P. T. Kuruganti, and Z. Liu, "Early Detection of Breast Cancer Using Thermal Texture Maps," *IEEE International Symposium on Biomedical Imaging*: *Macro to Nano*, Washington, DC, July, 2002.

[17.48] G. Ravichandran and D. Casasent, "Advanced In-plane Rotation-invariant Correlation Filters," *IEEE Transactions on Pattern Analysis and Machine Intelligence*, **16**(4), 1994.

[17.49] S. Reddi and G. Loizou, "Analysis of Camera Behavior During Tracking," *IEEE Transactions on Pattern Analysis and Machine Intelligence*, **17**(8), 1995.

[17.50] D. Ringach and Y. Baram, "A Diffusion Mechanism for Obstacle Detection from Size-Change Information," *IEEE Transactions on Pattern Analysis and Machine Intelligence*, **16**(1), 1994.

[17.51] J. Rocha and T. Pavlidis, "A Shape Analysis Model with Applications to a Character Recognition System," *IEEE Transactions on Pattern Analysis and Machine Intelligence*, **16**(4), 1994.

[17.52] Y. Sato, M. Moriyama, M. Hanayama, H. Naito, and S. Tamura, "Acquiring 3D Models of Non-rigid Moving Objects from Time- and Viewpoint-invariant Images: A Step Toward Left Ventricle Recovery," *IEEE Transactions on Pattern Analysis and Machine Intelligence*, **19**(3), 1997.

[17.53] D. Slater and G. Healey, "The Illumination-invariant Recognition of 3D Objects Using Local Color Invariants," *IEEE Transactions on Pattern Analysis and Machine Intelligence*, **18**(2), 1996.

[17.54] T. Sogo, H. Ishiguro, and T. Ishida, "Acquisition and Propagation of Spatial Constraints Based on Qualitative Information," *IEEE Transactions on Pattern Analysis and Machine Intelligence*, **23**(3), 2001.

[17.55] W. Sohn and N. Kehtarnavaz, "Analysis of Camera Movement Errors in Vision-based Vehicle Tracking," *IEEE Transactions on Pattern Analysis and Machine Intelligence*, **17**(1), 1995.

[17.56] F. Solomon and K. Ikeuchi, "Extracting the Shape and Roughness of Specular Lobe Objects Using Four Light Photometric Stereo," *IEEE Transactions on Pattern Analysis and Machine Intelligence*, **18**(4), 1996.

[17.57] G. Stiny and J. Gips, *Algorithmic Aesthetics: Computer Models for Criticism and Design in the Arts*, University of California Press, 1972.

[17.58] M. Swain and D. Ballard, "Color Indexing," *International Journal of Computer Vision*, **7**(1), 1991.

[17.59] T. Syeda-Mahmood, "Indexing of Technical Line Drawing Databases," *IEEE Transactions on Pattern Analysis and Machine Intelligence*, **21**(8), 1999.

[17.60] H. Takeda, C. Facchinetti, and J. Latombe, "Planning the Motions of a Mobile Robot in a Sensory Uncertainty Field," *IEEE Transactions on Pattern Analysis and Machine Intelligence*, **16**(10), 1994.

[17.61] M. Tichem and M. Cohen, "Submicron Registration of Fudicial Marks using Machine Vision," *IEEE Transactions on Pattern Analysis and Machine Intelligence*, **16**(8), 1994.

[17.62] S. Trika and R. Kashyap, "Geometric Reasoning for Extraction of Manufacturing Features in Iso-oriented Polyhedrons," *IEEE Transactions on Pattern Analysis and Machine Intelligence*, **16**(11), 1994.

[17.63] L. Tsap, D. Goldgof, and S. Sarkar, "Nonrigid Motion Analysis Based on Dynamic Refinement of Finite Element Models," *IEEE Transactions on Pattern Analysis and Machine Intelligence*, **22**(5), 2000.

[17.64] T. Wakahara, "Shape Matching using LAT and its Application to Handwritten Numeral Recognition," *IEEE Transactions on Pattern Analysis and Machine Intelligence*, **16**(6), 1994.

[17.65] R. Wallace, A. Stentz, C. Thorpe, H. Moravec, W. Whittaker, and T. Kanade, "First Results in Robot Road-Following," *Proceedings of the International Joint Conference on Artificial Intelligence*, 1985.

[17.66] C. Wang, "Collision Detection of a Moving Polygon in the Presence of Polygonal Obstacles in the Plane," *IEEE Transactions on Pattern Analysis and Machine Intelligence*, **16**(6), 1994.

[17.67] C. Wang and W. Snyder, "MAP Transmission Image Reconstruction via Mean Field Annealing for Segmented Attenuation Correction of PET Imaging," *17th International Conference of the IEEE Engineering in Medicine and Biology Society*, Montreal, September, 1995.

[17.68] C. Wang and W. Snyder, "Frequency Characteristic Study Of Filtered-Backprojection Reconstruction And Maximum Reconstruction For PET Images," *17th International Conference of the IEEE Engineering in Medicine and Biology Society*, Montreal, September, 1995.

[17.69] C. Wang, W. Snyder, and G. Bilbro, "Performance Evaluation of Filtered Backprojection Reconstruction and Iterative Reconstruction Methods for PET Images," *Computers in Medicine and Biology*, **9**(3), 1998.

[17.70] C. Wang, W. Snyder, G. Bilbro, and P. Santago, "A Performance Evaluation of FBP and ML Algorithms for PET Imaging," *SPIE Medical Imaging*, 1996.

[17.71] D. Weinshall and W. Werman, "On View Likelihood and Stability," *IEEE Transactions on Pattern Analysis and Machine Intelligence*, **19**(2).

[17.72] J. Weng, and S. Chen, "Vision-guided Navigation using SHOSLIF," *Neural Networks*, 1998.

[17.73] M. Wheeler and K. Ikeuchi, "Sensor Modeling, Probabilistic Hypothesis Generation, and Robust Localization for Object Recognition," *IEEE Transactions on Pattern Analysis and Machine Intelligence*, **17**(3), 1995.

[17.74] Z. Zhang, R. Weiss, and A. Hanson, "Obstacle Detection Based on Qualitative and Quantitative 3D Reconstruction," *IEEE Transactions on Pattern Analysis and Machine Intelligence*, **19**(1), 1997.

18 Automatic target recognition

Luke, you've switched off your targeting computer. What's wrong?

George Lucas

This is the principal application chapter of this book.[1] We have selected one application area: Automatic target recognition (ATR), and illustrate how the mathematics and algorithms previously covered are used in this application. The point to be made is that almost all applications similarly benefit from not one, but fusions of most of the techniques previously described. As in previous chapters, we provide the reader with both an explanation of concepts and pointers to more advanced literature. However, since this chapter emphasizes the application, we do not include a "Topics" section in this chapter.

There are lots of transducers which can provide information about targets. In this book, we only consider imaging sensors.

Automatic target/object recognition (ATR) is the term given to the field of engineering sciences that deals with the study of systems and techniques designed to identify, to locate, and to characterize specific physical objects (referred to as targets) [18.7, 18.9, 18.69], usually in a military environment. Limited surveys of the field are available [18.3, 18.8, 18.21, 18.66, 18.74, 18.79, 18.89]. In this chapter, the only ATR systems considered are those that make use of images. Therefore, our use of terminology (e.g., clutter) will be restricted to terms that make sense in an imaging scenario.

18.1 The hierarchy of levels of ATR

In this section, we define a few popularly used terms and acronyms in the ATR [18.57] world, starting with the five levels in the ATR hierarchy.

Detection. Identifying the presence or absence of a target in a given scene.

Classification. This term, at least in Army parlance, originally meant distinguishing between vehicles with tracks and those with wheels. Since this definition was

[1] The authors are indebted to Rajeev Ramanath, who assisted significantly in the generation of this chapter, and in fact wrote some sections, and to Richard Sims, and John Irvine who provided careful reviews and extremely helpful feedforward.

generated however, most ATR developments have bypassed classification with respect to performance requirements. Furthermore, the US Army is moving slowly away from tracked vehicles and this definition will certainly be obsolete when that happens.

Recognition. Distinguish targets from similar kinds. For example, distinguish tanks from front-end loaders, jeeps from automobiles, rocket launchers from school buses, etc.

Identification. Identify the type of target such as the type of tank (whether it is T90 or M1, etc.).

Characterization. Describe the identified target in more detail. In Army parlance, this level of process characterizes the target based on how many and what types of weapons are on board, e.g., a T90 tank with an extra 55 gallon oil drum attached to the back.

Each level of the ATR hierarchy is a refinement to the target description. Target characterization reveals the most detail of the target.

18.1.1 ATR terminology

There are a few other terms that are often used in the ATR literature. We give the definitions as follows.

Chip. A small image usually containing the image of a single target, extracted from a large image of a scene. Target cueing algorithms, which identify the likely presence of a target, often produce chips as output.

Detection rate. Fraction of targets correctly detected by the system.

Classification rate. Fraction of targets classified correctly, or more generally, the conditional probability of correct recognition given the target was detected.

Clutter. Objects that are imaged but are not targets. Clutter typically may be trees, houses, and other vehicles – anything that is in the picture but is not target.

Cultural clutter. Refers to man-made objects like buildings, as opposed to natural objects.

False alarm rate. Generally, the fraction of the number of detections that do not correspond to actual targets. However, this definition may be modified if the task is classification rather than detection. We observe that false alarm rate is not the same as probability of false alarm. The false alarm rate is usually given in false alarms per square kilometer. See section 18.3.

FLIR (Forward-looking Infrared). This refers to images formed in the midwave (3–5 μm) and longwave (8–14 μm) spectral bands. The term "forward looking" is no longer really meaningful, but the acronym persists.

IFF. Identify friend or foe.

18.2 ATR system components

The algorithmic components of an ATR system can be decomposed into preprocessing, detection, segmentation and classification (Fig. 18.1), although it is certainly possible that specific system implementations might not have one or more of these components. For example, a "blob" tracker can simply track the center of gravity of the infrared image, if there is just one hot region in the field of view, and no explicit segmentation is required.

ATR systems "see" images as their inputs. There is a variety of imaging modalities which have inherent advantages as each of them "see" different properties of these targets under consideration. For example, a visually well camouflaged tank in a field may be hidden to the visible band, but clearly visible in the infrared bands, simply because its engines are running! Fig. 18.2 illustrates images taken from two different imaging modalities. Fig. 18.2(a) shows what is captured by a regular video camera. Fig. 18.2(b) shows the images captured by a FLIR camera. Note how the engine of the tank and Hummer can be seen as "hot."

Table 18.1 lists some of the commonly used spectral bands. The boundaries in wavelength of some of these ranges may vary from one user to another [18.9].

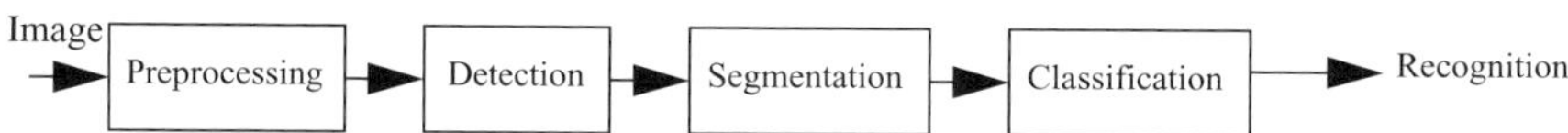

Fig. 18.1. Classical ATR system components.

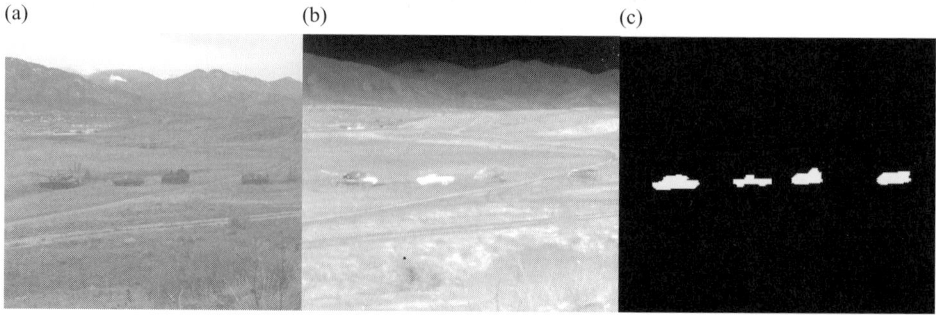

Fig. 18.2. Different imaging modalities. (a) Visible spectrum image. (b) Thermal IR image (notice hot tank engine on far left). (c) Ground truth (from [18.5]). Used with permission.

Table 18.1. *Popularly used spectral bands.*

Name assigned	Wavelength range	Source
Visible (V)	0.4–0.7 μm	Solar illumination
Near-infrared (NIR)	0.7–1.1 μm	Solar illumination
Shortwave infrared (SWIR)	1.1–2.5 μm	Solar illumination
Midwave infrared (MWIR)	3–5 μm	Solar illumination, Thermal
Thermal infrared (TIR) or Longwave infrared (LWIR)	8–12 μm	Thermal
Microwave, RADAR	1 mm–1 m	Thermal, Artificial

Given an image of the scene (which consists of a possible target and background), we need to detect the target. Target detection methods can be viewed as having two steps [18.86]. In the first step, appropriate measurements are extracted from an image using low-level image processing techniques. These measurements are then utilized to derive a primary segmentation of the image into regions. In the second step, higher-level descriptors of the segmented regions are used to determine the presence or absence of the target (detect) and possibly classify that target.

18.3 Evaluating performance of ATR algorithms

In this section, we consider some of the issues with evaluation of the performance of an ATR system. In this description, we use the term "classifier" assuming the job of the system is to correctly classify the targets in the scene. The same terminology – "false alarms," etc. – would be used if the objective of the system were any of the other levels of the ATR hierarchy.

For purposes of vocabulary, we will define these terms in the context of a pattern recognition problem which is to classify a result as "there" or "not there." Some applications of such classifiers are in automatic target detection (enemy target or not target), medical diagnosis (tumor or not tumor), and the digital communication channel (send a "1" or "0" at one end and receive it at the other). There will be several references back to Chapter 14, so you might want to glance at that first.

To illustrate a few key ideas in detection theory, we will begin with a simple example. Consider a digital communication system, sending a symbol a for "0" and b for "1".

We say we have hypothesis 0 (H_0) when a "0" is sent and hypothesis 1 (H_1) when "1" is sent. (In an ATR problem, we would have H_0 for a target absent and H_1 for a target present in the scene.) As the laws of nature state, there WILL be noise in the

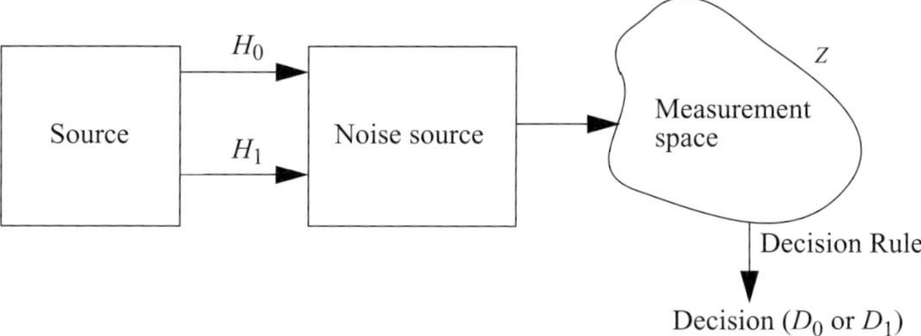

Fig. 18.3. Binary decision system.

system. So, when we send a "0" or in other words, a, we will receive $a + n$, where n is a noise sample. We have,

$$\begin{array}{ll} H_0 & Z = a + n \\ H_1 & Z = b + n, \end{array} \tag{18.1}$$

where Z is the received signal denoted a *binary hypothesis*.

Graphically we may summarize this decision-theoretic problem by Fig. 18.3.

Thus, D_0 and D_1 denote decisions, possibly incorrect decisions, made by the classifier, whereas H_0 and H_1 denote reality. We use $P(H_0)$ and $P(H_1)$ to represent prior probabilities or in a controlled experiment, *ground truth*.

There are four terms typically used in the evaluation of ATR algorithms:

True Positive. The object is there (e.g., the patient has a tumor; a "1" was transmitted) and our classifier says it is there (we decide the patient has a tumor; we received a "1").

True Negative. The object is not there (e.g., the patient does not have a tumor; a "0" was transmitted) and our classifier concludes no object (there is no tumor; we received a "0").

False Negative. The object is there (e.g., the patient has a tumor, a "1" was transmitted) and our classifier says otherwise (the patient has no tumor; we received a "0").

False Positive. The object is not there (e.g., the patient is healthy; a "0" was transmitted) and our classifier says there is an object (the patient has a tumor; we received a "1").

Clearly both "false" conditions are not favorable. False negative might be worse since we might miss a really dangerous target or overlook a malignancy. However, both types of error have associated with them different costs (refer to Chapter 14). The terms above are sometimes referred to by other names, such as "false alarms" (false positives) and "false misses" (false negatives). Based on these four values, two probabilities can be derived:

Probability of detection, P_{d}.

Sensitivity. Probability of a true positive, that is, the ratio between true positive and the summation of true position and false negative. $P(D_1|H_1)P(H_1)$. In the specific application of target detection (as distinct from classification, recognition, and identification), the sensitivity is referred to as the probability of detection and denoted P_{d}.

Specificity. Probability of a true negative, that is, the ratio between true negative and the summation of true negative and false positive. $P(D_0|H_0)P(H_0)$.

Observe that $P(D_i|H_i)P(H_i) = P(D_i, H_i)$. Now, the probability of a correct decision is

$$P(C) = P(D_0, H_0) + P(D_1, H_1) = P(D_0|H_0)P(H_0) + P(D_1|H_1)P(H_1).$$

18.3.1 ATR performance representation

In this section, we discuss how the performance of an ATR system may be characterized and estimated.

Once we have designed an ATR system, we invariably have some parameter we can adjust. One parameter might be a threshold on the brightness or pixels on target; in the quadratic classifier, it is the decision boundary; in the k-NN classifier (recall section 14.7), it is the number k, etc. How does performance vary with such parameters? Observe that it is not sufficient to design a system to have 100% sensitivity without considering specificity. A doctor who tells every patient seen "you have a tumor" would have 100% sensitivity. Instead, we wish to design systems which have as close to 100% sensitivity as possible while at the same time having as close to perfect specificity as possible. If we adjust the parameter(s), we can normally increase one performance measure at the expense of the other, as illustrated in Fig. 18.4.

In this figure, the *true positive fraction* (yet another term for $P(D_1|H_1)$) is graphed on the vertical axis vs. the *false positive fraction* on the horizontal axis. Each curve represents the performance of a particular classifier (or under a particular system setting), as some parameter is varied over its range. Of the three systems shown, the one with the sharpest bend, which passes closest to the upper left corner, is clearly the best. We can quantify this idea by calculating the area above the curve, and choosing the system with the smallest area.

Such curves are called "ROC" curves, a term which is an abbreviation for Receiver Operating Characteristic, a term from communication theory.

18.3.2 Generating ROC curves from training data

ROC curves can be generated by applying the classifiers to the training data. First of all, let's review the definition of probability of error.

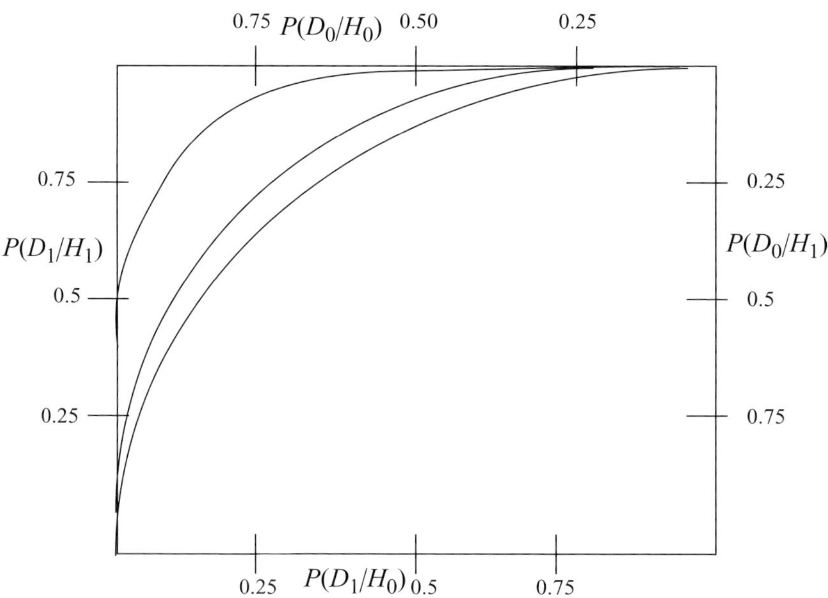

Fig. 18.4. Receiver Operating Characteristic (ROC) curve.

This comes from Bayes' rule. Take a look back at section 14.2.

In terms of decision regions, we can write the average probability of error as

$$\varepsilon = \int_{\Omega_0} P(H_1)p(z|H_1)\,dz + \int_{\Omega_1} P(H_0)p(z|H_0)\,dz, \tag{18.2}$$

where Ω_i is the region of measurement space over which we decide class i. $P(H_i)$ is the prior probability of seeing an example of class i, and $p(x|H_i)$ is the conditional probability density of whatever we are measuring (e.g., brightness) given that the example is from class i. Of course, we do not know the TRUE probability densities, we only know our estimates of those densities, determined from the training sets. Furthermore, we derived the decision regions Ω_i from those (estimated) densities.

We could try to determine the error rate by simply counting the number of elements in the training set that are misclassified: We call this the *apparent error rate*. Unfortunately, this leads to an optimistic result: It underestimates the error rate of the system when tested on data not in the training set. This occurs because the ATR has been designed to minimize the number of misclassifications of the training set, and unless the training set perfectly represents the true distribution of the data, the classifier will reflect characteristics of the training set which may not be true of the entire sample population.

True error rate is different from apparent error rate.

We must distinguish the apparent error rate from the *true error rate*. Although we have no way to determine the true error rate, we may get a better estimate using the two different approaches now described.

Separating the training set and the test set

The dimensionality of the problem is the number of measurements made (e.g., max. brightness, area, contrast).

This is straightforward. We simply divide our original training set into two parts (randomly of course) and build the classifier using half of the data. We then test the system on the other half. This approach works reasonably well if we have very large training sets (thousands of examples), or better said, something like 10^d where d is the dimensionality of the problem. (What does dimensionality mean here?) Unfortunately, such large training sets are unlikely in most problems.

The leave-one-out approach

Assume there are n points in the training set. Remove point 1 from the set, and design the classifier using the other $n - 1$ points. Then test the resulting machine on point 1. Repeat for all points. The resulting error rate can be shown to be an almost unbiased estimate of the expected true error rate for the classifier designed using all n points. Of course, this requires that we design n machines, which could be prohibitive. However, with such a result, we have numbers which we can put into the ROC curves.

18.3.3 ATR performance and system objectives

ATR algorithms can only be evaluated in context of the system objective. It would be silly to evaluate the performance of a missile (which only tracks hot spots) in its ability to distinguish tanks from trucks. Thus one is tempted to only evaluate each "block" of the ATR algorithm. But that introduces a problem as well. It is conceivable that each component of a system might function well in its own independent context but the entire system still not meet overall objectives. Therefore one tries to evaluate each component with respect to its own performance but also against the overall system objective [18.7].

Performance evaluation is a hard problem. It would be highly desirable to have an information-theoretic measure of the difficulty of a particular ATR scenario. Just as can be done with capacity of transmission channels, we would like to know: "What is the best performance we could reliably expect from *ANY* ATR algorithm operating on this class of scene?" Unfortunately we do not have any such measures available at this writing, although progress is being made [18.10, 18.61].

The preprocessing operations "correct" the input data improving image "quality" by performing denoising, deblurring, and other image corrections. For example, a denoising system may be evaluated by edge detection methods, comparing the number of edges detected, edge thinness or edge continuity, etc. Preprocessing steps generally increase the separability between the target and the background, hence, distance metrics in feature-space may also be used since the detection operation localizes potential target areas. Hence, probability of detection and false alarm may be used to evaluate this step. The segmentation operation extracts the target after it has been detected. We may therefore use measures like misclassified pixels, correlation coefficient between true and extracted target, etc.

18.4 Machine vision issues unique to ATR

The problems and issues of interest to the ATR community can be summarized as target signature variability, false alarm rate, segmentation, feature selection, performance degradation due to incomplete information, and performance evaluation [18.9, 18.71].

18.4.1 Target signature variability and false alarm rate

The "signature" of a target could mean a geometric signature and/or a spectral signature. Both are highly susceptible to change, given that we have a "smart enemy." Among all the affecting factors, variability in surface reflectance and natural illumination and occlusions of targets are most difficult to deal with.

Variability in surface reflectance and natural illumination

All images record signals received as a sum of emitted and reflected radiation:

$$f(x, y) = \xi(x, y) + \rho(x, y). \tag{18.3}$$

However, the amount emitted may be dramatically smaller than the amount reflected (in the case of visible light) or significant (in the case of longwave infrared). The *emissivity* is the ratio of emitted to total radiation.

$$\varepsilon(x, y) = \frac{\xi(x, y)}{f(x, y)}. \tag{18.4}$$

The emitted radiation is positively related to the temperature

$$\xi = \sigma A \exp(-T^4) \tag{18.5}$$

where A is the surface area, measured in m^2, σ is Stefan's constant, 5.67×10^{-8} W m^{-2} K^{-4}, and T is absolute temperature in degrees K.

Since the amount of radiation received from the sun and reflected off the target varies dramatically over a 24 hour period, and as parts of the object heat and cool at different rates, the same object may undergo contrast reversals as parts of it cool faster than others. This effect is illustrated in Fig. 18.5.

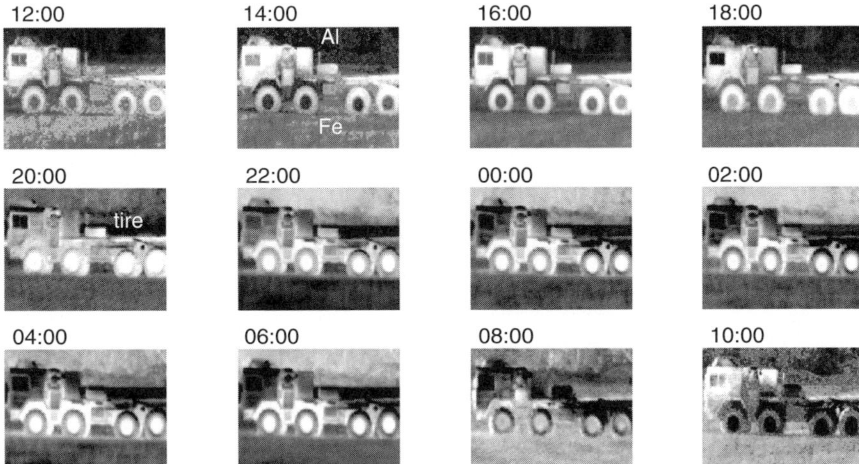

Fig. 18.5. Objects may exhibit contrast reversals over a 24 hour period. (From US Army Night Vision and Electro-optics Research Center, used with permission.)

Occlusions

Unlike industrial machine vision problems, where the setup of the manufacturing facility is specifically designed to minimize occlusion, not only do occlusions occur in ATR scenes, but targets are *usually* at least partially occluded. In fact, the opponent will be actively *trying* to have his equipment as occluded as possible [4.13]! An image of a truck occluded by a tree is shown in Fig. 18.7, later.

All these variabilities raise the question of "How well trained should the ATR be?" It is entirely too easy to over-train a system, and produce a system which performs very well on the data on which it has been trained, but poorly on data it has not seen, even though that data may (in the eyes of a human) be very similar. The problem is not to get the probability of detection high, but to do so while simultaneously keeping the false alarm rate low. The Neyman–Pearson test [18.53] provides a means to perform such a minimization with performance bounds on probability of false alarm.

18.4.2 Tracking

In ATR, many if not most applications require target tracking, and furthermore the tracking problems are less constrained and more challenging than tracking in the civilian domain. Centroid tracking is the simplest type of tracking algorithm, although there are ways to improve its sophistication [18.39]. The centroid tracker (usually) assumes there is just one target in the field of view, and that bright spot is much brighter than background. If those assumptions are true, the centroid of the target is the centroid of the field of view. More sophisticated tracking of moving objects is most often done using optimal filters like the Kalman–Bucy filter.

Haddad and Simanca [18.28] discuss the limitations of the Kalman filtering approach and propose a nonlinear tracking filter based on wavelets and the Zakai equation. Amoozegar *et al.* [18.3] provide a survey of fuzzy and neural techniques in tracking.

The process of tracking can be combined with the process of classifying vehicles [13.12, 18.22] as well.

18.4.3 Segmentation

In most ATR scenarios, the problem of separating clutter from target is fundamental. Clutter varies from one scene class to another and requires adaptive representations [18.38]. However, there currently is not even a uniformly accepted definition for "signal-to-clutter" [18.61, 18.68, 18.78].

Once a potential target is localized, it is extracted from the background as accurately as possible. However, every segmenter makes prior assumptions about the target and its neighboring pixels. These assumptions may not be valid for all viewing conditions. As we learned in Chapter 8, two common approaches to segmentation are edge or boundary formation and region growing [18.68]. Edge detection approaches are based upon recognizing dissimilarities in images whereas region growing utilizes similarity properties. Because edge detection techniques are quite sensitive to noise, successful edge detection usually depends on higher level semantic knowledge. Region growing techniques offer better immunity to noise and therefore do not have as much reliance on semantic knowledge. Qi *et al.* [18.63] propose an efficient segmentation approach to segment man-made targets from unmanned aerial vehicle (UAV) imagery using curvature information derived from an image histogram smoothed by Bezier splines. Experimental results show that by enhancing the histogram instead of the original image, similar segmentation results can be obtained in a more efficient way. In [18.87], a segmentation strategy based on the image pyramid data structure is developed, working its way from the top of the pyramid to the bottom, processing image detail hierarchically.

As we learned in Chapter 6, diffusion and diffusion-like processes [18.41, 18.42] provide excellent noise-removal steps as components of a segmentation process.

18.4.4 Feature selection

Most of the features used by researchers are geometric, topological and/or spectral [18.7]. The primary goal of feature selection is to obtain features which maximize the similarity of objects in the same class while minimizing the similarity of objects in different classes. The mathematics of feature selection are covered nicely in a textbook by Hand [18.30].

18.5 ATR algorithms

In the following, we relate a number of the topics covered earlier in this book to the specific application to ATR. This is done primarily through reference to the ATR literature. We will not attempt to survey all the relevant literature in ATR, since a great deal of that literature makes use of the methods of statistical pattern recognition, which deserves a textbook in itself. However, we cite a number of publications which will mention those methods.

The ATR application exhibits characteristics which are different from most other machine vision applications, such as industrial inspection. Key differences include the following.

(1) ATR systems must of necessity deal with *unstructured environments* – it simply is not possible to control the illumination, the viewing angle, the atmosphere, etc.
(2) *Occluded targets* are not just possible, they are likely.
(3) Only a *few pixels are likely to be on target*. The probability of correct classification strongly depends on the number of pixels on target. This is illustrated in Fig. 18.6 for a neural network classifier, but similar results, including especially the dramatic change around 50 pixels-on-target, occurs for any system, including a human.

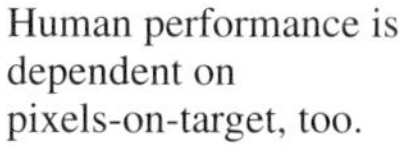
Human performance is dependent on pixels-on-target, too.

Interestingly, the first studies relating pixels on target to P_d were done prior to digital images, and considered scan lines rather than pixels. In infrared images of military targets, Johnson [18.36] found that target recognition by humans required at least four pixels across the critical (smallest) target dimension. Of course, such results vary with target complexity and recognition task details, but

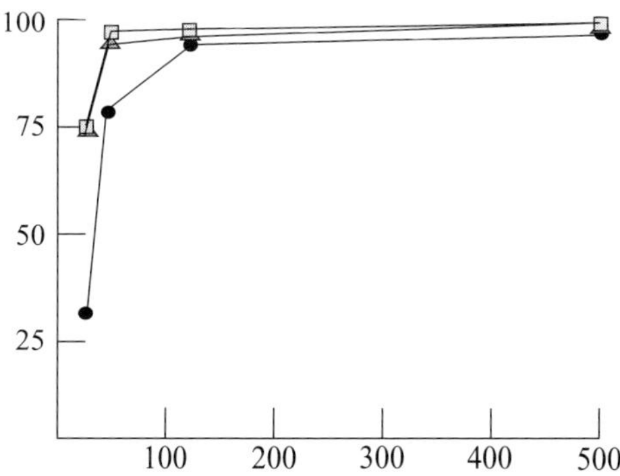

Fig. 18.6. Probability of detection (circles), probability of classification given detection (triangles), and probability of identification given detection (squares) as a function of pixels-on-target (redrawn from [18.84]).

the range of 3–5 pixels across, corresponding (for a square target) to roughly 20 pixels-on-target seems to hold fairly well. See also [18.37].

Clearly, the ATR system observer needs to remain as far as possible from the target, and therefore the number of pixels-on-target is always too small! In fact the observer would really prefer to not be there at all, which should strongly motivate the development of robotic forward observers [18.35].

(4) *3D information* must be considered. By 3D here, we mean that the target may be observed in a number of possible aspects. (See Chapters 9 and 12.)
(5) Any ATR system must *consider clutter and confusers*, which are other objects, accidently or deliberately introduced into the scene and which resemble targets. The most obvious confuser is friendly equipment of the same type as the target.

The requirements set out above dramatically affect the design of ATR systems. For example, one might consider a system which identifies targets by using a variety of different sized templates, one for each of many possible aspects. Or, a system might simply extract a variety of different sized windows, and do dimensionality reduction (using the K–L transform) and pass those results on to a classifier [18.14].

ATR philosophies of approach can be categorized using different taxonomies. In one way, the philosophies are divided into two groups [18.7]. First is the classical pattern recognition approach mentioned earlier, which uses statistical techniques. This is very popular due to ease in implementation and speed, provided it is possible to appropriately and effectively extract useful features. The other approach is AI-based and requires an additional step of symbolic manipulation.

Depending upon how the ATR systems function, other specialists in this area classify them as *geometry-based* (mostly single sensor) and *spectral-based* systems [18.71]. The basis for the *geometry-based* systems is the observation that, by preserving as much as possible the information that is received from an object and by using intelligent reasoning, a better interpretation of the scene containing the objects of interest can be made [18.70]. Thus, by better understanding the forward image formation process, the *inverse problem* – the task of recognizing objects from their received signatures – becomes more feasible. However, the integration of these models into a cohesive end-to-end model by considering the effects of their interactions is still very preliminary.

The geometry-based approach may be subdivided in various ways. For example, one could divide geometry-based systems into those that use a large collection of templates (an aspect graph) and those that start with a 3D model [18.77].

Spectral-based systems recognize targets based on spectral analysis and spectral matching [18.33]. Intuitively, the more spectral bands used, the more details about the scene can be revealed. The finer details provided in the spectral signature can substantially increase the detectability of targets, especially small targets, at pixel or even subpixel level [18.16, 18.47].

Another type of ATR system is based on the premise that the more sensory data that is available from the target of interest, the better the system performance. This is intuitively obvious for sensors that have complementary properties. Due to numerous limitations of single-sensor ATR systems, there has been a move toward multisensor targeting systems and, hence, the problem of correlating and combining data generated from *multiple sensors*. This is also referred to sometimes as *multisensor fusion*, however, the information sources may be different sensors (sensor fusion) or different algorithms (algorithm fusion) [18.32].

Finally, some researchers break the set of ATR algorithms into *model-based* methods, *statistics-based* methods, and *template-based* methods. These three categorizations are discussed in more detail below.

18.5.1 Model-based techniques

Most model-based techniques are geometry-based. They pose the question: "Given a certain viewing angle, what should the target look like?" [18.1, 18.6, 18.12, 18.13]. This is potentially a powerful philosophy, as it provides information on portions of the target that may have been occluded due to its position – e.g., from a certain viewpoint the barrel of a tank may be missing. However if we have a 3D model of the target, we could generate all possible views and perform a combinatorial search [18.65] to get a match. Model-based techniques are readily combined with different data types, especially range (laser radar) images [18.87]. However, like almost everywhere in machine vision, some optimization problem must be solved, using neural networks [18.29], genetic algorithms [18.10], or other optimization methods.

Usually, descriptions that correspond to scene structure and geometry alone are obtained as opposed to scene physics (heat, light, material properties, etc.). Matching is then the process of hypothesizing and verifying matches between model and image points. This process produces a 3D to 2D transformation which brings the 3D model points into correspondence with the 2D image points. The best match is then the transformation which best explains the scene. Solution of the 3D to 2D correspondence is basically solution of the perspective equation. Errors between projected model points and the corresponding image points help verify how good a match is.

Fig. 18.7. Image of a truck occluded by a tree [18.65]. SPIE, used with permission.

These methods are powerful, but require a lot of processing and a large database. They perform poorly when targets are occluded [18.73] as this results in a case of "incomplete information." To this end, a lot of work is being done in obtaining the "actual geometric shape" of the target from occluded views [18.75].

Occlusions, in general, fall into two distinct categories – contiguous (buildings or whole trees as shown in Fig. 18.7) and distributed (tree branches). The first type is easier to deal with as it is possible to have enough information on the nonoccluded portion to solve the problem. In [18.73], Sadjadi presents an approach to detection under occlusion. The boundary of the segmented region of the image is converted to a chain code. The chain code is then matched using histogram intersection (described in the next section) with a set of models. The match is used as a measure of the confidence of the system. If, however, the degree of confidence is still low, the system assumes that the object is partially occluded and now the object is matched with occlusion models in a database.

A histogram may be constructed of the chain code directions.

18.5.2 Statistics-based techniques

The ideas behind statistics-based techniques are the same as those described in Chapter 14: (1) Obtain features; (2) develop statistical measurements which characterize different classes; (3) make decisions which optimize some measure such as minimum cost, or maximum probability of correct decision. In this section, we consider only multispectral measurements.

Multispectral matching

One technique for multispectral analysis uses the concept of *histogram intersection* introduced by Swain and Ballard [18.82] for the realm of color images. The idea is simply to compare the histograms of two images and determine any overlap factor (how many pixels in the data base histogram match those in the histogram of the new image). Specifically, given a pair of histograms, I (from new image) and M (from database), each containing n bins, the intersection is defined to be $\sum_{j=1}^{n} \min(I_j, M_j)$. The result is the number of pixels that have the same color in the two images. This number may be normalized to obtain overlap factor. The color of an object is subject to significant change depending upon varying lighting conditions; and in this situation, this simple algorithm will clearly not give us a good match. To overcome this problem, Funt and Finlayson [18.24] combined the idea of histogram intersection with a concept referred to as "color constancy" [18.23], which removes effects of varying illumination conditions and in effect, normalizes the image to a standard illuminant. Now that we have a data base also in standard illuminant conditions, we can "compare apples with apples" and use the histogram intersection method described above. We have not put any restriction to the dimensionality of the histogram, and hence can extend this concept to higher dimensions (more sensors), obtaining a more robust system.

Another measure of spectral match between a known target signature and an observed signature is obtained by treating the signatures as vectors and finding the inner product between the two vectors [18.93]. The better the match, the closer the angular separation to zero. In other words, if we have two d-dimensional measurements in the spectral signatures, X and Y, then the distance between these two measurements can be represented by an angle between these two spectra

$$\cos\theta = \frac{X \bullet Y}{|X||Y|}.$$

Small θ indicates that the two spectra are quantitatively similar; similarly large θ indicates dissimilar spectra. In [18.93], Weisberg *et al.* use this measure to perform a clustering procedure, segmenting the image into multiple regions of interest. There are many other measures of similarity that are in use [18.34, 18.62].

In [18.86], Trivedi introduces the use of relative spectral information rather than absolute information about the target for remote-sensing purposes. This introduces some robustness into the system. For example, a certain object may be brighter than the background in a particular channel and darker in some other channel.

18.5.3 Template matching in ATR

Template matching is the simplest and most often applied algorithm in the ATR application domain. Template matching is most effective when the aspect can easily be established. For example [18.51], when the target is a ship, it is easy to find the center of gravity, and therefore to eliminate the translational degree of freedom; in addition, ships are long and thin, and their orientation can be determined by computing the major axis. Having eliminated rigid body motions, only a few template comparisons are required to get precise pose and identification.

In those applications where the pose of the target is not easily estimated, sets of *focused filters* can be used, with the strongest response indicating the target pose [18.2, 18.40].

Deformable templates of Chapter 9 are well suited to detecting and classifying approximately located objects [18.44, 18.76] like the silhouettes of aircraft [18.65], where the optimization problem is posed as a MAP estimation.

The problem of identifying and tracking roads in satellite images is discussed in [18.26, 18.48]. Barzohar and Cooper [18.4] also pose this as a MAP estimation problem.

18.6 The Hough transform in ATR

Since many man-made objects have straight-line properties, the Hough transform pops up often in ATR applications. For example, it may be used [18.19] to identify the track of a (subpixel) missile as it is observed from space. By taking differences in time, the target track becomes visible, along with a great deal of noise. The track is, however, more-or-less a straight line, and comes out very clearly in the Hough

Fig. 18.8. Hubble space telescope. Photo: NASA.

transform. Cowart *et al.* [18.19] also consider the use of a parametric transform which allows for maneuvering targets.

Viewed from above, a ship is a long, straight (unless the ship has had an unfortunate encounter) object. When viewed with SAR,[2] the image of a ship consists of spots and dropouts. One may estimate [18.25, 18.50] the orientation of a ship by simply taking the Hough transform. Surprisingly, it appears [18.25] that the Hough transform is less sensitive to noise than using principal axes. If the ship is moving, its wake is even longer, straighter, and may be more easily found [18.18].

A satellite, when viewed from a terrestrial telescope, has straight edges (see Fig. 18.8). The Hough transform can also be used to identify and characterize those edges [18.20].

18.7 Morphological techniques in ATR

The special needs of ATR affect the types of morphological techniques reported in the literature. For example, in many ATR applications, the targets are observed from above and have few pixels on target. Therefore, it is important that noise removal operations such as morphological opening be invariant to rotation. For example [18.60], the traditional opening

$$(f o B)(n, m) = \max_{i,j\in B}\{\min_{i,j\in B}\{f(n+i, m+j)\}\} \tag{18.6}$$

[2] SAR is an abbreviation for Synthetic Aperture Radar.

Table 18.2. *Result of modified opening with ring-shaped structuring element.*

L < R	Object erased
R < L < 2R	Some parts removed, some preserved
2R < L	Object preserved

Fig. 18.9. A ring-shaped structuring element and a rectangular object (redrawn from [18.60]).

can be replaced with

$$\gamma(f_B)(n,m) = \min\{\max_{i,j\in B}\{f(n+i,m+j)\}, f(n,m)\} \tag{18.7}$$

where B is the ring-shaped structuring element of Fig. 18.9. The behavior of this structuring element is different according to the relative size of the object, L, and the radius of the structuring element, R. Such methods [18.58] lead to morphology-based algorithms which are more specific to target recognition [18.10] or target tracking [18.92]. The result is shown in Table 18.2.

Shape recognition using morphology has also been addressed by Pham *et al.* [18.58].

18.8 Chain codes in ATR

Template matching and feature-based approaches have their own shortcomings. Most such algorithms depend on some prior knowledge about the object. Features are needed which are invariant with respect to size and orientation to the object in the field-of-view of the sensor. Similarly, approaches based on template matching require extensive data bases, with large search times. However, silhouettes may be matched [18.72] using the chain code of the segmented object, and then processing the histogram of that chain code (Fig. 18.10). This strategy has two very useful properties:

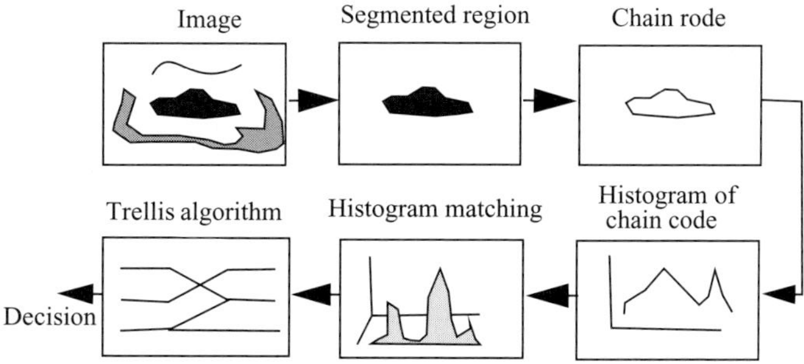

Fig. 18.10. Schematic of Sadjadi's proposed method [18.71–18.74] (redrawn).

(1) Scale variation in the image domain is equivalent to a vertical shift in the chain code histogram domain.
(2) Changes in orientation are equivalent to horizontal cyclic shifts in the histogram.

It is possible that two different objects have the same chain code histogram. To this end a trellis algorithm may be used to distinguish between such objects [18.81]. A trellis structure of a distorted pattern is created using each row vector as a "fractional pattern" (a node in the trellis). A large collection of observed data is used to provide a statistical basis for this trellis, to which the Viterbi algorithm (see section 2.4.2) is applied. Although the use has been demonstrated only for handwritten characters, this method may be universally applied to any category of distorted patterns.

18.9 Conclusion

A wide variety of papers are available in the public literature that have comparative studies on different techniques. For example, Li *et al.* [18.43] compared a number of neural, statistical and model-based approaches and concluded that: "At least for FLIR images, the neural network-based approaches gave better results than the PCA (principal components analysis) and the LDA (linear discriminant analysis)-based approaches." They found that methods based on the Hausdorff distance also performed "well."

Often, papers in the literature present conclusions that make the reader think the ATR problem is nearly solved when, in reality, either these systems have not been tested on real military data or their scope is so limited to one specific application and set of conditions that their "actual" performance may be doubted. We need to understand that the problem being dealt with here is not only nontrivial, it gets substantially more difficult with every new invention made by the enemy. What is "state-of-the-art" today may not be tomorrow.

Nonetheless, from an engineering point of view (which is almost always practical!), the authors believe that the following are needed to produce ATR systems that may be benchmarked:

- a large set of *standard* real-world images, with ground truth information available to the research community,
- a tool to optimally evaluate and incorporate the best of a large number of ATR techniques into a system with "optimal" performance.

Several questions arise when we take these issues into consideration. How do we come up with a *standard*? What objectives are we trying to reach – a powerful system that has the best ROC curves or developing a system that is portable, using minimal hardware? Clearly all these cannot be met at the same time and someone has to strike a balance.

General trends are seen, however, in the development of ATR systems. The use of more than one sensory system, portability of the system, its ability to be truly an automated system with minimal human interference, the increased use of mathematical tools to provide a sound basis for the system are clearly where research seems to be headed. In this chapter the authors hope to have communicated the vastness of this problem while also presenting the achievements of science in approaching this problem.

Bibliography

[18.1] J. Albus, "Applications of an Efficient Algorithm for Locating 3D Models in 2D Images," *Automatic Target Recognition VIII, Proceedings SPIE*, **3371**, April, 1998.

[18.2] K. Al-Ghoneim and B. Kumar, "Combining Focused MACE Filters for Target Detection," *Automatic Target Recognition VIII, Proceedings SPIE*, **3371**, April, 1998.

[18.3] F. Amoozegar, A. Notash, and H. Pang, "Survey of Fuzzy Logic and Neural Network Technology for Multi-target Tracking," *Automatic Target Recognition VIII, Proceedings SPIE*, **3371**, April, 1998.

[18.4] M. Barzohar and D. Cooper, "Automatic Finding of Main Roads in Aerial Images by Using Geometric-stochastic Models and Estimation," *IEEE Transactions on Pattern Analysis and Machine Intelligence*, **18**(7), 1996.

[18.5] J. Beveridge, D. Panda, and T. Yachik, "November 1993 Fort Carson RSTA Data Collection Final Report," Colorado State University Technical Report, CS-94-118, August, 1994.

[18.6] J. Bevington and K. Siejko, "Ladar Sensor Modeling and Image Synthesis for ATR Algorithm Development," *Automatic Target Recognition VI, Proceedings* SPIE, **2756**, April 1996.

[18.7] B. Bhanu, "Automatic Target Recognition: State of the Art Survey," *IEEE Transactions on Aerospace and Electronic Systems*, **22**(4), 1986.

[18.8] B. Bhanu and T. Jones, "Image Understanding Research for Automatic Target Recognition," *Proceedings of the DARPA Image Understanding Workshop*, 1992.

[18.9] B. Bhanu and T. Jones, "Image Understanding Research for Automatic Target Recognition," *IEEE Aerospace and Electronics Systems Magazine*, **8**(10), 1993.

[18.10] M. Boshra and B. Bhanu, "Predicting Performance of Object Recognition," *IEEE Transactions on Pattern Analysis and Machine Intelligence*, **22**(9), 2000.

[18.11] M. Bullock, D. Wang, S. Fairchild, and T. Patterson, "Automated Training of 3-D Morphology Algorithm for Object Recognition," *Automatic Target Recognition IV, Proceedings SPIE*, **2234**, April, 1994.

[18.12] J. Burrill, S. Wang, A. Barrow, M. Friedman, and M. Soffen, "Model-based Matching using Elliptical Features," *Automatic Target Recognition VI, Proceedings SPIE*, **2756**, April, 1996.

[18.13] L. Carin, L. Felsen, and C. Tran, "Model-based Object Recognition by Wave-oriented Data Processing," *Automatic Target Recognition V, Proceedings SPIE*, **2485**, April, 1995.

[18.14] L. Chan, N. Nasrabadi, and D. Torrieri, "Discriminative Eigen Targets for Automatic Target Recognition," *Automatic Target Recognition VIII, Proceedings SPIE*, **3371**, April, 1998.

[18.15] B. Chen, S. Thomopoulos, and C. Lin, "Feature Estimation and Objects Extraction using Markov Random Field Modeling," *Automatic Target Recognition V, Proceedings SPIE*, **2485**, April, 1995.

[18.16] S. S. Chiang and C. I. Chang, "Subpixel Detection for Hyperspectral Images Using Project Pursuit," *EUROPTO Conference on Image and Signal Processing for Remote Sensing V, Proceedings SPIE*, **3871**, pp. 107–115, September, 1999.

[18.17] C. Chun, D. Fleming, and E. Torok, "Polarization-sensitive Thermal Imaging," *Automatic Target Recognition IV, Proceedings SPIE*, **2234**, April, 1994.

[18.18] A. Copeland, G. Ravichandran and M. Trivedi, "Localized Radon Transform for Ship Wake Detection in SAR Imagery," *Automatic Target Recognition IV, Proceedings SPIE*, **2234**, April, 1994.

[18.19] A. Cowart, W. Snyder, and H. Ruedger, "The Detection of Unresolved Targets Using the Hough Transform," *Computer Graphics and Image Processing*, December, 1982.

[18.20] X. Du, S. Ahalt, and B. Stribling, "Estimating and Refining Orientation Vectors for Sub-components of Space Object Imagery," *Automatic Target Recognition VIII, Proceedings SPIE*, **3371**, April, 1998.

[18.21] Dudgeon D.E. and Lacoss, R.T. "An Overview of Automatic Target Recognition," *MIT Lincoln Laboratory Journal*, **6**(1), pp. 2–10, Spring, 1993.

[18.22] D. Ernst, H. Gross, D. Stricker, and U. Thönnesen, "Improvement of Object Classification in Image Sequences," *Automatic Target Recognition VII, Proceedings SPIE*, **3069**, April, 1997.

[18.23] D. Forsyth, "A Novel Color Constancy Algorithm," *International Journal of Computer Vision*, **5**, 1990.

[18.24] B. Funt and G. Finlayson, "Color Constant Color Indexing," *IEEE Transactions on Pattern Analysis and Machine Intelligence*, **17**(5), 1995.

[18.25] L. Gagnon and R. Klepko, "Hierarchical Classifier Design for Airborne SAR Images of Ships," *Automatic Target Recognition VIII, Proceedings SPIE*, **3371**, April, 1998.

[18.26] D. Geman and B. Jedynak, "An Active Testing Model for Tracking Roads in Satellite Images," *IEEE Transactions on Pattern Analysis and Machine Intelligence*, **18**(1), 1996.

[18.27] R. Ghamasaee, F. Amoozegar, H. Pang, Y. Chin, and S. Blackman, "Survey of Neural Networks as Applied to Target Tracking," *Automatic Target Recognition VII, Proceedings SPIE*, **3069**, April, 1997.

[18.28] Z. Haddad and S. Simanca, "Filtering Image Records Using Wavelets and the Zakai Equation," *IEEE Transactions on Pattern Analysis and Machine Intelligence*, **17**(11), 1995.

[18.29] M. Hamilton and T. Kipp, "Model-based Multi-sensor Automatic Target Identification for FLIR Fused with MMW," *Automatic Target Recognition V, Proceedings SPIE*, **2485**, April, 1995.

[18.30] D. J. Hand, *Discrimination and Classification*, New York, Wiley, 1981.

[18.31] H. Hayes, C. Priebe, G. Rogers, D. Marchette, J. Solka, and R. Lorey, "Improved Texture Discrimination and Image Segmentation with Boundary Incorporation," *Automatic Target Recognition V, Proceedings SPIE*, **2485**, April, 1995.

[18.32] Y. Huang and C. Suen, "A Method of Combining Multiple Experts for the Recognition of Unconstrained Handwritten Numerals," *IEEE Transactions on Pattern Analysis and Machine Intelligence*, **17**(1), 1995.

[18.33] R. L. Huguenin and J. L. Jones, "Intelligent Information Extraction from Reflectance Spectra: Absorption Band Positions," *Journal of Geophysical Research*, **91**(B8), pp. 9585–9598, 1986.

[18.34] D. Huttenlocher, D. Klanderman and A. Rucklige, "Comparing Images Using the Hausdorff Distance," *IEEE Transactions on Pattern Analysis and Machine Intelligence*, **15**(9), 1993.

[18.35] X. Jiang and H. Bunke, "Vision Planner for an Intelligent Multisensory Vision System," *Automatic Target Recognition IV, Proceedings SPIE*, **2234**, April, 1994.

[18.36] J. Johnson, "Analysis of Image Forming Systems," *Proceedings of the Image Intensifier Symposium*, October 6–7, 1958.

[18.37] J. Johnson and W. Lawson, "Performance Modeling Methods and Problems," *Proceedings of the IRIS Specialty Group on Imaging*, January, 1974.

[18.38] A. Katz and P. Thrift, "Generating Image Filters for Target Recognition by Genetic Learning," *IEEE Transactions on Pattern Analysis and Machine Intelligence*, **16**(9), 1994.

[18.39] A. Kumar, Y. Bar-Shalom, and E. Oron, "Precision Tracking Based on Segmentation with Optimal Layering for Imaging Sensors," *IEEE Transactions on Pattern Analysis and Machine Intelligence*, **17**(2), 1995.

[18.40] B. Kumar, A. Mahalanobis, and A. Takessian, "Optimal Tradeoff Correlation Filters with Controlled In-plane Rotation Response for Target Recognition," *Automatic Target Recognition VIII, Proceedings SPIE*, **3371**, April, 1998.

[18.41] A. Lanterman, M. Miller and D. Snyder, "Implementation of Jump-diffusion Algorithms for Understanding FLIR Scenes," *Automatic Target Recognition V, Proceedings SPIE*, **2485**, April, 1995.

[18.42] A. Lanterman, M. Miller, and D. Snyder, "Representations of Thermodynamic Variability in the Automated Understanding of FLIR Scenes," *Automatic Target Recognition VI, Proceedings SPIE*, **2756**, April, 1996.

[18.43] B. Li, Q. Zheng, S. Der, R. Chellappa, N. Nasrabadi, L. Chan, and L. Wang, "Experimental Evaluation of Neural, Statistical and Model-based Approaches to FLIR ATR, *Automatic Target Recognition VIII, Proceedings SPIE*, **3371**, April, 1998.

[18.44] J. Li and C. Kuo, "Automatic Target Shape Recognition via Deformable Wavelet Templates," *Automatic Target Recognition VI, Proceedings SPIE*, **2756**, April, 1996.

[18.45] Lincoln Labs, "Special Issue on Automatic Target Recognition," *Lincoln Laboratory Journal*, **6**(1), Spring, 1993.

[18.46] A. Mahalanobis and D. Kelly, "High Value Target Recognition using Correlation Filters," *Automatic Target Recognition VI, Proceedings SPIE*, **2756**, April, 1996.

[18.47] D. Manolakis, G. Shaw, and N. Keshava, "Comparative Analysis of Hyperspectral Adaptive Matched Filter Detectors," *Algorithms for Multispectral, Hyperspectral, and Ultraspectral Imagery VI, Proceedings SPIE*, **4049**, pp. 2–17, 2000.

[18.48] N. Merlet and J. Zerubia, "New Prospects in Line Detection by Dynamic Programming," *IEEE Transactions on Pattern Analysis and Machine Intelligence*, **18**(4), 1996.

[18.49] V. Mirelli, D. Nguyen, and N. Nasrabadi, "Target Recognition For FLIR Imagery Using Learning Vector Quantization And Multi-layer Perception," *Automatic Target Recognition V, Proceedings SPIE*, **2485**, April, 1995.

[18.50] S. Musman, D. Kerr, and C. Bachmann, "Automatic Recognition of ISAR Ship Images," *IEEE Transactions on Aerospace Electronic Systems*, **32**(4), pp. 1392–1404, 1996.

[18.51] Y. Nakano, Y. Hara, J. Saito, and Y. Inasawa, "Radar Target Recognition System using 3-D Mathematical Model," *Automatic Target Recognition VIII, Proceedings SPIE*, **3371**, April, 1998.

[18.52] E. Natonek and C. Baur, "Model based 3-D Object Recognition using Intensity and Range Images," *Automatic Target Recognition IV, Proceedings SPIE*, **2234**, April, 1994.

[18.53] J. Neyman and E. S. Pearson, "On the Problem of the Most Efficient Tests of Statistical Hypotheses," *Philosophical Transactions of the Royal Society London Series A*, **168**, pp. 268–282, 1933.

[18.54] C. Paiva, "Theater Targets Plume Edge Extraction and Hardbody Aimpoint Selection," *Automatic Target Recognition VII, Proceedings SPIE*, **3069**, April, 1997.

[18.55] T. Patterson, "Radiometrically Correct Sharpening of Multispectral Images using a Panchromatic Image," *Automatic Target Recognition IV, Proceedings SPIE*, **2234**, April, 1994.

[18.56] A. Pears and E. Pissaloux, "Using Hardware Assisted Geometric Hashing for High Speed Target Acquisition and Guidance," *Automatic Target Recognition VII, Proceedings SPIE*, **3069**, April, 1997.

[18.57] J. Perez-Jacome, "Automatic Target Recognition Systems with Emphasis on Model-Based Approaches," CSIP TR-97-01, ECE-Georgia Tech., 1996.

[18.58] Q. Pham and M. Smith, "A Morphological Multistage Algorithm for Recognition of Targets in FLIR Data," *Automatic Target Recognition VI, Proceedings SPIE*, **2756**, April, 1996.

[18.59] Q. Pham, T. Brosnan, and M. Smith, "Sequential Digital Filters for Fast Detection of Targets in FLIR Image Data," *Automatic Target Recognition VII, Proceedings SPIE*, **3069**, April, 1997.

[18.60] Q. Pham, T. Brosnan, M. Smith, and R. Mersereau, "A Morphological Technique for Clutter Suppression in ATR," *Automatic Target Recognition VIII, Proceedings SPIE*, **3371**, April, 1998.

[18.61] M. Phillips and R. Sims, "A Signal to Clutter Measure for ATR Performance Comparison," *Automatic Target Recognition VII, Proceedings SPIE*, **3069**, April, 1997.

[18.62] K. Plataniotis and A. Venetsanopoulos, *Color Image Processing and Applications*, Berlin, Springer, 2000.

[18.63] H. Qi, W. E. Snyder, and D. Marchette, "An Efficient Approach to Segmenting Man-made Targets from Unmanned Aerial Vehicle Imagery," *Optical Engineering*, **39**(5), pp. 1267–1274, 2000.

[18.64] H. Ranganath and R. Sims, "Self Partitioning Neural Networks for Target Recognition," *Automatic Target Recognition IV, Proceedings SPIE*, **2234**, April, 1994.

[18.65] K. Rao, "Combinatorics Reduction for Target Recognition in ATR Applications," *Automatic Object Recognition II, Proceedings SPIE*, **1700**, 1992.

[18.66] J. Ratches, C. Walters, R. Buser, and B. Guenther, "Aided and Automatic Target Recognition Based upon Sensory Inputs from Image Forming Systems, *IEEE Transactions on Pattern Analysis and Machine Intelligence*, **19**(9), Sept. 1997.

[18.67] A. Reno, D. Gillies, and D. Booth, "Deformable Models for Object Recognition in Aerial Images," *Automatic Target Recognition VIII, Proceedings SPIE*, **3371**, 1998.

[18.68] E.M. Riseman and M.A. Arbib, "Computational Techniques in Visual Segmentation of Static Scenes," *CCGIP, 7, Target Recognition VIII, Proceedings SPIE*, **3371**, April, 1998.

[18.69] R. Robmann and H. Bunke, "Towards Robust Edge Extraction – a Fusion Based Approach using Greylevel and Range Images," *Automatic Target Recognition V, Proceedings SPIE*, **2485**, April, 1995.

[18.70] F.A. Sadjadi, "A Model-Based Technique for Recognizing Targets by Using Millimeter Wave Radar Signatures," *International Journal of Infrared and Millimeter Waves*, **10**(3), 337–342, 1989.

[18.71] F.A. Sadjadi, "Automatic Object Recognition: Critical Issues and Current Approaches," *Proceedings SPIE*, **1471**, 1991.

[18.72] F.A. Sadjadi, "Automatic Object Recognition: Critical Issues and Current Approaches," 1991. Selected SPIE Papers on CD-ROM, Volume 6. Automatic Target Recognition SPIE: 1 PO Box 10, Bellingham, WA 98227-0010, USA.

[18.73] F.A. Sadjadi, "Automatic Object Recognition: Critical Issues and Current Approaches," 1991. Selected SPIE Papers on CD-ROM, Volume 6. Automatic Target Recognition SPIE: 1 PO Box 10, Bellingham, WA 98227-0010, USA.

[18.74] F.A. Sadjadi, Special Section on ATR, *Optical Engineering*, **31**(12), 1992.

[18.75] F.A. Sadjadi, "Application of Genetic Algorithm for Automatic Recognition of Partially Occluded Objects," *Automatic Object Recognition IV, Proceedings SPIE*, **2234**, 1994.

[18.76] R. Samy and J. Bonnet, "Robust and Incremental Active Contour Models for Objects Tracking," *Automatic Target Recognition V, Proceedings SPIE*, **2485**, April, 1995.

[18.77] R. Sharma and N. Subotic, "Construction of Hybrid Templates from Collected and Simulated Data for SAR ATR Algorithms," *Automatic Target Recognition VIII, Proceedings SPIE*, **3371**, April, 1998.

[18.78] R. Sims, "Signal to Clutter Measurement and ATR Performance," *Automatic Target Recognition VIII, Proceedings SPIE*, **3371**, April, 1998.

[18.79] R. Sims and B. Dasarathy, "Automatic Target Recognition using a Passive Multisensor Suite," Special Section on ATR, *Optical Engineering*, **31**(12), 1992.

[18.80] A. Srivastava, B. Thomasson, and R. Sims, "A Regression Model for Prediction of IR Images," *Proceedings of SPIE Aerosense ATR XI*, Orlando, 2001.

[18.81] L.B. Stotts, E.M. Winter, L.E. Hoff, and I.S. Reed, "Clutter Rejection using Multispectral Processing," *Proceedings SPIE Signal and Data Processing of Small Targets*, **1305**, pp. 2–10, 1990.

[18.82] M. Swain and D. Ballard, "Color Indexing," *International Journal of Computer Vision*, **7**(1), 1991.

[18.83] H. Tanaka, Y. Hirakawa, and S. Kaneku, "Recognition of Distorted Patterns Using the Viterbi Algorithm," *IEEE Transactions on Pattern Analysis and Machine Intelligence*, **4**(1), 1982.

[18.84] W. Thoet, T. Rainey, D. Brettle, L. Stutz, and F. Weingard, "ANVIL Neural Network Program for Three-dimensional Automatic Target Recognition," *Optical Engineering*, **31**(12), December, 1992.

[18.85] M.M. Trivedi, "Detection of Objects in High Resolution Multispectral Aerial Images," *SPIE Applications of Artificial Intelligence* II, 1985.

[18.86] M. Trivedi, "Object Detection Using Their Multispectral Characteristics," *Proceedings SPIE*, **754**, 1987.

[18.87] M. M. Trivedi and J. C. Bezdek, "Low-Level Segmentation of Aerial Images with Fuzzy Clustering," *IEEE Transactions on Systems, Man, and Cybernetics*, **16**(4), 1986.

[18.88] A. Ueltschi and H. Bunke, "Model-based Recognition of Three-dimensional Objects from Incomplete Range Data," *Automatic Target Recognition V, Proceedings SPIE*, **2485**, April, 1995.

[18.89] J. Wald, D. Krig, and T. DePersia, "ATR: Problems and Possibilities for the IU Community," *Proceedings of the DARPA Image Understanding Workshop*, IUW, 255–264, San Diego, January, 1992.

[18.90] B. Wallet, D. Marchette, and J. Solka, "A Matrix Representation for Genetic Algorithms," *Automatic Target Recognition VI, Proceedings SPIE*, **2756**, April, 1996.

[18.91] B. Wallet, D. Marchette, and J. Solka, "Using Genetic Algorithms to Search for Optimal Projections," *Automatic Target Recognition VII, Proceedings SPIE*, **3069**, April, 1997.

[18.92] S. Wang, G. Chen, D. Sapounas, H. Shi, and R. Peer, "Development of Gazing Algorithms for Tracking Oriented Recognition," *Automatic Target Recognition VII, Proceedings SPIE*, **3069**, April, 1997.

[18.93] A. Weisberg, M. Najarian, B. Borowski, J. Lisowski, and B. Miller, "Spectral Angle Automatic Cluster Routine (SAALT): An Unsupervised Multispectral Clustering Algorithm," *Proceedings of IEEE Aerospace Conference*, 307–317, 1999.

[18.94] D. Xue, Y. Zhu, and G. Zhu, "Recognition of Low-contrast FLIR Tank Object Based on Multiscale Fractal Character Vector," *Automatic Target Recognition VI, Proceedings SPIE*, **2756**, April, 1996.

[18.95] C. Zhou, G. Zhang, and J. Peng, "Performance Modeling Based Adaptive Target Tracking in Multiscenario Environment," *Automatic Target Recognition VI, Proceedings SPIE*, **2756**, April, 1996.

Author index

Index

CD-ROM

The CD-ROM accompanying this book contains seven folders.

Doc

The documentation for the IFS (Image File System) software system. This includes the principal documentation on IFS, ifsmanual.pdf. In addition, ifsflip.pdf describes a collection of special subroutines written for processing floating point images and optimized for speed; qsyn.pdf describes the quadric synthesizer, which allows synthesis of range images; and IFSIndex.pdf is a very brief description of all IFS, including special installation instructions for Macintosh, Solaris, Windows, and Linux.

Images

A rather extensive collection of images is included in this folder, including industrial, medical, and "standard" images. Each image is accompanied by a documentation file (same name, but with extension.txt), which can be read and displayed by wxIFSView.

Source

This directory contains a couple of example programs and includes the files required to compile IFS programs.

Platform-specific releases of IFS: Linux, MacX, Sun, Win32

Each of these directories contains a collection of programs (in the ifsbin subdirectory) and several libraries (in the ifslib directory). With the exception of wxIFSView, all programs are written to be run from the command line (the "terminal" in Mac OS-X, the "cmd" in Windows).

This CD-ROM does NOT contain software development tools such as a C-compiler, a linker, a debugger, or an editor. For Windows, we have had good luck with Microsoft Visual C++; on the Mac, ProjectBuilder works well with IFS; and there is an example using both of these in the source/sample directory. On Windows machines, other, much lower-cost systems are available such as Lcc. On the Mac, ProjectBuilder comes with the developer package download from Apple; however, ProjectBuilder is not required, as IFS can be used with the standard Unix cc command.